A PLUME BOOK

SPYCRAFT

ROBERT WALLACE is the former director of the CIA's Office of Technical Service and lives in Virginia. A recipient of the CIA's Intelligence Medal of Merit, Wallace founded the Artemus Consulting Group in 2004, providing management and intelligence counsel to corporate and government clients. He is also a contributor to the oral history program of CIA's Center for the Study of Intelligence.

H. KEITH MELTON is an internationally recognized author, historian, and expert on clandestine devices and technology. He is the technical tradecraft historian at the Interagency Training Center in Washington, D.C. He has assembled the world's largest collection of espionage devices and lectures widely throughout the U.S. intelligence community and abroad. He resides in Florida.

HENRY ROBERT SCHLESINGER is an author and journalist who has covered intelligence technologies, counterterrorism, and law enforcement. His work has appeared in *Popular Science*, *Popular Mechanics*, *Technology Review*, and *Smithsonian* magazine. He lives in New York City.

The authors can be contacted through their Web site at www.ciaspycraft.com.

Praise for *Spycraft*

"Their tales will leave you shaken, if not stirred."
—*Discover* magazine

"Full of neat gadgets."
—*Chicago Tribune*

"Hollywood's early depiction of spy gadgets is not far from reality."
—*Pittsburgh Post-Gazette*

"A fascinating and often funny compendium of spook contraptions."
—*The Miami Herald*

"007 and actual spies do have one thing in common: Q, the gadget maker, who in real life was played by Wallace."
—*New York Post*

"History-making, patent-never-pending objects of espionage."
—*Boston Herald*

"*Spycraft* portrays the ingenuity of the CIA, the success of its operations, and the bravery of its officers."
—Ronald Kessler on Newsmax.com

"Mind-boggling, high-tech spy gear."
—*Knoxville News-Sentinel*

"Exciting content and slam-bang style."
—*Publishers Weekly*

"Well-written account . . . The details of operational activity are as engrossing as the descriptions of the equipment, military and otherwise—e.g., miniature cameras and radios, obscure drugs, tiny weapons, secret compartments, and forged documents—depicted here in 100-plus fascinating diagrams and photographs." —*Library Journal*

"Regaling readers with the paraphernalia CIA case officers use in running their agents—audio devices, miniature cameras, secret writing, disguises, codes, dead drops, etc.—Wallace and his coauthors well capture the spy-versus-spy dynamic."
—*Booklist*

"Wallace and Melton provide an in-depth account of not only the 'wizardry' of OTS technology but also the innovative and often heroic application of this technology to support clandestine operations. For me, it has been a distinct honor and privilege to have had the opportunity to serve with the men and women of OTS. Their creativity and dedication to service continues to significantly contribute to U.S. national security."
—James R. Gosler, former director, Clandestine Information Technology Office, Central Intelligence Agency

"*Spycraft* is one of the best, if not the best and most revealing books on intelligence ever written. It is a must-read for anyone interested in intelligence, whether the reader is an historian, an 'aficionado,' or someone seeking an understanding of the profession. If only we had had some of those tools when I was a young ops officer."
—Larry Devlin, author of *Chief of Station, Congo* and former chief, CIA Africa Division

"This book is absolutely the best I've ever read about the CIA's 'spy-techs' and the critical role they have played. . . . A must-read for anyone interested in how the clever use of technology gives America's intelligence services a decisive advantage in the espionage wars."
—Pete Earley, author of *Comrade J: The Untold Secrets of Russia's Master Spy After the End of the Cold War*

"Just amazing! Page after page of jaw-dropping revelations about incredible cases and amazing technology. There has never been anything like this book."
—Richard Gid Powers, author of *Secrecy and Power: The Life of J. Edgar Hoover*

"A must-read for anyone interested in the world of CIA clandestine operations. The authors open a door on a hidden area that even those of us who have served in the agency rarely see. . . . Incredible research and great writing make this a fun ride through the history of this until now overlooked secret world deep inside the CIA. . . . The authors are finally able to bring the long overdue story of this critical side of the agency operations to light. Let our enemies around the world see why they can run, but they cannot hide."
—Gary C. Schroen, author of *First In*

"This is a story I thought could never be told. The CIA's super-secret gadgets and technical operations were the difference maker in the espionage wars. . . . Behind all of us who did the front line spying for the CIA stood some remarkable and unsung heroes, the scientists and engineers of OTS. It was a beautiful partnership. Don't miss this book. Nothing like it has been written before."
—James M. Olson, former chief of CIA counterintelligence and author of *Fair Play: The Moral Dilemmas of Spying*

"Stuffed with stories about chemical taggants, forged documents, physical and psychological disguises, software beacons that reveal the location of a cell phone or a laptop, about long-range surveillance cameras and ivory letter-opening knives, this extraordinary, detailed, accurate book tells more about what spies really do, the risks they run and their schemes to avoid them, than all the James Bond stories put together. Essential for any serious student of spycraft."
—David Kahn, author of *The Codebreakers*

SPYCRAFT

The Secret History of the CIA's Spytechs,
from Communism to Al-Qaeda

Robert Wallace and H. Keith Melton

with Henry R. Schlesinger

A PLUME BOOK

PLUME
An imprint of Penguin Random House LLC
375 Hudson Street
New York, NY 10014

First Plume Printing, June 2009

 REGISTERED TRADEMARK—MARCA REGISTRADA

The Library of Congress catalogued the Dutton edition as follows.

Wallace, Robert.
    Spycraft: the secret history of the CIA's spytechs from communism to Al-Qaeda / Robert Wallace and H. Keith Melton with Henry R. Schlesinger.
        p. cm.
    Includes bibliographical references and index.
    ISBN 978-0-525-94980-0 (hc.)
    ISBN 978-0-452-29547-6 (pbk.)
    1. United States. Central Intelligence Agency. Directorate of Science and Technology—History. 2. Intelligence service—United States. I. Melton, H. Keith (Harold Keith), 1944– II. Schlesinger, Henry R. III. Title.
    JK468.I6W35 2008
    327.1273—dc22   2007046734

Printed in the United States of America
Set in Times New Roman • Original hardcover design by Amy Hill
20 19 18 17 16 15 14 13

For the families of TSS, TSD, and OTS
who served their country with patience,
courage, and honor through quiet,
unheralded support of the Spytechs

Contents

Office of Technical Service crest, 2001

Foreword

by GEORGE J. TENET
Director, Central Intelligence
1997–2004

Minutes before I was to deliver the keynote speech at CIA Headquarters recognizing the fiftieth anniversary of the Office of Technical Service (OTS) on September 7, 2001, I was unavoidably called away to a meeting downtown. What I had prepared to say to the several hundred OTS officers gathered that morning would seem prescient ninety-six hours later when al-Qaeda struck the American homeland. Those words remain appropriate today as our nation confronts terrorism on every continent. For five decades, OTS officers and their wondrous devices had played a vital role in virtually every major CIA clandestine operation. In the words of Jim Pavitt, our Deputy Director for Operations during my tenure, OTS was the "technical problem solver for what appears impossible."

CIA's Office of Technical Service was established in 1951 at the beginning of the Cold War to meet a threat very different from the one America faces today. Throughout the next fifty years, OTS fashioned a history of adapting brilliantly to the challenge of applying new technology to the intelligence needs of each era. Whether CIA operations required an ultraminiature camera, a battery the size of a fingernail, or travel documents with unique inks and printing, OTS became the organization that did not just make magic, it made magic on demand.

In 1951, the future of the United States and Western democracy was confronted by an ideology of advancing communism sponsored by a nuclear-armed Soviet Union. In those uncertain times, leaders of the four-year-old CIA, DCI Walter Bedell Smith, his deputy Allen Dulles, and a promising operations officer, Richard Helms, envisioned technology as the means to secure a decisive intelligence advantage over the Soviet Union and its client states. Drawing on their collective World War II experiences in the Office of Strategic Services, they concluded that the best operations were a partnership between technical specialists and operations officers. The concept they enacted was simple and brilliant—CIA would apply the full force of America's technological ingenuity, whether sponsored by government or industry, to solve the problems of clandestine operations. From that idea, the Technical Services Staff emerged and its successes became legendary.

Now some of the previously untold stories of the impact on our intelligence history by this remarkable collection of people and technology can be told. Every CIA director confronts the tension between secrecy and the American public's right to know what its government is doing. Secrets are the necessary currency of the intelligence profession and protection of confidential sources and special methods is a solemn duty of every CIA officer. Regrettably, there have been instances when secrecy was invoked to deny knowledge of information that has long since lost sensitivity but is vital for public understanding and consideration. Such misuse of secrecy can result in flawed policy decisions, wild speculation about the CIA's activities, and a misleading historical record. For the CIA to maintain the public trust, responsible and accurate presentation of information on intelligence subjects is both wise and necessary.

The thousands of books and news articles produced about the CIA's operations have generally concentrated on large technical programs, such as the U-2 spy plane, satellites and communications intercepts, or spies who worked for the CIA or against American interests. In spy episodes, the technical equipment, an essential element of agent operations, is often obscured by the drama and action surrounding the human characters. Little attention had been given to the technological origins of the gadgets or the people who made them. *Spycraft* now presents this well-told story in a long overdue tribute to the previously unrecognized contributions to U.S. intelligence of an eclectic, talented collection of scientists, engineers, craftsmen, artists, and technicians. *Spycraft* is a history of the CIA's fusion of technical innovation with classic tradecraft and, equally, a call to young men and women with similar talents to enlist in the battle against America's new enemies.

The authors of *Spycraft* have been my associates and friends for many years. Together they bring a career of operational experience and a lifetime of study of intelligence to this work. During my tenure as DCI, Robert Wallace headed the Office of Technical Service after serving more than twenty-five years as an Agency operations officer and manager. Throughout his thirty-two-year career, he received multiple awards for operational success and leadership. H. Keith Melton has been a friend of the CIA for more than two decades. He is a frequent lecturer throughout the intelligence community, a bestselling author, and an internationally recognized collector and interpreter of historical intelligence technical devices and artifacts.

My draft of September 7, 2001, remarks concluded with the the observation that "the twenty-first century will present major challenges to our Agency and OTS ingenuity will be put to the test in the years ahead." The test began four days later. In the months that followed 9/11, the CIA again turned to OTS for technical innovation to build the gadgets that would detect and defeat a different and deadly enemy operating in an information environment revolutionized by the Internet and digital technology. *Spycraft*, whether detailing Cold War operations or those directed against terrorists, offers a somber warning to our adversaries and fresh encouragement to those who cherish freedom that we will prevail.

Preface

The CIA's new Deputy Director for Operations, David Cohen, called me to his office in August 1995. "I want you to apply for the job of Deputy Director, Office of Technical Service in the DS&T [Directorate of Science and Technology]," Cohen directed. "We need a DO [Directorate of Operations] person there and I think you're a good candidate."

He might as well have suggested that I apply for NASA's astronaut program. I had been an operations officer for almost twenty-five years, but for the past eighteen months I was assigned to the Comptroller's office. At age fifty-one, I was out of operations, doing the type of staff and budget work that motivated me to plan for early retirement.

"I've never worked in the DS&T. I'm a history–political science major, an operations type. I'm an analog guy in a digital world. I don't even change the oil in my car," I objected.

"I know what your skills are and this is a good assignment for you." Cohen left no doubt about the answer he wanted.

"Okay, I'll apply, but I can't imagine I'll be competitive if there are other candidates."

"There'll be other candidates and you'll do fine. We need a DO officer in OTS who knows senior ops people and is someone I have confidence in. You've been looking at the budgets in every DO division almost two years, so you know the key players and they know you. I have to make sure technical service and operations stay linked."

The interview was over. I had worked for Cohen several years earlier and recalled his frequent admonition that once a decision was made he had no patience for further discussion. This was not the first time that he had directed me to a job I had not sought, and the others had worked out pretty well. In 1988, he sent me to a large station that had never been on my assignment wish list. That put me into position to join the ranks of the CIA's Senior Intelligence Service. Three years later he ordered me back to Headquarters to serve as a division-level resource manager. Since I had spent the previous eighteen years in various field stations, this responsibility introduced me to a previously unknown world of billion-dollar budgets and the Agency's senior leadership.

Six weeks following the conversation with Cohen, after separate interviews with the Executive Director, Nora Slatkin, and the Deputy Director for Science and Technology, Dr. Ruth David, both recent appointees to their positions, I became Deputy Director, OTS.[1] Evidently, they agreed with Cohen that the job required a breadth of operational and management experience more than a technical degree.

"OTS is America's 'Q,' sort of," said "Roy," in welcoming me to the office and offering no apology for the reference to the gadget master of James Bond movies. Roy had spent his first ten years at OTS in the forgery shop working as a "document authenticator," making certain that CIA-produced travel and alias-identity documents were flawless in print type, color, design, and paper texture. Now, as a senior staff officer for Robert Manners, the Director of OTS, he had drawn the task of providing the new guy with a much-needed *CliffsNotes* version of the office. "I say 'sort of,' " Roy continued, "because, unlike the movies, if one of our visas doesn't pass muster at an immigration checkpoint, or one of our concealments accidentally opens and spills its contents, we can't reshoot the scene. If people are arrested or get killed because of our mistakes, they stay in jail for a long time or they really die."

Roy also made it clear that America's "Q" consisted of not one scientific genius or a handful of eccentric inventors, but a large contingent of technical officers, engineers, scientists, technicians, craftsmen, artists, and social scientists deployed throughout the world and cross-trained in operational tradecraft. OTS had a hand in every aspect of the CIA's spy gear from design and development through testing, deployment, and maintenance.

"Now, this is what's really important," Roy said, beginning the comprehensive briefing with a slow, deliberate delivery that conveyed no-nonsense seriousness. "We consider ourselves part of the Directorate of Operations as much as a part of the DS&T. Whatever the DO stations and case officers need for technical support, we do everything in our power to deliver. When we go to the field to do a job, there's no question who we work for—the chief of station."

Roy explained that the techs did much more than build and deliver spy gear. "Usually we are right there with the case officer or the agent, at the user's side in the operation. We train agents, install equipment, test systems, and repair stuff that breaks. We take the same risks as case officers—share the same emotion of accomplishment or otherwise. Over the course of his career, the tech becomes involved in more operations and meets more agents than many case officers."

Roy described OTS's five primary organizational elements, or "groups," as these were designated. The largest was a covert communications, or "covcom," group with a name that described its function. This group developed systems for agents and case officers to communicate covertly and securely. Secret writing, short-range radio, subminiature cameras, special film, high-frequency broadcasts, satellite communications, and microdots were all included in covcom. A second OTS group designed and deployed audio bugs, telephone taps, and visual surveillance systems. These techs were often on the road up to fifty percent of the time, traveling from country to country, as their services were required. The third group, called on for special missions that may include support to paramilitary operations, included a mixture of technical and "soft science" capabilities. This group produced tracking devices and sensors, conducted weapons training and analysis, analyzed foreign espionage equipment, performed operational psychological assessments, and built special-use batteries. Roy came from a fourth group that made disguises and "reproduced" documents. Its work in creating counterfeit travel documents could be traced directly back to a predecessor organization in the Office of Strategic Services. Rounding out OTS were the concealments and electronics fabrication laboratories, known collectively as Station III, and a field structure with regional bases in South America, Europe, and Asia.[2]

Roy's briefing supplemented my prior knowledge of OTS from two recent assignments, one operational and the other administrative. For two years in the early 1990s, I served as Deputy Chief for the CIA's nonofficial cover (NOC) program. There I worked with OTS officers who support NOC officers with documentation, covert communications, disguise, identities, and concealments to assure the NOCs were never identified with the U.S. government. The OTS provided the equipment and documents that enabled

NOCs to live a "normal" life as, say, a businessman, freelance photographer, scientist, or rice merchant while engaging in their clandestine work for the Agency.

In the Comptroller's Office, I encountered OTS from the perch of a budget weenie.[3] Beginning in 1991, after the collapse of the Soviet Union, the Comptroller had the unenviable task of managing a declining CIA budget at a time when operations officers were, in reality, being pressed by demands for new, better, and faster intelligence on counterproliferation, counterterrorism, and counternarcotics. OTS, like other components of the CIA, struggled to absorb the impact of reduced budgets without any reduction in demands for spy gear.

For the next three years as OTS Deputy Director or Acting Director, I would deal firsthand with the damage that the budget cuts of the 1990s did to the CIA's countersurveillance systems, advanced power sources, technical counterintelligence capabilities, and paramilitary-related weapons and training.[4] Then, beginning in 1999, as new resources began to be available, I would have the opportunity as Director to lead OTS in creating and reconstituting capabilities for the twenty-first century.

From its formation in 1951, OTS concentrated its efforts on creating devices and capabilities to improve the CIA's ability to identify, recruit, and securely handle clandestine agents. Whether the operational requirement needed research, development, engineering, production, training, or deployment, OTS responded. Motivated by a philosophy of limitless possibility, a few hundred technical specialists gave American intelligence its decisive technical advantage in the Cold War, a conflict that continues today in the worldwide battle against terrorists.

Collectively, the stories that form the OTS history convey a level of dedication and commitment by officers whose pride in their service to America was more important than personal wealth or individual acclaim. At their best, these experiences are models for successive generations of intelligence officers who would apply technology to agent operations. I cannot imagine a more rewarding responsibility or an honor greater than working with this remarkable cadre of technical officers, successors to the rich heritage of OSS General William Donovan and his chief technical genius, New England chemist Stanley Lovell.

The genesis of *Spycraft* occurred during an afternoon-long conversation with John Aalto, a retired case officer, in San Antonio in February 1999. I had been appointed Director of the CIA's Office of Technical Service three months earlier. John had joined the CIA in 1950 and spent the next five decades in Soviet operations.

John took note of my recent appointment and with unexpected seriousness asked, "Do you have any concept of what OTS and its predecessor, the Technical Services Division, accomplished for operations?"

Before I could respond, John continued. "I tell you," he began, "it is because of the techs and TSD that we in Soviet operations eventually won the intelligence war against the KGB in Moscow. And to my knowledge, no one has ever recorded that story, officially or unofficially."

Over the next three hours John described a remarkable inventory of TSD devices, technologies, inventions, gadgets, and tricks that he and others used in Moscow and throughout the Iron Curtain countries during the forty-year Cold War. He recounted fascinating tales about the leadership of Dr. Sidney Gottlieb, the cleverness of the TSD engineers, the inventiveness of the field techs, and the determination shared by TDS and Soviet Division case officers to break the stranglehold of the KGB on the CIA's operations in Moscow.

"You should do something," John urged, "to get this story recorded before all of us who were involved are gone and the inevitable organizational changes at CIA obscure this history."

Two years earlier I had met H. Keith Melton, a lifelong student of intelligence history and private collector of espionage devices and equipment. Keith lent the Agency hundreds of artifacts from his private collection of espionage equipment for display during the CIA's fiftieth anniversary in 1997. Subsequently, I assisted Keith in transforming the display into a permanent Cold War exhibit in CIA's Original Headquarters Building. On September 7, 2001, OTS celebrated its fiftieth anniversary with a gala dinner highlighted by Keith's presentation of the international history of spy gadgets and technical espionage.

Shortly after I retired, Keith and I had dinner during one of his visits to Washington. As we shared our admiration of the creativity and courage of the engineers and technical officers whom we had come to know, Keith asked if I had considered writing an account of my tenure as OTS Director. I had not, but his question reminded me of John Aalto's admonition four years earlier and sparked the idea of writing a public history of OTS from the accounts of retired technical officers. It would be a true espionage story that, combined with Keith's wealth of knowledge and images of historical spy gear, could be a valuable addition to intelligence literature. Keith agreed, and *Spycraft* was born.

We understood the obligations from my CIA employment to submit writing about intelligence subjects to the Agency for prepublication review to preclude the inadvertent release of classified information. I anticipated no particular difficulties with such review. Before beginning the project, I met with the CIA's Publications Review Board, outlined the concept, and received encouragement to proceed. In July 2004, the board approved a detailed outline of a proposed "popular account of OTS adventures and contributions to U.S. intelligence" along with the two sample chapters we had submitted. Relying on that approval, we contracted with Dutton, an imprint of Penguin USA, for publication, with full expectation of delivering a properly Agency-reviewed manuscript in late 2005. We submitted our 774-page manuscript under the title *An Uncommon Service*, to the board on September 6, 2005. Agency regulations specify manuscripts are to be reviewed within thirty days.

After six months, on March 13, 2006, the board issued us a letter stating: "except for Chapters 1–3 your manuscript is inappropriate for disclosure in the public domain." The Agency had approved only the first thirty-four pages, all of which discussed equipment from the Office of Strategic Service (OSS) World War II inventory. The 740 "inappropriate" pages included the previously approved detailed outline and sample chapters. No specific classified material was identified. Rather, the Agency applied a previously discredited "mosaic theory" of redaction, contending that a compilation of unclassified information becomes classified when written by someone at my senior level. The board's letter asserted that "in the aggregate the manuscript provides so much information . . . it would be of immense value to our adversaries." There seemed to be no awareness that adversaries read English and have the same Internet access and Google tools we used in our research.

During my previous seven years with OTS, I reviewed several books and articles as part of the Agency's prepublication review process. In preparing this manuscript, I exercised the same conscientious judgment regarding potentially classified information as I had done as a government employee. In its attempt to prevent the authors from publishing *Spycraft*, the March 2006 letter revealed the Agency's apparent unwillingness to distinguish between responsible writing on intelligence subjects and unauthorized leaks of classified information.

With the assistance of attorney Mark Zaid, we filed an appeal two weeks later. Such appeals, according to Agency regulations, would be adjudicated by the CIA's Executive Director within thirty days of receipt.

We received no response to our appeal for eight months. Mid-level officers of the bureaucracy took no action in what appeared to be an attempt to deny publication by causing an indefinite delay. Faced with the unwillingness of the Agency to conduct a review consistent with its prepublication policies, we prepared to seek relief in federal court. In the opinion of our legal counsel, the Agency's refusal to honor its own regulations, coupled with the capricious deletions of unclassified material from the manuscript, constituted a violation of First Amendment Constitutional rights.

Before taking the legal step, we made a personal request to the CIA's Associate Deputy Director in December 2006 for intervention. As a result, on February 8, 2007, we were advised that another review had reduced objections to approximately fifty of the manuscript's pages. Further, the board offered to reconsider the remaining deletions if the authors could demonstrate the material was not classified. Although we believe none of the disputed material is classified, as an accommodation, we revised certain passages and deleted some terminology that the CIA considered operationally sensitive. On July 18, 2007, we received approval to publish virtually all of the original manuscript.

The best that can be said of the experience is that Agency management eventually recognized a need to reform its prepublication policy and repair the broken review process. A historical irony is that William Hood encountered a similarly recalcitrant bureaucracy in 1981 when writing *Mole*, an account from the 1950s of the Soviet spy Pytor Popov.[5] "Every word in this manuscript is classified," said the initial CIA review. Twenty-five years later, *Mole* is now recognized as an espionage classic.[6]

The first five sections of *Spycraft* recount remarkable stories of ingenuity, skill, and courage throughout the first fifty years of OTS history. Section VI presents the doctrine of clandestine tradecraft from the perspective of espionage historian H. Keith Melton and includes a chapter devoted to the revolutionary changes digital technology has brought to spy work.

We wrestled from the beginning with the difficulties of when to present necessary explanations of the operational doctrine behind the technical topics that appear in the text. The impracticality of repeating explanations each time a technical topic appeared became quickly obvious. Lengthy footnotes also seemed more likely to distract rather than enlighten the reader.

Therefore, we consolidated into Section VI the five essential elements of clandestine operations used by every intelligence service regardless of nationality or culture. These chapters, drawn from Melton's widely acclaimed lectures, writings, and exhibits, set out the basic principles underlying technical support to operations. These principles transcend any specific service and represent knowledge common and available to intelligence professionals and civilians alike from print, electronic, and film media. The individual chapters will aid the reader in understanding the basic philosophy and principles of assessment, cover, concealments, surveillance, and covert communications as practiced by professional services. Readers have the option of diving directly into the OTS story and the development of CIA's clandestine spy gear in Chapters 1–19 or first immersing themselves with the doctrine and terminology of espionage operations presented in Chapters 20–25.

Spycraft combines the experiences and lore of the techs based on the authors' personal interviews and correspondence with nearly one hundred engineers, technical operations officers, and case officers. We verified specific details to the extent possible by collaboration with public material and multiple primary sources. The names of several individuals quoted by the authors throughout the book are changed as a matter of security, cover, or requested privacy. Appendix E provides a list of pseudonyms the authors assigned to these officers. Otherwise, we use true names throughout.

We did not seek access to, or use, classified files. At times, the fallibility of memory may produce less than a perfectly accurate account of events many years past. In a few instances, we purposefully obscured facts to protect operational information, or omitted sensitive details for the same reason. For example, the locations of operations, except those in Moscow, the former Soviet Union, and other denied area countries, are regionalized. Some operational terms and Agency jargon that appear in works by other authors not bound by secrecy agreements have not been used at the request of the Agency.

Why do history? Two thousand years ago Cicero observed, "To be ignorant of what occurred before you is to remain always a child. For what is the worth of life unless it is woven into the lives of our ancestors by the records of history?" A twentieth-century view, as expressed by G. K. Chesterton, is: "In not knowing the past we do not know the present. History is a high point of vantage from which alone we can see the age in which we are living." Richard Helms, who headed CIA operations in the early days of the Cold War and served as Director of Central Intelligence from 1967 to 1973, explained that he wrote *A Look Over My Shoulder* because it is "important that the American people understand why secret intelligence is an essential element of our national defense."[7] Our hope is that *Spycraft* becomes a part of that legacy.

—RW

INTRODUCTION

Future Technology Challenges in Espionage

by MICHAEL J. SULICK
*Former Director of CIA's National
Clandestine Service*

Technology has revolutionized every facet of human existence and espionage is no exception. Some of the most stunning developments are occurring in nanotechnology and information technology, two areas that are particularly well suited to the world of espionage. Nanotechnology is the science of manipulating materials on a molecular scale, especially to build microscopic devices. This technology already plays a role across a broad spectrum of the human enterprise, aiding in the development of smaller and faster computers, lightweight spacecraft, and even implants in the human body to dispense medicine more effectively to sick patients.

The ability to miniaturize, of course, is essential to espionage, where concealment, one of the "five pillars" discussed in Chapter 22, is crucial to the security of clandestine collection, storage, and transmission of secret information. Smaller, more easily concealable devices are especially applicable to collecting intelligence on two of the gravest national security threats that will continue to confront the United States in coming years—terrorism and the proliferation of weapons of mass destruction. Gathering intelligence on these threats is extremely difficult. Terrorist hideouts and training camps are often located in remote areas inaccessible to Westerners, and weapons research and storage sites of interest are heavily guarded facilities in regimes closed to the outside world. "Clandestine intelligence collection," as CIA Director Allen Dulles noted, "is chiefly a matter of circumventing obstacles in order to reach an objective."[1] To overcome these obstacles, miniaturization and its impact on concealment will play a critical role in infiltrating an array of devices against these targets.

As one example, the U.S. intelligence community relies heavily on sensors to collect data on the unintended emissions from targets such as weapons systems, nuclear tests, and missile launches. Thanks to nanotechnology advances, future research and development will reduce the size of sensors for cover enhancement while also increasing power capacity to ensure long-term collection.[2]

Nanotechnology will also have a dramatic impact on the burgeoning science of robotics. So-called "nanorobots" are under development to search out and destroy pathogens and cancer cells in the human body. The Defense Advanced Research Projects Agency (DARPA) is already working on Unmanned Aerial Vehicles (UAVs) the size of an insect.[3] The miniature size, maneuverability, and potential cover they afford also make these nanorobots ideal candidates for espionage. Nanorobots masked as insects or small reptiles could eventually crawl or slither undetected into an adversary's weapons facility or the inner councils of a hostile regime. Nanotechnology has also facilitated the integration of various data technologies into a single small device, and thus an ostensible "insect" could remain unnoticed in a corner of a room simultaneously collecting and transmitting voice, image, and other types of data, adding a futuristic twist to the traditional practice of "bugging."

The convergence of nanotechnology and information technology also will have critical applications in espionage. "Real agents," as military historian John Keegan indicates, "are at their most vulnerable when they attempt to reach their spymasters. The biographies of real agents are ultimately almost always a story of betrayal by communications failure."[4] Advances in information technology have already resulted in smaller, more easily concealable devices storing larger volumes of data transmittable at higher speeds, thus improving both the security and efficiency of clandestine communications. The slang term "sticks and bricks" for the older method of "impersonal communications" discussed in *Spycraft* could be replaced nowadays by "bugs and bots."

The ever shrinking size of electronic devices will increasingly facilitate the secure collection and transmission of secrets. Spies like Jonathan Pollard packed satchels full of secrets and smuggled them from classified facilities. Pollard, in fact, first came under suspicion when a colleague noticed him lugging batches of materials out of his office building. This risky practice has been virtually eliminated now that secret information resides in computers and can be downloaded to easily concealed flash drives and memory cards capable of storing large volumes of data. The storage capacity of these portable devices is expected to increase to two terabytes or more of data, the equivalent of two thousand copies of the *Encyclopedia Britannica* (ten terabytes could hold the print material of the entire Library of Congress).[5] Soon a spy will be able to download the complete holdings of entire ministries into such devices. Beyond this practice, the imperfect human elements in using an agent are also eliminated from cyber-espionage when computers can be penetrated remotely and information from their hard drives exfiltrated without the knowledge of the owner.

One goal of future information technology is "pervasive computing," which envisions integrating data technologies and embedding them in everyday items such as clothing, appliances, and even inside human skin at the same time as they are connected to a vast array of other networks.[6] In the world of human intelligence, this development would significantly enhance the cover and thus the security of an agent even more and facilitate his ability to communicate with his case officer on a real-time basis.

This integration of information technologies will prove essential to the quality and effectiveness of intelligence collection and analysis. The U.S. intelligence community has been severely criticized in the past, often rightfully so, for its "stovepipes," the isolation of its various collection methods that has impeded the sharing of information and integration of data across its disciplines, old-fashioned human spying, electronic intelligence, and photo reconnaissance, to name but a few. Information technology has significantly enhanced the integration of data and already resulted in a truly "all-source" approach to solving difficult intelligence questions.

The future holds even more promise for further integration. Current data technologies already available on devices such as cell phones and computer tablets will be supplemented by newer voice, image, text, mapping, and numeric software applications. Imagine this scenario: A well-placed agent in a volatile country learns of an impending military coup that could have adverse impact on U.S. national security. Thanks to pervasive computing, the agent has ready—and more secure—access to his communication device and can swiftly alert his case officer and send him documents, audio of actual conversations among the plotters, and maps and images of military preparations.

The case officer, linked directly with analysts at his headquarters, relays the information to them. In return, the analysts, armed with more advanced text mining and link analysis software and databases, provide results of voice recognition tests, initial assessment of the documents for authenticity, information on the plotters gleaned quickly from a host of databases. Finally, they provide the case officer with an initial assessment of the information and additional questions to obtain an even more thorough and up-to-date picture of the crisis.

One caveat, however. Technology is a double-edged sword. A characteristic of technological advances is their widespread availability. Thus the same technology to improve espionage capabilities are and will be used against us by nations rich and poor and non-state actors like terrorists. As one example, the authors of *Spycraft* note that the availability of personal database information can significantly facilitate an intelligence service's spotting of potential spies. The proliferation of social media networks also provides similar opportunities for spotting and assessing potential recruits. Foreign intelligence services, however, will undoubtedly exploit these opportunities in the future to dispatch "virtual" double agents to feed false information, discover our modus operandi, or simply waste resources on a useless operation.

The electronic barrier between interlocutors inherent in these social networking sites is ideally suited for deception and has already been exploited by criminals and predators of every stripe assuming a range of false identities with their targets on the other end of the network. In espionage, the development of a cyber-relationship without direct human contact prevents the vetting and assessment of a potential spy that can only be accurately obtained by the interaction of a case officer and his target in face-to-face encounters.

Another form of information technology that appears far less sensational than other James Bondish wizardry currently portrayed in film and fiction is machine translation. The ability of computers to quickly and accurately translate foreign language documents for use by analysts will save hours of time of translators poring over texts, especially material on advanced technical topics. At the same time, this will undoubtedly make the study of foreign languages even less appealing and exacerbate the chronically woeful state of Americans' ability to communicate with foreigners.

Biometrics, measures of unique physical or behavioral characteristics that can authenticate an individual's identity, is another area of information technology with applications in espionage. Biometrics includes a vast array of measures ranging from digital fingerprint identification, iris and retinal scans, voice and facial recognition, to scanning the veins in one's hand and gauging the gait of one's stride. Some of these measures are already deployed to control access to facilities and provide a reliable defense against thieves, criminals, and terrorists.

Despite the defensive advantages, biometrics will entail risks as well as benefits in future espionage. The ability to determine identity beyond a shadow of doubt will present daunting challenges to intelligence officers operating in alias and disguise. False passports and old-fashioned disguises of eyeglasses and a beard plastered on with spirit gum are already ineffective and vulnerable to exposure. The same disadvantage will increase regarding global positioning systems (GPS) and other tracking and beaconing devices such as RFIDs (radio frequency identification devices), which are in everyday use in a number of areas. The very technology that already enables the tracking of stolen goods or the electronic payment of highway tolls can also be deployed against intelligence officers to simplify the efforts of adversaries to track their movements and activities and frustrate their ability to collect secrets for the nation's security.

The technological advances cited above are merely a few of those that already are and will continue to have applications in the world of espionage. Sun Tzu, the Chinese philosopher and military strategist, emphasized the importance of the "divine skein" of espionage, the threads of a fishing net that must be pulled together to obtain the most effective intelligence picture of a situation. Technology has increasingly become one of the primary cords pulling the strands of intelligence together in a more integrated fashion. Despite these advances, in the foreseeable future technology alone will not overcome all the barriers to the collection and analysis of intelligence. Technology, no matter how well developed, is still invented and operated by human beings. Machines malfunction. Human beings make mistakes.

Our adversaries are also human beings and can also counter our technology with their own measures. A hostile dictator can deny access to the Internet during a crisis just when a spy in his circle must rely on that electronic backbone to communicate with his case officer—sophisticated technology is rendered useless precisely at the moment when it is most needed. Intelligence services must consider such scenarios and be prepared to rely on old-fashioned, nontechnical solutions as emergency backup plans.

U.S. intelligence will also have to rely on human beings, those spies inside the enemy's camp who can reveal the plans and intentions of terrorists and leaders of closed regimes. No search engine can unearth the motives that may impel a person to spy or the thoughts of a hostile dictator. No software can yet replicate the complex human interaction between a case officer and his agent. There is still no substitute for the case officer huddled alone with his agent, judging the subtle changes of mood that may be affecting his cooperation, his reliability, and the quality and veracity of the information he provides.

Some believers in so-called "techno-utopias" predict that in the distant future technology will enable us to peer into the human mind and divine the inner thoughts of others. Scientists are already investigating possibilities of "whole brain emulation" or "mind transfer," a hypothetical process in which the neural network of a human brain is scanned into a computer.[7] The project would clearly have a decisive impact on espionage and possibly render the practice irrelevant. At this point, however, such visions are in the realm of science fiction. Still, the history of technology is littered with the comments of disbelievers who underestimated the human imagination. In 1946, Darryl Zanuck, famed film producer and studio executive, claimed, "Television won't be able to hold on to any market it captures after the first six months. People will soon get tired of staring at a plywood box every night."[8] Three decades later, in 1977, Ken Olsen, the cofounder and CEO of the Digital Equipment Corporation, claimed, "There is no reason anyone would want a computer in their home."[9]

So anything is possible. But, until these hypothetical marvels occur, U.S. intelligence will continue to rely on the imagination and ingenuity of our "spytechs" to apply technological advances to our national security. Based on their history thus far, the nation should be well served.

Official Message
from the CIA

The Central Intelligence Agency requested the following message be included in *Spycraft*. To provide the reader a sense of the reality of covert communications, the authors have presented the message using a page from a one-time pad issued to Aleksandr Ogorodnik (*TRIGON*) in 1977. Chapter 8 presents the *TRIGON* story. Use the one-time pad on page 97 and the instructions in Appendix F to decipher this message.

25886	14155	75126	50200	19082	18193	73799	86932
21351	10043	47273	79962	35859	31419	67511	71466
74048	43427	79468	17464	21551	05369	23777	57954
39206	81440	75115	34678	27628	64265	95474	68273
07454	30545	57041	64491	84617	37194	65182	32028
10856	42127	98147	08212	80461	19159	76906	96350
06801	03739	79616	59897	74718	88039	57655	43996
96548	60171	10516	80703	42355	55453	18959	54960
48072	34595	24879	89432	74811	30669	49194	06105
86431	91706	51389	41559	57081	45856	29817	88628
45609	60007	85961	33296	91619	73179	04316	16318
78511	63202	26270	04975	57067	87112	06824	18890
23476	00497	12853	10704	85157	82625	80302	39568
98740	89702	62880	27515	01159	00782	10019	09324
76309	62253	29920	93879	79588	50325	30160	63686
36758	94379	96557	55805	16400	36597	45151	17432
68270	84821	11592	28099	35403	73705	90023	31866
41596	83244	40964	59866	92175	01481	85834	93496
68589							

AT THE BEGINNING

7 September 1951

1. Effective immediately the Operational Aids Division is redesignated the Technical Services Staff.

CHAPTER 1

My Hair Stood on End

<hr>

The weapons of secrecy have no place in an ideal world.

—Sir William Stephenson, *A Man Called Intrepid*

On a quiet autumn evening in 1942, as World War II raged across Europe and Asia, two men sat in one of Washington's most stately homes discussing a type of warfare very different from that of high-altitude bombers and infantry assaults. The host, Colonel William J. Donovan, known as "Wild Bill" since his days as an officer during World War I, was close to sixty. A war hero whose valor had earned him the Medal of Honor, Donovan was now back in uniform.[1] Donovan responded to the call to duty and put aside a successful Wall Street law practice to become Director of the Office of Strategic Services (OSS) and America's first spymaster.[2]

Donovan's guest, for whom he graciously poured sherry, was Stanley Platt Lovell.[3] A New Englander in his early fifties, Lovell was an American success story. Orphaned at an early age, he worked his way through Cornell University to ascend the ranks of business and science by sheer determination and ingenuity. As president of the Lovell Chemical Company, he held more than seventy patents, though still described himself as a "sauce pan chemist."

Donovan understood that the fight against the Axis powers required effective intelligence operations along with a new style of clandestine warfare. Just as important, he appreciated the role men like Lovell could play in those operations. "I need every subtle device and every underhanded trick to use against the Germans and the Japanese—by our own people—but especially by the underground in the occupied countries," he had told Lovell a few days earlier. "You'll have to invent them all . . . because you're going to be my man."[4]

The wartime job offered to the mild-mannered chemist was to head the Research and Development (R&D) Branch of the OSS, a role Donovan compared to that of Professor Moriarty, the criminal mastermind of Sir Arthur Conan Doyle's Sherlock Holmes stories.[5] Lovell, although initially intrigued by the offer, was now having doubts and came to Donovan's Georgetown home to express those reservations.[6] He had been in government service since that spring at a civilian agency called the National Defense Research Committee (NDRC). Created by President Roosevelt at the urging of a group of prominent scientists and engineers, the NDRC's mission was to look into new weapons for what seemed to be America's inevitable entry into the war. Lovell had joined the NDRC to act as liaison—a bridge—between the military, academics, and business.[7] But what Donovan proposed now was something altogether different.

The mantle of Professor Moriarty was, at best, a dubious distinction. An undisputed genius, the fictional Moriarty earned the grudging respect of Holmes by secretly ruling a vast criminal empire of London's underworld with brutal efficiency and ingenuity. In his role as Professor Moriarty of the OSS, Lovell would oversee the creation of a clandestine arsenal that would include everything from satchel concealments to carry secret documents and subminiature spy cameras to specialized weapons and explosives. These were the weapons to be used in a war fought not by American troops in uniform, but by soldiers of underground resistance movements, spies, and saboteurs.

Spying and sabotage were unfamiliar territory for both America and Lovell, who had made his fortune developing chemicals for shoe and clothing manufacturers. America, Lovell believed, did not resort to the subterfuge of espionage or the mayhem of sabotage. When the United States looked into the mirror of its own mythology, it did not see spies skulking in the shadows of back alleys; instead, it saw men like Donovan, who faced the enemy in combat on the front lines.

"The American people are a nation of extroverts. We tell everything and rather glory in it," he explained to Donovan. "A Professor Moriarty is as un-American as sin is unpopular at a revival meeting. I'd relish the assignment, Colonel, but dirty tricks are simply not tolerated in the American code of ethics."[8]

Donovan, as Lovell would later write, answered succinctly. "Don't be so goddamn naïve, Lovell. The American public may profess to think as you say they do, but the one thing they expect of their leaders is that we be smart," the colonel lectured. "Don't kid yourself. P. T. Barnum is still a basic hero because he fooled so many people. They will applaud someone who can outfox the Nazis and the Japs. . . . Outside the orthodox warfare system is a great area of schemes, weapons, and plans which no one who knows America really expects us to originate because they are so un-American, but once it's done, an American will vicariously glory in it. That is your area, Lovell, and if you think America won't rise in applause to what is so easily called 'un-American' you're not my man."[9]

Lovell took the job. Donovan knew what he wanted, but even more important, he knew what was needed.[10] He had toured the secret labs of Great Britain that created just such devices. He also maintained close ties with the British Security Coordination (BSC), England's secretive intelligence organization in North America, through which the United States was already funneling weapons to assist in the war effort. Even the mention of Sherlock Holmes's ruthless criminal adversary may not have been a chance literary allusion. Two years earlier, in 1940, British Prime Minister Winston Churchill signed into existence the Special Operations Executive (SOE) with the instructions "Now go out and set Europe ablaze!"[11] SOE's mandate was unconventional warfare, including the arming of resistance fighters in the war against Germany. Its London headquarters was an undistinguished office building on Baker Street, the same street as Sherlock Holmes's fictional address.

Although Donovan eventually persuaded Lovell to join the OSS, the chemist's initial assessment of the American public's dim view toward espionage was not unfounded. From the beginning, the idea of an American intelligence service was controversial. One senator proclaimed, "Mr. Donovan is now head of the Gestapo in the United States."[12] In the best tradition of Washington's bureaucratic infighting, the person in charge of the State Department's Passport Office, Mrs. Ruth Shipley, insisted on stamping "OSS" on the passports of Donovan's personnel traveling overseas, making them perhaps the most well-documented secret agents in the history of espionage. To remedy the situation, which had reached a deadlock between the OSS and the State Department, FDR himself had to intervene on the young agency's behalf with the stubborn Mrs. Shipley.[13]

The media of the day was no more charitable, often treating the OSS dismissively. The Washington columnist Drew Pearson called the nascent spy agency "one of the fanciest groups of dilettante diplomats, Wall Street bankers, and amateur detectives ever seen in Washington."[14] More colorful phrases were penned by Washington's *Times-Herald* society columnist, Austine Cassini, who breathlessly wrote:

> *If you should by chance wander in the labyrinth of the OSS you'd behold ex–polo players, millionaires, Russian princes, society gambol boys, scientists and dilettante detectives. All of them are now at the OSS, where they used to be allocated between New York, Palm Beach, Long Island, Newport and other Meccas frequented by the blue bloods of democracy. And the girls! The prettiest, best-born, snappiest girls who used to graduate from debutantedom to boredom now bend their blonde and brunette locks, or their colorful hats, over work in the OSS, the super-ultra-intelligence-counter-espionage outfit that is headed by brilliant "Wild Bill" Donovan.*[15]

Cassini made it all sound like good clean fun. A bastion of pampered blue bloods, the OSS seemed no more dangerous than a country club cotillion. But at a time when less privileged sons and husbands were fighting and dying in the South Pacific and North Africa, the levity in the words "gambol boys" and "dilettante detectives" was almost assuredly bitter reading for many. Not surprisingly, the organization's acronym was soon transformed into the less than flattering "Oh So Social" by career military officers and draftees alike. The fact that an early OSS training facility was based at the plush Congressional Country Club, located just outside Washington, only served to reinforce the notion of privilege and elitism.[16]

If OSS seemed a bastion of aristocrats and bankers, it was not without reason. Donovan worked on Wall Street in the days leading up to World War II. When he became Coordinator of Information (COI), an OSS predecessor, in 1941, Donovan staffed the organization from circles with which he was familiar—the New York legal, business, and financial worlds—along with graduates from the nation's finest universities. However, there was more to this than simply establishing an "Old Boys' Club" of espionage. Prior to World War II, travel opportunities for abroad and learning foreign languages were largely limited to the privileged. As a result many of those recruited came with intimate knowledge of the European landscape, including the cities and towns of France, Germany, and Italy, from past travels. Others had done business in Europe before the war and could reestablish contacts.

Less visible than the privileged blue bloods were the refugees, those recent immigrants and first-generation native-born Americans (many of them academics) who also joined the ranks of the OSS. Unlike the Wall Street bankers and ex–polo players, these recruits brought day-to-day knowledge of foreign cultures, along with clothing, identity papers, and language skills.[17]

Even as it became the target of Washington infighting and attracted the derision of newspaper columnists, Donovan's organization expanded rapidly.[18] If the United States was going to enter what Rudyard Kipling called "the Great Game" of international espionage Donovan needed to move quickly. Spurred on by the urgency of war, the OSS would share clandestine responsibilities with the Allies. The London Agreements, negotiated in 1942 and 1943,[19] established a protocol for clandestine cooperation between OSS and the SOE, defining each side's role, down to the development of weaponry and financial responsibilities. Theaters of secret operations were divided between the United States and Great Britain. OSS had responsibility for China, Manchuria, Korea, Australia, the Atlantic Islands, and Finland, while SOE covered India, East Africa, the Balkans, and the Middle East. Western Europe would remain primarily British, with U.S. representation.[20]

As "junior partner" in this joint wartime venture, Donovan needed to build not only America's first spy agency, but one capable of waging a global intelligence war. This was no easy task. Whatever espionage legacy remained from previous wars was largely out of date or forgotten. He would have to assemble the organization from the ground up with assistance from the British. The United States provided technology while Britain offered experience and counsel, training Americans in the craft of intelligence.

The blue bloods, so easily dismissed by the society columnists as frivolous playboys and genteel sportsmen, learned quickly from their British tutors.

"Ah, those first OSS arrivals in London!" wrote veteran British intelligence officer Malcolm Muggeridge. "How well I remember them arriving like *jeune filles en fleur* straight from a finishing school, all fresh and innocent to start work in our frowsty old intelligence brothel. All too soon they were ravished and corrupted, becoming indistinguishable from seasoned pros who had been in the game for a quarter century or more."[21]

As the British schooled that first generation of American spies, American ingenuity was about to transform espionage. Lovell's new R&D unit was officially established on October 17, 1942. General Order No. 9 in early 1943 described its mission as the invention, development, and testing of "all secret and special devices, material and equipment for special operations, and the provision of laboratory facilities." R&D was divided into four divisions: Technical, Documentation, Special Assistance, and Camouflage. Each would work closely with Division 19 (originally codenamed Sandman Club) of the Office of Scientific Research and Development (OSRD), which served as their link with contractors in the private sector. Division 19 maintained its own testing laboratory at the Maryland Research Laboratory (MRL), located on the site of the Congressional Country Club.

At the time Donovan and Lovell were sipping sherry in Georgetown, the OSS in its infancy was already showing evidence of American character, differing from its SOE cousin in subtle but significant ways. While the British had kept SOE separate from the country's traditional intelligence-gathering arm, the Secret Intelligence Service (SIS), the OSS combined espionage and unconventional warfare into a single organization. Whereas the SIS was a civilian agency, OSS was a military organization, functioning with relative independence under the Joint Chiefs of Staff (JCS).[22]

The new agency also differed from its British counterpart in the way it acquired its clandestine technology. Great Britain created government laboratories for the scientific and technical work in espionage, scattering them throughout the country. These highly secretive "Stations," as they were called, operated largely independently and with defined responsibilities. Station VIIa, for instance, responsible for covert radio production, was located in Bontex Knitting Mills in Wembley, while a part of the camouflage section, Station XVa, was housed in the Natural History Museum in London.[23] England's best scientific and engineering minds had been recruited to work at these top-secret government labs and used whatever limited wartime resources they could muster.

By contrast, Lovell, rather than recruit engineers and scientists into government service and build laboratories from the ground up, sought out private companies with the technical expertise and manufacturing capabilities to produce the needed gear, either from all original designs or by modifying existing consumer products for clandestine work. Traditionally clever artisans turned out one custom-made clandestine device at a time. Under

Lovell's leadership the new generation of spy gear would be engineered and produced using modern manufacturing techniques.

American industry and Lovell were particularly well suited for the mission. The advances in science and engineering since World War I were broadly integrated into the nation's manufacturing and technical infrastructure and Lovell offered OSS far more than just management and technical expertise. As a scientist and businessman of the post–World War I generation, he arrived at his task with a lifetime of business and research contacts. These personal relationships with executives and scientists would prove invaluable for OSS.

Producing clandestine devices required a mind-set on the part of the designer and motivation on the part of the manufacturer quite different from other wartime industries. Work on spy gear was highly secretive, specialized, and the dollar value of the production runs relatively small. Compared to wartime contracts for millions of canteens or boots, the OSS might require only a few hundred clandestine radios or few thousand explosive devices. To recruit contractors and their technical talent, Lovell would need to appeal to an owner's patriotism and personal history, more than profit.

In the months following his meeting with Donovan, Lovell and his OSS/ R&D branch developed an arsenal of special weapons and devices with which to "raise merry hell," along with increasingly inventive schemes.[24] Time-delay fuses for explosives were needed, so agents or saboteurs could safely leave the area before detonation. Building on the work of the British SOE, Lovell's engineers developed the *Time Delay Pencil*, a copper tube containing a glass ampoule of corrosive liquid and copper wire connected to a spring-loaded firing pin, which could also be used to ignite incendiary devices. Small and reliable, the *Pencils* were color-coded to indicate different timing intervals.[25] A pocketable cylinder called a *Firefly*, developed by Lovell's team, mated a small explosive incendiary device with a self-contained time-delay fuse for a saboteur to drop into a car's gas tank.[26]

Another explosive device called a *Limpet*, named after the mollusk that fastens itself to rocks, was specifically designed to attach to the sides of ships beneath the waterline and blow a twenty-five-square-foot hole through either steel plates.[27] The *Limpet* featured an delay detonator that could be set for hours or days or rigged to set off multiple detonations sympathetically with the concussion of one timed explosion triggering the others nearly simultaneously.[28]

OSS scientists discovered that explosives in powder form could be mixed with wheat flour and safely shipped, shaped, and even baked until needed for sabotage operations. The "explosive flour" could pass inspection as ordinary flour except under microscopic examination.

The *Limpet*'s delay relied on acetone to eat away a celluloid disk and trigger the detonation. While the timing of the explosion varied with the water temperature, it still offered a marked improvement over a British version that used aniseed balls—a traditional British hard candy—dissolving in water as a fuse.[29]

Several Lovell-inspired devices relied on the environment or the target's natural function to set them off. The *Anerometer*, a small barometer-activated device designed to sabotage airplanes, triggered an explosion when the aircraft reached an altitude of 1,500 feet above its starting elevation.[30] A sabotage tool intended for trains featured an early version of a photosensitive "eye." Called the *Casey Jones* or *Mole,* the eye reacted to the sudden absence of light.[31] When attached to the undercarriage of a train, it ignored gradual light changes, but exploded in dark tunnels, derailing the train. Clearing a train wreck from within a tunnel compounded the effectiveness of the sabotage. Explosives were also disguised as coal for sabotaging a locomotive's firebox or a power plant. Since the enemy often left stocks of coal unprotected the disguised explosive coal was simply tossed onto the pile.[32]

In one exceptional example of camouflage, Lovell's engineers began work in November of 1942 on a new type of high explosive disguised as flour. Eventually, DuPont produced fifteen tons of the granular explosive,

*The OSS experimented with a cigarette pistol
supplied to them by the SOE. The close-
range, one-time-use, .22 caliber weapon
was intended to improve one's chance for
escape and evasion.*

nicknamed *Aunt Jemima*, for use by OSS in China. Designed to match the
gray color of Chinese wheat flour, *Aunt Jemima* could be safely used to bake
pancakes or biscuits indistinguishable from the real thing in appearance and
taste, other than a slightly gritty texture.[33] With the proper detonator attached, however, the biscuit contained sufficient explosives to become a
small bomb.

Other devices provided by Lovell and his men were less subtle. The *Liberator* pistol fired a single .45 caliber bullet. General Motors mass-produced
this inexpensive but deadly weapon from sheet metal in its Guide Lamp Division.[34] For airdrop to resistance forces behind enemy lines, the *Liberator*'s
packaging included ten rounds of ammunition, pictorial firing instructions,
and a stick to poke out the empty shell casing after firing.[35] With an effective
range of twenty-five yards, it became increasingly inaccurate beyond six
feet, and was "the gun to get a gun." Due to its low cost and Spartan design,
the firearm soon acquired the unflattering nickname "Woolworth Gun."[36]

A more substantial weapon was the silenced .22 caliber automatic pistol
Lovell's team created by modifying the commercially available *Hi-Standard*

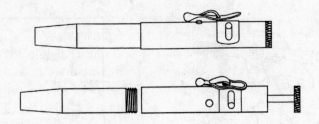

The Tear Gas Pen *was a commercially produced personal defensive weapon designed for carrying in a pocket or purse. The pen had an effective range of six feet, firing strong tear gas to incapacitate the target or attacker long enough to allow an escape.*

pistol to add a silencer. The silencer reduced ninety percent of the weapon's noise, so its gunshot would be drowned out by traffic noises, closing doors, and other activities of everyday life. It was ideal for use inside closed rooms or when eliminating sentries.[37]

A third weapon, the *Stinger*, was a small single-shot disposable .22 caliber pistol about the same size as a cigarette and intended for use at close range. Inexpensive to produce in large quantities, the *Stinger* was concealable and could be fired from the palm of a hand at a person sitting in a room or passing in a crowd.[38]

Lovell's wartime efforts also included spy gear and gadgets for agents to conduct conventional espionage. When unable to obtain Minox subminiature cameras in sufficient numbers, OSS joined forces with Kodak to develop America's first spy camera. Small enough to be camouflaged as a European-style matchbox, the tiny *Match Box Camera* or *Camera-X* held two feet of 16mm film, enough for thirty-four exposures. The lens design allowed agents to capture distant images of enemy installations, while documents could be photographed with a special attachment. Easily concealed, the camera was operable with one hand and could be requisitioned with a choice of camouflaged matchboxes that included Swedish or Japanese origins.[39]

OSS printers counterfeited currency and reproduced identity documents with "official" seals and forged signatures.[40] Beginning in 1943, they issued hundreds of virtually perfect German stamps, pay books, identity papers, ration cards, and even Gestapo orders.[41] OSS tailors created clothing so flawless the stitching resembled the genuine article from the country of supposed manufacture.[42]

No idea seemed too far-fetched for Donovan, whose motto became "Go ahead and try it." The R&D lab created a soft metal tube with a screw cap that projected a thin stream of liquid chemical with a repulsive and lasting odor as a psychological harassing agent. When squirted directly on the body or clothing of a person, it engulfed them with the odor of fecal matter. The plan called for Chinese children in occupied cities to squirt the liquid at Japanese officers. Lovell dubbed it "*Who Me?*"[43]

When a civilian dentist suggested to President Roosevelt that one million bats with tiny incendiary devices attached to them could be released over Japan to ignite a firestorm among houses constructed almost entirely of wood and paper, experiments leading to what would become known as *BAT* or *Project X-ray* were undertaken.[44] Bats were clandestinely collected from Carlsbad Caverns in New Mexico and transported to a test site. Developers designed a parachute container to house the bats during their descent from a highflying airplane, while Division 19 engineers produced tiny (15 grams) incendiary and *Time Delay Pencil* devices.[45] The initial testing at Carlsbad Air Base was both a high and low point for the project. The armed bats successfully, but accidentally, burned down a hangar after crawling into the rafters of the newly constructed building.[46]

For a brief time the plan seemed to have potential. In large quantities, the price of the incendiary device and time-delay fuses were less than four cents per unit and the bats could be obtained at no cost during their hibernation cycle. The separate elements necessary for the project to work were all in place and tested, but military planners would not authorize a bat operation, declaring insufficient data existed about the processes needed to arm and transport one million bats for an air strike. The project was cancelled in March of 1944.[47]

Additional experiments were undertaken to use a larger animal, the common Norwegian rat, to deliver bigger payloads than the tiny bats. Tests showed that a rat could carry up to seventy-five grams of explosives attached to its tail. The rats, which normally live in buildings, factories, and warehouses, were thought to provide a way of introducing explosives into guarded installations.[48] But, like the bat attack, this project also floundered in military planning.

Another unconventional project that failed, although it had been supported by the Chairman of the Senate Appropriation Committee, was the *Cat Guided Bomb*. The idea was to harness a cat to the underside of a bomb in such a way that the feline's movements would steer the explosive to its target. In theory, when a cat was dropped over open water with a ship in sight, it would steer itself, and the bomb, toward the safety of the ship's deck. Initial tests proved cats were ineffective and the concept died as quickly as the first test subjects.[49] Another failed idea included plans to poison Hitler with female hormones by injecting them into the vegetarian Fuhrer's vegetables.[50]

Some programs that approached the edge of America's ethical standards were accepted as the price for winning an unconditional surrender from Germany and Japan. Botulism and toxins were toyed with, along with the possibility of using germs and nerve gas, although such projects never represented a major effort by the OSS.

There were also some experiments with truth drugs and hypnosis but these never progressed very far.[51] The idea of a truth serum was not new. Law enforcement had been searching for such a magic elixir for years with little success. Nevertheless, Lovell budgeted a modest $5,000 for the project but it turned up nothing substantial. "As was to be expected, the project was considered fantastic by the realists, unethical by the moralists, and downright ludicrous by the physicians," Lovell wrote in a preliminary report.[52]

In May 1943, after less than a year on the job, Lovell visited David Bruce, the OSS chief of station in London, where the New England chemist captured Bruce's attention. The day after the meeting, Bruce wrote to General Donovan: "Stanley Lovell arrived yesterday, and he and I have just had a long talk at lunch, in the course of which he made my hair stand on end with his tales of the new scientific developments on which he has been working." Clearly taken by Lovell's ideas, Bruce continued: "His [Lovell's] arrival has been anxiously awaited and I have put him in touch immediately with various people [at SOE] who are engaged in similar work."[53]

One of the most forward thinking projects undertaken by Lovell's team was *Javaman*, a remote-controlled weapon consisting of a boat packed with four tons of explosives. Using early television technology, a camera mounted on the boat's bow broadcast images to a plane circling fifty miles away where a crew member watching a monitor guided the boat to its destination, then triggered the explosives by remote control. Despite encouraging tests, the project was eventually dropped.[54] According to Lovell, the Navy abandoned the idea because it judged the explosive load as too dangerous to carry either by ship or submarine.[55]

By the summer of 1944, with bases of operations established throughout the world, OSS printed an expanded Sears and Roebuck–style catalog of espionage and sabotage devices, listing the specifications of each piece of equipment along with pictures.[56] Station chiefs could peruse the catalog and choose whatever device they required. At war's end in 1945, OSS had produced—in less than thirty-six months after its creation—more than twenty-five special weapons and dozens of sabotage devices, along with scores of other gadgets, including concealments, radios, and escape and evasion tools.[57]

Mirroring the accelerated wartime production schedules that turned out ships, canteens, boots, and bombs in record time, it was a remarkable achievement. With initial guidance from the British, the OSS progressed in two years from offering a handful of basic tools of the spy trade to the design, manufacture, and deployment of an astonishing array of devices. The OSS officer corps developed at a similar frenetic pace, establishing intelligence networks throughout Europe, the Middle East, and Asia. Yet, in the autumn of 1945, the fruits of America's dramatic entry into the international spy game were nearly lost in the wake of America's rapid military demobilization.

CHAPTER 2

We Must Be Ruthless

We cannot afford methods less ruthless than those of our opposition.

—John Le Carré, *The Spy Who Came in from the Cold*

With the end of the war growing near, Donovan remembered the lessons of Pearl Harbor and the value of intelligence in occupied Europe and other theaters of war. At the behest of President Roosevelt, he prepared a detailed memorandum calling for the creation of a permanent postwar agency to act as a central clearinghouse for intelligence. In the covering letter of this 1944 memo, Donovan wrote: "When our enemies are defeated, the demand will be equally pressing for information that will aid us in solving the problems of peace . . ."[1]

However, Washington politics during the last days of World War II eroded Donovan's influence along with his dream of forming a civilian central intelligence service. Many in government considered the OSS a temporary wartime agency, not needed in peace time any more than the Office of Price Administration, which oversaw the rationing of sugar and car tires. For them espionage was an inconvenient wartime necessity like gas coupons and war bond drives. Unable to see future challenges to national security, they believed America's involvement in spying should end with the war.

Donovan's memo, intended for the private consideration of the President and the Joint Chiefs of Staff, was leaked to the press. Columnist Walter Trohan, leading the charge against a standing intelligence agency, wrote in

February 1945, in the *Chicago Tribune* and *New York Daily News*: "Creation of an all-powerful intelligence service to spy on the postwar world and to pry into the lives of citizens at home is under serious consideration by the New Deal. The unit would operate under an independent budget and presumably have secret funds for spy work along the lines of bribing and luxury living described in the novels of E. Phillips Oppenheim."[2]

This was a direct policy and class attack on Donovan and his "blue-blooded" operatives, even down to the mention of Oppenheim. A popular and prolific British spy novelist of the day, Oppenheim pioneered the genre that would eventually became known as the international thriller, rarely missed an opportunity to have his characters revel in extravagant luxury. The message was clear: spying was elitist, unsavory, and un-American.

After Donovan's confidential report remained unacted on by Roosevelt, a second negative report made its way to President Truman's desk. This one, prepared by a Roosevelt aide, Colonel Richard Park, Jr., offered a devastating review of the OSS and with it, Donovan's proposed peacetime intelligence agency.[3] Truman accepted Park's position and wasted no time acting. Within weeks of V-J Day in mid-August, the President signed an order on September 20, 1945, abolishing OSS and directing it to disband by October 1, 1945.[4] Providing only ten days for the dissolution of the agency, the executive order left no time for a political counteroffensive by Donovan and OSS supporters.[5]

Two days prior to its official termination, the OSS staff gathered in Washington at the Rock Creek Park Drive skating rink (near the present-day Kennedy Center for the Performing Arts) to bid farewell to one another. Addressing the assembled crowd, Donovan said, "We have come to the end of an unusual experiment. This experiment was to determine whether a group of Americans constituting a cross-section of racial origins, of abilities, temperaments and talents, could risk an encounter with the long-established and well-trained enemy organizations."[6]

Closing down of OSS did not completely dissolve its capabilities. Bits and pieces of the organization were seen as valuable and absorbed into other government entities. Research and Analysis was moved to the State Department, and other sections were incorporated into the War Department (later the Department of Defense) under the name Strategic Services Unit. Those transferred included overseas OSS stations and a skeleton crew from operations and technical support made up of a few experts in wireless communications, agent documentation, and secret writing (SW).[7] However, the majority of OSS engineers, scientists, and craftsman assembled for wartime duty, returned to the private sector, taking with them their expertise in producing the specialized equipment required for intelligence operations.

America was without a functioning centralized intelligence agency, though not for long. In January of 1946, two months before Winston Churchill warned of the coming Soviet challenge in his historic "Iron Curtain Speech" in Fulton, Missouri, President Truman signed the Central Intelligence Group (CIG) into existence. The occasion became a jovial ceremony where attendees were supplied black cloaks, black hats, and wooden daggers.[8]

The CIG's two basic missions were strategic warning and the coordination of clandestine activities abroad. Absorbing the Strategic Services Unit along with its officers, agents, files, overseas stations, and unvouchered funds, the new agency's overseas component was named the Office of Special Operations (OSO), with responsibilities for foreign intelligence, counterintelligence, covert action, and technical support. However, without independent funds, the CIG did not function well and, within the first year and a half, had three directors.[9]

With the Cold War intensifying and with the CIG underperforming, government leaders recognized that without independent statutory authority, the structure could not carry out the required mission. As a response, Congress passed the National Security Act of 1947 that created the Central Intelligence Agency. Like the CIG, the new Agency focused on providing early warning and preparation for any Soviet invasion of Western Europe. On the military front, weapons were cached, agents infiltrated into Eastern European countries, stay-behind resistance groups organized, and plans for counterattacking Soviet invaders drawn up.

The more traditional job of spying fell to the OSO, which had been absorbed into the CIA intact. With more than one-third of its officers drawn from the OSS, the OSO proved effective, but technical support could not keep up with operational demands. As a result, in September 1949, OSO created an Operational Aids Division staffed by officers with prior experience in OSS Cover and Documentation Division. The "operational aids" included agent authentication and documentation papers, secret writing, photography, and audio surveillance.[10]

A year earlier, in September 1948, a separate organization known as the

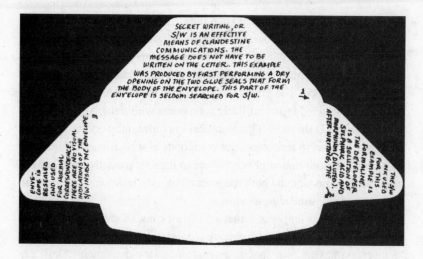

*The underside of flaps of an envelope, shown here unfolded
and after developing, were often used for secret-writing
messages during World War II and afterward.*

Office of Policy Coordination (OPC) was formed to conduct aggressive
paramilitary and psychological warfare operations against the Soviet Union
and Eastern Europe. Between 1948 and 1952, OPC grew from 302 employees, with no overseas stations, to a staff of more than 2,800 staff and 40
overseas stations.[11] OPC had its own small R&D shop and staff, inherited
from OSS, that conducted research in chemistry, applied physics, and mechanics.

The two offices operated independently and competed for the limited
resources available to produce the clandestine devices needed by agents and
officers. With little quality control and without a coordinated research and
development program, early CIA technical equipment was often in short
supply and of uneven quality.

In October of 1950, President Truman, dissatisfied with the CIA's intelligence following North Korea's invasion of South Korea, appointed General Walter Bedell Smith as Director of Central Intelligence. Smith, in turn,
appointed Allen Dulles head of clandestine operations, giving him the title
Deputy Director for Plans (DDP) in 1951. All of the Agency's operational
components came under the DDP in January 1952.[12] Dulles appreciated the
value of technical equipment for clandestine operations through firsthand
knowledge. As an OSS case officer, he had used devices supplied by Lovell's

R&D branch. He also understood that the CIA faced a problem of applying emerging postwar technologies to improve clandestine gear and deploy the equipment to field operatives.

Dulles first turned to Lovell, who had returned to the private sector, for advice in early 1951. The Professor Moriarty of the OSS responded by proposing a centralized technical R&D component within the Agency similar to the OSS/R&D division. This technical organization, working under the DDP, would develop technology for operations as well as conduct research on new capabilities that might contribute to intelligence gathering. The engineers would understand both the potential of new technology and how to apply it to clandestine requirements.

"Warfare is no longer a matter of chivalry but of subversion," Lovell wrote to the man who would dominate U.S. intelligence for the next decade. "Subversion has its own special arsenal of tools and weapons. Only Research and Development is capable of creating such an arsenal."[13] Lovell also advised that a central R&D component for the CIA should begin with a minimum staff of several hundred scientists and engineers.[14]

The recommendation found a receptive ear with Dulles and he assigned his special assistant, Richard Helms, to study the issue of technical support. In turn, Helms tasked Colonel James H. "Trapper" Drum, then head of the OAD, to produce a report with recommendations that addressed the problem. Four months later, Drum, a West Point graduate, who left the military a full colonel to join the Agency, produced a lengthy report that formed the foundation for a new approach to providing operations with technical support.

Known as *Drum's Bible,* the report advised combining all technical elements responsible for supporting operations into a single organization directly under the DDP. Drum wrote to Dulles in August of 1951 that the proposed new office would "provide the tools of the trade required to support the operating components of the Clandestine Service."[15] As Lovell had recommended several months earlier, Drum envisioned a new organization with two primary responsibilities: centralized technical support to deliver gear needed by field operations, and research and development to improve collection capabilities.

Dulles accepted the recommendations and created a Technical Services Staff (TSS) with "powers and authorities" equivalent to those of the other operational offices in the CIA.[16] On September 7, 1951, the DDP formally announced establishment of TSS, a small component numbering about fifty

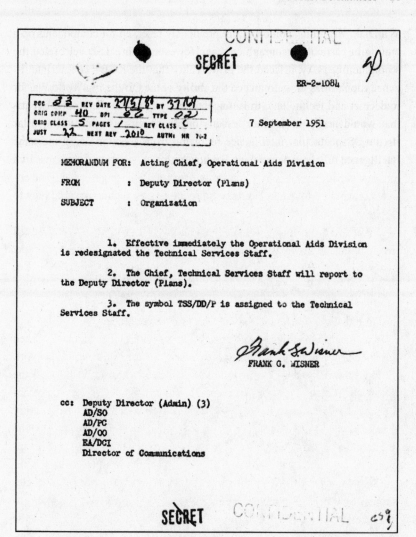

The OTS "birth certificate." This official memo authorized formation of the Technical Services Staff on September 7, 1951.

officers, with Drum at its head. Explosive growth followed. Within two years, the demand for TSS "products and services" was so strong that the staff expanded more than fivefold. TSS existed until July of 1960, and was then renamed the Technical Services Division (TSD).

It took nearly a decade for technical services to gain formal recognition as a DDP "division," nomenclature previously reserved for components operating in particular geographic areas. However, before TSD celebrated its second anniversary, it faced the grim reality that the KGB's counterintelligence capabilities far outmatched the ability of the CIA, using World War II tradecraft and technology, to run agents securely inside the USSR. Events that would teach those bitter lessons began promisingly in 1961 with a stream of spectacular intelligence reporting from a senior Soviet military intelligence officer, Colonel Oleg Penkovsky.

PLAYING CATCH-UP

CHAPTER 3

The Penkovsky Era

Each man here is alone. —Oleg Penkovsky quoted in the *Penkovskiy Papers*

B ad news, like every secret communication from Moscow, arrived at CIA Headquarters encrypted. The news that arrived mid-morning on November 2, 1962—as the Cuban Missile Crisis was winding down—was particularly bad. Colonel Oleg Vladimirovich Penkovsky, a career Soviet military intelligence officer and the Agency's most spectacularly successful spy, was, in all likelihood, lost. Penkovsky had held a senior position in the Glavnoye Razedyvatelnoye Upravlenie (GRU), the Chief Intelligence Directorate of the Soviet General Staff while secretly reporting to U.S. and British intelligence. In the colorful parlance of espionage, he had almost certainly been "rolled up."

At the new Agency compound at Langley, Virginia, the paint was barely dry on the walls when the Communications Center on the ground floor—Headquarters' sole secure link to Moscow personnel—received the super-enciphered message. It arrived as an "IMMEDIATE" cable, a long, narrow strip of paper snaking out of a bulky machine, much like a price quote from an old-fashioned stock ticker. The encoded message was contained in an intricate pattern of perforations that ran along the paper's length. When the transmission was complete, the paper was torn off by the communicator, and then run through a printer that produced a neat array of seemingly random numbers and letters on a sheet of standard letter-sized paper. A second level of decryption was needed to render the message into plain text. This phase of decryption guarded against the potential for security failures along the transmission path, whether over the air or via land lines. Like placing a strong, small safe inside a larger safe, this last layer of decryption could be performed only by one of a handful of authorized officers from the Soviet Russia Division (SR) of the CIA's Directorate of Plans.[1]

Although the DDP sounded like the dullest of bureaucracies, its name veiled the most secretive directorate in the Agency. Hidden beneath the vague acronym resided the responsibility for the CIA's "cloak and dagger" work. Within the DDP, SR was particularly shrouded with "cloak."

If asked about their job by neighbors or friends, SR personnel would repeat a carefully rehearsed cover story of working for one or another government department, but never the CIA. It was not unusual for DDP operations officers to remain undercover even after retirement, and maintain their cover stories until their deaths. Even the top-secret clearance, required for employment at the Agency, did not authorize someone to know rudimentary details regarding SR or its personnel. If an Agency colleague asked about an SR staffer's job, they would receive only generalized replies and most knew better than to probe for details. Secrecy within the Agency was both enforced by official policy and expected as part of professional etiquette.

Virtually no one, with the exception of SR personnel, was allowed into SR spaces. A no-nonsense secretary immediately confronted any visitor who opened the unmarked, always closed, hallway doors that led into the division's suite and friends of SR officers from other parts of the agency did not drop in to plan weekend activities or for office gossip. When SR officers left the area, even for a short time, security procedures mandated that desks be cleared and all work secured in one of the division's high-security 500-pound black steel safes.

SR Division applied strict need-to-know compartmentation through BIGOT lists that restricted access to what many would consider routine information coming out of the Soviet Union. Within the division information was distributed like pieces of a jigsaw puzzle. Only a very few ever saw an entire operational picture. Those outside of SR could only assume that a puzzle existed. Within CIA's instinctively tight-lipped security environment, SR's added multilayered security cloak created a mystique that some viewed as arrogant and unnecessarily obsessive.

The term "BIGOT list" existed—and still exists—as a holdover from World War II when the most prized stamp on the orders of personnel traveling from England to Africa was "TOGIB," meaning "to Gibraltar." To reach Africa, the majority of personnel made the dangerous journey by ship through seas controlled by German U-boats. However, for a select few, there were the highly prized seats on a flight to Gibraltar. For these lucky individuals, the stamp on their orders was reversed to read BIGOT and the term thus acquired its special meaning in intelligence circles, carrying with it the inference of not only rarity, but also safe passage and a valued mission.

There were other levels of compartmentation as well. A top-secret clearance did not provide automatic access to specific operations or programs. TS, a security clearance level required for all CIA staff employees, only made one eligible for potential access to a compartmented program. The BIGOT access was granted based on responsibilities and an individual's demonstrated need to know about the operation.

SR's security policies extended to written communications within Headquarters. SR did not rely on the CIA's usual interoffice mail couriers nor were its officers permitted to use the 1960-era state-of-the-art pneumatic tube system that carried classified documents to every corner of the 1.4 million-square-foot building.[2] Everything regarding Soviet operations was hand-carried from office to office by either an SR operations officer or one of a dedicated cadre of women known as Intelligence Assistants.

It was standard operating procedure for the communicator to place the encrypted message in a heavy manila security envelope, securely seal it, and call SR to advise that a cable had been received from Moscow. On the morning of November 2, the young SR officer who walked to the communications vault, accepted the sealed envelope, and, without opening it, retraced his three-minute route to SR's small warren of offices, could not have known that he now had a role in one of history's most significant espionage events.

At his desk, the officer opened the envelope, removed the single sheet of paper, and, with painstaking care, began deciphering the message by hand. He used a one-time pad, or OTP, whose printed columns of numbers and letters exactly matched those used by the person who had composed the brief message. After the message was deciphered, the page of the one-time pad used was destroyed. The Soviet Union paid a heavy price during World War II when they reused one-time-pad pages for communicating with agents in different parts of the world. This seemingly innocuous error provided an advantage to U.S. code breakers who were able to unravel many Soviet ciphered communications that had been intercepted from Washington, D.C. and New York City. This secret would become known as VENONA and remains one of the notable achievements of the Army Security Agency and later the National Security Agency.[3]

The cable did not mention Penkovsky by name. Rather, it reported that Richard Jacob, a CIA officer in Moscow, was apprehended while clearing a dead drop. After a nerve-shattering but relatively brief interrogation, the message continued, Jacob was released to the custody of the U.S. ambassador and returned to the safety of the U.S. embassy. Because he was a diplomat, Jacob could not be formally charged with a crime. Instead, he was "PNG'ed," declared persona non grata by Soviet authorities and ordered out of the country.[4]

Penkovsky's arrest by the KGB was not confirmed during those first few hours, but it did not seem realistic to hold out much hope for the agent. As in the immediate aftermath of any roll-up, there were more questions than facts, but for those few who knew about the case, it required no imagination to conclude that Penkovsky either was dead or would be very soon.

The officer delivered the decrypted cable up the chain of command to the SR Division Chief. The Chief took the bad news to the Deputy Director for Plans who in turn briefed John McCone, the Director of Central Intelligence. Within twenty-four hours, McCone would personally inform President Kennedy. That so few understood the enormous impact Penkovsky's arrest would have on America's national security was partially due to the extraordinary secrecy surrounding the nearly eighteen-month operation and the care given to the handling of the remarkable intelligence he single-handedly supplied.[5]

Intelligence reports based on Penkovsky's information had been structured to suggest that the intelligence originated from multiple sources. To reinforce this illusion, the Penkovsky product circulated under two code names, IRONBARK for that material that was scientific or quantifiable and CHICKADEE for material that included his personal observations.[6] For anyone outside the small group who knew the truth, the vast quantity of intelligence flowing from the Soviet Union looked like the work of an extensive spy network, coupled with mysterious and advanced technical collection, rather than the efforts of a single spy.

A small team of CIA and British intelligence officers ran Penkovsky. He was alternately known as *HERO* to his American handlers and *YOGA* to the

British.[7] Jacob had been chosen to service the dead drop because he had recently arrived in Moscow and had a strong cover in a traditionally non-alerting, low-level administrative position. As such, he was less likely to be identified as a CIA officer and draw KGB surveillance.

According to later accounts, Jacob entered the dingy hallway of an apartment house at 5/6 Pushkinskaya and removed an ordinary matchbox wrapped in a short length of wire that formed a hook to secure it behind a radiator. As Jacob was placing the matchbox in his pocket, the KGB team jumped him from their hiding places in the vestibule. During the ensuing scuffle, he managed to drop the matchbox to the floor through a slit in the lining of his raincoat pocket, ridding himself of incriminating evidence and avoiding the nasty legal and diplomatic problems arising from having Soviet state secrets on his person. The technicality did not matter to the KGB team, since it was obvious why the American was in the building. Once subdued, Jacob was hustled into a waiting car and whisked off to a nearby militia station.[8]

The final act of the Penkovsky drama had begun that morning with two voiceless phone calls—silent calls—to a phone answered by a U.S. official. The silent call was a signal activating the communication plan issued to Penkovsky by his handlers when they had met outside the Soviet Union. Arguably the most critical piece of any operation, the commo plan provided agents, such as Penkovsky, with precise contact instructions and schedules to establish secure communication under both ordinary and extraordinary circumstances.

Because the CIA assumed that the KGB monitored all telephone calls to and from the U.S. officials, the silent call represented a clever piece of tradecraft that allowed a message to be sent, even if the call was monitored. Penkovsky had been instructed to go to a remote public telephone and call a specific number. When the phone was answered, he said nothing, but waited ten seconds before hanging up. The call to the specific number and the length of silence before hanging up were the message that directed intelligence officers to a telephone pole marked with a symbol written in chalk, an X.[9] The simple chalk mark announced that the dead drop site at the Pushkinskaya apartment house had been loaded.

These standard pieces of tradecraft—the silent call, followed by a signal site marked with an X and dead drop—were part of a commo plan, codenamed DISTANT, designed specifically for Penkovsky to provide an early warning of imminent Soviet attack on the West.[10] The small matchbox that

Jacob found tethered by wire behind the radiator might have contained information signaling the start of World War III.

With the silent call, Penkovsky, who had not been heard from or seen since early September, had apparently, reemerged.[11] It was possible that nothing serious was wrong. If it was a trap—a provocation on the part of the KGB—then it was worth the chance. "We had been worried about him, it had been quiet for quite a while," said the case officer who decrypted the message and whose memories are still vivid after more than four decades. "But in the past he had come up again. To my knowledge we had no warning, nothing to indicate they'd caught him."

Now, with Jacob's arrest, whatever glimmers of hope that might have existed with Penkovsky's reemergence, seemed far-fetched. It was possible that a bystander had seen Penkovsky suspiciously fiddling behind the radiator as he loaded the dead drop and called authorities who then laid in wait. It was also possible that the KGB had not been fooled by Jacob's cover and defeated his countersurveillance maneuvers en route to the dead drop site.[12] Any number of other scenarios about Penkovsky's fate was possible, but only a single distressing conclusion was probable.

Penkovsky's handlers had grown increasingly troubled by recent events surrounding the operation. Penkovsky had vanished from operational sight for several weeks prior to the silent call and his GRU superiors abruptly canceled his scheduled trip to Seattle in the autumn of 1962.[13] Additionally, the sheer volume of intelligence he was providing on his Minox film cassettes suggested a level of clandestine activity that could not continue undetected indefinitely. So voluminous was Penkovsky's productivity during the first half of 1962 that his handlers decided to discontinue temporarily tasking him for new intelligence collection.

The operation would refocus on supporting his work for the GRU by providing comprehensively written technical articles to be published under his name and supplying harmless intelligence products he could take back to Moscow from trips to the West. The intent was to strengthen Penkovsky's credibility among superiors, raising him above suspicion and moving him into circles of even greater access to Soviet secrets.

During a three-month period between October 1961 and January 1962, Penkovsky met with his contact in Moscow, Janet Chisholm, the young wife of British MI6 officer Roderick Chisholm, eleven times in public locations. During these brief encounters, she received thirty-five rolls of film containing hundreds of images of top-secret Soviet documents.[14] In January, Penkovsky reported what he believed was surveillance on Mrs. Chisholm but showed no personal alarm. Rather, he suggested that dead drops replace their contacts "on the street."[15] Early successes, it seemed, emboldened Penkovsky but, in his handlers' opinion, the agent's level of productivity was alarming as well as gratifying.

Had Penkovsky dropped his guard or grown careless as the inherently dangerous work became routine? It was possible. Had he grown to feel invulnerable and above suspicion? That, too, was possible. It only became known much later that George Blake, an MI6 officer who spied for the Soviets, alerted the KGB that Janet Chisholm was actively supporting her MI6 husband in operations. Consequently, when the couple arrived in Moscow, KGB surveillance teams were waiting for them.

Confirmation of the disaster arrived a few hours after the first message with news of the arrest of Greville Wynne, a British businessman traveling in Hungary. A sometime contact between Penkovsky and his handlers, Wynne was arrested by a KGB team in Budapest, also on November 2, and flown back to Moscow.

The final curtain fell a month later. On December 12, a notice in the Soviet newspaper *Pravda* announced Penkovsky's arrest in late October, more than a week before Jacob's apprehension. Six months later, on May 7, 1963, Penkovsky stood in a courtroom before the same judge who had presided at the trial of Francis Gary Powers, the American pilot whose U-2 spy plane had been shot down in May 1960 over Sverdlovsk.

The trial lasted four days. Penkovsky, in an attempt to save his life, admitted that he had passed secrets to the Americans and British. Prosecutors cited "moral degradation" among the reasons for his traitorous acts, while a witness bolstered this claim by testifying that he had seen the defendant sipping wine from a woman's shoe during a night of heavy drinking.[16]

On May 17, a public notice appeared that Penkovsky had been executed.

Rumors about his death eventually began to leak out. While the Soviet press announced an execution by firing squad, another, unconfirmed report, claimed that he had been burned alive in a crematorium and the grisly episode filmed as a warning to new GRU officers who might someday consider cooperating with the West.[17]

Wynne was also tried, found guilty and sentenced to eight years in prison. He was released in 1964 as part of a spy swap for Gordon Lonsdale, a Soviet spy convicted in Britain.

Like a silent explosion, the capture, trial, and execution of Penkovsky

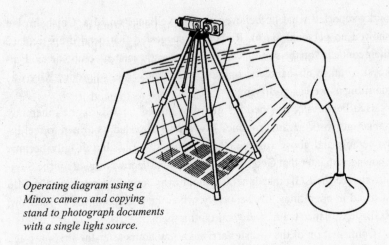

*Operating diagram using a
Minox camera and copying
stand to photograph documents
with a single light source.*

sent shock waves of uncertainty, recrimination, and retribution through American, British, and Soviet intelligence circles. While the badly burned Soviets restructured the GRU, the British and Americans, uncertain about when and how Penkovsky was first identified, faced a flood of questions.[18] If Penkovsky was under KGB suspicion as early as December of 1961, or January of 1962, did this mean the Soviets manipulated the information he provided? If so, when did he begin reporting controlled information designed to mislead American and British analysts? For that matter, could *anything* he reported be trusted?

Material long disseminated by analysts to policy officials was recalled and painstakingly reexamined. The eventual conclusion was that the Soviets had not played Penkovsky back against the Americans and British, but that left unanswered the mystery of why, if Penkovsky was suspected as early as December 1961, the Soviets continued to allow him access to secret files and materials.

Over the next several years, the Penkovsky case would become a cottage industry within the CIA as every aspect of the operation was analyzed to determine what was accomplished and what went wrong.

The Penkovsky operation had produced an astonishing amount of material. During his year and a half as an active agent, he supplied more than a hundred cassettes of exposed Minox film (each containing fifty exposures or frames). The more than 140 hours of debriefings in London and Paris produced some 1,200 pages of transcripts and reams of handwritten pages.

He identified hundreds of GRU and KGB officers from photos, and provided Western intelligence officials with their first authoritative view of the highest levels of the post-Stalin Soviet Union. In fact, he supplied so much information that both the CIA and MI6 set up teams dedicated exclusively to processing the material, which resulted in an estimated 10,000 pages of intelligence reports.[19]

More than the quantity, the substance of the documents on the Minox film and his knowledgeable debriefings impressed both CIA and MI6. Penkovsky appeared at a crucial time during the Cold War when tensions and the potential for nuclear war between the Soviet Union and the West were at an apex. This volatility was heightened by a lack of certainty on each side about the intentions and capabilities of the other.

The failed Soviet attempt to isolate the British-, French-, and U.S.-controlled sections of Berlin by blocking all ground and rail transportation and shipments into the city during 1948 and 1949 was still a fresh memory when the United States was caught off guard by unpredicted assertive Soviet technological, military, and political actions beginning in 1957. The USSR launched Sputnik in 1957; shot down Francis Gary Powers's U-2 reconnaissance plane on May Day, 1960; and built the Berlin Wall in 1961. So anemic was U.S. intelligence access to the plans and intentions of the Kremlin that the text of Nikita Khrushchev's famous speech denouncing Stalin at the Twentieth Party Congress in 1956 came to the CIA via a third party, an Israeli source operating behind the Iron Curtain.[20]

Through the late 1950s, Khrushchev's seeming obsession with the United States was rising to dangerous levels. His fixation with U.S. objectives was fueled first by an alarmist 1960 KGB report that falsely described the Pentagon's intention to initiate war against the Soviet Union "as soon as possible" followed by a failed attempt to overthrow Castro in 1961. Then, in 1962, two erroneous GRU intelligence reports warned of an imminent nuclear first strike on the Soviet Union by the United States.[21]

"Our production of rockets is like sausages coming from an automatic machine, rocket after rocket comes off the assembly line," bragged Khrushchev.[22]

Penkovsky's assignment to the State Committee for the Coordination of Scientific Research Work granted him access to the highest levels of military circles. He, in turn, provided the West with a contrasting view of both Soviet capability and Khrushchev's belligerent stance. "His [Khrushchev's] threats are like swinging a club to see the reaction. If the reaction is not in his favor, he stops swinging," Penkovsky explained to the team in a Paris hotel room in 1961.[23]

For the Kennedy administration, Penkovsky's reporting put the lie to the Soviet leader's braggadocio, while the intelligence he provided, combined with overhead intelligence, influenced downward revisions of Soviet missile production in National Intelligence Estimates.[24]

Penkovsky also revealed the real dangers of diplomacy without independent and timely intelligence. As the Cuban missile crisis heated up, Soviet Ambassador Anatoly Dobrynin used back-channel communication through Attorney General Robert Kennedy, Adlai Stevenson, and other White House officials to assure President Kennedy that only short-range defensive, rather than offensive, missiles were going into Cuba.[25] Similar false assurances also flowed through the back channels of diplomacy from GRU Colonel Georgi Bolshakov, working under cover of the TASS news agency, through Robert Kennedy.[26]

However, the technical manuals provided by Penkovsky for the Soviet SS-4 medium-range ballistic missiles allowed CIA photo analysts to identify and match the deployment pattern or footprint with U-2 reconnaissance photos taken over San Cristobal, Cuba. Far from being defensive and of short range, the missiles were armed with 3,000-pound nuclear warheads and a range of some 1,000 nautical miles, and were more than capable of reaching Washington, D.C., and New York City.[27]

Finally, Penkovsky's information provided analysis of the Soviets' overall lack of preparedness for war, allowing President Kennedy to face off against Khrushchev during the crisis. His insights, derived from personal access to Kremlin leaders, added independent weight to technical evidence that Soviet military threats were overstated, if not hollow. The American President was emboldened to act and denied the Soviets a nuclear missile foothold in the Western Hemisphere. For that brief and critical moment in time, history turned on the material provided by one man, Oleg Penkovsky.

In the wake of the Penkovsky case, the CIA undertook the unprecedented measure of bringing to press in 1965 *The Penkovskiy Papers* [sic]. The Agency, working with journalist Frank Gibney and the publisher Doubleday & Company, publicly exposed many of the operational aspects of the GRU revealed by Penkovsky. An immediate bestseller, the book presented most Americans with one of the first in-depth looks at Soviet intelligence operations in the West.

The Penkovskiy Papers offered remarkable details of Soviet tradecraft, from tips on American personal grooming and social customs ("Many Americans like to keep their hands in their pockets and chew gum") to evading surveillance and selecting dead drop sites. One section warned of the dangers presented by squirrels running off with small packages left at dead drop sites in New York's Central Park.

For American readers, the book confirmed their worst suspicions that Soviet spies were active and successful in the United States. It may have also implied an equally aggressive and thriving U.S. espionage capability in the Soviet Union. Unfortunately, this was not the case. The few who understood how dependent American intelligence had been on *HERO*'s production knew the time had come to change the game plan. The case had revealed grave deficiencies in the tradecraft needed to handle long-term agents inside the Soviet Union. America's technology and the CIA's Technical Services Division would become key players in a new operational strategy.

CHAPTER 4

Beyond Penkovsky

Soviet intelligence is over-confident, over-complicated, and over-estimated.

—Allen Dulles, *The Craft of Intelligence*

In nearly every respect, the Penkovsky case was a traditional agent operation. Relying more on the professionalism of the agent and handler than gadgets, the tradecraft employed differed little from what was used during World War II, and some methods, such as signal sites, dated back to the Old Testament.[1] Penkovsky was briefed and debriefed in hotel rooms in London and Paris. These were cordial working sessions that lasted hours, filling the rooms with cigarette smoke, and then ending with the optimism of chilled wine and gracious toasts. In Moscow, Penkovsky used dead drops and brush passes to deliver his intelligence and, in one instance, used the overhead water tank of a toilet during a diplomatic function as a dead drop.[2]

Well-designed and properly executed dead drop exchanges are among the most secure means for agent communication. Brush passes, though less secure, are still relatively safe. However, Penkovsky conducted an excessive number of personal exchanges between October 1961 and January 1962, all with Mrs. Chisholm. Even more alarming, eleven of these exchanges were in public view and some were poorly executed and transparent to surveillance teams.[3] The KGB's Seventh Directorate surveillance officers later commented that while surveilling Mrs. Chisholm and her children in a park off

Tsvetnoy Boulevard in 1961, they observed an elderly man approach one of the children and proffer a small box of chocolates. The young girl took the present to her mother who, without opening the box, placed it inside the baby carriage.[4] To the KGB the act was suspicious, and the elderly man was later identified as Penkovsky.

Whatever shortcomings might have existed were not entirely without reason. There simply were no suitable devices on Agency shelves for this type of operation. For instance, as late as 1962 the CIA had yet to develop a small, reliable document copy camera for agents. Rather, Penkovsky relied on the commercially available Minox Model IIIs camera.[5] Small enough to conceal inside a man's closed fist, the Minox boasted an excellent lens that easily captured images of letters, memos, and pages from a book but could not be used covertly. The sliding shutter release required two hands, making it impossible to use inside an office or archive with anyone else present. Good pictures required even lighting, proper photo technique, and privacy.

The only item Penkovsky used that could properly be called advanced tradecraft was his "agent-receive" communications through a one-way voice-link. These encoded messages, known as OWVL, were broadcast over shortwave frequencies at predetermined times from a CIA-operated transmitter in Western Europe. Penkovsky listened to these messages on a Zenith TransOceanic radio—strings of numbers read in a dispassionate voice—and then decoded them using a one-time pad. Although foreign consumer technology, such as a Panasonic radio, was rare in the Soviet Union, Penkovsky could display his openly in the small study of his apartment since the radio raised no questions of disloyalty for a senior officer in his position. However, the system only received messages and left him without a means to send a reply.

Penkovsky secreted his spy gear—one-time pads, Minox cameras, film, and commo plan—inside a clever homemade concealment built into a wooden desk in his study. All of this was eventually displayed in open court as proof of his covert activities.

In sharp contrast to Penkovsky's basic spy gear was the sophisticated KGB technical surveillance operation that enveloped him after coming under suspicion. The KGB established three key points of observation to monitor his activities inside his home. The first was in the apartment directly above his, from which a KGB audio surveillance post monitored all

Steady the Minox against your face,

tuck both elbows well into your side,

and stand with your feet slightly apart.

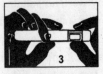

1) *Right;* 2) *Right;* 3) *WRONG: Don't allow finger to cover lens.*

COAT POCKET

SUB-MINIATURE CAMERA

SPECIALLY DESIGNED SLOT IN POCKET WITH A LENS TUBE FOR PENETRATING IT WHEN NEEDED.

CAMERA HELD BEHIND OPEN FOLD OF COAT

SUB-MINIATURE CAMERA IN FOLD OF GLOVES

CAMERA HELD IN POCKET

Minox subminiature concealment drawings. Covert photos with Minox cameras could be taken in a variety of circumstances but required user precision and experience to acquire high-quality pictures.

conversations. A pinhole opening was drilled through the ceiling of his study from a monitoring post overhead and photographed Penkovsky with a special 35mm camera (codenamed LINOCK) while he worked at his desk.

A second camera system, also from the apartment above, was mounted in the small balcony that overhung his window. Hidden in the concrete floor of this balcony was a camera and a small, remotely controlled trapdoor rigged to open and capture images as he photographed documents at the windowsill. The even light coming in through the open window, which overlooked the Moscow River, produced excellent images and he never imagined anyone could observe his activities.

The KGB set up a third observation post on the other side of the river in an apartment building facing Penkovsky's apartment, at *Naberezhanaya Maksima Gorkogo* (Maksim Gorky Embankment) 36, Flat 59. From there, KGB cameras with telephoto lenses produced high-quality images of his photo sessions, even capturing Penkovsky at his desk as he monitored shortwave broadcasts and copied down the transmitted numbers.

After-action assessments following Penkovsky's roll-up focused attention on the absence of effective spy gear, particularly in the area of agent communications. While in Moscow, his communication channels to his handlers were limited to dead drops and brief meetings. Silent phone calls were no more than prearranged emergency signals.

Technology did little to enhance either Penkovsky's production or security. His remarkable success was achieved not because of technology, but despite the lack of it. His official position allowed him periodic travel outside the USSR and opportunities for extensive debriefings. Without these personal meetings, Penkovsky would not have been successful.

What became clear was that the CIA in the 1960s did not have the operational methodology, clandestine hardware, or personnel to run secure agent operations inside the USSR. The absence of a secure and clandestine means of communicating in Moscow forced both the agent and his handlers to take risks that eventually played into the hands of the KGB surveillance methods. The recruitment of agents inside the Soviet Union meant little if the KGB could quickly identify them or if they could not securely communicate the secrets to which they had access.

For agents to be handled clandestinely in-country, the CIA needed the means to detect and counter pervasive KGB surveillance before conducting an operation; to conduct secure impersonal communications; and to pass and receive materials securely from the agent. These were no small matters.

In fact, a decade would pass before TSD and OTS engineers created the covert devices needed to conduct multiple sustained clandestine operations inside the USSR. Although neither side realized it at the time, Penkovsky's capture marked the start of a fifteen-year span during which the technological advantage would swing decisively in favor of the CIA.

When Penkovsky volunteered to spy for the West, the CIA lacked the ability to handle him in Moscow. In contrast, the KGB and GRU packed Soviet embassies, consulates, trade organizations, trading companies, the United Nations, international organizations, and press offices around the world with intelligence officers and co-opted civilians, often to the annoyance of genuine diplomats.

For the few U.S. intelligence officers who could get into the Soviet Union, operational success was nearly impossible. While the Soviets could not accurately identify every U.S. intelligence officer, the KGB erred on the side of caution and assumed that all Americans worked for the CIA until proven otherwise.

The KGB's Second Chief Directorate opened a file on each American and profiled each against age, gender, job description, personal activities, and possible intelligence roles.[6]

The KGB also charted "expected activity patterns" to go along with the profile that included predictable routes of travel to and from work each day. Even noted were the wives' favorite place to shop or visit. Added to these were the necessary travel to sporting and cultural events, sightseeing, and social activities outside of work. They knew that activities outside the regular travel pattern could mask a more sinister purpose, such as an agent meeting, casing a site, or servicing a dead drop, and thus identify the suspect as an intelligence officer.

Americans also quickly learned that, contrary to spy novels and James Bond movies, obvious attempts to evade surveillance only proved counterproductive. Any aggressive action to escape the surveillance teams set off alarms among the watchers of the Seventh Directorate.

The KGB penalized its surveillance officers for carelessness, which included losing a person under close observation. So taunting, antagonizing, or making the surveillance teams' job more difficult could result in dog feces being rubbed on car door handles or a smashed windshield. Particularly unnerving for most American drivers was a technique called bumper lock, in which the surveillance car remained literally inches from one's rear bumper.

Provocations were commonly employed. Soviets posing as disaffected or greedy officials volunteered information in an effort to engage the CIA. These individuals, aptly called "dangles," made it both difficult and essential to verify the authenticity of potential agents. Fortunately for the West, some of the most significant agents proved remarkably persistent in attempts to establish contact after their initial efforts were rebuffed.

If such caution toward volunteers was understandable, it could also lead to disaster. In 1963 a former officer in the KGB's Second Chief Directorate,[7] Aleksandr Cherepanov, handed a package to a pair of American tourists visiting the Soviet Union. Agency opinion differed on whether the material was genuine or part of a provocation. At the time, there was simply no way to tell.

The contents of the package, which provided details on KGB surveillance methods, were photographed and eventually handed back to the Soviets through diplomatic channels. Cherepanov, learning of the betrayal, fled Moscow. Eventually captured, he was tried in secret and executed in 1964.[8]

"It is not possible to determine why the Americans betrayed Cherepanov," a KGB assessment observed. "Either they suspected that his action was a KGB provocation or they wanted to burden the KGB with a lengthy search for the person who had sent the package to the embassy."[9]

Americans were closely observed even inside their own Tchaikovsky Street embassy. A ten-story structure erected in the 1950s as an apartment complex in a Russian version of beaux-arts style, it was typical of Soviet Union architectural design. The interior represented Soviet construction of the day, featuring a claustrophobic maze of narrow halls and small rooms.

American diplomats had been in the building since 1952, when Stalin ordered a move from Mokhovaya Street near the Kremlin and the National Hotel to the more remote location. Had the Americans stalled, as the British managed to do, the move might have been unnecessary, as Stalin died a short time later.

Extensive renovation by the American occupants produced only limited improvements. American construction crews discovered that the walls and floors were insulated with sawdust, ash, and other debris from the original construction. Seemingly installed as an afterthought, the electrical wiring would have been state-of-the-art in the 1920s, but was inadequate for the power requirements of modern appliances.

Soviet citizens employed as workers by the U.S. Embassy had access to all but the most sensitive areas of the building. Working in low-level administrative, maintenance, or service positions, they reported on personal habits, personality types, and office gossip to the KGB. During the 1960s and 1970s, Russian nationals equaled or outnumbered American citizens working in the Moscow embassy.[10] Conversely, the Soviet Embassy in Washington, D.C., did not employ a single American citizen.

The presence of so many Soviet citizens, a large percentage of them no doubt co-optees or informers, was not without its amusing moments. For two decades, a vivacious woman known as Valentina ran the barbershop and beauty salon in the embassy's basement. No one doubted she reported to the KGB, but the situation became untenable when information regarding a KGB operation was traced back to her. Valentina was swiftly fired but returned to the embassy one last time when a group of her American customers hosted a going-away party for her.[11]

In addition to the ubiquitous informers among the embassy staff, pervasive technical surveillance within the embassy itself was discovered in 1963. A defector reported that the embassy was riddled with listening devices and the assertion carried enough credibility that an Agency sweep team was dispatched to find the bugs. For the majority of American diplomats, the embassy compound served as both home and office space. If the embassy was bugged, there was no limit to what secrets, personal and professional, the KGB's microphones captured.

After the sweep team's initial hunt for listening devices turned up nothing, Navy Seabees were flown in to conduct a physical search that included demolition of a sample office. Walls, floors, and ceilings were torn up without finding any trace of covert wiring. It was only after removing the cast iron steam radiator that squatted at one end of the room and dismantling the wall behind it, that the first listening device was discovered. Standing amid the wreckage of the demolished room, a tech pointed to an inch of protruding wood and asked, "Now, what do you suppose this is?"

Cleverly hidden behind the radiator, the device consisted of a hollowed-out wooden dowel positioned so its center was flush against a pinhole in the wall's plaster. About a foot long, the dowel provided a clear air passage for sound to travel to a microphone concealed in an oversized brick on the building's exterior. Wires from the microphone were not run through the interior walls, where they would have been more easily detected, but through the mortar of the exterior stucco façade and into the basement, eventually trailing off to a listening post.

Members of the sweep team marveled at the ingenuity. Low-tech dowels had defeated the West's advanced metal detectors by placing the metallic microphone beyond range. Positioning the bug behind the radiator not only minimized the possibility of discovery, but also reduced the risk of the air passages being sealed by paint or plaster.

Given such realities, the psychological pressure on Agency personnel and their families throughout the sixties and seventies was especially intense. At times it seemed that the embassy took on a "through the looking glass" atmosphere. "You just assumed your apartment was bugged," said the wife of a TSD tech. "The KGB came with the apartment, like the nanny." Families needing privacy could go to "the bubble," a Plexiglas-like box measuring ten feet by ten feet by six feet high in a sealed and shielded room in a section of the embassy off limits to Soviet nationals. The bubble, although facilitating personal conversation, served as a stark reminder of the extraordinary measures required to evade KGB listening devices.

The KGB capabilities extended to breaking into the safes of foreign embassies. Surreptitious entry teams used a portable x-ray device positioned over a safe's lock to view the tumblers falling into place. The clever device came with one design flaw, the emitting of high levels of radiation that slowly poisoned its users. Within the KGB, members of these teams were known as *bezzubyye,* which roughly translates into "the guys with no teeth."

Especially alarming was a case in the mid-1960s involving a foreign diplomat recruited by the Agency for a single mission: to load one dead drop site in Moscow. For the concealment package, TSD engineers fabricated a four-inch hollow anodized pointed aluminum alloy spike to hold a one-time pad and the agent's commo plan. The cylindrical concealment was designed for quick planting at a precise location by simply stepping down on it, driving the spike into the ground and then covering the head with dirt.

However, the agent proved unreliable. Not only did he fail to load the drop but he also ignored security instructions. "We told him never to let the spike out of his possession because we had reliable information that KGB teams were going in and out of the safes in many of the embassies, including his," said the officer who directed the operation. "But naturally, people didn't believe us. They think this stuff only exists in the movies. So our agent stored the device in his embassy safe before returning it to his American contact."

When the spike, still loaded with the one-time-pad and commo

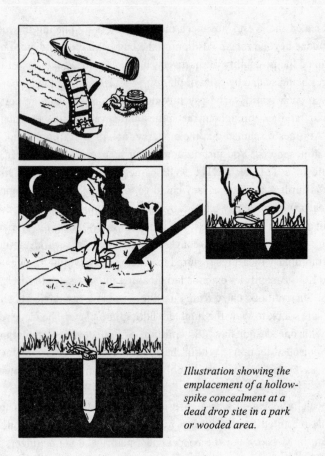

Illustration showing the emplacement of a hollow-spike concealment at a dead drop site in a park or wooded area.

instructions, arrived back at Langley, the receiving CIA officer put it in his office safe. Several months passed before he got around to returning it to TSD. As he walked into the lab, a Geiger counter sitting on a nearby shelf sounded. The spike registered as highly radioactive.

The investigation that followed concluded that the KGB had entered the diplomat's safe, removed the spike, extracted the contents, and impregnated the one-time pad with cobalt 60. It was estimated that the OTP contained enough radioactive material that a standard Geiger counter could register its presence through a brick wall. "That kind of experience makes counterintelligence very real," said the officer in whose safe the spike had been stored. "When all of a sudden you realize you've been sitting eighteen inches away from a device emitting radiation for months, you understand what the Soviets are capable of. Anyway, the Agency's Office of Medical Services gave me a lot of close attention for the next ten years."[12]

For a decade after Penkovsky's death the combined intense pressure from the KGB and scrutiny by the Agency's own Counterintelligence (CI) staff caused a virtual cessation of agent operations in the Soviet Union. Headquarters placed severe limitations on recruiting and handling agents inside the USSR. Field officers could not instigate or engage in any operational activity without prior Headquarters review and approval. While officers could express opinions, by saying, for instance, "We don't like that drop site because . . . ," Headquarters made all final decisions.

Given such restrictions, recruiting agents in countries bordering the Soviet Union became the priority focus, but even those opportunities were so infrequent that any reasonable lead merited immediate attention. In 1968, a Russian-speaking case officer assigned to Headquarters received orders to make an immediate, unexpected a trip to Helsinki. There was a possibility that a Soviet target would become available. The officer waited in vain for a month in Helsinki for the Soviet to arrive, then returned home empty-handed. There was no other choice. With so few prospects, any potential opportunity received urgent attention.

Consequently, defectors, émigrés, and legal travelers to the USSR became important sources of intelligence. But these assets were, almost by definition, often far removed from the political and military centers of power or technical institutes. Only a spy near the center of power—given the ability to communicate securely with his handlers—offered the potential for a reliable stream of quality intelligence.

A tightly held secret among the elite of the CIA's Soviet Russia Division and Counterintelligence staff was the reality that neither the United States nor its allies could confidently recruit and securely handle Soviet agents unless they were able to travel outside the USSR. Frustration over Moscow's severely constrained operational environment remained with officers long after they retired. "I was in Moscow for two years in the mid-sixties following the loss of Penkovsky, and to my knowledge, we unloaded only one dead drop during that entire period," said a veteran case officer. "In those twenty-four months I never had a 'sit-down' dinner or a private visit with a nonofficial Soviet. I spoke good Russian but was never invited to a Russian's home. I traveled all over the country, and the only contacts I had were with people who, the minute they found out I was with the American government, either literally ran or turned their back on me and walked away out of fear."

Those occupying senior positions at Langley shared the frustrations felt by CIA officers stationed in Moscow. "Our operations were criticized both by people who thought we were being duped and those who thought we were too timid," recalled a Moscow hand. "There were others who thought it [running agents] wasn't worth doing because the U-2 and satellites could gather intelligence just as well."

"Soviet intelligence is over-confident, over-complicated, and over-estimated," wrote Allen Dulles in his 1963 book, *The Craft of Intelligence.* Published the year following Penkovsky's arrest, the assertion was more bravado than fact, as Dulles no doubt had full knowledge of the situation in Russia at the time.[13]

Dulles, however, was not blind to the potential of technology. Nearly a decade earlier, in the winter of 1954, a twenty-seven-year-old Technical Services Staff (TSS) officer received an odd proposition from the TSS chief, Willis "Gib" Gibbons. "He asked whether I was game for taking on an unusual job. I asked for a better description than that and, of course, didn't get one," the officer remembers.

The assignment turned out to be that of a technical tutor for Dulles, who had been appointed CIA Director by President Eisenhower in February of 1953. Dulles was not an unknown to the young tech. The two had met the previous autumn when the tech worked on another unusual job for the DCI's security detail. As part of the construction of the new DCI suite, the tech installed several covert audio devices, including microphones in the ceiling hardwired to recorders in the security offices. He also installed a secret button on the DCI's desk to summon a secretary should a visitor outlast his welcome.[14]

An intelligence officer of the "old school," in which clandestine activity was conducted "nose-to-nose" or using easily understood devices, such as dead drops and concealments, Dulles realized that he was now involved in a technically complex world. Increasingly surrounded by engineers, including his deputy, Air Force General Charles P. Cabell, Dulles was determined to keep his knowledge relevant. He seemed to sense, even at this early stage, the advances in technology would shape Cold War intelligence as well as the CIA itself.

"Apparently Cabell stepped to the fore whenever anything technical came up, and Dulles didn't like to be overshadowed," the staff officer recalled. "The DCI wanted a technical education and needed it quickly. Basically, he was uncertain and somewhat afraid of the jargon—that sort of thing."

As with many of his generation, Dulles was ill adept, if not ill at ease, with even simple technology, including his office telephone and intercom system. Born in 1893, he was part of the generation that bridged the nineteenth and twentieth centuries and witnessed the unfolding of technological miracles of a "modern world." His generation was the first to experience the emergence of technology in daily life, in which the underlying science is neither intuitively apparent nor easily understood without some technical acumen.

The young engineer assigned as Dulles's tutor was only three years out of college, and had a degree in physics and electro-mechanical systems. As he had done several times when making the DCI's audio installations, he walked the two miles from the Technical Service Staff's covert building on 14th Street near the Department of Agriculture to Dulles's office on the second floor of South Building at 2430 E Street. Perched on a modest rise of land called "Medicine Hill," the facility was a hand-me-down from the U.S. Navy. Several scattered buildings had sprung up over the years within the small fenced compound, which had also been the final Headquarters for the OSS.[15] First home to the U.S. Naval Observatory, it then served as the Naval Museum of Hygiene, and the Naval Medical School with hospital facilities for sick officers. Now, in the mid-fifties, the complex was again pressed into service for a "spy agency," though, by current standards, security was surprisingly relaxed.

"When I first went to see Dulles, I remember particularly the women who worked for him. They were an odd assortment. They struck me as being very grandmotherly old biddies but, in fact, were the sharpest creatures you would ever meet," the technical tutor recalled. "They gave this very matronly appearance. They seemed thirty-five or forty years older than me. I was twenty-seven at the time and had an idea that they were somehow frail, elderly ladies. But once a man came to the office whom they knew intended to do bodily harm. He was one of St. Elizabeth's [the nearby Washington Psychiatric Hospital] 'walk aways.' I watched them disarm the fellow, who had a pistol. They talked to him very calmly, very politely, and all of a sudden a big hulking security guy was behind him; then it was over quickly." Inside the inner office, the tech found Dulles sitting behind an imposing desk. Bespectacled, gray-haired, and dressed in rough Scottish tweeds, he looked very much like the headmaster of a very good boarding school or a Wall Street lawyer at a prestigious firm (a job he had, in fact, once held). "I went in and introduced myself," the TSS staffer recalled. "I asked him what he wanted to know, exactly, and he said, 'I don't have the slightest idea, just start broad.' So, we talked about physics, and we talked about chemistry."

Over the next nine months, the tech and the DCI engaged in nearly two dozen sessions. Dulles, ever the spymaster, was effectively debriefing the junior engineer. What started out with principles of pure science soon narrowed to specific questions or requested tutorials regarding particular technologies, such as Doppler radar or sonar. "In hindsight, I know this was at a time when he and the Air Force were arguing about the U-2. I don't doubt that his deputy, General Cabell, was pushing the Air Force view, rather than Edwin Land's view, which was more aggressive as to what could be done," said the engineer. "Perhaps Dulles was beginning to feel that he was being cut out."[16]

Land, founder of Polaroid, was heading an intelligence subpanel of distinguished scientists on long-range missile development. He, along with Massachusetts Institute of Technology president James Killian, proposed a technologically ambitious overhead reconnaissance role for the CIA while the Air Force was advocating a more conservative approach. In the end, President Eisenhower approved plans for the more advanced plane championed by Land and Killian, eventually code-named U-2. The project was put under CIA control and the aircraft designed by the legendary Clarence "Kelly" Johnson at the Lockheed "Skunk Works" outside Los Angeles, California.

Dulles began his scientific push by forming the CIA Research Board in 1953. Comprised of prominent scientists and business leaders, members of the Research Board included Land and Rear Admiral C. M. Bolster (ret.) of General Tire and Rubber Company, with Navy Rear Admiral Luis de Florez serving as chairman. "They'd come in for a day or so, and Dulles would entertain them at the Alibi Club," said the Advisory Board's secretary at the time. "Many of the sessions were very informal. He—Dulles—liked the informality. Once we had oysters, and I mean they brought in bushels, put them on the table and everyone pitched in, with a trash can next to the table and plenty of beer."

It is not difficult to imagine Dulles's likely methodology. As the case officer, Dulles was debriefing and building networks—essentially conducting an intelligence collection operation on the giants of industry and the "technologists" driving the aerospace and electronics revolutions.

If Dulles's pursuit of technical briefings seemed prescient, he was not alone. Other prominent national security experts saw substantial opportunities to use technology for intelligence objectives. In 1955, Air Force general and World War II hero James H. Doolittle, working at the request of President Eisenhower, led a small team in preparing a confidential report on the state of America's intelligence capabilities.[17]

The sixty-nine-page report took just eight weeks to complete, and its conclusions sounded an alarm:

> *The usable [intelligence] information we are obtaining is still far short of our needs . . . [Therefore] the U.S. should [utilize] every possible scientific and technical avenue of approach to the intelligence problem . . . We must develop effective espionage and counterespionage services and must learn to subvert, sabotage, and destroy our enemies by more clever, more sophisticated, and more effective methods than those used against us.*

Whether Doolittle sensed that technology could transform human spying, or concluded, as did a growing number of scientific thinkers, that technology applied to intelligence collection could replace traditional spying, is moot. The intelligence strategy of America would swing toward big technology programs. Technical collection—early satellite photography, spy planes, and signal monitoring—born in the 1950s and nurtured in the subsequent decades, soon became the focus for America's intelligence investment, beginning with the Corona satellite program. This was "Big Technology" done in a big way, and with big budgets.[18]

Corona, a photoreconnaissance satellite conceived in 1946 by the Rand Corporation, was launched in February 28, 1959. The first flight failed, as did the next eleven attempts. On the thirteenth test launch, the low-orbiting satellite was a success and its engineering payload recovered. Then, on August 18, 1960, the fourteenth Corona test launch took images of the USSR from space and, the following day, successfully ejected the film canister over the Pacific Ocean for retrieval by plane in midair.[19]

Carried within the canister were more than 3,000 feet of exposed film that captured a million square miles of the Soviet Union, providing intelligence officials with their first look at vast outlying areas of Russia. There could be little doubt that U.S. intelligence had dramatically changed since August of 1949 when the Soviets detonated their first nuclear device. Then, intelligence analysts rummaged through Herbert Hoover's Presidential archives at Stanford University, a collection dating back to the former President's days as a mining engineer, looking for a map of the Ural Mountain region where the blast occurred.[20] Now, with the satellite pictures, analysts had current images of precise areas of interest.

As Big Technology, with its big budget satellites and aircraft began operating against the Soviet Union, classic tradecraft struggled for relevance. The Big Technology programs attracted scientists, inspired technical creativity, and pushed engineering, literally and figuratively, to new heights. Satellites in the skies were viewed as less susceptible to the kinds of risks and unpredictability that plagued spies on the streets. In the minds of some, "technical collection" was untainted by the moral, ethical, and diplomatic entanglements associated with human espionage.

Satellites would not be arrested in the hallways of Moscow apartment buildings nor were they likely to cause international incidents. Satellites had no motive for betrayal and did not require reassurance and flattery. Moreover, if the satellites that delivered images cost billions of dollars, it was not because of some foreigner's personal greed. Yes, satellites might have mechanical failures, but they did not quit working out of fear or disillusionment. They did not violate the territorial integrity of the Soviet Union while photographing facilities whose guards would shoot a trespasser. As long as they had fresh film, fresh batteries, and a cloudless sky, they delivered the otherwise unattainable: nonpoliticized data.

There were, to be sure, limitations. A satellite could capture the image of missiles deployed to remote areas, but seemingly endless spools of film and powerful lenses could not divine the intentions of Soviet leadership. It could see submarines in their pens at the Severodvinsk naval base, but could not penetrate the roofs of government labs in Moscow and Leningrad to record images of the future weapon systems spread out across the drawing boards of engineers. Nor could it see into the minds of the Politburo or capture the complex internecine dynamics of Kremlin leadership. Only an agent inside the Kremlin could do that.

Pictures neither lie nor reveal the complete story. With the early successes of the photoreconnaissance satellites, American leadership also desperately needed to know more about what Soviet leaders were thinking and planning. At no time was this more publicly apparent than during the 1960 U.S. presidential election.

Democratic presidential nominee John F. Kennedy charged that Republicans were insufficiently attentive to national defense. How could the Republican administration, asked the Democrats, have allowed the United States to fall so woefully behind in this critical area? Backed by imprecise Pentagon estimates and Soviet Premier Nikita Khrushchev's tough talk, the issue touched a nerve with American voters. Eisenhower, of course, was basing his moderate Soviet policy on secret U-2 photographs, which would support his position, if made public. The U-2 photographs appeared to provide convincing rebuttal to the argument that the Republicans were weak on national defense, but without credible corroboration, their interpretation was hotly debated. National Intelligence Estimates of Soviet missiles showed an alarming increase in capability and numbers, with the U.S. lagging behind.[21] The phrase "missile gap" entered the national vocabulary.

What the American public heard were Khrushchev's exaggerated claims and Kennedy's allegations against the Republicans. Two years later, intelligence provided by Penkovsky, combined with satellite photographs, prompted a downward revision of the official estimate of Soviet missiles during Kennedy's presidency.[22]

Knowing Soviet capabilities with certainty was impossible. A totalitarian state such as the USSR possessed a large advantage over an open society in its centralized control of media and citizens. Within the Soviet Union, even road maps and railroad schedules were routinely falsified. Conversely, in any edition, *The New York Times,* the *Washington Post,* or the *Wall Street Journal* provided more reliable accounts for the Soviets on the thoughts and motives of the American leadership than U.S. intelligence could tell the American President about events within the Soviet Union. Farm reports, stock prices, economic statistics, and dozens of other sources of information, freely and widely available in the United States, revealed data that the USSR either held as a state secret or purposely distorted.

The Iron Curtain was a geopolitical one-way mirror. Soviet leadership could see out if they wanted, but American leadership, desperate for the smallest glimpse, could not see in. During the darkest days of the Cold War, the intentions of the Kremlin leaders remained obscure at best. The placement of Soviet officials on the reviewing stand atop Lenin's mausoleum in Red Square during May Day festivities, combined with whatever grainy photos military attachés could get of the Soviet Army's parade of military equipment, became objects of intense analysis for Western intelligence organizations. Anxious for any information, analysts considered nothing too trivial for scrutiny. The practitioners who studied such minutia had a professional name, *Kremlinologists.*

However, a small but growing cadre of officers emerged within the CIA who argued that new tradecraft based on advanced technology could be applied to operations on the streets of Moscow just as had been done in the skies above the USSR. These officers, subjected to Soviet counterintelligence tactics behind the Iron Curtain for more than a decade, had acquired a substantial body of operational knowledge that could be used to counter the seemingly invulnerable KGB.[23] They argued that if new tradecraft methods, combined with yet-to-be-invented spy gear, were developed and applied selectively, then the KGB's surveillance stranglehold in Moscow could be broken. This post-OSS generation of case officers found eager allies in Seymour Russell, Chief of the Technical Services Division (TSD), and his operationally oriented scientists and engineers.[24]

Bring in the Engineers

Warfare is no longer a matter of chivalry but of subversion, and subversion has its own, special arsenal of tools and weapons. Only Research and Development is capable of creating such an arsenal . . .

—Stanley Lovell writing to Allen Dulles in 1951

When Seymour Russell took the helm of TSD in the summer of 1962, no one doubted his disappointment with the assignment. For Russell, a highly regarded and ambitious Clandestine Service operations officer, an assignment outside one of the geographic divisions was almost certainly a detour in a fast-tracked career. After a series of impressive successes in the field as a case officer and station chief, Russell had every reason to expect an assignment as Division Chief overseeing operations in Western Europe or Asia, or even heading up all CIA operations as Deputy Director for Plans.

"Seymour Russell lived operations," said a TSD officer who became one of his top lieutenants. "He made no secret that he didn't want the TSD job. He wanted a senior job in the DDP." While Technical Services was a "division" in the DDP, it did not have the status of the six area divisions with their geographic responsibilities, such as the Far East, Africa, or Soviet Russia.

Joining Russell on the new TSD management team were his deputy, Dr. Sidney Gottlieb, and Richard Krueger as Chief of Research and Development. A chemist by training, Gottlieb was known internally for directing some of the Agency's most sensitive research under the MKULTRA program. Gottlieb arrived at Langley through a circuitous career path. He entered government service in 1944 by way of the Department of Agriculture, followed by a stint at the Food and Drug Administration, and then a spot at the University of Maryland before joining the Agency in 1951. After running a small, dozen-person chemistry division inside the Technical Services Staff for six years in the mid-1950s, he took a two-year assignment in Germany before returning to Washington in 1959 to head the TSS Research and Development program.

Krueger, who had been the young tech who installed the secret microphones and recorders in Dulles's office and served as the DCI's technical tutor, then moved on to the CIA's U-2 and radar programs. Now, steeped in the science of big technology, he was returning to the basics of espionage tradecraft.

Barely a decade old, TSD had expanded from fewer than fifty technicians in 1951 to an office of several hundred engineers, craftsmen, scientists, psychologists, artists, printers, and technical specialists. After 1962, with the formation of a separate CIA Directorate of Research, TSD existed solely to support operations, with 20 percent of its staff assigned to a network of forward-deployed bases overseas. With the exception of "denied areas" such as the USSR and China, these dispersed technical specialists could be summoned to any part of the world to provide immediate support to an operations officer. If an operation called for concealing a camera, secreting a microphone in the office of a target, or installing a phone tap, the tech could provide it, install it, or fashion a custom version from "off-the-shelf" parts. If something didn't work, the tech could repair it. Moreover, if it still did not function, the tech could figure out a "work around" from whatever materials were at hand.

Among newer case officers, the techs were becoming prized for their ingenuity as well as their engineering skills in the field. But there were serious limitations. Many other officers, while not exactly technophobes, did not fully embrace technology. Operations were generally conducted using World War II tradecraft refined by personal experience. When TSD techs were brought in for their expertise or gadgets, their assistance was often not viewed as critical for either the successful day-to-day agent operations or an ops officer's career. "It was nice to have the techs there when we needed them, but if we didn't have their gadgets, we still ran the operation," said one case officer active during the early 1960s. The necessary technologies for miniaturization, low power, and reliable electronics were in their infancy. Both the scientific advances of the 1960s and the leadership of Russell and Dr. Gottlieb would be required before technology and operations would become interdependent.

Russell, who lived for operations, was also one of a handful of senior case operations officers of his generation who understood the potential advantage technology could give to operations. "When you went to one of Russell's stations overseas, you saw good technical skills meshing with ops. He really sucked the techs dry in terms of wanting to know what they could provide and how it was going to happen," said one TSD veteran. "A lot of station chiefs literally didn't want to know that 'tech stuff,' but Russell wanted to know it all."

Over the next four years, the Russell-Gottlieb-Krueger team proved to be an inspired, if unlikely, trio. This combination of a case officer with no scientific experience, a scientist of limited operational experience, and an engineer steeped in Big Technology programs would transform TSD. Eventually, this new organization would play a major role in nearly every significant CIA operation for the remainder of the twentieth century.

Russell wasted no time in exerting his influence on TSD. He conveyed a sense of operational reality and urgency that made the case officer's concerns for recruiting and handling spies a concern for his engineers, while daily advances in technology began to influence his vision of operations.

However, Russell and TSD faced a problem unrelated to technology, that of the tightly compartmented world of the Agency itself. Although TSD was a "global division," its technical officers were rarely privy to the details or scope of the operations they supported. Compartmentation sealed off all but the basic facts of an operation to anyone who did not have an absolute need to know. This constraint was of little consequence if the requirement was to secretly photograph a document or prepare a dead drop container. However, with more sophisticated and flexible technology becoming available, the more the techs knew, the better they could match TSD expertise and technology to the operation.

"This was a place where compartmentation and need to know were at odds," said one case officer. "To perform the task perfectly the tech should know everything. But in our world the techs weren't allowed to know everything. Compartmentalization is a necessary fact of life."

Along with this situation, which separated the tech and the case officer, there also existed a subtle cultural divide. In the DDP, the case officer was the star player. The culture of the DDP had evolved from the OSS. The "Ivy League" image that once prompted derision in the press and the "Oh So

Social" sobriquet carried over with the former OSS officers who were among the founding members of the CIA. DCI Allen Dulles had an Ivy League background (Princeton, class of 1914) and an association with New York's powerful white-shoe law firm of Sullivan and Cromwell. Richard Helms, later a DCI himself, had attended boarding schools overseas, including the prestigious Swiss prep school Le Rosey, before attending Williams College. OSS head William Donovan, although not from a wealthy family, attended Columbia University (class of 1905) and its law school (class of 1907). DCI William Colby, the son of an army officer, graduated from Princeton and Columbia University Law School. DCI Bill Casey, who graduated from Fordham University, Catholic University School of Social Work, and Brooklyn Law School, represented the exception that proved the rule.

President Lyndon Johnson may have been thinking of the historical characterization of the so-called blue-blooded case officer when he said, "The CIA is made up of boys whose families sent them to Princeton but wouldn't let them into the family brokerage business."[1]

Conversely, TSS and later, TSD, included few famous family names or Ivy League bona fides, with the notable exception of Cornelius "Corney" V. S. Roosevelt (grandson of President Theodore Roosevelt), who served as director of TSS/TSD from 1959 to 1962. The reason for this was quite simple. For the most part, the OSS technical and engineering staff returned to their corporate or university laboratories after the war ended in 1945. By 1951, when Allen Dulles authorized the formation of TSS, the Agency turned to state universities, technical colleges, and institutes, where engineering programs were emphasized, to hire its first wave of technical officers.[2]

Typically, these technical recruits had shown a childhood penchant for tinkering that eventually turned into engineering and hard-science degrees. They were often the first or only member of their family to attend college and many came from rural communities in the Midwest and Southwest. They arrived at the CIA seeking technical opportunities and adventure.

It did not take long before these newly minted engineers began delighting in calling operations officers "liberal arts majors." For engineers, this less than flattering term summed up both a case officer's educational background and the imprecise, unscientific nature of agent recruiting and handling.

Case officers, for their part, had their own traditions. In theory, spies are acquired through a methodical process of spotting, assessing, developing, recruiting, and handling. All of it was usually done person to person. While living in Switzerland during World War II, Allen Dulles met with agents in his well-appointed study.[3] Penkovsky was met and debriefed in hotel rooms. Face-to-face meetings between the agent and case officer were common practice. Agents were briefed, debriefed, and tasked in safe houses and out-of-the-way restaurants over leisurely dinners.

These meetings built rapport, mutual trust, and personal relationships that often approached friendship. Despite the inherent manipulation, deception, and potential for fatal consequences, the handler and spy worked as a team, with the best case officers also playing the role of a psychologist, cheerleader, banker, confidant, or best friend, depending on the needs of the agent.

Given the limited capabilities of most counterintelligence services after World War II, this process worked well until the 1960s. Highly valued spies such as Pyotr Popov and Penkovsky could use commercial cameras, pass film to their handlers via dead drops, receive messages in the form of strings of random numbers over a standard shortwave radio, and decrypt those numbers with OTPs like those used by the French or Polish underground in Occupied Europe.

As a result, case officers were generally passive toward technology's potential in operations. "We didn't comprehend what was there [technically], and either took a defensive position of operational arrogance or retreated into a shell of saying, 'Look, there's no way I will ever understand what you tech types are doing and if the success of this operation depends on me understanding, then we're just not going to make it,'" remembers one case officer.

For case officers, lacking a clear understanding of technology meant not grasping its potential, while techs ran the danger of either misapplying technology or failing to capitalize on some special advantage. This attitude was especially frustrating for the eager engineers of TSS seeking insight into what was really needed in the field. Highlighting the separateness between operations and technology was the fact that TSD did not receive space in the new Langley CIA Headquarters building where the DDP operational components resided. Six miles separated TSD's downtown Washington location from Langley, a distance that precluded the professional lubrication of drop-in meetings, cafeteria luncheon conversations, or office gossip.

Russell and Gottlieb, both with operational experience, understood this divide and undertook the task of bridging it. "Clearly there was a cultural division. No question about that. When Sid Gottlieb came back from Germany in 1959 to head TSD's research and development work, his approach was, 'Yes, there is this divide, but it needn't be there. And TSD has to bridge it, because the DDP won't,'" explained a TSD engineer. "Gottlieb was right on the mark." Ultimately, the DDP controlled the operations, the dollars and the manpower, so if TSD was to get the funds and requirements needed to be successful, it had to become part of operations.

One of Gottlieb's first moves was to bring research and development closer to the techs directly supporting operations.[4] Less obvious at the time was the significance of the creation of the Directorate of Research in 1962.[5] With this decision, R&D supporting the Big Technology of aerospace and satellite programs became independent from the DDP. TSD remained in the DDP with the single mission of operational support, specifically technical support to stations, case officers, and agents.[6]

Gottlieb and Russell saw a future in which TSD technology would enable operations through new tradecraft devices and techniques. The Big Technology of satellites and photoreconnaissance had proven successful on a large scale in the relatively predictable environment of space. Now TSD had the opportunity to demonstrate how sophisticated, scaled-down technology could expand the scope of what was possible in the unpredictable environment of street operations.

This ambitious strategy first had to address some immediate and not altogether pleasant realities. Much of the equipment in TSD's inventory was woefully outdated and the technical staff seemed imbalanced toward technicians at the expense of engineers. As late as 1960, most electronic gear available for field deployment was too big, too cumbersome, too unreliable, too complex, and too power hungry. In the decidedly blunt terminology of one scientist of the day, the equipment was "junk."

What Russell and Gottlieb had was largely a "craft and special service boutique" for operations. TSD could provide excellent forged documents, quality clandestine printing, and well-made concealments. But that was insufficient to meet the DDP's demands for conducting clandestine operations against the Soviet Union.

"TSD leadership had two mountains to climb. One was the technology, which was pretty bad," said a TSD staffer from the era. "For instance, with secret writing, we were issuing systems that Caesar could have used during the Gallic Wars. We used systems that were developed during the First

World War. We weren't taking advantage of post–World War II chemistries; and the opposition, the Soviets, certainly had the capability to detect our existing systems. So, we had to get better technology. Second, we had to convince working level case officers that we had something to contribute. But it [the technology] wasn't on our shelves, so we'd have to develop the new capabilities and the new equipment across the board."

In chemistry, Gottlieb's field, the few academically degreed chemists directing research were trained prior to World War II. Other TSD "chemists" and techs supporting secret writing in the field were former military medics with no professional training. To remedy this problem, Gottlieb and Russell began recruiting university graduates—scientists and engineers, rather than technicians—specifically for R&D. Their strategy was to exploit the current science from university and research centers and package it quickly into viable covert systems.

Russell broke ties with the Agency's Office of Personnel, TSD's traditional source of "new hires," at the urging of his R&D chief, and began sending senior officers to universities to talk to engineers and recruit new graduates. A co-op program was started in which college sophomores and juniors were hired for a summer or school term and put to work in a lab. An instant success, the co-op program offered a view into the newest research at the universities and allowed TSD to assess a potential employee before making a long-term commitment. In a bit of clever bureaucratic maneuvering, the division skirted personnel ceilings by hiring engineers as "contract employees" on two-year contracts. Not counted as permanent CIA staff, they fell outside the personnel ceiling limitation. The contracts could be renewed every two years as long as sufficient funds were available, and TSD always found the dollars.

"I came in as part of Gottlieb's program to hire these young people with fresh ideas just out of graduate school. I remember sitting in my first SW [secret writing] course. Several of us newly hired chemists started giggling at the 1930s technology the instructor was giving us," said one chemist. "We'd sneak in a question, 'Does this d-orbital . . . ?' The instructor didn't know the term 'd-orbital' [an advanced chemistry term related to the subatomic properties of certain substances, such as crystals and metals].[7] I'm picking on one guy on one point, but that illustrates the level on which the chemical technology existed."

The new hires had a profound impact on TSD's technology. Young chemists improved formulas and processes for SW that had remained unchanged for decades. The SW chemists referred to themselves as "lemon squeezers" in acknowledgment of one of the oldest SW ingredients—lemon juice.

"In World War II secret messages were prepared with a wooden stick and a little bit of water-based ink," explained an SW chemist. "The agent would dissolve the ink chemical, stir it around, take a small piece of cotton, wrap it around the end of the stick and dip it in. He had to first steam the paper, then write the message, re-steam the paper, and then press the paper flat. Finally, he had to write a cover message over the top of the secret writing."

Although not particularly complex, the laborious multistep process required time to complete, and given the limited privacy in many apartments behind the Iron Curtain, was not very practical. "About the time I was hired, we understood the Russians and the British did it a little bit differently and much more securely. My guess is that's when management finally realized we were far behind the curve in SW chemistry," said the chemist. "Why are we using this liquid stuff? Why couldn't we do it dry?"

If the operations officers were not immediately aware of the changes taking place, there were good reasons. TSD had not been moved into CIA's new Headquarters Building at Langley in 1961 with the other DDP divisions. Rather, in 1965, TSD consolidated many, but not all, of its functions into three buildings, Central, East, and South, at the original CIA complex in Washington, D.C., on E Street on Medicine Hill next to the State Department. The TSD chief and deputy chief occupied offices formerly used by Dulles and other CIA directors. The consolidation improved communications among the techs but required a six-mile trip from South Building to Langley for the techs to meet with case officers.

By the late-1960s, Gottlieb's focus on hiring engineers and scientists, combined with adequate funding from the DDP and revolutionary technology, had transformed TSD. In audio surveillance and secret writing, the technical advances produced new capabilities to meet technical support requirements in Africa, Latin America, the Middle East, and much of Asia. The exceptions were China, the Soviet Union, the Soviet Bloc, and Cuba—"denied area operations" countries where direct access to targets was nearly impossible and internal security strict. Yet, outside of the Vietnam War, these were the countries of highest priority for U.S. intelligence. A major initiative to make technology that would work in the toughest environments was demanded.

The Soviet Union presented a special set of operational problems. One involved the very technology that could help operations. The KGB, under its chief Yuri Andropov, fielded one of the most pervasive counterintelligence services imaginable. By virtue of its primary mission "to protect the Revolution," the KGB regarded the Soviet citizenry, foreigners, and emerging consumer technologies with deep suspicion.

For the KGB, even simple technology in the hands of the public was a potential threat to the government and "state security." Virtually every typewriter sold in the Soviet Union, for example, had its fonts sampled on a sheet of paper that was then filed away should the need ever arise to trace the origins of a suspicious document. Complex procedures granting access to copy machines in government offices included signed authorizations and meticulously kept logs of the copies produced.[8] Even consumer items commonly available in the West, such as Kodak point-and-shoot cameras, electric appliances, and battery-powered transistor radios, could not be purchased in the USSR. A typical Soviet possessing such Western-made items would assuredly draw KGB attention.

For agents to operate in this security-obsessed environment, with its deep suspicion of technology, TSD would need to develop an array of special cameras, communications equipment, concealments, and countersurveillance devices.

CHAPTER 6

Building Better Gadgets

The Game is so large that one sees but a little at a time.

—Rudyard Kipling, *Kim*

Russell saw a new role for technology, particularly in its potential for enhancing agent communications. For a spy the greatest danger usually is not stealing a secret, but rather, passing it to his handler. Covert communication (known as "covcom") dominated operational planning. Without the means to transfer information securely between agent and handler, espionage could not exist. The most secure covert communication system was an impersonal exchange that separated the agent and case officer by distance, time, place, or some combination of the three.

The tradecraft lexicon is filled with colorful phrases for impersonal exchanges of information. The best-known method, and most widely used, is called dead drop by the CIA, *taynik* by the Soviets, and dead letter box by the British.

Another personal exchange, the brush pass, requires the agent and handler to walk close enough to each other so that a note or package can be dropped or passed quickly and discreetly. The drop might be made into an open shopping bag or handed off folded into the morning's newspaper. The car toss, a variation of the brush pass, involves throwing a package through the open window of a slowly moving vehicle.

What these pieces of tradecraft have in common is the goal of minimizing the time agent and handler spend in the same space at the same time.

With some techniques, such as with the brush pass, the time is reduced to a fraction of a second. Nevertheless, even the seemingly insignificant half second required to make a successful brush pass brings the agent and handler to the same place at the same time. In hostile areas, such as Moscow, mere proximity of two individuals could arouse suspicion. Was it possible that in Moscow, a city of millions, an American would, by chance, bump shoulders with a leading scientist on a streetcar? The KGB would not believe in such chance encounters. Their view might well have been derived from the legendary New York Yankees catcher, Yogi Berra, who is reputed to have said, "That's too coincidental to be coincidence."

Dead drops, a preferred means of covert communication in denied areas, separate the agent and handler by time, but carry the risk of leaving the package unattended in an environment that could change without warning. A concealment package left at the site could be found by an unwitting passerby or buried by an unexpected snowstorm. The act of loading or clearing the drop site if it appeared unnatural could draw suspicion.[1]

Russell, the case officer, now surrounded by engineers, proposed a solution for improved agent communications that was both ingenious and technologically elegant. A TSD audio surveillance engineer remembered receiving an unexpected call from the chief. "Russell called me up one day in early 1963 and said, 'I was just thinking last night: If you compromise your audio operation by telling the agent where a hidden microphone is, he can talk near it and you have a one-way communication system.'"

Soon afterward, techs began suggesting to case officers that audio devices or bugs, previously used exclusively for audio surveillance, could become a one-way communication system if an agent knew he could be heard through a concealed mic and transmitter.[2] The result became known as an audio dead drop. In one western European city, TSD techs planted a microphone in a tree in a park. To communicate with his handler, the agent "talked to the tree." "I remember one time, we bugged the exterior of a building, so our guy could pause at the corner of the building, say whatever he had to say and keep on walking," a TSD staffer recalled. "We really got involved in that. Audio dead drops took off like gangbusters once we started, and it all began with Russell."

Imagining a dignified diplomat pausing and muttering a few words into a tree trunk seems comical. Yet, the humor is overshadowed by considering how dramatically this new capability expanded the options for communication beyond the chalk marks for signals or loading and unloading dead drops, the level of tradecraft employed by Penkovsky. However, even with the clever audio dead drop, two-way impersonal covert communications inside the Soviet Union remained the prize, a necessary weapon to counter the massive security apparatus of the KGB's Second and Seventh Directorates.

Taking on the KGB inside the USSR began modestly when TSD launched operations to identify the postal censorship techniques used by the Soviets to monitor and examine both internal and international mail. In one basic method of covert communication, an ordinary letter with standard text could also contain a hidden message in secret writing. Mixed in with millions of pieces of mail, cover letters with nonalerting descriptions of vacations and family news could be virtually undetectable. Since World War II agents working for U.S. intelligence had routinely written and received hundreds of secret writing messages from most areas of the world. But the Soviet Union was different from "most of the world."

The KGB watched the mail going in and out of the USSR assiduously. Soviet postal censors were well aware of SW techniques, and the KGB unapologetically opened and inspected the mail of its citizens and foreigners alike. However, since even the KGB could not open, read, and test every single letter, the TSD staff theorized that the Soviets must have censorship protocols. If TSD could understand the systematic organizational sieve that captured and flagged suspicious letters, then they could defeat it.

"For us the question was always: What is the decision process that gets a particular letter routed to the KGB's chemist inside Moscow's Central Post Office? Once that happened, once the letter is suspect, and your guy, whether sender or receiver, is in trouble," said a TSD staffer. "Their chemist may not have confirmed it yet, but there was something, an anomaly, that the first-line postal censor, who is not a chemist, sensed or saw. Why did he pull that letter aside? Why was that one sent over there to the chemist?"

In an exercise called probing, TSD staffers directed the mailing of hundreds of test letters in and out of the Soviet Union with a seemingly endless number of permutations including: date and time of mailing, site of postal box, country of destination, type of letter or postcard, and whether it was written or typed. Probing continued for several years with the letters varying in language, size, and style. Letters were sent from the United States to East European and Russian addresses. Letters were sent from those denied areas to accommodation addresses, known as "AAs," in Europe and the United States.[3] Many AAs were the homes of ordinary citizens recruited for the sole purpose of receiving mail from unknown parties. Once the mail was received, the recipient would call a number alerting the Agency of its arrival and requesting pickup.

The letters were delivered to TSD for examination and analysis. Envelopes were screened for markings, fingerprints, and opening techniques, as well as traces of chemicals that could have been used by the KGB to test for secret writing. Small details, such as the positioning of fingerprints along the perimeter of the paper, revealed valuable clues about the level of scrutiny given a particular letter.

"I traveled to Leningrad and then to Prague, just looking at transit times. One item, like a postcard, came through in about two days; sealed items came through in about two weeks," recalled one member of TSD's probe team. "We began to get a real good feel for what various countries were doing with censorship. The project gave us something solid to take to the Soviet Russia Division officers and recommend, 'Use this technique for mailing from these cities.' We had real postal data that the case officers wanted to hear and could use."

Transit times of letters, postcards, and other commonly mailed items may seem a prosaic detail of intelligence, yet in this way TSD began chipping away at the massive KGB security apparatus.

It took years of efforts by the engineers to make even modest progress. George Saxe had an engineering degree when he was recruited into the Agency directly out of college in 1951. Like most new CIA officers of the time, his career path was not something he anticipated, although the happenstance manner that launched his twenty-five-year service in espionage remains one of his favorite stories.

As an engineering student in the Southwest, Saxe had high hopes for a corporate career in a solid company like Westinghouse or General Electric. It was during his senior year, facing a tight job market, that George spotted a recruitment notice on a campus bulletin board. Turning to a friend, he asked, "What do you think the Central Intelligence Agency is?"

President Truman had established the Agency four years earlier, but the organization had little visibility outside of Washington. George, with no better employment prospects on the horizon, signed up for the interview. When he arrived at the interview, George had no hint that the man seated across the table from him had introduced himself with an alias or that it was not George's engineering skills that made him an attractive candidate.

"The guy had scars all over his face. And I've got papers with my grades and the courses I'd taken, and I'm ready to talk about engineering—what I've been studying for the last four years," remembered George. "The first question was, 'I understand you're on the pistol team?'" George, as it happened, was the captain of the team and regularly posted the highest scores.

"The next question was about whether I ever handled a small boat. Now, this is in a setting where I'm the young engineer looking for my future career," George said with a laugh. "Then he looked at me kind of intently and asked, 'What do you think about jumping out of airplanes?'"

Truthfully, George had not thought about it all that much, but answered that he imagined he could do it. After a few more questions, the interviewer invited George to the campus hotel for a follow-up conversation. "I got there and the first thing he does is haul out a bottle of bourbon, which was not allowed on campus. So, that was my introduction. This was not exactly interviewing for a position at Westinghouse."

Nor was Saxe interviewing for an engineering job. The CIA's interest in George came from its covert paramilitary work to counter a potential Soviet invasion of Western Europe. Soviet-U.S. tensions had not yet coalesced into their four-decade Cold War standoff and all forms of military action seemed possible, if not likely.

"In the early 1950s the Joint Chiefs of Staff and National Security folks—everybody who thought they knew anything about the strategic situation in the world—believed the Red army was going to cross the Rhine River," said George. "So my first tour in Germany had nothing to do with recruiting spies and everything to do with the Soviet Union. I was caching—burying—arms and demolitions for stay-behind teams to wage covert operations against the invading Soviets, like the French resistance during World War II. After a year, I had to fill out a reassignment questionnaire: Where did I want to go next? A fishing buddy of mine said, 'Wouldn't it be neat to go to Alaska?' We got what we wanted. Alaska, not yet a state, was one of Soviet-Russia Division's field locations."

Assigned to Alaska with his fishing buddy, George became the most supervised employee in the CIA. As the only case officer in Anchorage, he was managed by both the Chief of Station and the Deputy Chief of Station. The three were there because Alaska, like Germany, was seen as another critical point into which the Soviet Union could launch hostilities. From a base just across the Bering Strait, Soviet pilots would regularly probe the

Distant Early Warning line of U.S. radar and other air defenses protecting the U.S. western and northern borders. However, when budget cuts hit the tiny intelligence outpost a year later, George was ordered back to Washington where, on that early November day in 1962, he deciphered the message that signaled Penkovsky's capture.

George spent the next two years in SR Division, running a modest number of peripheral operations and learning passable Russian before securing a Moscow assignment where he worked for two stressful but operationally uneventful years. Returning to Headquarters in the mid-sixties Saxe found a changed attitude in the SR Division. Despite the paralysis that gripped operations against Soviet targets, there was a new determination within the ranks of the SR officers to challenge the KGB on its own turf. The idea was to begin taking some risky, but carefully calculated initiatives that would, with luck, lead to productive operations.

Shortly after Penkovsky's arrest, with Moscow operations all but dried up, a Soviet engineer had walked into an American Embassy outside of the USSR and offered his services. By way of bona fides, he brought with him images that detailed Soviet missile capability. With pressure not to make a mistake, the station provided the "walk-in" with a basic commo plan that included instructions on how to receive shortwave coded messages via OWVL, but no follow-up contact was authorized.

The lead had grown cold by the time George learned of the case. Sensing that an opportunity to reactivate the proposal remained, Saxe received approval to initiate an operation. No one gave him much chance of success, and the passage of time had only complicated an already complex situation. Even if the walk-in wasn't a provocation, even if he could be recontacted, and even if he responded, the Moscow office had little ability to sustain communications.

Added to these problems was the fact that the volunteer had specified that he could provide detailed technical intelligence, specifically engineering drawings. It was difficult enough to pass printed or handwritten documents securely inside the USSR, but large blueprints presented special problems. Technical drawings could not be paraphrased or readily copied by hand, and they could not be removed from the facility for any significant length of time without triggering security alarm bells.

Out of necessity, the plan George developed departed markedly from operational tradition. First, the operation would be run not from the Moscow

office, but out of SR Division at Headquarters. Second, there would probably be no face-to-face meetings. In all likelihood, the agent would never
again talk with an American.

George was determined that the operation be handled exclusively
through impersonal communications to lower both the political and security risks. "What would happen if this turned out to be a provocation? We
could lose one of our few case officers in Moscow. And then what happens?
It goes up to the Secretary of State who calls the Ambassador," explained
George. "What does the Ambassador do? He raises holy hell. He yells at
the office chief, 'You CIA cowboys are out there upsetting Soviet-U.S. relations! We have enough trouble with the Soviets without you jackasses
going out and doing something on the street with a guy you know nothing
about.'"

Aside from the risks of a diplomatic flap, there was also the safety of the
agent to consider. The material was so specialized that if the KGB intercepted it, the agent could quickly be identified as the source.

George, although a case officer, was still an engineer at heart with a
natural affinity for TSD. Working with the techs, he devised a one-of-a-kind
communications plan. "For two years I spent half my time working on this
one guy and working with TSD people," George recalled. "The agent had
access to a laboratory where blueprints for missiles were reproduced, and
he could get 35mm reproductions of their engineering drawings of missiles."

The plan that TSD engineers and George eventually devised was not
simple, but ensured the safety of the agent.

TSD mail probes of the Soviet postal system had shown that a nonpoliti-
cal, inoffensive message from an American tourist on a postcard back to the
U.S. attracted little attention from the censors. Conversely, a postcard or letter from a Soviet citizen going overseas warranted a more careful look.
Without revealing the source or purpose of the request, encouragement
went out to U.S. officials in Europe to buy black-and-white picture postcards, common across the Soviet Union, whenever they were traveling in
the country. The postcards, of the variety favored by tourists, featured images of Russian landmarks, such as the Hermitage, the Kremlin, and Red
Square. The postcards were sent to Langley and filled out by staffers with
messages typical of an American tourist visiting the Soviet Union. The
cards were then returned to the Soviet Union and dead dropped to the agent
for use with his new commo plan.

Addressed to accommodation addresses in the West, the recipient's only connection to the Agency was a "sterile" phone number to call when a card arrived.

The dead drop for the agent also contained a long portion of what TSD engineers called "stripping film." Originally created for satellites, the high-resolution film was eventually rejected for the space-based program because the thin, light-sensitive emulsion layer was easily peeled away from the thicker plastic backing. Once separated from the backing, the film resembled the type of clear plastic wrap used to keep leftovers fresh in the refrigerator. Big Technology's excess film became TSD's operational treasure.

To create a covert image, the agent would transfer a standard 35mm image onto the larger-format stripping film by making a "contact print." This involved placing 35mm negatives of the missile plans firmly against the stripping film and briefly exposing them to light. After developing the large-format film, the agent then bleached the image to a nearly transparent white and stripped away the backing. If done with precision, one postcard-sized transparency would hold up to nine 35mm images. In the final step, the agent fastened the clear plastic film to the front where the detailed diagrams, now bleached on the film's emulsion, vanished against the images of Russian tourist attractions. The finished product, which looked like any other postcard printed on glossy paper stock, would attract little notice from the postal censors. However, the agent had to accomplish all this in his tiny Soviet apartment that offered little or no privacy.

Since mating a covert picture with a postcard was a complex procedure, George, the engineer, spent two weeks learning the process before writing up step-by-step instructions in Russian.[4] He went back and forth with native Russian linguists on every word to confirm that the message would be clearly understood. Satisfied that he had both mastered the technique and accurately described the process in Russian, George turned his attention to the trusted one-way voice link radio. A brief OWVL message told the agent that a package had been put down at a secure dead drop site. Inside the package, the agent found his instructions:[5]

IN THIS PACKAGE WE HAVE INCLUDED A ROLL OF SPECIAL "STRIPPING" FILM OF APPROXIMATELY 90MM. OPEN THIS FILM PACKAGE ONLY IN A DARK ROOM USING A SAFELIGHT. USE THIS FILM TO

MAKE CONTACT COPIES OF YOUR 35MM NEGATIVES WHICH YOU WISH TO SEND TO US.

CUT OFF A PIECE OF THE FILM, SLIGHTLY LONGER THAN A PHOTOGRAPHIC POSTCARD. ARRANGE AS MANY OF YOUR NEGATIVES AS POSSIBLE ON THIS PIECE OF FILM (EMULSION TO EMULSION WITH THE 35MM NEGATIVES ON TOP). BE SURE THE 35MM IMAGES DO NOT EXTEND BEYOND THE SURFACE OF THE POSTCARD.

USING A 100 WATT LIGHT BULB POSITIONED ONE METER ABOVE THE 35MM NEGATIVES, THE EXPOSURE TIME SHOULD BE A FEW SECONDS. WE HAVE PROVIDED YOU WITH EXTRA FILM SO THAT YOU CAN EXPERIMENT TO OBTAIN THE BEST EXPOSURE TIME FOR MAKING GOOD COPIES.

DEVELOP THE EXPOSED FILM USING ANY HIGH QUALITY FILM DEVELOPER AVAILABLE TO YOU. ONCE YOU HAVE DEVELOPED, FIXED, AND WASHED THE FILM CONTAINING COPIES OF YOUR 35MM NEGATIVES, THEN PROCESS THE FILM IN A BLEACHING SOLUTION UNTIL THE 35MM IMAGES ARE TOTALLY BLEACHED OUT AND THE FILM IS AGAIN COMPLETELY CLEAR AND THERE IS NO SIGN OF THE LATENT 35MM IMAGES. WASH THE FILM THOROUGHLY.

PRACTICE THIS PROCEDURE USING A BLANK CARD (NOT ONE OF THE PRE-WRITTEN POSTCARDS WE HAVE PROVIDED YOU) UNTIL YOU ARE FAMILIAR WITH THE ENTIRE PROCESS AND ARE SATISFIED WITH THE FINAL APPEARANCE ON THE PRACTICE POSTCARDS.

CAREFULLY DAMPEN THE PICTURE SIDE OF A POSTCARD WITH WATER TO SOFTEN THE PHOTOGRAPHIC EMULSION, TAKING CARE NOT TO WET THE OPPOSITE SIDE OF THE POSTCARD WHERE THE HANDWRITING WOULD BE. THEN WHILE THE PIECE OF SPECIAL FILM IS STILL DAMP, GENTLY SEPARATE THE EMULSION FROM THE FILM BACKING. THIS "STRIPPING" FILM IS MANUFACTURED TO ALLOW REMOVAL OF THE EMULSION WITHOUT DAMAGE. PLACE THE EMULSION SIDE OF THE POSTCARD SO THAT THE SURFACE OF THE EMULSION WHICH WAS FORMERLY IN CONTACT WITH ITS FILM BACKING IS NOW IN CONTACT WITH THE POSTCARD EMULSION. STARTING AT THE CENTER OF THE POSTCARD AND WORKING TOWARD THE EDGES GENTLY PRESS OUT ANY AIR BUBBLES OR EXCESS MOISTURE TO ENSURE COMPLETE CONTACT. PLACE THE POSTCARD ON A SMOOTH, FIRM, DRY SURFACE, AND THEN PLACE

CLEAN BLANK SHEETS OF ABSORBENT PAPER ON TOP OF THE POST-
CARD AND ADD A BOOK OR SOMETHING SIMILAR TO ALLOW THE
POSTCARD TO DRY WHILE PRESSED FLAT.

WHEN THE POSTCARD IS COMPLETELY DRY, TRIM OFF EXCESS
FILM EMULSION ALONG THE EDGES OF THE POSTCARD AND CARE-
FULLY INSPECT THE POSTCARD TO ENSURE THAT IT APPEARS NOR-
MAL IN EVERY WAY AND THAT THERE IS NO SEPARATION OF THE
FILM EMULSION FROM THE POSTCARD.

WHEN YOU HAVE PREPARED ONE OF THE PRE-WRITTEN POST-
CARDS, ADD THE APPROPRIATE STAMP, MAIL IT IN AN AREA THAT
IS VISITED BY FOREIGN TOURISTS.

UPON RECEIVING ONE OF YOUR POSTCARDS, WE WILL RECOVER
YOUR 35MM IMAGES USING ANOTHER PROCEDURE.

Each element of operational planning and preparation had required
time-consuming, exacting attention to detail, but the work paid off. Images
of top-secret Soviet missile blueprints began arriving in U.S. mailboxes in
large cities and rural communities. The recipients, who had agreed to help
the Agency out of a sense of patriotism, could not have suspected that some
of Moscow's greatest secrets were passing through their hands.

At Langley, the bleached images of engineering drawings were reconsti-
tuted at TSD using specialized equipment to produce high-quality prints. The
first batch of images convinced both counterintelligence and Soviet-weapons
analysts of the agent's access, and as production continued, the operation was
declared a major intelligence success.

The stream of technical information on Soviet missile design came at a
point in the Cold War when knowledge of missile capability was a major
requirement for the intelligence community. Yet the most serious operations
have lighter moments. During one briefing, George sat in a conference
room with ten analysts studying a copy of missile design blueprints. "I remember we couldn't read every digit. The analysts were looking for serial
numbers and parts numbers and those types of things. As hard as they tried
they couldn't be confident of every letter and number. Finally in frustration
one of the analysts blurted out, 'I hate to be critical of you guys in operations, but for Christ's sake go out and buy the guy a better camera.'"

George recalled, "The analyst is thinking that we handed this guy a cheap
Pentax camera. I remember what we went through to get those images back
and what the techs had to come up with technically and operationally. Think

about what the agent had to do and the pressure he was under. So all I could do is laugh when this analyst offered a solution to the part of the problem he saw. 'Spend an extra three hundred bucks and get us a little bit clearer image.'"

Later, as the operation matured, the agent communicated that he would be visiting a missile test site and offered to try to recover a piece of a spent missile. Soviet military materials were veritable "gold" for Department of Defense and CIA weapons analysts since the material's composition could yield otherwise unobtainable intelligence about a weapon's capability, design, and production processes.

The agent advised that officials of his rank traveling on government business were allowed to buy quantities of goods and foodstuffs from regions outside the cities and bring those purchases back home. This made it common practice for officials to travel with large empty suitcases to fill with local items such as meats, cheese, fish, vegetables, and other hard to obtain delicacies. The area of the missile test range, the agent noted, happened to be famed for its herring, exactly the type of fish that his family enjoyed. Since he would take two large cases to fill with herring for family and friends, there would certainly be enough room in one of them for a small piece of rocket assembly.

So important was the acquisition of a fragment from an operational missile that Langley approved a high-risk clandestine moving-car delivery on the streets of Moscow. The plan instructed the agent to arrive at a predetermined site in an alley that also served as a through street. The alley had no lights and the meeting would be scheduled for a moonless night. The agent, carrying a shopping bag, would hide the missile part under whatever was available at the market that day and remain at the location no longer than five minutes.

On the appointed afternoon, an American whose pattern of activity frequently involved early-evening shopping and a drink at one of the hotels catering to Western businessmen drove away from his house. As usual, KGB surveillance fell in behind the car, maintaining a polite distance.

For more than three hours, the American attended to routine activities with surveillance trailing behind. After finishing a nightcap consisting of more tonic than gin, he headed home at an unhurried speed. With surveillance hanging back at a steady fifty meters, the American assured himself there was ample distance for what would come next. After five more minutes of driving, surveillance did not close the gap. Apparently they were not going to "bumper lock" this evening, but neither were his KGB watchers going to abandon their surveillance. With both cars maintaining their respective speeds, the American concluded the surveillance vehicle was far enough back. This was the moment. If wrong, he would be signing the agent's death warrant.

Making an abrupt right turn down an alley that served as a shortcut to his house, the American's car was shielded temporarily by three-story buildings on either side for a few seconds. Surveillance could see neither the brake lights blink nor the car's three-second pause in the darkened alley. As the KGB team rounded the corner, the American was driving just a bit slower due to the narrowness of the alleyway and a few minutes later parked at his residence.

The next morning's report by the surveillance team no doubt included details of an uneventful evening. No mention would have been made of a darting shadow that appeared from a hidden doorway at the very instant the American's car turned the corner or that an old shopping bag was dropped through the open window on the vehicle's passenger side.

That same morning George smiled as he deciphered an ops cable advising that a courier had departed Moscow that day for Washington with a special delivery. His package weighed somewhat more than normal for hand luggage and had a slightly fishy odor.

In part, because of his work in developing the sophisticated commo plan that yielded the valuable missile diagrams, Saxe was appointed special assistant to the chief of Soviet Operations in 1967. This new post not only took advantage of Saxe's skill at creating commo plans, but also his ideas about using technologies that could finally make operations possible inside the USSR.

Sid Gottlieb, now heading TSD, recognized in George exactly the kind of ops officer that would assure the long-sought "relevance" of TSD to operations. George was one of the few case officers in the Soviet Division willing to spend time on technology and agent communication. While most of his colleagues wanted to make their career in recruiting agents, George possessed no aspirations to become the DDP or the DCI. He genuinely liked the techs and saw the value in what they could do for operations.

Six miles away from Langley, across the Potomac River at the TSD headquarters, George was able to translate the basic concepts of denied-area operations to TSD engineers who possessed little, if any, operational experience. The quandary was how to reveal the needed information about an operation without violating compartmentation. Seemingly minute details, such as an agent's military rank or nation of origin, could breach security.

But without those basic facts, how could TSD techs know what type of camera to propose? Issuing a $1,200 camera in 1970s to an agent inside the USSR would surely attract unwanted attention. Suspicions might also arise if the agent, who had not been abroad, suddenly acquired equipment not available on the Moscow market. These were precisely the small but significant details TSD needed to understand. The techs needed to know what types of equipment, such as cameras or radios, an agent, based on his salary and status, could easily own in his country. Conversely, from the case officer's point of view, high-quality images required better cameras, but only an informed tech could explain the necessary technical and security tradeoffs. SR Division officers would need to reveal more about the ops to the techs, and the techs would have to honor that trust.

Mastering operational requirements for equipment was no small thing, particularly for operations behind the Iron Curtain. To the case officers it sometimes seemed engineers operated from principles of design that were in conflict with covert operations. Engineers are schooled in the design of industrial and consumer products so that form usually follows function. The can opener in a kitchen and the ratchet wrench in the workshop look the way they do because engineers chose the most logical design solution. Spy gear inverts that concept. For clandestine use, function must often adapt to forms that disguise the true nature of the device. The challenge for TSD engineers was to design a can opener to look like a shoe, a vase, or a tube of toothpaste—anything but a can opener—and they had to do it without sacrificing any of the can opener's functions or reliability.

One of SR Division's early concealment requirements came to TSD during 1967 when operational planners needed a dead drop container for passing money to an agent. Moscow officers collected brick fragments from the dead drop area to match the color and texture of the local masonry. Even the brick collection operation required careful scripting, since Americans in Russia did not stop their cars and jump out to pocket a few random stone fragments from a construction site without prompting questions from KGB surveillance.

While the Moscow office concerned itself with finding brick fragments, another officer was dispatched to Switzerland to obtain nontraceable and well-circulated small denomination rubles. The TSD lab worked on fabricating a hollow "brick" and compacting a wad of money to go inside.

Its work completed, TSD called George to the lab for a look at the brick. What he found was a beautifully constructed concealment, matched in color, texture, and dimension, thanks to the brick fragments. Every detail was perfect, until George picked it up. It felt light as Styrofoam.

"This won't work. Go back to the lab. Get a real brick and weigh it. Your brick doesn't have to be precisely the same, but it has to be close," he told the TSD staff. "I don't care if you put lead in it or whatever you have to do. I know you have to make it hollow and large enough for the wad of rubles, and that paper is lighter than brick material, but whoever picks it up has to believe they're holding a brick."

What George wanted—and what the operation required—was a "brick" that fit into the environment in every way possible. It was not good enough that it simply looked like a brick. It had to *be* a brick to anyone who accidentally came across it. The critical question was: If you put it down in the playground, would a ten-year-old boy come over and pick it up and say, "Boy that's a light brick," or would he say, "That's just an old brick," and throw it away? Would a construction worker who picks up rubble all day say, "That doesn't look or feel right"? This was the degree of protection required to handle agents inside the USSR and the product quality demanded from the TSD.

Getting money to agents in denied areas was, in fact, another ongoing problem. Operational security dictated that higher denominations of rubles were more likely to arouse the suspicions of shopkeepers or bartenders, who could report the unusually large notes to the KGB. However, for a dead drop concealment to remain inconspicuous there was a limit to its size. George took TSD a stack of twenty-ruble notes with the requirement to figure out a way to fit the Soviet currency into the smallest cubic inches possible. At the TSD lab, the engineers devised a combination shrink-wrap, vacuum-packing technique that compressed hundreds of ruble notes into a roll that felt like a stone of solid paper. That single process eventually enabled SR officers to pass millions of rubles in small concealments to the agents.

A few months later, a top-level Soviet scientist, acting on instructions from his OWVL message, approached a high-voltage transmission tower outside of Moscow. There, as instructed, he picked up a brick at a specific location that matched the dead drop description he was given. Something must be wrong, the agent concluded, because this was identical to all the

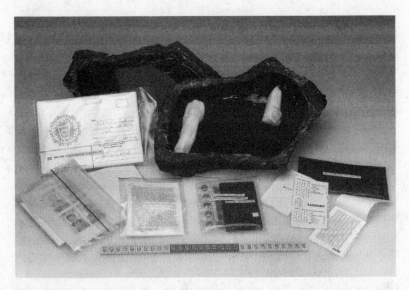

Russian counterintelligence image of a CIA Cold War dead drop rock container and contents, late 1970s.

Closed CIA rock concealment of the type used to pass instructions, cameras, and money to agents, late 1970s.

other bricks scattered around the area by workers who erected the tower. Discarding the brick, he headed to the bus for the journey home wondering what had gone wrong.

Several anxious days passed as the case officer awaited confirmation that the agent had successfully "unloaded" the dead drop. Then came word that the agent had indeed gone to the site, but no "special" brick had been seen. An OWVL broadcast quickly followed, reiterating the instructions with assurances that the ordinary-appearing brick he had discarded was indeed a very special one. It was a modest, but important success.

"The guys in TSD were technically adept, but they were coming over into the operational lion's den," remembered George. "We who did Moscow operations were the most security conscious part of the whole SR Division, which was the most buttoned-up part of the CIA. I mean it was tight and TSD never had the benefit of having a tech assigned in Moscow or dealing closely with denied-area case officers in operational planning. At first they didn't realize all the little points we were always thinking about, worrying about."

A new way of thinking about the operational environment and technology in denied areas was required of both case officer and engineer. For case officers this meant understanding that technology used in satellites could be adapted and applied for operations on the streets of Moscow or Leningrad. For TSD's engineers this meant experiencing the realities of operating in the denied area and integrating that knowledge into the design of clandestine gear. Success demanded that much of the traditional compartmentation of laboratory development from the field operations end.

SECTION III

IN THE PASSING LANE

CHAPTER 7

Moving Through the Gap

Adversity is a good teacher. —Russian proverb

The U-2 spy planes and Corona satellites that produced spectacular new intelligence from their successful first missions overshadowed the slow evolution of technology used by human spies in the early 1960s. TSD chief Sid Gottlieb, who replaced Russell in 1966, remained unwavering in this belief that technology would become an integral role in agent operations. During Russell's four years in the division, the two officers had become lifelong friends and shared the view that the KGB's massive security apparatus was vulnerable to U.S. technology. In the predigital, analog environment of the mid-1960s, this confidence in technology was not routinely accepted in the DDP.

Gottlieb continued to rely on George Saxe to bridge the gap between the operations and technology. As case officers began to use the increasing store of TSD equipment available, new problems arose. One recurring dilemma was communications between case officers and engineers. The cultural divide between the pragmatic engineers of TSD and the "liberal arts types" in the DDP was one not only of background, but of language.

Technology brought with it a vocabulary that was not always clear to the outsider. The potential for confusion was compounded by the already colorful espionage vernacular. This was illustrated when a case officer asked TDS to fabricate a "phone tap." What he wanted was a device that logged the numbers dialed for outgoing calls to identify the contacts of a target. Hearing the words "phone tap," TSD built a system that covertly taped the target's conversations but did not collect the numbers dialed.

Gottlieb understood the necessity for operational compartmentation, but technical requirements had to be precisely translated if TSD was to build the right gear. Misuse of words, misunderstanding of technical concepts, lack of clearly defined operational requirements, and excessive application of compartmentation all contributed to operational failures. George's job was to make sure that SR's operational needs were well defined and clearly communicated to the engineers of TSD.

One significant step that Gottlieb took to bridge this gap was to invite George, as a representative of the SR Division, to TSD's annual retreat. Convened at a covert testing and training facility on an island off the East Coast, the event was a chance for senior scientists, engineers, and craftsmen to let their hair down. A seemingly small thing in retrospect, George's presence caused a stir at the time. "They'd sit in an auditorium in a group and say things like, 'What I don't like is that this group over there is not providing me the kind of support I need to do my job' or 'We have a new idea with a contractor, and we need fifty thousand bucks, but can't get the money,'" recalled George. "So when I showed up, guys were pointing at me and saying, 'Who let him in? An operations guy listening to us talk about our problems?' Gottlieb told them, 'We need to have more trust with operational elements that we support. If he goes back and tells his people we have some problems, what else is new?' That was an effort by a very smart director of TSD to break down the wall and get a better flow of information. I can't overemphasize how revolutionary that was to people in TSD."

The assignment of two senior DDP officers, Everett C. O'Neal and Quentin Johnson, as Gottlieb's deputies, also served to build bridges between the technical and operational. Johnson had firsthand knowledge of the dangers surrounding denied area operations, having served as one of the principal CIA case officers handling Penkovsky a decade earlier.

As head of TSD, Gottlieb began shaking things up inside the Division. He instituted a program of daily operational briefings in a conference area that became known as "the situation room." "Every afternoon beginning at about four, the bosses had to go to the situation room," recalled a chemist. "On one wall hung a huge map of the world with pins representing a tech, somewhere, doing something operational. Those responsible for secret writing would not normally hear about audio because the individual operations and targets were compartmented. But with this briefing, TSD chiefs were forced to think about the problems and requirements faced in other disciplines."

Gottlieb had little tolerance for personality differences or rivalries within the Division. When long-simmering tensions between two TSD chiefs showed no signs of dissipating, he put them in the same office. "They shared a small office, their desks faced each other, head to head," recalled a TSD staff member. "Sid was quoted as saying, 'They may refuse to talk to each other, but by God, they're going to sit there and look at each other all day.'" Gottlieb's personal attention to the TSD "family" became legendary. He called staff and officers on their birthdays and remembered spouses and hobbies. "It sounds hokey, but he had a touch with that kind of thing," said a TSD chemist. "It came across as, 'The boss knows me.'"

As TSD and Soviet Russia Division were beginning to mesh internally, a bureaucratic and political turf war raged among the CIA's senior officers. From its inception, TSD had been a part of the Agency's operational directorate, but with the formation of the Directorate of Research in 1962, the organizational position of TSD became a matter of debate.[1] DCI John McCone believed that all Agency technical capabilities should be centralized. Conversely, Richard Helms, Deputy Director for Plans at the time, opposed moving TSD to the new directorate and argued successfully that operations needed a technical component "as their right arm." Helms then became DCI, and, for the next decade, TSD remained in the operational directorate.

President Nixon moved to replace Helms by nominating James Schlesinger in December 1972 as the successor. Schlesinger became DCI in February 1973 and almost immediately initiated a major reorganization of the Agency. TSD was realigned from the DO to the DS&T.[2] The move also brought TSD a new name: Office of Technical Service (OTS) and a new chief from the DS&T, John McMahon. Gottlieb retired in May 1973.[3]

The internal turbulence was soon matched by controversy on Capitol Hill. When OSS veteran William Colby followed Schlesinger as DCI, political clamor about the CIA's activities in the 1950s and 1960s erupted.[4] In December of 1974, *The New York Times'* investigative journalist Seymour Hersh revealed evocative CIA "crypts" (cryptonyms), like MHCHAOS and MKULTRA, and described past operations within the United States.[5] One of the most damaging revelations was the Agency's involvement, along with the FBI, in opening the mail of U.S. citizens.[6] As a result, both Congress and the Ford administration conducted investigations. The Church Committee in the Senate, the Pike Commission in the House of Representatives, and the presidential-appointed Rockefeller Commission each examined past CIA activities deemed illegal, improper, or misguided.[7]

Intent on making a full disclosure, Colby released sensitive and previously closely held operational details referred to as the Agency's "family jewels."[8] Ordered by Schlesinger, the "family jewels" documents had been hastily compiled in 1973 during the Watergate inquiries. Colby made the highly classified material available to a Senate committee headed by Frank Church. He then unintentionally provided an impromptu visual coup for the Church Committee on September 16, 1975, by displaying an ominous-looking pistol called a *Nondiscernible Microbioinoculator*. The press dubbed the weapon "the CIA dart gun." In fact, it was not a CIA device, but the result of a Fort Detrick research and development program. The weapon, along with others developed by the Army, had been sent to Langley and other elements of the intelligence community for evaluation and comment.[9] Colby had taken the pistol to the committee meeting, thinking it would be of interest as a curiosity, a miscalculation that inadvertently and permanently linked the Agency to the weapon.[10]

Information about CIA activities from reports of the Church Committee and the Rockefeller Commission provided a more complete picture of post–World War II U.S. intelligence than had ever been seen. Several OTS officers were investigated or subpoenaed as a result of their participation in drug testing projects, assassination planning, mail opening, or support to Nixon's "White House plumbers" in the Watergate break-in. Eventually all of OTS's activities were found to have been part of approved operations, and not a single OTS officer was found guilty of any wrongdoing.

While Washington was preoccupied with the politics of scandal, technology was advancing at lightning speed. Personal computers, called "micros" in the parlance of the day, were entering the mainstream. Once the domain of hobbyists and large organizations, these new systems, with their unwieldy five- and eight-inch floppy disks, pointed toward an unexplored world of digitally stored information. The future could also be glimpsed in the hands of teenagers playing a video game called "Pong" from a new company with the strange name: Atari. In U.S. research labs, scientists were laying the digital groundwork for the "wired world." ARPANET, a Department of Defense–distributed computer network, was quietly expanding into a communication system that would evolve into the Internet within two decades.

For OTS, the question was, how quickly could viable and reliable spy gear be built integrating this new technology? Like their OSS R&D counterparts, OTS engineers recognized that technology flowing from private industry could meet intelligence requirements. Technology for espionage seemed poised to match the imaginations of screenwriters who dreamed up fictions such as *The Man from U.N.C.L.E.* and *Mission: Impossible.* A "pen communicator" or a "self-destructing" taped message seemed plausible. The transistor, which had revolutionized audio surveillance operations a decade earlier when it supplanted the vacuum tube, was now being replaced by the microchip. The reliable, affordable Xerox copier ended the labor-intensive need for OTS techs to photograph, develop, and print copies of sensitive documents agents secretly lent to case officers. For OTS, the question became which one of the technologies to pursue.

Early in 1975, an OTS scientist was invited to the lab of an engineer in another part of the Agency. "I have a technology you guys really ought to look at," explained the engineer. At the lab, the OTS scientist saw an experimental setup that allowed for storage and retrieval of relatively large amounts of digital information in a very small sphere. Aptly named "bubble memory," the storage technology could be used to create a new short-range agent communications (SRAC) device.

At the time, SRAC systems could store and transmit only a limited number of characters. With bubble memory, it might be possible to store and transmit entire pages of data. The scientist proposed the project as a practical solution to the communication problem. A few days later, he received an estimate of fifty thousand dollars to build a bench model of a bubble memory module for a SRAC device. Even more quickly, the answer came back—"That's too much money"—and the project wilted. Eighteen months later, funds appeared, but too late for bubble memory, which was already overtaken by the inexpensive and adaptable Read Only Memory (ROM). Literally, technology was advancing faster than the government's funding process.

Another compelling technology quietly emerging was the Charge-Coupled Device (CCD) developed by researchers at Bell Labs in the late 1960s. Originally conceived as a memory storage device, each CCD chip is made up of an array of light-sensitive capacitors. As photons hit the CCD, an electron is dislodged, creating a small electrical charge to form a pattern that varies in degree to the intensity of the light. By focusing light through a lens, the pattern becomes well defined, similar to the chemical reaction of photographic film. A software program that "remembers" where individual charges are located creates the picture. Instead of the image's resolution being determined by the size of the silver grains on the film, the number of capacitors (or pixels) defines detail. In 1974, OTS began building its first digital imager. Rather than copying documents with film cameras, the idea was to replace film with a linear array of imaging sensors from the emerging CCD technology. With a modest investment through a classified contract, OTS engineers worked with a team of scientists at a leading American electronics company to develop a "camera" that would work as well in an agent's hands as the KH-11 imagers (cameras) worked from space.[11]

It would take more than ten years for a product to emerge, a remarkable black box called a "filmless camera" that captured and stored digital images. More important for clandestine operations, the "black box" contained a feature for digital transmission of the electronic images, turning the camera into a two-way covcom device. By 1989, OTS had a piece of spy gear that worked like a cell phone's digital camera.

Advances in technology both radically reduced size and increased capabilities of spy gear. Size reduction expanded possibilities for concealment, minimized power requirements, and improved an agent's ability to conceal it, carry it, dead drop it, and use it. "Can't we make it smaller?" and "Why is this so big?" may have been the most frequently asked questions of OTS engineers.

CHAPTER 8

The Pen Is Mightier Than the Sword (and Shield)

Q: What is a Soviet trio?
A: A quartet returning from an overseas tour.

—1970s underground Soviet humor

I n 1973, a Soviet diplomat stationed in Colombia entered the steam room of the Bogotá Hilton. A few minutes later, another man casually joined him and struck up a conversation in Spanish. The Soviet was Aleksandr Ogorodnik, a member of the Ministry of Foreign Affairs, and the other a CIA case officer. What appeared to be a chance meeting in an unlikely location was actually part of a finely coordinated plan to recruit Ogorodnik to spy inside the Soviet Union.

An economist with a specialty in Latin America, Ogorodnik had access to information about Soviet policy through his diplomatic status and assignment. Here was a chance for the United States to learn what the Soviet leadership was thinking and planning for its Latin American policy. If successful, this operation could provide sustained, detailed intelligence on Soviet plans and intentions not available from satellites.

Soviet officials were more accessible in countries outside the Soviet Union, but recruiting them was still no easy task. As much as possible, the Soviets tried to create a security cordon around their diplomats. Soviets living abroad were watched carefully by their own security officers stationed in the embassy, who, ever alert for a hint of the smallest political crime, would make note of something as harmless as attending a foreign movie. Soviet missions were laced with *stukachi* (informers) eager to curry favor with superiors by reporting any trivial transgression. Soviet diplomats were required to report even a casual conversation with Americans to the security officer at the KGB *rezidentura* (station) inside the embassy. Most complied with the restrictions because, compared to conditions in Moscow, foreign living was luxurious. These diplomats had prospered under the Soviet regime and guarded their elite status jealously.

However, when the case officer entered the steam room that day, he had every reason to believe the Soviet economist could be recruited. According to CIA assessments, Ogorodnik was different. Unconventional by the standards of the Soviet diplomatic community, he was known to like "the good life" and enjoyed a Western lifestyle that included a nice car and a poodle.

Ogorodnik also had three problems that made him vulnerable to recruitment. The first was that a KGB officer in the embassy was trying to recruit him to work as an informer. Such a role would place additional demands on the diplomat's life, while turning down the invitation could generate questions of loyalty. Ogorodnik's second dilemma was that, although married, he had a Colombian mistress, and she was pregnant with his child. This led to the third problem. He would be compelled to return to Moscow when his tour ended. Trapped in a diplomatic soap opera that involved a failing marriage, pregnant mistress, career ambitions, and the KGB, Ogorodnik faced a difficult situation.

Once contact was established, it became evident to the case officer that Ogorodnik had both strong motivations and character traits to become a spy. He hated the Soviet system and was prepared to work against it, though was not foolish. He required compensation and precautions. In exchange for his commitment, funds would be deposited into an escrow account to support his mistress and child in the short term. In the long term, after an undetermined period, the Agency would assist him in defecting. The CIA gave Ogorodnik a code name, *TRIGON,* and insisted that as few people as possible be involved in the operation, since the more people who knew about his secret work, the greater the risk that one of them would betray him. However, to be securely handled inside the Soviet Union, *TRIGON* needed intensive tradecraft training before returning to Moscow.

George Saxe got the call. The instructions were precise. He was to close up everything else he was working on and concentrate all his efforts on *TRIGON.* He could talk to no one about the new assignment in which he was to train the agent and create a communications plan for passing photographs of documents to the CIA in Moscow.

Although the training would take place in Colombia, there were still security concerns. The KGB had a strong presence in the country and maintained close ties with local sources, including police, journalists, and government employees. Ogorodnik's Western lifestyle made him a high-profile personality within the diplomatic community, and local KGB officers were monitoring him as part of their own recruitment campaign.

It took George a month to work out a training and commo plan. Since he spoke and wrote excellent Russian, he would conduct the training himself in Colombia with an OTS tech available for technical consultation and assistance. Unlike Penkovsky, *TRIGON* was not a professional intelligence officer. He would require basic instruction in tradecraft and operational techniques, including the use of dead drops, signal sites, brush passes, car tosses, and accommodation addresses. This would be followed by a series of more advanced covert technology and tradecraft techniques, such as document photography, receiving OWVL broadcasts, using one-time pads for encryption and decryption, secret writing, and microdot reading.

Even for professionals, comprehensive operational training takes months of study and years to perfect. Now, in a room in the Bogotá Hilton, George had the daunting task of schooling a spy to operate in the toughest counterintelligence environment in the world in a matter of weeks.

Among the devices issued to *TRIGON* was a new OTS ultraminiature camera. Work on the subminiature camera that began in early 1970 had a direct connection to Penkovsky. Quentin Johnson, during his assignment to TSD, pressed for development of "a camera that an asset could use to photograph documents while inside a KGB *rezidentura*." Initially, the technical requirements seemed nearly impossible. In addition to being able to capture high-resolution images of a full page without distortion at the edges or benefit of flash, the camera needed a film capacity of at least a hundred frames and a silent shutter system. Added to this, the camera had to be small enough to conceal in an item that a person would normally carry with them into guarded and secure facilities.

OTS responded with a subminiature camera design that carried the designation "T-100."[1] Just one-sixth the size of the Minox issued to Penkovsky a decade earlier, its small size and cylindrical shape allowed the T-100 to be integrated into a wide array of personal items, such as pens, watches, cigarette lighters, or key fobs.

A jewel of watchmaking mechanical precision and optical miniaturization, the camera's 4-millimeter diameter lens was made up of eight elements. Tiny, precisely ground glass elements, some only a bit larger than a pinhead, were exactingly stacked, one on top of another, to achieve clarity in photographing a standard 8½-by-11-inch page.

"The craftsmanship and the technology that went into making the lens assembly was something that may never be repeated," said George, more than three decades after the camera was first introduced.

The T-100's film, lens, and shutter mechanism were housed in a single aluminum casing that measured one and a half inches long and three-eighths of an inch in diameter. As each picture was snapped, the film automatically advanced from one tiny spool inside the cassette to another, making it the world's smallest "point and shoot" camera. Under optimum conditions, the camera's 15-inch filmstrip could hold approximately 100 exposures.

Built under tight security to OTS specs by a precision optical contractor, the T-100 was designed specifically for document copying. An agent could appear to be studying a technical manual, engineering drawings, or a policy paper and noiselessly snap photos by holding the camera in his closed fist eleven inches above the target. Since the lens design allowed some tolerance in the focus distance, most users could place their two elbows at normal shoulder width on a table, and with the document between them, conceal the camera in clasped hands at the apex of the triangle.

In other document copy operations, agents could mount a 35mm camera on a tripod, frame the document, snap away, and be assured of quality pictures. With the T-100, the agent became the tripod and needed to position the camera precisely for each image. Although the film advanced automatically with each exposure, there was no autofocus function and without a viewfinder it was difficult to be certain the document was centered.[2]

Nothing about the T-100 was ordinary, right down to the film it required. Due to of the size of the cassette's spools and operational realities that favored packing as many images as possible in a single cassette, extremely thin film with high resolution was needed. OTS engineers found the solution not in custom-made film, but in retired stocks of Kodak 1414 film used in early satellite photography programs. Because of the sensitivity to "payload" weight in satellite launches, the film was designed with ultrathin emulsion and backing. OTS sliced the film into 5mm-wide 15-inch strips to fit on the T-100's spools. Big Technology had once again assisted with the smallest of devices.

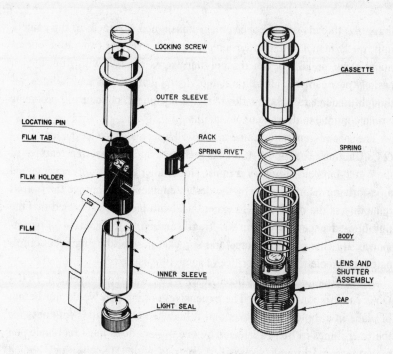

*Drawing of the tiny, intricate T-100 camera assembly that
could be hidden in a fountain pen or a cigarette lighter, 1972.*

Loading the film into the subminiature cassettes required a skill that few could master. The small lengths of film were loaded by hand and wound around the tiny spool either in total darkness or with the aid of an infrared viewer. Then, once the film was loaded, questions always remained as to whether it was loaded correctly. "It's kind of like testing flashbulbs," said one tech. "The only way to test it is to run it through a camera, develop the film and see what the images look like. If they look okay, you say, that was good, but now I have to load another one. You had to have someone who knew what they were doing to load these things in a reliable way. If somebody did it all the time, the loading required about fifteen minutes. Since you couldn't test each one, the best you could do was treat and package each one with care."

The exacting work required to mount a loaded camera in its concealment and the precision needed to load the film was a job for the techs. Attempts to instruct agents on how to remove the camera from the concealment and insert a replacement proved to be extraordinarily difficult. TSD routinely found the threads of the concealment misaligned, stripped, or the camera improperly mated with the tension spring. After several frustrating operational failures, TSD began resupplying agents not just with film, but an entirely new camera. This eliminated virtually all mechanical failures, although inadequate light, shadows over a document, or focusing problems would continue to degrade photo quality.

The second-generation camera, the T-50, was issued to George for *TRIGON*'s training. The T-50 had all the technical and engineering features of the T-100, but held only fifty frames. The smaller film capacity represented a performance compromise by the design engineers to improve the overall reliability of the camera. Field experience with the T-100 revealed that the tiny film-advance mechanism was fragile and prone to inconsistent performance. By changing the clutch and sprocket design, the engineers eliminated the problem, at the sacrifice of some film capacity.

An internationally recognized luxury pen was reproduced for *TRIGON*'s camera concealment. The expensive-looking pen would not be out of place in a diplomat's pocket and it matched Ogorodnik's fondness for the finer things in life. Fabricated by one of America's most reputable pen manufacturers through a classified contract with OTS, the thick-bodied pen looked and worked like its commercial namesake, though a smaller ink sac and imperceptibly shorter base for the nib created a cavity for the spy camera.[3]

Before *TRIGON*'s training began, George perfected his own expertise with the T-50 by spending hours practicing clandestine photography techniques around Agency Headquarters and in the local library. He would carry the pen in his pocket, select a book or magazine, sit at a desk with others, and take covert photos. Repeatedly, George positioned his elbows on a table, forearms at an angle, hands together, and tried to establish a comfortable position. At home, with a ruler, he practiced using different postures to develop a "feel" for the exact 11-inch distance from lens to document. No one seemed to notice. After shooting several rolls, George returned the film to OTS where his work was developed and critiqued.

After an intense month of practice, George was finally confident of his own expertise with the camera. Flying into Colombia under an alias, he checked into the Bogotá Hilton to begin a Spartan existence. He minimized interaction with other Americans and purposely avoided the embassy and government officials. The CIA officer handling the case arranged clandestine meetings with George to coordinate activities and pass instructions, but made no effort to lessen the pressure. George knew the importance of *TRIGON* to SR Division, and with the operation based in Columbia, the Latin America division chief was also watching closely. The case officer had done his job recruiting *TRIGON*; now everything depended on the performance of the camera and operational training.

The first thing George would need to determine was *TRIGON*'s aptitude for learning the clandestine skills to operate the T-50. His first impression of *TRIGON* was "This guy is smart." The training was in Russian, and George, conscious that his Russian was not native, repeatedly confirmed the instructions were understood and that he was using the correct verbs and sentence structure.

The training began with 35mm cameras to familiarize the agent with basic photographic techniques, equipment, and film. The camera selected was the Pentax OM-1, a $200 product that a Russian diplomat could reasonably acquire during an overseas assignment. *TRIGON* quickly grasped the instruction, and his practice photos produced nearly perfect results.

George then demonstrated the modified pen, emphasizing the critical importance of discreetly getting the body geometry of hands and elbow positioned and taking the picture without compromising the camera. Recalling his month-long self-training process, George knew just how hard this could be while also consciously maintaining situational awareness. Awareness of the environment while performing a clandestine act was a fundamental lesson *TRIGON*'s life depended on.

"When I did this," George recalled telling *TRIGON*, "the hardest part for me was sitting in an office. Suddenly, either I'd catch myself paying no attention to the camera or no attention to what was going on around me. It's difficult to maintain that split personality."

Training progressed fitfully over several weeks, with *TRIGON* sneaking away for small blocks of time that did not disrupt his normal pattern of activity or catch the notice of the KGB. On one occasion, he left a few minutes early for an appointment at the Cultural Society, taking a route that included a visit to the Hilton. On another day, after giving a talk to the Bogotá Chamber of Commerce about Soviet assistance to Latin America, he stopped at the Hilton before returning to the embassy.

These unscheduled training sessions could last between fifteen minutes and two hours and George never knew when his pupil might arrive. Confined to his hotel room and fighting boredom, he waited for *TRIGON* to knock on the door and say, "I've got fifteen minutes" or "I've got an hour."

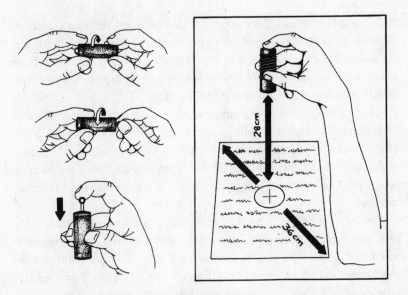

Diagrammed instructions for one method of using the OTS T-100 ultraminiature camera to photograph text on a full-sized sheet of paper in a single frame.

TRIGON proved a fast learner, but George understood it was one thing to perform flawlessly in the safety of the training environment with the instructor and quite another to operate spy equipment alone, with no support. However, whatever concerns George had about the agent's courage were soon put to rest, when during one of the rushed training sessions, *TRIGON* dropped a bombshell. "There's a new, highly restricted Soviet policy paper on China that has just come into the embassy and I should be able to have access to it," he told George one day.

George did not encourage *TRIGON* to photograph the document, since his training on the camera was not completed and top-level policy papers were not readily accessible, even to diplomats. *TRIGON* had the security clearance for the information but the document itself was closely guarded. The policy paper was being held in the *referentura* and its control involved signing out the document from a custodian, then reading it while a guard observed the room through a small viewing port.

The instruction continued, and after three more successful practice sessions, *TRIGON* departed the room with the loaded pen saying, "I may be able to do this."

Twice he returned to George, describing the security in the area and showing the strain of the risk he was taking. "After I get to the room, a guy walks behind me. I couldn't use the camera," he said. "You know what will happen if I make a mistake."

George sympathized with the agent's anxiety. "We would like you to do this," he replied. "And I can tell that you would like to do it, but it's not worth your life. Just keep thinking about it. This is good practice for when you are back in Moscow, although it won't be as difficult there, since you should be able to take documents to your own office."

A few days later, *TRIGON* reappeared at the hotel room door smiling. "I think I've got it," he said.

After *TRIGON* departed, an elated George went to a public phone outside the hotel, called the tech standing by in another part of town, and passed the verbal parole for an emergency meeting. Securing the pen, still loaded with exposed film, in a money belt, George headed out. Fearful of taxi muggings, he walked for more than an hour to pass the pen to the tech who then caught the next plane to Washington, hand-carrying the treasure.

The report George received was astonishing. When the film was processed, only two frames out of fifty were illegible. All significant contents of the policy paper had been captured. With more than twenty years in Soviet operations, this was, to his knowledge, the first time that top-secret documents had ever been photographed by a CIA agent inside a Soviet embassy's *referentura*. *TRIGON* had more than proven he could use the T-50 operationally.

The report added that the information had gone to the "seventh floor" at Headquarters from whence the DCI hand-carried it to Secretary of State Henry Kissinger, who was quoted as saying that the copied document was "the most important piece of intelligence that he had read as Secretary of State."[4]

TRIGON had yet another surprise in store for George at the end of the next training session. He said, "Oh, and by the way, please get me something to kill myself, in case I think I'm going to be caught."

Despite the fiction in espionage novels, lethal substances, called L-pills, were rarely deployed and not available as an off-the-shelf stock item from OTS. Only when an agent could not be dissuaded and after approval by the DDP himself would an L-pill be produced.

George reported the request to the case officer, who cabled it to Headquarters. An immediate answer firmly stated that the request had been reviewed by the Soviet Division chief and the Deputy Director for Operations and the answer was "No way."

Since this was an "agent handling" matter, the case officer broke the news to *TRIGON* that his request had been denied. An L-pill would not be issued.

"Okay, fine, I won't be working for you anymore," *TRIGON* replied.

A furious field-to-Headquarters cable exchange followed. Headquarters asked if the agent was bluffing. The case officer asked George's assistance in preparing a considered response. *TRIGON* and his handler, even prior to the steam room recruitment, had spent considerable time together in Bogotá. And while *TRIGON* did not speak English, the case officer, who was fluent in Russian and Spanish, used both languages to build confidence and trust. Their relationship included informal evenings sitting on street corners at night drinking and talking about politics, philosophy, and personal interests. The case officer was certain he knew *TRIGON* the person as well as *TRIGON* the agent, and George had just spent several intense weeks training the diplomat in espionage. The two agreed the cable should read: "EITHER HE GETS AN L-TABLET OR WE DON'T HAVE AN OPERATION."[5]

OTS was instructed to produce an L-pill and conceal it in the barrel of a pen identical to the one that held the camera.

With his tour in Colombia at an end, *TRIGON* returned to Moscow in 1975 as part of the normal pattern of diplomatic rotation. From the Agency's perspective, *TRIGON* could not have received a better Moscow assignment. Appointed to a key post in the American Department of the Soviet Foreign Ministry, *TRIGON*'s job gave him access to read and photograph reports from Soviet ambassadors around the world.[6]

Since Soviet officials returning from overseas were watched carefully by the KGB for signs of corruption, no immediate contact was attempted. Then, after a cooling-off period of several months, *TRIGON* recovered a dead drop containing new one-time pads, a commo plan, and the T-50 spy camera. Thereafter, he began providing a steady stream of documentary intelligence detailing Soviet policy, confirming the quality of both his training and equipment.

To maintain operational security, with a single exception, *TRIGON* never met his Moscow case officer. The operation relied on communications conducted using OWVL and written instructions passed through dead drops.

95 1100

ДЛЯ РАСШИФРОВКИ

```
24765 93659 55146 09380 18882 67898 69598 95436
25341 88038 31282 39057 21708 51305 66499 20567
65096 02819 74377 27960 20471 53361 18687 06458
19226 31329 55134 83869 26588 24850 81322 67478
01334 80225 37061 13995 88627 07293 53021 81129
90865 91712 80927 18799 71311 57151 71976 06245
98890 61224 59636 08076 65747 36834 49525 92576
95428 50476 06584 38399 37155 75549 11968 12962
43041 83175 29737 88523 76769 29465 47144 75691
77230 19601 57378 51440 48030 63857 15846 37829
32548 48508 71999 22399 86499 22365 91365 74317
57311 83798 06280 74855 58916 46616 07784 57382
10464 00582 08702 30607 80017 50120 76361 88759
93610 38382 57828 27710 00947 00977 02927 89429
53217 20255 20839 63759 74408 60213 32159 73481
31617 14857 97505 25301 14258 36792 42161 05427
52190 32626 07392 88180 32382 22884 82072 81263
39585 92345 44974 09467 88114 50678 84634 02982
44347 73204 49702 60171 56691 11969 32188 62818
06460 37447 02998 93679 05391 96625 21874 88256
85784 28585 57163 61054 85038 41729 76885 51723
12105 61287 69331 72620 98079 56863 59622 96951
94389 88086 36174 39492 54706 56234 49308 07472
79967 13807 72543 07594 89680 63806 18102 32416
65413 91747 01977 31100 62600 78129 31020 07515
09685 11575 35283 37365 15236 28014 82731 07629
35772 51501 01308 09111 40637 41959 81825 82217
69421 13874 28982 52087 95908 43908 06689 55318
64308 31000 08437 64768 79907 58033 78288 44541
39151 31450 44942 53264 04459 19196 33063 68732
57000 78066 10301 31438 87160 08879 10617 39947
41192 47297 79960 45748 24756 60210 83200 78918
91761 48988 10844 64704 86812 61530 69324 30482
03174 79631 96669 88017 31989 32177 73058 80287
94449 59824 50666 22217 36665 78788 88951 51139
92675 67604 01497 28710 65505 37546 76036 64619
84157 68553 92307 42962 21660 78980 52154 40531
57646 07563 92053 84974 34262 59764 68318 44568
65986 82656 13413 64402 77821 46528 50300 34720
43525 90572 90038 01483 75550 94795 48699 55418
```

Page from a covert communications one-time pad for CIA agent TRIGON. The OTP's random numbers were arranged in five-digit groupings. Only two copies were produced—one for the agent, the other for the handler. The number groups on each page differ from those on every other page. The size of the page was determined by the method by which it would be transported and concealed.

TRIGON never met the case officer who loaded and unloaded the drops. Had he done so, he likely would have been shocked.

Martha Peterson arrived in Moscow in 1975. She worked a regular eight-hour daily cover job, improved her Russian language skills, and kept a low profile. For the KGB watchers, she merited little attention. Only on her scheduled breaks, lunch hours, and nights did she assume the responsibility of handling a spy she would never meet, the CIA's highest-placed agent in Moscow.

A young woman, dressed in the latest fashions and doing a woman's administrative job was not how the KGB envisioned a CIA officer. If Peterson did not fit the KGB profile of Agency personnel in the USSR, she did not fit the traditional CIA denied-area case officers, either. She had won the coveted assignment due to the insistence of the office chief who had been impressed by her earlier operational work.

From the time of his recruitment in Columbia, *TRIGON* had proven himself to be a productive agent. Then, in the spring of 1977, the operation took a turn for the worse. The first indication that something was wrong surfaced when communication schedules between the agent and CIA broke down. Finally, an early July OWVL radio message was sent to *TRIGON* instructing him to put out the standard meeting signal by parking a car in a specific location at a specific time. Uncharacteristic of his normal responsiveness, no car appeared at the site. However, a second request for an alternate signal, a red lipstick mark on a "child crossing" pole, did elicit the appropriate response—the red mark appeared on the pole the next day. The signal indicated *TRIGON* would pick up a package at a predetermined time and designated dead drop site.

On July 15, Peterson made a point to leave work on time. For the next four hours, she walked, rode, drove, and used public transportation, executing a carefully constructed surveillance detection run. Her single operational act that night would be to load a dead drop for *TRIGON*. She had conducted more than a dozen similar operations during the previous eighteen months. As in the past, she encountered no problems or indication that KGB surveillance might be following her.

The dead drop site was a window crevice in one of the arched pillars along the footpath on the Krasnoluzhskiy Bridge that spans the Moscow River. Peterson carried an OTS concealment fabricated to look like a piece of asphalt debris in her shopping bag. The relatively flat 6 × 8 × 4-inch "black rock" was sealed with reversed screws and rubbed with soil and mud to give it a dirty look. The cavity was filled with small-denomination Russian rubles held tight with a rubber band, a resupply of six loaded T-50 camera bodies, some jewelry, a pen concealment, new commo schedules, one-time pads, contact lenses, and a personal note on Kalvar film.[7]

Included in the package was a cautionary note in Russian that read:

ONE-TIME PADS

DESCRIPTION: The One-Time Pad is a printed pad of random numbers, usually arranged in five-digit groupings. The random numbers on each page differ from the random numbers or every other page. The size of the pad is determined by the method in which it will be transported and concealed.

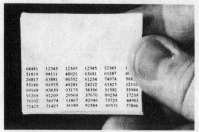

64481	12345	12345	12345	12345	1
51819	94111	48921	63681	60387	46
26817	63881	90552	61254	74974	588
53140	01955	49281	24212	61825	12116
99168	93839	03175	38356	51582	55984
91269	91200	29569	37670	09254	17239
76102	74074	13867	82946	75725	44961
72425	71425	36189	92584	09531	77896

A One-Time Pad measuring 2" × ¾".

PURPOSE: The One-Time Pad is used to encipher and decipher messages that are absolutely unbreakable. There are two steps involved in using the pad:

a) the letters of the plain text message are converted into numbers; and

b) these numbers are added to the numbers from the One-Time Pad using noncarrying math.

Since the numbers are absolutely random, each must be used only once and thrown away to make the message unbreakable.

Many ciphers use mathematical keys that can be generated from an algorithm memorized by the agent. The resulting keys produced from the algorithm are not random, and can be broken using modern computers. The weakness of the One-Time Pad system is that replacement pads must be conveyed to the agent as the old keys are used. For that reason, One-Time Pads are usually very small.

ACTUAL SIZE

A single leaf from a One-Time Pad designed to be concealed behind a postage stamp and mailed to the agent; ⁵⁄₁₀" × ⅛".

HOW TO USE A ONE-TIME PAD

Each letter of the alphabet is assigned a numerical equivalent:

A B C D E F G H I J K L M N O P Q R S
01 02 03 04 05 06 07 08 09 10 11 12 13 14 15 16 17 18 19

T U V W X Y Z
20 21 22 23 24 25 26

Substitute the numbers for the letters in a plain text message:

S E R V I C E   D E A D   D R O P
19 5 18 22 9 3 5   4 5 1 4   4 18 15 16

Use the random numbers shown on the pad in the illustration.

Separate the One-Time Pad numbers into pairs and write them across the page (for this example we have begun with the last row of numbers on the pad). To these we add the numbers from the message.
***(Note:* When adding, use noncarrying arithmetic.)**

S	E	R	V	I	C	E	D	E	A	D	D	R	O	P
19	05	18	22	09	03	05	04	05	01	04	04	18	15	16
72	42	57	14	25	36	18	99	25	84	09	53	17	78	96
81	**47**	**65**	**36**	**24**	**39**	**13**	**93**	**20**	**85**	**03**	**57**	**25**	**83**	**02**

The message is then arranged into five-number groups and transmitted.

81476   53624   39139   32085   03572   58302

To decipher, the receiving agent, who has previously been given an identical One-Time Pad, separates the numbers into pairs, and subtracts the same set of random numbers used for encipherment:

81	47	65	36	24	39	13	93	20	85	03	57	25	83	02
72	42	57	14	25	36	18	99	25	84	09	53	17	78	96
19	05	18	22	09	03	05	04	05	01	04	04	18	15	16

These numbers are reconverted into our original message:

19 05 18 22 09 03 05 04 05 01 04 04 18 15 16
S E R V I C E D E A D D R O P

Official name . . . Cipher Pad, One Time; (special sizes and concealments must be specified)
Circa 1948

Instructions for use of a one-time pad in covert communications, 1975.

CIA-prepared dead drop instructions for TRIGON *at Moscow bridge, 1977.*

COMRADE! YOU HAVE PENETRATED INTO ANOTHER'S SECRET. TAKE THE MONEY AND THE VALUABLES, THE REST THROW INTO A DEEP PLACE IN THE RIVER. AND FORGET ABOUT ALL OF THIS. YOU HAVE BEEN WARNED.[8] Should someone other than *TRIGON* stumble across the container, the note, with its threatening message, might persuade them to dispose of the package instead of reporting the find to the KGB.

Dusk was darkening into night when Peterson finished her surveillance detection run. Confident she had no surveillance, she loaded the dead drop at 10:30 PM, pushing the concealment as far back as her arm could reach into the crevice, and then walked a few steps to a long staircase that descended from the bridge. Just before she reached the bottom, three men suddenly appeared in front of her and grabbed her arms. Immediately, a van pulled out from under the bridge and a dozen more KGB officers piled out. Sensing the KGB's surprise at finding a woman loading the dead drop, she took advantage of the momentary confusion and screamed, "Provocation!" If *TRIGON* was in the area, the shout and the ruckus might warn him away. In a brief struggle that followed, Peterson's green belt Tae Kwon Do instinct flared and she landed one painful kick to the groin of a Russian before being subdued.

The KGB recovered the dead drop container and its contents. Peterson was also searched and the KGB found, Velcroed to her bra, an OTS-developed frequency scanner used to intercept surveillance radio transmissions. Peterson's "necklace" was the scanner's induction coil antenna. Thinking the scanner was a communication device, the KGB officers talked into the black box in an attempt to elicit a response from another party. Throughout the ordeal, the small receiver Peterson wore remained undiscovered.

Peterson was then driven to Lubyanka, headquarters of the KGB's Second Directorate, where questioning began. Within a short time, calls were made to the U.S. Embassy with news that an American citizen had been arrested.

The State Department representative who arrived at Lubyanka was as surprised to see Peterson in custody as the KGB was to see her at the bridge. By 2:00 AM, she had been released. The next day, the Soviet government declared her persona non grata and ordered Peterson out of the country. She left on the first flight out of Moscow, without ever returning to her apartment.

The CIA later learned that *TRIGON* had been dead for at least a month before Peterson's apprehension, compromised by Karl Koecher, half of the Karl and Hana Koecher husband-and-wife spy team. The Koechers were Czech nationals sent to the United States in 1965 under the control of the Czech intelligence service—the Stani tajni Bezpecnost (StB). Claiming to have fled their homeland in search of freedom in America, they posed as virulent anticommunists. Karl earned a degree from Columbia University, and then landed a translator job at the CIA. The StB shared reports from its agents with the KGB and whatever information Koecher gleaned from his translation work about a Soviet diplomat working for the CIA in Colombia was enough for the Soviets to launch an investigation that eventually identified Ogorodnik.

Precise details of *TRIGON*'s death remain clouded, but his early insistence in having an L-pill was prescient, at least according to an account of the death of "agent Trianon" written in 2000. "Trianon" is clearly *TRIGON*. The author, a retired KGB officer, Igor Peretrukhin, who claimed he led the investigation, described "Trianon" sitting in his apartment surrounded by KGB officers at 2:00 AM. *TRIGON* requested paper and a pen to "write an explanation addressed to the KGB leadership." He then requested his own fountain pen that was laying on the table and which one of the KGB officers had inspected. The pen received another, more thorough examination before being given to *TRIGON*. As he began writing, *TRIGON* slowed down and fiddled with the pen. When no one was near the table where he was writing, *TRIGON* managed to disengage the L-pill and get it into his mouth. Suddenly he quivered, leaned against the back of his chair, and began to wheeze. The KGB officers rushed to him and with a metal ruler tried unsuccessfully to open his firmly clenched teeth to find the suspected poison ampoule. Foaming blood began coming out of *TRIGON*'s mouth. He never regained consciousness.[9]

Karl Koecher was arrested in New York City on November 27, 1984, and charged with conspiracy to commit espionage. He served less than two years in prison before a swap for imprisoned Soviet dissident, Anatoly Scharansky, allowed him and his wife to return to Czechoslovakia.[10]

CHAPTER 9

Fire in the Arctic

And ye shall know the truth and the truth shall make you free. —John 8:32

—Inscription at the entrance of CIA's Original Headquarters Building

The death of *TRIGON* and arrest of Martha Peterson were bitter victories for the KGB. The Soviets may not have known *TRIGON* had been an active agent for more than four years, but they certainly would have understood the significance of the information he provided. *TRIGON* did not hold the most senior of Soviet government ranks but had access to documents containing vital national secrets.

Unlike Penkovsky, there would be no public trial during which *TRIGON*'s pet poodle and South American mistress would be submitted as evidence of "moral degradation" and "individualism." Worse for the KGB, *TRIGON*'s suicide precluded any kind of interrogation and follow-up damage assessment of what he may or may not have handed over to the Americans. They would have to assume the worst, and the worst was significant. Throughout his tenure in Moscow, *TRIGON* had access to some of the most sensitive policy and planning documents in the ministry, including those pertaining to the Soviet negotiating positions during the Strategic Arms Limitations Talks.[1]

TRIGON supplied the type of vital intelligence that fed directly into the poker game of *realpolitik* diplomacy and military threat assessment at a critical juncture in the Cold War. It was intelligence that could not be gained from the technology of satellites, the open sources of Soviet media, or a defector who fled across the border at night. It was also perishable information, the sort of intelligence that required continuous refreshing to track shifts in the dynamics of international relations. Only a human agent with the tools to copy documents, record conversations, prepare secret correspondence, and communicate regularly with his handlers could acquire it.

The KGB did not provide accurate details of its investigation beyond those printed as propaganda in *Pravda* and other government-controlled media outlets. However, from the KGB's counterintelligence perspective, it was now painfully apparent that the CIA was capable of running agent operations in denied areas, including Moscow. It was as if a banker, supremely confident in his massive safe, alarm system, and armed guards, had suddenly realized a doorway had been cut at the rear of the vault.

There could be little doubt that secrets in Moscow—and the entire Soviet Bloc—were now vulnerable to American intelligence, and technology was playing an essential role. The level of American communications and collection technology evident in the subminiature cameras and the surveillance receiver confirmed a new level of technical capabilities available to American agents and case officers in the field.

For the KGB, early evidence of America's technical sophistication occurred in 1974 with the discovery of a ground sensor. Concealed as a tree stump near an air force base, the device was crammed tight with electronics capable of capturing and transmitting radio data from the airfield.[2] Now *TRIGON*'s roll-up confirmed that advanced technology was also in the hands of agents. Dead drops, signal sites, and one-time pads were still used by the Americans, but operations were evolving. It did not require great imagination by KGB leadership to envision that soon these time-honored pieces of tradecraft would soon be supplemented or supplanted entirely with a new generation of ingenious devices that would move agent communications from the street to the airwaves.

Peterson presented yet another problem for the KGB officers of the Second Directorate. Because she was not a stereotypical case officer by KGB standards, Peterson's cover work fit cultural and professional expectations of Western women in Moscow. For nearly two years, the immense Soviet security apparatus apparently had not focused on her activities. "When they caught Marty Peterson, I think that opened their eyes a lot," observed an officer intimately involved with the case.

The KGB had done its best to create a cold, dark, inhospitable arctic-like environment for agent operations in Moscow. For years, both the reality and myth of the KGB's blanket surveillance capability contributed to caution bordering on paralysis when it came to aggressive operational efforts. However, instructions given the new Moscow chief by his superiors in early 1973 were to "go out there and shake things up."

TSD chief Sid Gottlieb conveyed a similar message to the techs. "You are not there to fiddle around with the technical gadgets," he directed. "I'm sending you there to become part of their operations and use our gadgets wherever they think will help." Over the next two years, the combination of new tradecraft and technical gear would revolutionize the operational environment in Moscow. Fire was being carried to the Soviet arctic.

Traditionally, new CIA officers received a block of basic technical training during the six-month-long intensive tradecraft course required by the DO. This instruction included photography with a 35mm, Minox, and some subminiature cameras; principles of developing and printing 35mm film, lock picking, sketching of operational sites, secret writing, fashioning an improvised concealment, and "burying" microdots. The objective was to introduce case officers to a variety of technical tools and familiarize them with the operational capabilities available through OTS.[3]

OTS technical operations officers, known as TOOs, who were trained specifically for field assignments, came from all disciplines, including documents, disguise, audio, and secret writing. After acquiring proficiency in each of the other areas, their primary expertise was then matched with the overriding need of an office. In the early 1970s, the prevailing requirement in Moscow was support for covert photography used for operational casing.

Depending on circumstances, Moscow officers would use commercial 35mm or special subminiature cameras to photograph persons or equipment of interest as well as locations for meetings, dead drops, and signal sites. A favorite small camera was the Tessina, which took a half-format image using 35mm film to produce seventy-two exposures instead of the standard thirty-six and featured a spring-wound motor that allowed for one-handed operation and ten exposures before rewinding, eliminating the need to remove the camera from its concealment after each shot. The techs also made concealments for these casing cameras, incorporating them into purses, little leather pouches, and books.

Case officers returning from an operational photography mission, called a "casing run," would hand over the camera to the tech who would develop and print the film, and then return the camera, loaded with fresh film and in a concealment, to the case officer.

The TOO in Moscow could also fabricate small concealments for dead drops out of fabric, leather, pieces of wood, pipes, dirty work mittens, plastic tubing, or discarded cardboard milk or juice cartons. What the TOO could not do on-site with his supply of hand tools and ingenuity, he accomplished by playing the role of consultant, relaying requirements for covert devices—such as audio transmitters—to a larger OTS tech base outside the Iron Curtain or to Headquarters. A TOO working in a denied area experienced the same pressures as case officers, including drawing surveillance when they shopped in the local market or drove their children to the international school.

The KGB may not have realized the degree to which the Americans were monitoring and exploiting the radio transmissions of its surveillance at the time of Peterson's arrest. Certainly they had been confounded by the small receiver, no larger than two packs of cigarettes, she carried.[4]

Development of the SRR-100 began in the early 1970s after a communications tech intercepted transmissions on known KGB frequencies and correlated those with the movement of American personnel. Just as TSD had probed Soviet Bloc mail in the 1960s to understand postal censorship patterns, the CIA now began orchestrating a series of "rabbit runs" to probe KGB surveillance transmissions. A set of scanners identified signals and recorded the transmissions while a dilapidated plotter charted frequencies.

Receiver packages, small enough to be carried under a coat, included a time marker capability so the "rabbit" would push the button when he left the office, then five minutes later, as he took a street to the left, push the button again. Analysis of the marked times and locations were correlated with KGB radio transmissions to determine when the rabbit had surveillance and what frequency was used. Over time, a picture emerged of the type and intensity of surveillance Americans faced and how the KGB coordinated its efforts. Patterns surfaced that defined a standard operating procedure by the KGB, and detailed analysis identified types of behaviors that would likely draw surveillance and which Americans were most closely watched.

Recordings showed that some KGB conversations were in shorthand combinations of terms and numbers. For instance, the term "twenty-one" meant "I have the target in sight," while some code words described specific people or activities. Brief conversations or even single words would indicate surveillance, while extended silence became a reliable sign they were free of their KGB watchers.

The CIA discovered that their KGB watchers would follow an individual with several different teams. One set would conduct overt surveillance, while the second team would hang back, invisible. Sometimes the surveillants would walk, changing articles of clothing to avoid recognition. At other times, they would rotate colors and models of cars. Surveillance could be in fixed positions located in apartment buildings and offices. "Warming rooms" were provided during winter months where inactive surveillants were put on stand-by should their services be needed.

For the operations officer this meant that surveillance could appear and disappear at almost any time. To know with certainty when one was free of surveillance, even momentarily, was the key to conducting a clandestine act. To take advantage of these transmissions, OTS was tasked to build a concealable, body-worn monitor. Conceptualized by an engineer in the Office of Communications and produced by OTS, the SRR-100 enabled the wearer to eavesdrop on KGB surveillance transmissions.

The first models of the SRR-100 could pick up only a single frequency. To alter frequencies, the TOO had to change out the two-pronged crystal, though once the monitor's effectiveness was demonstrated, OTS developed a multichannel receiver to keep pace with the KGB's communication systems.[5]

The second half of building an operational receiver for clandestine use was designing the scanner to be covert as well as functional. The scanner had to be small enough not to attract attention on the street when worn under either summer or winter clothing. While transistors could solve this problem by powering a three-quarter-inch-thick receiver that was no larger than a pocket radio, the question remained of how the case officer could covertly hear the transmissions.

Thirty years after the SRR-100's introduction, it is now commonplace to see the young and elderly walking with headphones and earpieces coupled to devices such as an iPods and cell phones. However, in Moscow, in 1973, an ear bud or headphones—particularly worn by an American—would surely have attracted attention and suspicion. A telltale trace of wiring, no matter how cleverly disguised, trailing from the ear to the shirt pocket of a foreigner on the streets of Moscow would be noticed and reported by the KGB watcher teams.

The solution the techs eventually arrived at was an ingenious use of existing technology, known as an "induction loop." Based on an electrical phenomenon whose applications may be traced to the mid-nineteenth century, the "induction loop" operates on the principle that electrical currents sent through a wire (loop) generate an electromagnetic field that can be picked up by another nearby wire. Electromagnetic induction is similar to the way vibrations in one prong of a tuning fork cause the other prong to vibrate as well.

Engineers created an induction loop that could be worn around the neck under clothing. The loop in turn was connected to a monitoring receiver worn in a shoulder harness under the armpit. For women, the device could also be placed in a special purse with the shoulder strap containing the induction loop. The loops performed double duty, acting as both an antenna receiving surveillance intercepts and a transmitter that sent intercepts to the receiver. The receiver included another, much smaller, loop mounted in an earpiece resembling a hearing aid. Barely larger than the head of a cotton swab, the commercial device was manufactured by the Swiss hearing aid company Phonak.

Although the earpiece was small, it was not small enough to be worn on the street without the possibility of attracting attention. "The earpiece had an obvious problem," said one OTS staffer who was involved with the design. "You couldn't be seen wearing a piece of plastic in your ear without drawing attention." So OTS disguise specialists produced a "Hollywood solution." After taking a casting of a case officer's ear, they fashioned a false, silicone ear that fit over the Phonak receiver. Realistic down to the last detail, the covering was sculpted and tinted to duplicate the shadow of the ear canal. Each case officer received four earpieces, two for the right ear and two for the left ear. Officers could insert the receiver into the ear canal and place the ear mold in front to cover the device.

In addition to masking the earpiece, the sham ear exterior offered another benefit. The sculpting was done with such precision that it not only held the listening device firmly in place without adhesive, but also blocked out ambient street noise, rendering the Russian transmissions more intelligible.

Over time, OTS experimented with other methods for transmission surveillance. In one design, the smaller induction coil was placed in a smoking pipe, called "the Tooth Fairy." The case officer could hold the pipe in his teeth and "hear" through bone-conducted vibrations transmitted along his jaw to his ear canal. Another engineering concept using bone conduction called for incorporating the Phonak circuitry into the bridge of an officer's set of false teeth.

Within two years, the technology had progressed from correlating KGB radio communications with surveillance to being able to identify locations and activities of surveillance teams. A case officer could now walk out on the street, monitor the transmissions, and know with certainty whether or not he had surveillance. "When I heard that transmission and knew I'd been called out, I knew I was, for whatever reason, of some interest to them on that day," said a tech about his Moscow experience. "I didn't know whether or not they were calling me out to a surveillance team that was waiting around the corner or because they needed instructions about whether I was a target that day. I just knew that if the transmissions continued, they were looking at me. If the transmissions ceased, I knew there was a good chance I was free. And if the transmissions resumed later, I knew the KBG had me back on their active list."

The CIA's accumulated operational experience combined with OTS technological countermeasures revealed that the KGB surveillance apparatus, while daunting, was by no means perfect. A key to operational success became patience, as case officers learned that weeks, even months, of routine activity in pattern and profile, was often necessary to set the stage for a single clandestine act.

In time, case officers discovered that even under surveillance they could sometimes *go black*—vanish from sight—for relatively brief periods without setting off alarm bells. Soviet-style clothing, for instance, might be enough to blend into the population for relatively brief periods of time—just long enough to perform an operational act—and pop up again in view of the watchers, who no doubt breathed a sigh of relief. Personnel in Moscow called this "operating through the gap." Such risky acts depended on a well-established pattern of travel, so that when the officer briefly disappeared the KGB surveillance teams would assume that it was their error to have lost him.

By the early 1980s, those skeptical that the CIA could operate in Moscow had been silenced with several remarkable clandestine successes. Viktor Sheymov, a brilliant engineer from the KGB's Eighth Directorate (communications security and signals intelligence) had been smuggled out of the USSR with his wife and daughter in May 1980.[6] A. G. Tolkachev was reporting regularly about advanced Soviet aviation developments during clandestine meetings in Moscow.[7] Although *TRIGON* had been lost in 1977, he, along with another agent, code-named *AEBEEP*, a GRU general, had been handled successfully inside, and new technical collections systems were being deployed. Technology was melting some of the iron in the Iron Curtain.

CHAPTER 10

A Dissident at Heart

I have chosen a course which does not permit one to move backward, and I have no intention of veering from this course.

—A. G. Tolkachev, as quoted in *Studies in Intelligence*

C landestine operations inside the Soviet Union through the 1970s steadily increased along with reliance on new spy devices. These operations, although relatively small in number, were growing in frequency and yielding successes that would have been unimaginable in the Penkovsky era. Agent operations, once wholly dependent on the traditional tools of secret writing, signal sites, and dead drops were undergoing a technological revolution and defeating the KGB counterintelligence apparatus. The new generation of equipment focused on three areas critical for running agent operations: copying documents, communicating with agents, and countering surveillance.

Prior to 1970, techniques for agent communications were limited to a small number of proven techniques, primarily secret writing, microdots, radio broadcasts, and dead drops. Now materials, electronics, chemistry, and miniaturization were transforming agent operations. Technical capability was becoming integral to operational planning and execution.

Whether it was an electronic surveillance package such as the SRR-100 or the T-100 subminiature camera, denied-area operations received the first run of every new device that flowed out of OTS—or at least got first crack at it. A new piece of spy gear provided the element of surprise, since the longer an item was in use, the more likely it would be exposed by defecting agents and subsequently susceptible to technical countermeasures.

Moscow, where operations were largely dependent on the new gadgetry, was to become the proving ground in modern espionage. High-tech spy gear was new territory for agent operations and critical questions had to be answered. For example, would an agent accept "impersonal" handling? How will the equipment be delivered? How are agents trained? Can the agent be trusted with gear that cost millions of dollars to develop? Can the agent reliably operate the new technology? If a device malfunctions, how will it be repaired? Where can the agent hide obvious spy gear?

A second, more subtle change had also taken place among case officers. Almost all Agency personnel working in Moscow were baby boomers, just a few years out of college. Thirty years later, photographs of the era surprise even those who posed for them. The now-fading pictures show grinning youths, relaxed and dressed in the casual American fashions of the day. This change in appearance by the "'60s generation" confounded the KGB, making it more difficult for them to pinpoint the intelligence officers who now dressed in the casual fashions of the 1970s rather than official-looking suits and ties.

In part, youth was also an operational requirement, since the KGB monitored all Americans in Moscow. Once an American was identified as being CIA, he was flagged throughout all subsequent postings and into retirement. A hint of suspicion would earn the officer additional surveillance.

Officers began building their official covers months, sometimes years before leaving for an assignment. They learned the procedures, lingo, and customs of their cover jobs, so by the time they arrived in the Soviet Union, they were virtually indistinguishable from their nonintelligence colleagues. Years after returning home to a Washington suburb, one TOO remembered, with some pride, being approached by a former colleague who inquired whether his wife, and not he, had been a "spy."

Youthful officers provided another, though unexpected, operational benefit. The baby boomers were, compared to the earlier generation of officers, comfortable with the pace of technological innovation. They had grown up with consumer products being regularly introduced, updated, and drifting into obsolescence. In their experience, one technology always supplanted another. Every gadget got smaller, more reliable, and less expensive. If a transistor was better than a vacuum tube, then an integrated circuit was superior to a transistor, and another advance could be expected within a few years. This expectation carried over from consumer products to a constant demand for newer, smaller, and more reliable spy gear.

Just as the class of scientists that entered OTS in the 1960s found technology in the lab lagging behind what existed in private research centers, the new case officers entering the DO in the 1970s discovered that their expectations of "spy gadgets" outpaced the reality. The fantasy of 1960s TV spy shows such as *Get Smart* and *Mission: Impossible*, together with the popularity of the James Bond movies, had changed expectations about the role technology played in supporting operations. Ops officers began to believe that the "magic" of engineering and spy gadgetry they saw in Q's laboratory might be, at least on some level, real.[1] Senior OTS officers recall temporarily reassigning techs to telephone duty to handle inquiries on the day following the airing of new episodes of *Mission: Impossible*. Most of the calls were from operations officers who wanted to know, "Could OTS do *that*?"

Case officers did not need to understand the physics that put Neil Armstrong and his crew on the moon. It was enough to know it was possible. The same was true about the science behind a dry secret writing carbon paper or a battery that would last twenty years in an eavesdropping device. All that mattered was that it worked, was reliable, and met the operational requirement at hand.

As technology advanced, CIA scientists began designing large-scale, highly sophisticated technical collection platforms. In many cases, these plans were so imaginative they led to a conflict between the technically possible and operationally realistic. A technically viable operation from the perspective of a Langley laboratory could represent an unacceptably high risk for the case officer facing KGB surveillance in Moscow. Another problem arose when sophisticated equipment that performed flawlessly in the lab proved impractical to maintain and service in the field.

To resolve these conflicts between technology and operations in Moscow, the TOO became the trusted intermediary between the cadre of scientists at Langley and the case officers in Moscow. The TOO's role was that of translating case officers' requirements into the technical language of engineers and making technical constraints understandable to the ops planners. Case officers needed to appreciate the limitations, as well as the capabilities these advanced systems offered while the design engineers had to recognize the clandestine realities of the denied area.

"Back in the late seventies and early eighties the DO and OTS regularly received proposals for technical collection operations in denied areas," recalled one tech who served in Moscow. "A scenario could require getting someone fifty kilometers outside the city, carrying eighty pounds of equipment, then climbing a tree to a height of about a hundred feet, putting a collection package up there and aligning the antenna to within a degree or two, while evading surveillance the entire time. Well, as valid as the target or the operation might be, the likelihood of ever being able to do it was slim to none."

The culture of the DO added yet another hurdle for technical operations. Large-scale technical collection systems transformed the traditional roles between case officer and technology. Essential technology, such as secret writing, document photography, and agent communications, historically had served the DO as an aid to agent operations. However, technical collection operations in a hostile country were new. When technology became the means for collecting intelligence, the role of the DO sometimes shifted to supporting the collection operation rather than managing it.

Some DO case officers felt that they were being asked to carry all of the operational risks in denied areas, while any successes were credited to the technology. And the risks were high. Like agent operations, a compromised technical program could expose methods of collection, result in the arrest of officers, create an international incident, and jeopardize other ongoing, but unrelated intelligence activities.[2]

Dmitri Fedorovich Polyakov, a career Soviet intelligence officer who eventually rose to the rank of Lieutenant General within the GRU, began life as a spy for the United States at almost the same time as Penkovsky in 1961 and continued his clandestine work through the 1970s.[3] During his service as an agent for U.S. intelligence, Polyakov was a pioneer in the transition from the CIA's reliance on traditional "low-tech" to innovative high-tech tradecraft.

Polyakov was recruited in New York in 1961 by the FBI as a counterintelligence source. He provided identities of illegals working for Soviet intelligence within the United States as well as the names of several Americans who were Soviet penetrations of the U.S. government.[4] After his reassignment in 1966, Polyakov was handed off to the CIA and continued reporting from a series of postings in Burma and India. Rising within the ranks of the GRU, he accumulated a long list of CIA code names, including *GTBEEP, TOPHAT,* and *BOURBON.*

TOPHAT requested little money from his case officers, and accepted only a few small gifts from the Agency, such as woodworking tools and a couple of shotguns for hunting. Motivated primarily by his hatred of the Soviet system, he saw himself as a proud Russian, but a reluctant Soviet. One case officer who knew Polyakov well described the agent as capable of "both intense moments of pride in the Soviet military while simultaneously despising the system it served."

He was the consummate professional. During a turnover meeting in India in the late 1970s, when a departing CIA case officer introduced his replacement, Polyakov noted the new officer's neatly trimmed beard. "We don't allow beards in the GRU," commented Polyakov. At the next meeting, when the two case officers arrived at a hotel room safe house in advance of Polyakov, the senior officer asked the replacement why he had not shaved the beard.

"Why should I?" asked the younger man.

"What our friend was telling you," explained the senior officer, "is that a GRU general is not comfortable being with someone wearing a beard. It could raise questions."

The young officer got a razor from the front desk, went into the bathroom, and shaved off his beard. Shortly thereafter Polyakov arrived and immediately complimented his new case officer on neatness and appearance. The turnover went smoothly, although the officer's wife found it curious that, after wearing a beard for several years, her husband, without warning or explanation, had decided suddenly to go clean-shaven.

Understandably, the Agency and FBI were eager to keep open this pipeline of counterintelligence, which had grown over the years to encompass Soviet espionage operations outside the United States as well. Polyakov also reported on the Red Army and its armaments, including biological and chemical weapons programs. If one were to judge the value of intelligence by the number of customers who used the information, then Polyakov's product was priceless. His information flowed from the CIA to the Pentagon, White House, and State Department.

In the assessment of officers who worked with him, Polyakov was a nearly perfect spy. Not only was he highly placed within the military structure, he possessed the training and discipline of a skilled intelligence officer.

Because he understood KGB counterintelligence tactics, Polyakov proved to be an extremely cautious agent. Methods of covert communications were changed frequently to lower the risks as much as possible. Initially he communicated using OTS's latest secret-writing techniques and conventional dead drops. Outside the USSR, he employed brush passes and received signals via personal ads placed in *The New York Times,* under the name "Donald F."[5]

These methods required significant time and planning, and carried with them varying amounts of risk. None permitted real-time or near real-time exchanges with the case officer in case of an emergency. "We had an agent providing us with critical counterintelligence and positive intelligence about military policies, technical information on Soviet equipment, and penetrations of the American government. These are invaluable reports. But whatever he had, it had to be condensed into short messages of a few hundred words laboriously written down and ciphered using an OTP," said one of his ops officers. "Because Polyakov didn't have a private place to work, this was all done while hiding in a closet or sitting on the john. His family didn't know about his secret work, and his children, wife, and mother-in-law were living in his apartment. He had to sit and work with these miniature one-time pads and encipher all this information prior to sending it to us and decipher what we sent him. Then, Polyakov had to go outside, take a deep breath, put down a dead drop in a public place and hope he was not seen and that the case officer found it before anybody stumbled across it." The operation needed a device enabling the Agency to communicate quickly and securely with an agent while lowering the risk of compromise.

The dilemma of securely communicating with Polyakov was solved while he was posted to India (1973–76), when the CIA developed its first electronic short-range agent communication system for use in a denied area. The new SRAC device, a form of "burst transmitter," carried the code name BUSTER. Measuring approximately 6 × 3 × 1 inches and weighing just over half a pound, the unit was small enough to conceal easily in a coat pocket. BUSTER had a tiny single-digit display with a Cyrillic type font and a keyboard that was no larger than an inch and a half square. To load a message, Polyakov would first convert his text into a cipher using a one-time pad, then poke at the tiny keyboard one character at a time to store up to 1,500 characters. After the data was loaded, Polyakov would contrive a reason to go within the transmitter's thousand-foot range of the base station receiver and press the SEND button.

The receiver base station was a larger unit, measuring approximately $8\frac{1}{2} \times 11 \times 5$ inches thick, and typically sat in one of several windowsills of

Agency residences or in parked cars. Because base stations were maintained in multiple locations, Polyakov could vary his transmission points making his pattern of movement around the city difficult, if not impossible, for any KGB watchers to discern. BUSTER's burst signal minimized transmission time, thereby limiting the KGB's ability to detect the signal and pinpoint its source.

When Polyakov returned to Moscow, he could communicate while riding in a car, streetcar, or bus, or when walking or riding a bicycle, simply by pressing a button on the device in his pocket during the few seconds he was within range. He could now send electronic messages at the time and location of his choosing. Better still, the communication link was two-way. Once the base station received the message, it replied with a confirmation signal and transmitted its own preloaded, though more limited character set, back to the agent's unit. All this occurred in less than five seconds.

"He would poke in his message, then go out for his walk, or on an errand. He had been informed of the general area, but not specific location, of the base station. Once he got into range, he'd push a button," explained one case officer. "His message went to the case officer's machine, which received the information and sent a message back automatically. Polyakov looked at BUSTER and saw a red light flashing that indicated the transmission was successful. Then he returned to his apartment and read our message." In a primitive form, BUSTER possibly represented the world's first text message exchanges.

BUSTER was a technical leap in covert communications equivalent to the telephone in public communications. The distance protected the identity of the communicators, while the short burst and encryption protected the communication itself. The major drawback was possession of the SRAC device, which would conclusively identify the owner as a spy.

The cost of BUSTER's development required expenditures exceeding what OTS alone could afford. To cover the budgetary burden of such projects, OTS teamed up with other offices in the Directorate of Science and Technology, most often the Office of Development and Engineering and the Office of Research and Development. Many of ORD's original scientists who had been part of TSS and TSD now focused on long-term research programs and technology with potential value to the Agency. OD&E engineers were responsible for satellite and overhead programs and the myriad technologies associated with their platforms, cameras, and sensors.

The consulting role OTS played with ORD and OD&E was not an

entirely new function. George Saxe had performed a similar task a few years earlier, translating DDP operational requirements for technology to OTS engineers. Now, the OTS engineers were supplying operational requirements to scientists in ORD and OD&E—merging new technology with the stringent operational requirements of denied areas.

Just as a low-tech dead drop disguised as a brick had to come as close as possible to looking, feeling, and weighing the same as a real brick, similar types of pragmatic, operational design features had to be integrated into this new generation of high-tech devices. Whatever came out of the lab needed to be practical for an agent whose life could not be risked. It had to be easy and uncomplicated to use even under enormous stress.

These requirements were both obvious and not so obvious. For instance, BUSTER had to be small enough to conceal while transmitting. Two buttons to send a message was one too many, because the device had to be activated covertly with one hand inside a coat pocket. Its short-range, line-of-sight transmission would need to be the equivalent of a whisper, rather than a shout, since weaker transmissions were more difficult for counterintelligence receivers to detect.

Powering the device presented unanticipated problems. With commercial batteries frequently in short supply in Moscow, BUSTER's batteries could be either rechargeable or resupplied via dead drop. The decision was made to go with rechargeable batteries, since every operational act, including loading and clearing a dead drop, was inherently risky. However, this meant the agent would need a battery charger and a way of concealing yet another piece of spy equipment at his home or office.

OTS provided the concealment for Polyakov's BUSTER inside a stereo unit he purchased and shipped home before returning to Moscow in 1977. Other concealments were provided when he was outside the Soviet Bloc, together with spare BUSTERs. To attract less attention these concealments were often designed to fit inside existing objects the agent already owned.

There were logistical as well as operational challenges in Moscow to be solved. From what locations could the agent make a transmission without attracting suspicion? The TOOs ran covert signal path surveys to identify specific spots around the city suitable for transmissions and evaluated apartments as possible base stations. In order not to attract KGB attention while conducting these surveys, the casing work was divided among all the office personnel who, with their concealed cameras, would covertly photograph the streets and the views outside apartments to identify potential send and receive locations.

The TOO processed the film from casing runs and these street-level pictures were mated with Headquarters-supplied overhead photography. The distances, angles, and interfering structures between the potential base stations and an agent's send positions were then plotted. As usable sites were identified, a site sketch map was prepared and passed to the agent via dead drop with instructions, "Here are the places to go to make your transmission during the next time period."

From the vantage point of the twenty-first century, BUSTER is a primitive technology. However, compared to technical spy gear available to Penkovsky, it was a miracle advance. Not only did it link the agent and case officer more closely in real time than ever before possible, it also provided anonymity and security for the agent. Within little more than a decade after Penkovsky's arrest, covert communications advanced from a matchbox dangling behind a radiator to electronic messaging that was virtually impossible to detect.[6]

BUSTER was not only created for Polyakov, but also benefited from his suggestions regarding its design and functionality. "In my opinion, Polyakov drove the technology," said one of his case officers. "He would kind of teasingly say, 'You mean to tell me the best you can do is give me something this big?' Sometimes I think he would do it for the fun of it, his way of letting the case officer know they had more of an equal relationship than agent–handler. Not everyone gets to be a general in the GRU, and he knew he was very good at his job. Plus he had succeeded in working for us for years while leading a double life."

Polyakov's input proved invaluable, despite his teasing remarks. He understood the Soviet counterintelligence capabilities at home and abroad, as well as the consequences he faced should the device or its use be discovered. When he finally received the finished version of BUSTER, he told his case officer, "Tell your technical people this is great, I love this piece of equipment . . ." In 1980, Polyakov retired to pursue his passions for fishing, hunting, and woodworking. The story should have ended there, with a man of conscience quietly living out his remaining years.

The KGB eventually discovered Polyakov's clandestine activities, but as with *TRIGON,* the discovery did not come from any known deficiencies in technical equipment or tradecraft. Both were betrayed by Americans working for Soviet intelligence.[7] Two Americans compromised *TOPHAT*'s identity. The first, FBI Special Agent Robert Hanssen exposed Polyakov to the GRU in 1979 while serving as a counterintelligence officer stationed in New York. The revelation strained credibility within the GRU and was thought to be almost unbelievable.[8] By then Polyakov was a respected senior officer and even the somewhat timid proposal to launch a full investigation was turned down.[9] Within the ranks of Soviet military intelligence, casting the shadow of suspicion on a Major General was to gamble one's own career if the charge proved false, and few within the GRU were prepared to take that risk.

But in May 1985, CIA officer Aldrich Ames provided information to the KGB that implicated General Polyakov as a CIA asset. As part of its ensuing investigation, the KGB lured Polyakov away from his modest dacha outside Moscow, then arrested, interrogated, and executed him in 1988.[10]

Reliable accounts relate that during his interrogation he revealed details of what he handed over to his American contacts, and that he had refused the opportunity for exfiltration. In 1990, *Pravda* announced Polyakov's execution in a story detailing the account of a Soviet citizen who spied for the Americans, named "Donald." *DONALD*, one of Polyakov's code names used in personal ads in *The New York Times* years earlier, was the type of detail revealed only during a detailed confession.[11]

By publicizing the story, the Soviets intended to send a message to the CIA and FBI as well as to members of Soviet intelligence who might consider following Polyakov's path. From another perspective, the Soviet account confirmed Polyakov's status as, in the words of a case officer, "the most perfect agent one can reasonably imagine."

Adolf Tolkachev waited patiently on a snowy street near a Moscow gas station in January 1977. The station was frequented by foreigners and when an American-appearing driver stopped, Tolkachev asked, in English, if he was from the United States. When the driver answered that he was, Tolkachev calmly dropped a folded sheet of paper on the car seat through the open window.

Since neither individual looked like what they were, it's difficult to say which of them would have been more surprised to learn the truth. Tolkachev, a middle-aged, undistinguished Russian, was actually a top Soviet military engineer who had recently decided to become a spy. The young, casually dressed American in the car was the local CIA chief.[12] Both were carrying their own secrets and wary of the KGB watchers.

Tolkachev, then a systems engineer in the NIIR (Scientific Research Institute of Radio) building was, by his own description, a "dissident at heart." Inspired by writers like Solzhenitsyn and Sakharov, he was in a state of near anguish. In one of his letters about his decision to spy, he wrote:

> *Some inner worm started to torment me; something has to be done. I started to write short leaflets that I planned to mail out. But, later, having thought it out properly, I understood that this was a useless undertaking. To establish contact with dissident circles which have contact with foreign journalists seemed senseless to me due to the nature of my work. (I have a top-secret clearance.) Based on the slightest suspicion, I would be totally isolated or liquidated. Thus was born my plan of action to which I have resorted.*[13]

Startled by the brief encounter, the chief read the note, which asked for a meeting with an appropriate American official for a confidential discussion and suggested a choice of discreet meeting places in either the car of the American or a Metro station entrance. It proposed a response signal to confirm the meeting involving a car parked at a specific location. The envelope contained precisely drawn sketches of the locations along with a diagram showing how the car should be parked to send the correct signal.

Tolkachev could hardly have picked a worse time for the contact. Although a few operations, like *TRIGON*, within Moscow were progressing, there remained a deep suspicion within the CIA of any Russian volunteer, especially in Moscow. That chance had brought the Agency's Moscow chief face to face with Tolkachev on the initial contact attempt only amplified this distrust. What were the odds that a legitimate volunteer would hand a note to the highest-ranking American intelligence officer in Moscow? And the fact that Tolkachev's approach should occur just before a scheduled diplomatic visit by Cyrus Vance on behalf of a newly elected President, Jimmy Carter, made it all the more suspicious.

On the other hand, many of the Agency's most important agents, including Penkovsky, had volunteered in a similar manner. Penkovsky, for instance, had sent several messages in 1960 to two American students, a British businessman, and a Canadian businessman in an effort to establish a communications channel before the British followed up.[14]

However, if Tolkachev was a KGB "dangle," responding to his approach could allow the Soviets to pinpoint agency personnel, identify agent-handling tradecraft, disrupt ongoing Moscow operations, and embarrass a newly elected President.[15] Added to this was the fact that Tolkachev's notes failed to provide sufficient personal information to identify him or specifics about his access to information that might justify the risk in making contact. CIA Headquarters directed its Moscow officers not to respond to the note.

A month later, Tolkachev was back. This time he slid into the chief's car as it was parked. The two had a brief conversation and Tolkachev left another note. Headquarters again directed that no response be made. Two weeks later, Tolkachev returned a third time, leaving a note that provided additional personal and professional information. Headquarters considered the proposal again, but determined that counterintelligence concerns overrode any meeting. Then, in May, Tolkachev made a fourth approach, spotting the car and pounding on it to get the American's attention. He was ignored, and that summer the CIA chief left Moscow for another assignment.

Six months passed until December 1977 when an Italian national working for the Americans was approached by an unknown Soviet and handed a note at a local grocery store. In this note, Tolkachev volunteered to work as an agent and included two pages of technical data on Soviet aircraft electronic systems to establish his access to sensitive information.

Langley continued to forbid contact. There were now even more reasons to be cautious than before. A few months earlier in August of 1977—barely a month after the arrest of Martha Peterson and loss of *TRIGON*—a fire in the embassy, which some claimed was suspicious, destroyed three upper floors, including the roof. And other operations were compromised that autumn.

The fire had been a physical disaster for much of the embassy. Ron Duncan, a TOO, was having an evening drink at the bar in the Marine House when reports arrived about smoke in the embassy. A few moments later, a Marine cadre discovered the fire already burning out of control on the upper floors. As calls went to the Moscow fire department, Ron rushed to his post and, with water from fire hoses pouring through the ceiling, acted as a guard throughout the night to prevent documents and equipment from being taken by the Russian firemen swarming through the building.

"The Russian firemen, I have to commend them, they were super. They were superb firefighters," said Ron. "The Russians came in with a lot of fire trucks. They poured water on the fire from ten o'clock Friday night until eight Saturday morning. And I swear, they were so efficient with their aim that not one drop of water hit the sidewalk. It was all concentrated on the fire in the building. But the combination of fire, water, and smoke wiped out several offices. Almost everything was destroyed."

The fire presented a myriad of security problems. Damaged documents and photographs could not be thrown into the trash, which would likely be searched by the KGB. Nor, for the same reason, could the damaged furniture or office equipment be hauled to the Moscow dump. Security required that chairs, desks, typewriters, and other furnishings not locally procured had to be returned to the United States. A simple thing, such as the number of discarded chairs, might provide KGB's counterintelligence with an estimate of the number of Agency personnel.

It would, Ron knew, be a massive cleanup and reconstruction project. Virtually every piece of office furnishings was damaged beyond use and would have to be disassembled for shipment. The restoration work would go on seven days a week, eighteen hours a day, for more than a year.

So Ron Duncan found himself with a third full-time job—"ODA"—Other Duties As Assigned. Cleaning up, disassembling, packing, shipping, and reconstruction occurred alongside his cover work and his TOO responsibilities of building concealments and deploying technology. "At least I had some training," Ron recalled. "My tech skills, electrical, painting, carpentry, all came in handy."

Tolkachev, unaware of the disruptions caused by the fire, persisted in attempting to secure a meeting with an Agency officer. It was not until February of 1978 that Headquarters finally approved contact with the determined volunteer. Information on Soviet aircraft that he passed in his earlier note was found to be of such high value that the potential benefit was judged to outweigh the risks.

In his previous letter, Tolkachev had included a partial telephone number missing two digits. His instructions informed the CIA he would be standing in line at a designated bus stop at a certain time holding two pieces of plywood, each with one of the missing digits written on it. All the CIA officer had to do was drive by the bus stop at the designated time and copy the missing digits. The plan failed. An attempt to dial the completed number proved unsuccessful. However, if Agency personnel had any concerns that Tolkachev had given up, they were unfounded. The would-be agent was nothing if not dogged.

As the chief was walking with his wife in March, Tolkachev unexpectedly approached, and quickly handed over a package with eleven handwritten pages on airfield technology. Included in the package was detailed personal information regarding both himself and his family, along with an address, phone number, and other pertinent information. A case officer called the telephone number specified in the note, and this time made contact. If he was not a dangle, Tolkachev had taken an enormous risk in identifying himself and family while passing secret military information. Had that note fallen into the hands of the KGB, the consequences would have been dire.

Five months after the initial phone call, Tolkachev was directed to a Moscow dead drop site near his home. Ron had constructed a concealment using a locally obtained construction worker's mitten. To give the appearance of an unusably dirty, well-worn mitten, he ripped and dirtied the glove so that it resembled a piece of litter. Dropped beside a phone booth near the would-be spy's apartment, it would attract no attention, Tolkachev would find concealed within the worn fabric all the materials needed to begin his clandestine reporting.

The concealment contained special carbon paper for secret writing, three cover letters with nonalerting content whose blank reverse side would hold the SW message, accommodation addresses, a one-time pad to encipher and decipher exchanges, instructions for using the OTP, intelligence requirements, and operational instructions.[16]

A month later, three SW letters arrived at accommodation addresses outside the USSR. Once developed by OTS techs, their physical and chemical analysis revealed all three had been opened, but none showed signs that the SW was detected.[17] The letters contained tantalizing information from Tolkachev, including the assertion that he had ninety-one pages of handwritten notes on subjects such as the new Soviet airborne radar and reconnaissance system and the status of a new Soviet aircraft weapons system. Soviet specialists judged the information to be so important that, despite the *TRIGON* roll-up, Langley authorized a high-risk personal meeting to establish an in-country communications plan between Tolkachev and his handler.

The meeting was scheduled for New Year's Day 1979. Because it was a popular holiday, KGB surveillance tended to be light. A Moscow case officer, after determining he was "black"—that is, without KGB surveillance—called Tolkachev at his apartment to trigger the meeting at a predetermined location. They talked while walking outside in the frigid Moscow winter for less than an hour. Tolkachev passed nearly a hundred pages of highly technical aeronautical design data that included diagrams, electronic specifications,

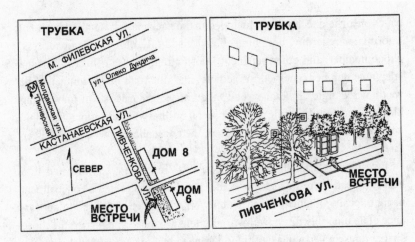

CIA operational sketches of covert communications sites
for use by A. G. Tolkachev in Moscow, circa 1984.

and material copied from official papers. In return, the case officer gave him a list of intelligence requirements along with "good faith" money. The case officer reported that Tolkachev was "calm, deliberate, and one of the few Russians sober on such a major holiday."

In Washington, Pentagon and Agency analysts marveled at the detail of the intelligence passed along at the meeting. The reporting was consistent with other verified information about Soviet aircraft technology and expanded the U.S. understanding of Soviet efforts in weapon systems design.

Seven months before that first personal meeting with Tolkachev, samples of his initial letters and notes were studied by an OTS handwriting analyst. The graphologist, who received no identifying information about Tolkachev, not even his nationality, or background on the case, returned a report that concluded:

The writer is intelligent, purposeful, and generally self-confident. He is self-disciplined, but not overly rigid. He has well-above-average intelligence and has good organizing ability. He is observant and conscientious and pays meticulous attention to details. He is quite self-assured and may plow ahead at times in a way which is not discreet or subtle. All in all, he is a reasonably well-adjusted individual and appears intellectually and psychologically equipped to become a useful, versatile asset.[18]

A year later Tolkachev would describe his motivation:

I have chosen a course which does not permit one to move backward, and I have no intention of veering from this course. My actions in the future depend on [my] health, and changes in the nature of [my] work. Concerning remuneration, I would not begin to establish contact for any sum of money with, for example, the Chinese Embassy. But how about America? Maybe it has bewitched me, and I am madly in love with it? I have not seen your country with my own eyes, and to love it unseen, I do not have enough fantasy or romanticism. However, based on some facts, I got the impression that I would prefer to live in America. It is for this very reason that I decided to offer you my collaboration. But I am not an altruist alone. Remuneration for me is not just money. It is, even to a greater extent, the evaluation of the significance and the importance of my work.

Tolkachev had studied "opto-mechanical radar training," graduating from the Kharkov Polytechnical Institute in 1954. He worked as a "leading systems designer" in a large open office with twenty-four other people. Earning 250 rubles a month plus a 40 percent "secrecy bonus," he lived with his wife and son in a ninth-floor apartment consisting of two rooms plus a kitchen, bath, and toilet. Though above average for the typical Soviet, the cramped quarters would later complicate his clandestine work.

Following the first personal meeting in January 1979, the CIA realized that Tolkachev, in contrast to Polyakov and Penkovsky, had no familiarity with basic tradecraft, such as dead drops or countersurveillance. His job and personal status precluded any secure way to provide the kind of training *TRIGON* had received outside the USSR. Under those circumstances, the decision was made to continue personal meetings in Moscow as the primary means for communicating.

For the next eighteen months Tolkachev was safely met every two or three months. During those meetings Tolkachev passed information while the case officer refined operational details of the elaborate communication system customized to accommodate the agent's circumstances at work and home. Every three months Tolkachev was given the opportunity to pass material to his handler via a dead drop. If the dead drop was to be used, there was a "ready to receive" signal left at a prearranged public site. The signal would have been as simple as a lipstick mark on a telephone pole or a colored thumbtack left on a wooden signpost, invisible to anyone who might pass by. Once the package had been received, the next day the case officer put up a "recovery" signal confirming that the dead drop had been "cleared."

Tolkachev could also initiate a dead drop any Monday by making a mark at a predetermined location. The case officer would reply that he was "ready to recover" the following Wednesday by using a "parked-car" signal.[19] The same night Tolkachev saw the signal he would "fill" the dead drop.

Despite the greater safety of dead drops, Tolkachev preferred personal meetings and argued they were no more risky than dead drops since the case officer had to be free of surveillance to either "make the meet" or "clear the drop."[20] His handlers acquiesced, in part, because new OTS technologies had greatly improved the CIA's surveillance detection tools.

Tolkachev had access to an unprecedented array of technical documents, but lacked a way to copy them.[21] A film-based, easily concealed camera was the best solution available. The Minox Model III had functioned well for Penkovsky twenty years earlier, but was best suited for copying documents in a studio or office where the user could work free from worries about security.[22] Except for their size, these commercial subminiature cameras were not designed for covert use. With their noisy shutters, no automatic film advance, and lack of capability to discreetly photograph a document, the commercial products did not make good spy cameras for high-risk operations.[23]

What Tolkachev needed was a camera that could be concealed and operated while seated at his desk, like the one used by *TRIGON*. Unfortunately, for an untested and untrained agent, the agency did not want to risk the technology. OTS came up with an alternative, constructing a small, concealable camera and light meter, which was passed to Tolkachev via a dead drop glove in February of 1979.[24]

While the camera had worked perfectly in the OTS laboratory, Tolkachev identified a number of operational limitations. The slow film speed, although faster than the original, commercially available Minox film, required more light than was readily available inside his office. The "click" noise of the shutter also caused him to worry about attracting unwanted attention. He found it awkward and impractical to hold the camera steady for the required exposure time. Of more than a dozen rolls of film that Tolkachev passed to case officers in April and June of 1979, most of the images turned out to be blurred and unreadable.

At Tolkachev's suggestion, until a better camera could be provided, he was given a Pentax ME 35mm SLR with a commercial "copy clamp" that could be affixed to the back of a chair to hold the camera steady over the document. The arrangement produced excellent images, but was suited for use only when he could find a few moments of privacy in his apartment.[25] Even given these restrictions, the camera didn't dampen his enthusiasm or slow production. In October and December of 1979, he passed 150 rolls of 35mm film that contained high-quality images taken with the Pentax.

After nearly a year of experience with Tolkachev, the CIA determined that he could be trusted with more advanced and sensitive gear. OTS then provided two of its best custom 4mm-lens subminiature cameras in October of 1979 and four more in December. The cameras, which produced fifty frames, were "packaged" in active concealments and could be disguised as almost any commonly carried item, such as a pen, a cigarette lighter, or a tube of lipstick.

More problems arose in December of 1979 when Tolkachev's institute introduced new security procedures. Under the new guidelines, he could no longer check out an unlimited number of classified documents from the institute library promising they would be returned by the end of the workday, then run home for an impromptu photo session. With the new procedures, classified documents could still be checked out, but only if the borrower's building pass—*propusk*—was left in the library. This ended his ability to photograph documents at home.

Improvising, Tolkachev began photographing documents inside a bathroom stall during working hours. By February he had filled four of his six subminiature cameras with over 200 exposures, but remained dissatisfied with the small camera, complaining that it "still required too much light," was difficult to hold steady, and sometimes malfunctioned.

In order to return to his home photography, Tolkachev suggested the CIA fabricate an exact reproduction of his *propusk* (building pass) to leave in the library while he used his genuine pass to enter and exit the building. To this end, he provided a color photograph and description of the pass to create the replica. However, before the fake pass could be fabricated, the new security restrictions were abruptly cancelled because record keeping under the guidelines became a burden for the library staff.

The impact of the change was immediate as Tolkachev was again able to take documents home to photograph with his Pentax. Weather proved to be an unexpected, though important factor. During the winter, his heavy outer clothing allowed him to conceal more documents than during the summer. In June of 1980, more than 200 rolls of 35mm film were passed, the largest exchange of film in one meeting.

Despite the surprise change of security requirements at the institute, graphic artists at OTS continued work on the *propusk* reproduction. In October of 1980, Tolkachev then asked the CIA to reproduce the institute's document sign-out card containing a running list of all documents he accessed. He was justifiably worried, since a leak could trigger an investigation leading to the institute and document cards. However, a clean sign-out card would sever all links between him and the compromised documents.

In the fall of 1980 Tolkachev was issued an emergency back-up communication system based on the SRAC technology originally developed for Polyakov. This second-generation BUSTER could transmit an entire type-written page of text and partially eliminate the need for riskier personal meetings. The new SRAC, called DISCUS, consisted of two identical units—for case officer and agent—the size of two cigarette packs laid end to end. Each unit had a small detachable antenna, charger, battery pack, Russian or English key plates, and instructions. In advance of an exchange, the agent and case officer would key in their respective messages one character at a time. As each character was entered, the unit automatically converted it to a cipher before storing it. These new units exchanged messages in "burst transmissions" of less than three seconds. The DISCUS system also had a longer line-of-sight transmission range than BUSTER and, once received, a message was automatically deciphered and readable on a small screen built into the face of each unit.

Tolkachev received a DISCUS in March of 1981, but a malfunction necessitated a replacement unit and, some months later, a successful exchange of messages occurred.

Using this new technology required all the elements of disciplined trade-craft perfected for Moscow operations. For Tolkachev, the communications plan involved initiating the exchange by leaving a signal, such as a white chalk mark on a specific telephone pole along the route normally taken by CIA officers. Once the signal was laid down and acknowledged, the agent and case officer would move into that prearranged ELD (electronic dead drop) area at the same time for the exchange.[26]

OTS completed the fake sign-out card and passed it along with the duplicate *propusk* to Tolkachev in March of 1981. Wasting no time, he quickly swapped the new "clean" document sign-out card for the original, confident he was safe, at least for the immediate time being. The outside cover for the forged *propusk,* however, was a slightly different color than the original and returned. Fortunately, with the relaxed security procedures, he was still able to check out documents without having to surrender his building pass.

To ensure continued contact in case of emergency, Tolkachev was also given a commercially available shortwave radio, an OTS demodulator, and two OTPs. Part of an "Interim One-Way Link (IOWL)," the system was a technical upgrade version of what Penkovsky used twenty years earlier.[27] However, because of limited privacy at his apartment and problems related to transmission times, the system had only limited operational utility for Tolkachev.

In November of 1981, the institute reinstated the security restriction that required building passes be surrendered when checking documents out. Fortunately, OTS had a new fake pass waiting, but again Tolkachev judged the color match to be less than perfect. Recognizing the technical difficulty facing the OTS artists attempting to duplicate the precise color of a folder they had only seen in a photograph, he proposed loaning the CIA the original pass for two months during his vacation in January and February of 1982. However, the handler worried that the risk was too great if the return of the original was delayed. Tolkachev's safety was of paramount importance, even if it meant decreased production.

Then, in March of 1982 Tolkachev contacted the CIA for an unscheduled meeting and passed his handler a piece of the external cover of an original *propusk,* enabling OTS to match precisely the colors.

With the increased security and no longer able to copy documents at home during his lunch hour, Tolkachev's production decreased significantly. However, even the reduced level of production was significant given the constraints on copying. Once his case officer commented that photographing the documents at work was "dangerous." Tolkachev laughed and replied, "Everything is dangerous!"

Tolkachev preferred the face-to-face method of passing material to his case officers and receiving new gear. Ten such clandestine encounters occurred between October of 1980 and November of 1983. Each meeting first depended on the CIA officer's confidence that he had been able to evade KGB surveillance; if there was any doubt, the meeting was aborted.

Pervasive surveillance of suspected officers required creative as well as technical solutions by OTS to defeat the Soviets. The SRR-100 receiver that monitored surveillance teams' transmission had been upgraded since its discovery by the KGB when Martha Peterson was arrested. Although it remained a valuable tool, the Soviets had also changed procedures, making reliance on just the SRR-100 alone unwise. Case officers meeting Tolkachev needed other means to free themselves from surveillance.

One possible solution arose from the increased use of private automobiles in Moscow. In studying KGB surveillance tactics, case officers concluded there were moments when a passenger could exit the car unobserved by surveillance. If the passenger's profile could be replaced, the surveillance team would continue to believe they had their target in sight while the officer was free to "operate black."

Getting out of the car was a tradecraft problem left to case officers, but creating a device to replicate their profile became an OTS technical challenge. What followed evolved into one of OTS's stranger attempts to adapt consumer technology for clandestine use.

Two young OTS engineers were dispatched to a windowless store in Washington, D.C.'s seedy red light district on 14th Street. Inside they examined the stock of adult entertainment and sex enhancement products until they found what they needed: three inflatable human figures intended for use as sex dolls. Although the figurines were excessively detailed in anatomical respects, they also possessed all the necessary features to become an inflatable piece of tradecraft.

Technical requirements stipulated that the dolls needed to inflate quickly and sit up straight. The dummy occupant of the car had to appear in the passenger seat within a split second after the passenger exited since a car could be obscured from KGB surveillance for only brief moments, such as rounding a corner. This required, the techs reasoned, a burst of compressed air to inflate the mannequin, which would pop out of some type of container. Initial testing demonstrated that the figurines' construction could not withstand the pressure of a rapid inflation in less than a second. Seams blew out and slow air leakage was common, affecting the dolls' posture.

Discreet acquisition of additional "test dolls" became problematic, if not embarrassing. When the young techs returned to a store for more dolls, the proprietor's quizzical stare seemed to raise uncomfortable questions about their private lives. After all, they could not explain, "You see, we work for the CIA, and . . ."

Even with reinforced seams, stronger materials, and less air pressure for rapid inflation, the problems persisted. The inflatable mannequins continued to sag in the wrong places, resulting in a less than lifelike appearance. The sagging was corrected by installing a valve to provide continuous slow inflation, but the valve had a tendency to hiss. Quality control by manufacturers was another problem. All mannequins did not equally tolerate rapid inflation.

Once the figure was inflated and in place, the techs had to devise an equally rapid deflation and efficient storage process for the device that would allow the case officer to reenter the car and take his place in the passenger seat as quickly as he had exited.

It was then that the research veered to the "elegant solution." Was it necessary to create the most lifelike mannequin possible? KGB surveillance teams typically followed from behind or in front. Rarely did they pull alongside a car. Perhaps a three-dimensional representation was not required. Maybe all that was needed was a two-dimensional figure of the torso with a three-dimensional head. An alternative wooden, metal, and plastic mechanical pop-up prototype was constructed, eliminating inflatable dolls, air pumps, and difficult-to-explain trips to 14th Street.

This new piece of tradecraft acquired the nickname "Jack-in-the-Box," or JIB. The final product could be concealed in a medium-sized briefcase placed on the passenger seat. No installation was required. With one hand a driver unlatched the briefcase and the JIB popped instantly into place. When the passenger was set to reenter the car, the JIB could be pushed back into its resting position with one hand and the "briefcase" shoved to the floor.

In May 1982, personal meetings with Tolkachev were temporarily suspended when increased KGB surveillance forced CIA officers to abort several planned meetings. Eventually, using the JIB, an officer was able to reestablish direct contact. Tolkachev had remained active, using his fake building pass to evade security restrictions, and his home photo sessions continued.

Unexpectedly, another set of restrictions had been imposed at the institute involving a new style of *propusk*, so the fake building pass that had just begun to work was now worthless. Compounding the problem were other regulations that limited all but a few senior officers from leaving the institute without written permission except during lunch. Increasingly Tolkachev's only opportunity to photograph documents would be inside the institute.

CIA assessed the new security regulations as an indication that the KGB suspected the loss of information from the institute and attempts were under way to limit those losses until the leak could be identified. Tolkachev was urged to "stand down," but the persistent agent would not stop his spying.

Tolkachev next provided a photograph of the new building pass and a color strip from the cover to assist OTS in creating a duplicate. He reported smuggling his 35mm camera into the institute on three occasions to photograph documents at his desk and turned over a dozen rolls of film in March 1983 and another dozen in April. As a result, upgraded miniature document cameras were given to Tolkachev for use inside the institute, along with directions to stop taking documents home.

New apprehension about Tolkachev's security arose in April, after he passed information about a new Soviet fighter aircraft target recognition system. The institute's security department requested a list of all personnel with access to that specific information. Tolkachev, convinced that the leak could be traced back to him, destroyed all his spy gear and other potentially incriminating possessions. His SRAC device, Pentax camera, charred remains of documents were all tossed out of his car along a stretch of road during the drive back from his country dacha.[28]

Throughout the autumn of 1983, five attempted personal meetings were cancelled because of KGB surveillance or Tolkachev's scheduling problems. When a meeting did finally occur in mid-November, he had no film, but provided sixteen pages of handwritten notes. The case officer passed him two newly designed document cameras, a light meter, and a proposed schedule for future meetings.

Tolkachev's family situation required a change of communications plans. Since the inception of the operation, unattributable calls from pay phones located throughout Moscow to his home had been used by case officers to initiate unplanned meetings. This worked well until Tolkachev's son, in whose room the phone was located, became a teenager. Like most teens, at the first ring of the phone, the young man ran to it. The commo plan that successfully evaded KGB surveillance for four years now became vulnerable to a teenager who thought that every incoming call could be for him.

From the earliest stages of the case, the CIA began planning for a means of exfiltrating Tolkachev and his family out of Moscow. Ideally, Tolkachev would remain in place as long as possible, but there was always the potential of an emergency requiring immediate exfiltration. Tolkachev realized the consequences if he were ever caught. Four months after his initial meeting he asked his case officer for a "poison pill," stating, "I would not like to carry on a conversation with organs of the KGB." His request was refused by Headquarters, but he persisted and eventually wrote a personal letter to the DCI pleading his case. Tolkachev reasoned that the tasking required him to request highly sensitive documents that were outside his normal work. If he was willing to take this risk, then the Agency should be willing to provide him a poison pill. Additionally he suggested that this "means of defense" would limit any disclosures he might unwillingly make during an interrogation.

Tolkachev assumed that if he were to be arrested, the scenario would begin with a summons to the office of his boss where he would be seized. Now, as security continued to tighten at the institute, any time he was called to the office, he would conceal the poison pill beneath his tongue so he could immediately bite it. Given these urgent circumstances, Tolkachev stopped all document photography but continued to make notes on topics of known intelligence value. His Moscow officer was prescient when writing to Headquarters: "this is indeed a man who is driven to produce, by whatever means he deems necessary, right up to the end, even if that end is his death."

When the institute's new security regulations in early 1983 indicated to Tolkachev that Soviet authorities were suspicious that sensitive information was leaking, exfiltration planning was moved to the front burner. His case officer evaluated escape plans and concluded the "Leningrad Option" would be best. This involved getting his family to Leningrad, then smuggling them out of the USSR in an OTS-constructed concealment cavity built into a vehicle.[29] If the family was unable to reach Leningrad, a backup plan called for them to be picked up on the outskirts of Moscow where they would remain in hiding until exfiltration from the country by another means.

Tolkachev was offered details of the completed plan in April but refused to accept them "because of his current family situation." Both his wife and son knew people who left the USSR for Israel and later wrote back that they regretted the move. The CIA escape plans were no match for his wife's emotional bond to the Russian homeland or his son's attachment to his friends. Tolkachev later commented that he could not think about exfiltration "because I could never leave my family."

Headquarters determined the aviation information Tolkachev reported in March of 1983 had not been disseminated outside of CIA until June, leading to the assumption that no leak could have alerted the KGB in April. Nevertheless, the Agency decided to restrict future meetings with Tolkachev to twice a year and provided a new SRAC device to replace the one he destroyed. He was directed to use extreme caution, limit collection activities to writing notes, and to take photographs only if he felt completely secure using the subminiature cameras.

In April of 1984, Tolkachev passed two full cameras and thirty-nine pages of handwritten notes, twenty-six of which contained detailed intelligence. The images in the spy cameras were of excellent quality. In turn, he received two new cameras, a communications plan, and a hundred thousand rubles, but again refused an exfiltration plan. His spirits seemed high. No problems were reported from the security scare of the previous year. He apologized for destroying his Pentax 35mm camera and requested a replacement. The case officer concluded that from Tolkachev's perspective the operation was "back to normal."

The CIA's assessment of Tolkachev's situation was decidedly less rosy. He was not given a replacement Pentax camera because his handlers did not want him to risk removing documents from the institute. At an October meeting where he passed twenty-two pages of handwritten notes, he also pressed his request for a Pentax and insisted he was "ready to return to work." The Moscow office expressed concern that if Tolkachev was not given another 35mm camera, he would obtain one locally. As an alternative Tolkachev was given additional document copy cameras in an attempt to satisfy his compulsion to photograph documents.

In January 1985, Tolkachev returned three expended subminiature cameras and sixteen pages of handwritten notes. His resupply included five fresh cameras, new intelligence requirements, and another hundred thousand rubles. He explained he had been able to take better photographs inside the institute because the work was done from a toilet stall near a window. This gave him more light and he could arrange for another stop inside the building as cover for the twenty to twenty-five minutes he was away from his desk. Unfortunately, the images, taken on an overcast day, were underexposed and unreadable.

Tolkachev's behavior in the January meeting seemed normal. His written information was consistent in terms of subject matter, quantity, and quality of his previous offerings, and the case officer detected no change in surveillance activities before or after the meeting. Nothing indicated that Tolkachev was compromised or working under KGB control.

According to his notes, the unreadable images from the January meeting contained data about the design of a new frontline Soviet fighter aircraft scheduled to be operational in the 1990s. So significant was the intelligence that the Agency wanted his handlers to initiate an unscheduled meeting in March. The case officer signaled for the meeting, but the reply signal from Tolkachev—opening one of his transom windows between 12:15 PM and 2:30 PM—was not seen. (However, one of the other transom windows was open and later this was thought to have possibly been a danger signal.) Without a positive response from Tolkachev it was decided to wait until June to attempt another meeting.

In the first week of June Tolkachev signaled readiness to meet by opening the middle transom window in his apartment at the indicated time, but the case officer was forced to abort the operation when he detected heavy surveillance. On the next alternate meeting date, June 13, the readiness-to-meet signal was again seen at Tolkachev's apartment. This time, the case officer detected no surveillance en route to the meeting, but as he approached the site more than a dozen KGB personnel wearing camouflaged military uniforms and hiding in nearby bushes jumped him. Bundled into a waiting car, the case officer was taken to the Headquarters of the Second Chief Directorate at Lubyanka for questioning.

During interrogation, the case officer was accused of being a spy and confronted with the package he intended to pass to Tolkachev. Inside were five concealed doc copy cameras, books and drawing pens for Tolkachev's son, periodontal medicine, an OTS book concealment containing 250 pages of newspaper and magazine articles requested by Tolkachev, and an envelope with thousands of rubles. An accompanying note thanked Tolkachev for the "very important written information" he provided at their last meeting, and a description of a new low-light film. The arrest of the case officer was announced to the press, but with no mention of Tolkachev made until September of 1985. Even then his fate was unknown and would remain so until October 22, 1986, when the Soviet press announced that Tolkachev had been executed.

The compromise of the Tolkachev operation was not triggered by a lapse in tradecraft, but the treason of two CIA officers, first Edward Lee Howard and later Aldrich Ames. Howard, a disgruntled CIA officer, was dismissed in April of 1983, but had been briefed about Tolkachev in early 1983. Howard likely betrayed the invaluable agent to the KGB in September of 1984 in Austria. Ames confirmed Howard's information when he passed the names of several other CIA assets including Polyakov to the KGB in May and June of 1985.

In February 1990 the Soviet newspaper *Sovetskaya Rossiya* described the Tolkachev case in an article that was clearly the work of KGB officials. It contained a number of comments that offered grudging praise for the CIA:

CIA provided Tolkachev with a cleverly compiled meeting schedule. CIA instructors made provisions for even the tiniest of details . . . the miniature camera came with detailed instructions and a light meter . . . Let us give CIA experts the credit due them—they worked really hard to find poorly illuminated and deserted places in Moscow for meetings with Tolkachev. . . . Anyone unfamiliar with CIA's tricks would never imagine that, if a light were to burn behind a certain window in the U.S. Embassy, this could be a coded message for a spy. . . . Langley provided touching care for its agent—if he needed medicine, everything was provided. . . . In every instruction efficiently setting out his assignment, they checked up on his health and went to great pains to stress how much they valued him and how concerned they were for his well-being.

The inventory of OTS contributions to the operation involved virtually every element of spy gear and capabilities used by the CIA, including:

Secret writing materials
Concealments
One-time pads
Multiple models of subminiature cameras
Light meters
Disguises
Surveillance detection receivers
Commercial cameras
Clamps for holding cameras steady
Duplicates of security passes
Duplicates of library sign-out cards

Exfiltration containers
Short-range agent-communication system
Commercial shortwave radio
Demodulator unit for radio broadcast
L-pill
The Jack-in-the-Box (JIB)
Special low-light film

During more than five years, case officers and Tolkachev met clandestinely more than twenty times. Never before had this number of personal meetings with an agent inside the Soviet Union been contemplated or securely executed. During the five years of Tolkachev's active service, the fusion of tradecraft and technology demonstrated that there were no longer permanently "denied areas" for agent operations.

As early as March of 1979 a memorandum sent to the Director of Central Intelligence outlined the significance of the material beginning to flow from Tolkachev, concluding it to be "of incalculable value."

From the first rolls of film containing technical data on Soviet aircraft, Tolkachev's reporting was singular in both quality and detail. Especially significant was the window into future Soviet weapons systems that was unobtainable by technical collection programs. Tolkachev provided U.S. military analysts a perspective on Soviet capabilities a decade into the future. That information alone saved the U.S. government billions of dollars in military research and development.[30] Tolkachev's production in terms of volume and value was so significant that a special task force exploited it until 1990.

A senior CIA operations officer, after an exhaustive study of the case, characterized A. G. Tolkachev "a worthy successor to Penkovsky."[31]

CHAPTER 11

An Operation Called CKTAW

Secret intelligence has never been for the fainthearted.

—Richard Helms, *A Look over My Shoulder*

Running concurrently with the Tolkachev operation, a parallel technical collection operation known as CKTAW remained one of America's best-kept secrets.[1] Generating new insights into Soviet particle beam and laser weapons research, CKTAW's priceless intelligence did not come from an agent photocopying secret documents and loading dead drops in out-of-the-way locations. Neither did it arrive from the Big Technology of satellites with high-powered lenses. However, before it was over, the full spectrum of intelligence tradecraft and technology would come into play, from an orbiting satellite to hard won covert experience acquired over nearly a decade of successful operations handling agents in Moscow.

CKTAW was a wiretap on underground communications lines that linked the Krasnaya Pakhra Nuclear Weapons Research Institute in the closed city of Troitsk to the Soviet Ministry of Defense in Moscow.[2] Data flowing via phone, fax, and teletype between the two facilities was recorded as it passed through a seemingly secure section of cable running beneath the streets of a Moscow suburb.

Tunnels and tapped communications lines had been previously used against the Soviets. The Berlin tunnel operation (Operation GOLD) in 1955–56[3] was conceived at a time when Soviet communications were mostly unencrypted and carried by underground telephone and telegraph wires.

However, unprotected hardwired communications had fallen out of favor during the 1960s when governments and military commands adopted microwave links to communicate. These directional signals, transmitted from antenna to antenna in line-of-sight of each other, offered some security advantages. Though the towers were often visible, the directional transmission cone of the signals made them difficult to intercept without positioning an antenna between two towers, an espionage act that would have been impossible in Moscow.

In the mid-1970s, officers monitoring the RF (radio frequency) spectrum in Moscow discovered mysterious microwave signals with no clearly discernible origin. Eventually identified as data from a link connecting the MOD headquarters to the Krasnaya Pakhra lab,[4] the signals unexpectedly appeared during rainstorms and suddenly vanished when the rain stopped.[5] Engineers studying the phenomenon concluded that the temporarily exposed signals were caused by an atmospheric anomaly combined with one of the city's unique architectural features. Apparently, the rain caused enough diffraction to send the secret microwaves ricocheting off Moscow's tin-roofed buildings. The combination of rain and century-old roofs essentially turned a precisely targeted transmission into something resembling a broadcast. Rain and tuning to the right frequencies were all that was needed to listen in on some of the Soviet Union's most closely guarded weapons development secrets.

The intelligence windfall was short-lived. When the Soviets became aware their microwave transmissions were vulnerable to intercept, both from land-based and satellite collectors, the signals diminished and then, eventually, disappeared completely.

U.S. analysts were certain that another, more secure communications link had been established, but did not know where it was located. Eventually, analysis of images from the new KH-11 satellite revealed that the Soviet military was laying communications cables in a trench running between Moscow and Troitsk.[6] Launched in December 1976, the KH-11, the first space-based platform to use digital technology and transmit near-real-time images, provided eye-in-the-sky images from Moscow whenever it was in position and clouds did not obscure the target.[7] Corroboration of the image analysis came when CIA technical and operations officers in Moscow exploited information from every possible human and technical source to confirm that the trench did in fact carry the new Krasnaya Pakhra signals.

Closer examination of the route showed a series of manholes along the length of the buried communications link. The manholes were installed for routine repairs and security checks, but if the CIA could obtain access to the cabling from one of the manholes, perhaps the link could be compromised.

The job of identifying the best potential access point fell to CIA field officers who studied more than a dozen manholes along Varshavskoye Shosse (Warsaw Boulevard), a busy thoroughfare near Moscow's outer ring road. However, like all Moscow operations, any actions to survey the manholes required exacting planning. How many times could Americans travel to the target locations without alerting the KGB surveillance teams? Could clandestine photography of the sites be acquired? How closely could a manhole be examined without attracting attention?

In the final determination only a handful of ground reconnaissance and casing trips were authorized. Each trip—either on foot or by car—required detailed planning, disciplined tradecraft, and, perhaps most important, timing calculated to avert suspicion. For an American to drive along a particular stretch of road outside his normal pattern of travel five times in a week was certain to tip off the ubiquitous KGB watchers that *something* about the area was of interest to the CIA. The few authorized casing runs required the participants to create new patterns of travel and eventually took more than two years to complete.

The essential casing photography was acquired with OTS concealed Tessina cameras. With high-altitude images, casing photos, and sketches, the Moscow office worked up a three-dimensional view of the CKTAW target site, including close-ups of notable features. Target sites along the heavily trafficked Varshavskoye Shosse were narrowed down as one target manhole after another was rejected. The specific manhole eventually selected for an entry point was in the worst possible location, except for the fact it was better than all the others.

The area along the narrow treeline that bordered Varshavskoye Shosse became one of the most carefully plotted and studied pieces of real estate on the planet. Details of the manhole, located just off the roadway, were painstakingly examined. Among potential problems were the worrisome 120-acre open field opposite the target, and a hostile installation belonging to the Second Chief Directorate visible on a hill two kilometers away. One positive feature was the treeline that ran parallel to the roadway. Although less than fifty feet wide, its foliage offered good ground cover for approaching the manhole from May through October.

Depiction of proper technique for successful clandestine photography from inside a moving automobile, 1960. OTS continually improved spy cameras, but the techniques for taking quality photos remained unchanged.

Over several months, CIA officers covertly examined, photographed, and briefly entered the manhole. These operations allowed them to gauge the difficulty of removing its cover, measure the dimensions of the underground chamber, the depth of the ground water at the bottom, and confirm that the communications cables were accessible from a small steel utility ladder set permanently into one wall. The Soviets had not only shielded the cables in lead, but also installed sensors and alarms to detect any tampering. To tap a cable clandestinely meant not touching the actual transmission wires or compromising the integrity of the lead sheath encasing them.

Tapping the cable was a daunting challenge. So important was the potential intelligence carried over the line that in 1977, the CIA authorized an unprecedented, technically high-risk, and costly development program to do just that.

Technical expertise required coordination among multiple engineers to design and build the equipment and an interface among the developers and operations officers who would install and service the equipment. The operation included several DDS&T offices, a DO Division, and the National Security Agency. Management of CKTAW's technical development resided with the DDS&T's Office of Development and Engineering (ODE), the office responsible for satellites and other advanced technical sensor programs. The development team included several OTS engineers, with the operation eventually encompassing the Office of SIGINT Operations, the National Photographic Interpretation Center, and the DO's Soviet/East European Division.

The system's key component was a collar that surrounded the lead-sheathed cable and extracted the selected signals for capture by high-volume recorders at the site. The recorded signals were retrieved and returned to Langley where sophisticated equipment deciphered and analyzed the electronic pulses.

With the key technology still in development, the buzz of covert activity surrounding the target manhole continued. Memos, cable traffic, thick sheaves of casing photos, satellite images, operational proposals, and surveillance reports bounced between Langley and Moscow with particularly sensitive documents hand-carried by couriers. Refining the operational details required hours of discussions in the conference rooms at Headquarters and among case officers in Moscow.[8]

Over the course of the operation CIA officers would make periodic clandestine entries into the manhole in preparation for the installation of the collar and recording device. Officers trained for these assignments in a full-sized replica of the underground chamber at the CIA's covert facility called "The Farm." Rigorous sessions familiarized the officers with the operation's physical demands, allowed them to practice the tasks to be performed in Moscow, and assessed their capability for carrying out the assignment.[9]

Ken Seacrest was one of the OTS techs selected to enter the underground chamber. He would conduct a data sampling survey of individual communications cables to determine the one of greatest value. This critical phase of the operation would require Ken to remain alone, inside the manhole for up to two heart-pounding hours.[10]

The timing for an entry into the manhole was planned when Ken's activity was least likely to be observed by a passerby on foot, in a passing car, or on a streetcar. Current schedules and routes of the Metro, trolley buses, and the *electrichka* were collected and studied along with the sometimes wildly inaccurate official and unofficial maps of the area. The mileage to the target area from the office and potential drop-off points were precisely measured, as were the exact distances and drive times during different times of the day.

To ensure he did not raise suspicion when approaching the manhole, Ken would disguise himself as an ordinary Muscovite. However, given the dated style of fashion available in Moscow, an American purchasing new local clothing would attract suspicion and shopping for used clothing would have been even more alerting to the watchers. So, Eastern European apparel of the right size, style, and matched to the season, was purchased at flea markets and thrift shops in Vienna, East Germany, and Warsaw. The clothing was carefully inspected, cataloged, and packaged at Langley before shipment to Moscow, where it was stored in a secure area to preclude potential KGB attempts to tag and track the items.[11]

Ken arrived in the USSR in the summer of 1979, after completing a six-month crash course in Russian.[12] Among his fellow trainees was the senior operations officer slated to become CIA chief of Moscow. As part of his cover, Ken took an immediate interest in cultural activities of Moscow, never missed an opportunity to volunteer to show a visitor around the city, and spent as much free time as possible outdoors with his family. He filled their schedule with visits to tourist sites, cross-country skiing, and hiking. In short, he acted nothing like the intelligence officers depicted in the movies. Never an international man of mystery surrounded with beautiful women nor the toast of Moscow's small American community, Ken did not own a James Bond tuxedo. His car of choice was not an exotic Aston Martin, but a sensible Volkswagen bus. He rarely used his newly acquired Russian language, since Americans who spoke Russian came under closer scrutiny. Ken appeared to be the dedicated family man with a moderate taste for culture and a love of the outdoors, who enjoyed playing broomball during winter's short daylight hours and throwing darts in the evening.

In reality, Ken's carefully selected cultural excursions and interest in the outdoors worked toward a single purpose—to build a persona that established a predictable pattern of activity that began on the day he arrived and would continue until he departed. As artfully engineered as a piece of spy gear, Ken's job, hobbies, and interests—all of those elements that make up daily life—were meticulously designed before he stepped off the plane at Moscow's Sheremetyevo Airport.

Like all other Americans in Moscow, Ken would be a potential target for surveillance and subject to assessment by the KGB. He would be closely monitored during his first weeks in Moscow for inconsistencies in the carefully crafted profile. It would not take much to catch KGB attention. One or two anomalies outside the normal pattern of activity would label Ken an intelligence officer, bringing continuous surveillance and compromising his operational value.

Within a few weeks Ken concluded that he fell into the KGB's middle tier of surveillance priorities—an American who would be accorded periodic surveillance at random intervals. Even this would most likely lessen over time as surveillance observed him maintaining a consistent profile and pattern of activity in Moscow.

One critical piece of Ken's activity was frequent family outings doubling as well-planned surveillance detection routes (SDRs).[13] Whenever possible, Ken, accompanied by his wife and two small children, visited Moscow's *lesoparki* or "forest parks" on day trips. Large expanses of green—sometimes spanning hundreds of acres—these parks were a welcoming world away from the gray, soot-covered streets of Moscow's center and a logical destination for a young American family that loved the outdoors.

Ken and his wife, Sharon, spent hours in the parks with their children, who were four and seven at the time, picnicking, throwing Frisbees, and hiking the long trails through the forests. The rucksack Ken carried became a familiar accessory to those who might have him under surveillance. The rucksack contained food, water, toys, and blankets for a day in the park. These excursions also provided a logical reason to visit all areas of Moscow, allowing Ken to familiarize himself with the roads, geography, and traffic patterns as well as identify choke points, escape routes, and observation stops for planning future SDRs.[14] On days when they happened to have surveillance, Ken, with Sharon's help, could practice identifying KGB vehicles, license numbers, surveillance tactics, and team members.

On a spring morning in 1981, five years of operational planning, targeting analysis, satellite imaging, signals intelligence, and technical development culminated in one of history's most elaborately planned and expensive family picnics. Leaving home in the family's green and tan VW bus, Ken began a twenty-mile trip that appeared identical to his past travel patterns. But in this instance, each road, each turn, and each stop was designed to detect and identify surveillance.

In their ears, Ken and his wife wore small radio receivers for the OTS monitors secured in harnesses near their armpits and tuned to the KGB's 103.25 MHz primary surveillance frequency. Ken also carried a second receiver, a six-channel scanner that searched for any "near field" transmissions from militiamen and the ubiquitous Seventh Directorate.[15]

The inductive antennas looped around their necks were encased in soft fabric, concealed under their clothing, and transmitted the signals from the receivers to the earpieces no larger than the head of a Q-Tip. The small earpieces were concealed by latex sculpted by OTS disguise specialists to match natural ear contours and colored to blend with Ken's and Sharon's individual skin tone.

Each listened for the distinctive Russian word *dvadtsat odim* (twenty-one), which translated from KGB surveillance code meant, "I have the target in sight." If neither spoken numbers nor a series of clicks[16] were heard, Ken could conclude with some confidence that surveillance was ignoring him that day.[17]

As the family drove toward the park, no words were exchanged about how Ken would spend the majority of the day or the possibility of surveillance. "We wouldn't say much, particularly with kids there," Ken explained later, "because at those ages you never know what they're going to repeat or to whom. So they had to be totally unwitting about was going on."

Ken initiated a series of maneuvers designed to detect surveillance. Unlike action adventure movies, where the hero eludes surveillance through a series of spectacular high-speed stunts, the reality of espionage for Ken was both more complex and prosaic. He did not zoom through intersections or make dangerous last-second turns to lose the followers. His driving had to fit the profile of a family man taking his kids in a van to a picnic. To any surveillance team that might have been watching, Ken took a more or less direct route to the park with all the speed his four-cylinder family van could muster.

Once he abruptly pulled off the road, grabbed a kid out of the backseat, and rushed into the treeline for an emergency potty break. He also missed a turn-off and circled back to get on the right road. There was also the need to study the map when momentarily lost. If a surveillance team had been following as he performed these common maneuvers, they could not have stopped without drawing Ken's or Sharon's attention.

The surveillance teams of the KGB's Seventh Directorate were known to drive distinctive cars, usually the larger *Volga,* which resembled a mid-sized Volvo or the more compact *Zhiguli,* the Soviet version of a Fiat 124. Because only the KGB had automatic car wash equipment, their cars tended to be clean compared to other vehicles on the Moscow streets. Further, they boasted the luxury of windshield wipers, a valued commodity among those Muscovites lucky enough to own a car and often stolen if the car was parked unattended.

The SDR continued for more than an hour with Ken using a variety of seemingly routine stops and turns to confirm any vehicular surveillance.

Finally confident they were not being followed, Ken turned into a parking lot at the perimeter of the large park and pulled into a spot selected far in advance for the day's picnic. Ken purposefully avoided the section of the park popular among Americans, where KGB teams might already be watching other targets. If he was not being watched, he wanted to avoid walking into active surveillance of another target. The family walked a couple hundred meters into the park before Sharon spread out their blankets on the edge of a grove of trees. Ken and Sharon continued looking and listening for surveillance as he began a series of apparently innocent explorations into the woods. Like the vehicular maneuvers, these were also intended to force any watchers to betray their presence.

After half an hour trying to enjoy the picnic lunch with his family and becoming reasonably certain surveillance was not nearby, Ken put on his well-used and stuffed rucksack, nodded to Sharon, and slipped into the woods. The wordless gesture told her that if he did not return by a certain time, she should load the children into the van and drive directly home. Using a prearranged signal, she would alert the chief that Ken had not returned as scheduled and was likely in some kind of trouble. The chief would immediately take action to minimize potential collateral damage to other operations and prepare the U.S. ambassador for the inevitable diplomatic blowback from the Soviet press announcing, "another American spy had been caught attempting to destroy the peaceful relations between America and the Soviet Union."

As soon as Ken left the family, he began an extended surveillance detection run. There was need to reassure himself that he was clean before changing to Russian clothing.[18] As he well knew, the Seventh Directorate did not always detain or arrest suspected intelligence officers during an operation, but rather continued to observe in the hopes of being led to an agent or dead drop site. Even then, the officer was often allowed to continue on his way as if nothing was amiss.[19]

That had happened to Penkovsky in late 1961. The Seventh Directorate watchers first spotted a man suspected of passing material to Janet Chisholm, wife of MI6 officer Roderick Chisholm, inside a doorway just off one of Moscow's busy shopping streets. Rather than make an arrest, they increased surveillance and eventually spotted the same man a few days later in a public park making another exchange. The unidentified subject was then placed under surveillance and later identified as GRU Colonel Oleg Penkovsky.

The spy, not the intelligence officer, was the ultimate target.[20] However, once the KGB had identified an agent, arresting the foreign intelligence officer during an operation offered additional value. When the KGB arrested Martha Peterson on the Krasnoluzhskiy Bridge in 1977, although *TRIGON* was probably already dead, her detention brought negative international publicity to the United States and disrupted CIA operations in the USSR.

As Ken walked the paths of the park, the decision of whether to go to the manhole site was complicated by the number of people around him. Soviet citizens frequently used Moscow parks and on this early spring day, with the snow off the ground and the weather warming, the park was particularly crowded. Ken knew crowds could be used to his advantage, allowing him to blend into packs of visitors, but they also served a similar function for KGB watchers.

There was also danger in looking too closely at every person that crossed his way. Training and experience had taught Moscow personnel that if they looked hard enough, anyone could seem out of place. Since surveillance teams attempted to imitate everyday life, ordinary citizens can easily be mistaken for more than they really were. The old man wearing a cloth cap and walking slowly with a cane, the young couple strolling hand in hand along a path, a mother with children seeking a few hours' reprieve from a small apartment, or the stout, middle-aged lady with the ubiquitous plastic "perhaps bag" swinging at her side—any one of them could be members of the Seventh Directorate.[21] Then again, they could be just what they appeared to be, Muscovites enjoying a spring day outdoors.

Ken was well aware of this particular operational paralysis. Intelligence officers called it "seeing ghosts." The psychology of the phenomenon was rooted in the inherent anxieties of clandestine activity coupled with the real possibility of encountering hostile security. What made it particularly vexing was the uncertainty of proving a negative. If one observed surveillance, or heard transmissions over the monitor, it became a certainty, but if one did not see or hear surveillance, several possibilities came into play. There could be no surveillance, the officer had not identified surveillance, or surveillance could know where he was headed and be waiting there.

To combat the uncertainties, Ken relied on confidence that previously effective SDR techniques would work again. In the end, it came down to trusting experience, training, and whatever technology was at hand. When well executed, the plans, maneuvers, and detailed routes that had been so carefully constructed, studied, and practiced should reveal surveillance, but ultimately Ken would act on his instincts. Even if his elaborate vehicular

Case officers frequently donned light disguises, such as that of a Russian worker, for meeting with agents in Moscow, circa 1982.

and foot SDRs did not reveal surveillance, Ken, like every other officer, had the option to abort the operation based on nothing more than "gut feeling."

"You absolutely trusted the process. But at the same time, you developed an intuition about what was there and what wasn't," Ken explained. "After a while you began to get a feel for these things. Something might just not feel right."[22]

So, despite all the planning and preparation that came before, the final decision of "go, no go" was Ken's alone. With the park goers bustling around him, he paused, took several deep breaths of the chilly spring air, and decided, "It's a go." Entering an isolated section of forest, he quickly changed from his American clothes into the local clothing stuffed in the backpack.

The Russian clothing offered a fair degree of camouflage by blending in among the other drably dressed Soviet citizenry. Soviet clothing, in 1981, had little of the style found in Western fashions. The outfit Ken now sported featured well-worn grays and browns, a brimmed fedora-like hat, inexpensive Russian shoes, a knee-length overcoat, and rough, ill-fitting trousers. He would not have looked out of place in a World War II film among the costumed extras who populated the set of a crowded train station.

Casually draped on one shoulder was the rucksack with its unusual "picnic" supplies. Weighing nearly thirty-five pounds, it contained, along with his original clothing, a twenty-million-dollar investment in advanced U.S. eavesdropping equipment designed to extract and record signals from lead-sheathed communications cables the Soviets considered tamperproof. If discovered by the KGB, not only would the mission aimed at critical intelligence abruptly end, a technical collection capability that could be applied to similar targets in other parts of the world would be exposed.

Walking through the groves of birch that bordered the park, Ken took a lengthy, circuitous route leading him out of the park and immersing him in Moscow's general population on the public transportation system. Changing buses several times, his face set impassively, Ken became indistinguishable from the other passengers. The trolleys and buses made their way through Moscow traffic, carrying him far from the manhole on Varshavskoye Shosse and then, eventually, returning him to within walking distance of the target.

At each change of bus and trolley, Ken tried to be the last to step off and observe if anyone else was rushing to get through the closing doors. Still wearing the receivers, he listened for surveillance transmissions corresponding to his actions. Like the vehicular SDR, some elements of Ken's foot SDR were designed to compel improvisation by his KGB watchers, forcing them into the open by making mistakes. Other elements of the routine allowed him to pick out patterns or repetitions in surrounding faces. Was the young couple he noticed earlier in the park now getting on a bus far from where he first spotted them? Was it the same girl, only now dressed in a different coat and hat? Was the license plate of the car that trailed a respectable distance behind the bus the same one he had seen a half hour earlier?

Ken followed the procedures, moving steadily toward the time when he would make the second crucial decision: entering the manhole. The right choice meant the beginning of an operation that pried open the door to a potential treasure of intelligence. The wrong decision would result in his arrest, identification as a CIA officer, expulsion from the USSR, and collapse of a multimillion-dollar operation that had the attention of the highest levels of the U.S. government.

Exiting the bus at the final stop of his journey, Ken hiked three kilometers to the site. Although he had never been this close to the target before, the maps, casing photos, and satellite images that he had studied for months gave the surroundings a sense of familiarity. The site was "as advertised"—technically promising and operationally vulnerable. Partially obscured by the narrow treeline, the new spring's foliage offered marginal cover. A curious look in Ken's direction at the wrong time from the driver of the tractor working the nearby field or the unexpected presence of a Soviet citizen who had wandered off the beaten track in search of wild mushrooms—any of these or a thousand other unforeseen pieces of ordinary bad luck could wipe out years of planning and research.

The leafy trees along the roadway shielded Ken as he slipped into a pair of chest waders and pulled a specially made tool from the backpack to open the manhole. The operation required simple, as well as sophisticated technology. Once in the manhole, the waders would provide warmth and dryness. But first, the manhole's heavy metal cover had to be removed quickly, a task for which OTS had designed a special implement. Fashioned from aluminum, the 12-inch curved tool was compact and lightweight, but strong enough to lock into a utility hole and pry the heavy iron slab from its position.

Moving to the manhole, Ken inserted the pry bar and lifted, sliding the cover off the hole just enough to lower himself into the dank semidarkness. "That was easy," he thought, and made a mental note to thank the designers of the pry bar when he was back at the OTS lab.

Descending the ladder, Ken stepped into cold water. Looking up through the partly opened manhole, he could see only a small slice of gray sky and a thick cluster of treetops. Measuring four feet by six feet, the chamber was eight feet from bottom to top. Exiting from a matrix of conduits on one side of the manhole and reentering an identical conduit matrix on the opposite side were the lead-shielded cables that carried the target communications along with the rubber-encased civilian telephone lines.

Thigh-deep in cold water, Ken began the technical part of his clandestine work. His objective was to obtain samples of communications from the streams of high-speed data traffic flowing through the dozens of wires contained in each of the lead-sheathed cables. Although the CIA was confident that Krasnaya Pakhra communications were carried in one or more of the cables, Ken needed to identify the specific cable of primary interest by sampling signals from each of the dozen cables.

The sampling and monitoring device carried in the shabby backpack was engineered to pinpoint that cable. Designed by a private CIA contractor, the collection equipment had been shrunk down into a 25-pound package that looked like a rectangular portable radio with an array of indicator lights mounted on the top and leads set into one side to accommodate several probes simultaneously. The extracted data was recorded on what appeared to be a standard cassette tape, though capable of capturing high-quality signals on multiple tracks.

Pulling the device and various components from his backpack, he meticulously assembled the equipment before applying a probe to the first cable. The only indication of technical success for Ken, standing in the bottom of the manhole, would be if the tape moved and the light came on. If an indicator light failed or the tape malfunctioned, he had no way to troubleshoot the system and would need to abort the operation.

Ken placed the formfitting two-piece sensor, one section at a time, on either side of a cable and strapped it together with lengths of white fabric measuring a foot long by an inch wide. He then tightened and buckled the strap to hold the sensor in place. Ken's twelve buckling straps were white by design. If he dropped one, white would be easier to spot in the poorly lit manhole. Twelve was also an easy number to remember when he inventoried the gear prior to leaving the hole. No evidence of an unauthorized entry could be left behind.

A grid of more than a dozen underground cables entered the chamber, although Ken's attention focused on the well-protected, lead-sheathed, and gas-filled cables whose very design betrayed their significance.[23] Since the data flow was often irregular, he needed to keep the monitors on each cable long enough to get an adequate sample.

Above, the noise of traffic whizzing by only a few yards away was a constant reminder of Ken's vulnerability. If an inquisitive Soviet citizen crossed the road to look down the partially uncovered manhole, he would have come face to face with a CIA officer standing in ground water amid a thick cluster of cables.

It was tedious work. The position of each conduit and cable needed to be identified in a logical way. Working from a checklist, Ken collected samples, made identifying notations and then moved on to the next grouping. A small "bookkeeping" mistake of misidentifying a single cable could invalidate the sampling. Working methodically, he was conscious that each additional minute in the hole increased the chances of discovery, while haste could lead to unacceptable mistakes.

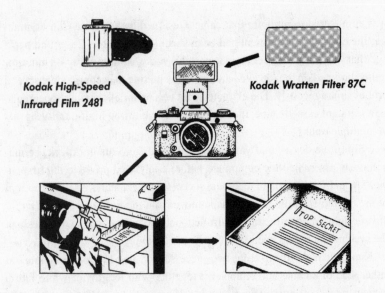

Kodak High-Speed Infrared Film 2481

Kodak Wratten Filter 87C

Infrared film combined with a Kodak Wratten 87C filter covering the flash allowed undetectable photography of documents or objects under low-light and dark conditions.

Ken had a second mission, as well. While waiting for sample collection, he "cased" the chamber, photographing the conduit and cable layout using a 35mm camera outfitted with an infrared flash unit.[24] This documentation would provide a current and complete record of the underground chamber, including details that could assist in any final design changes of the collar for permanent installation.

Finally satisfied that he had collected and recorded adequate samples of the data, Ken inventoried his tools, repacked the equipment, climbed up the ladder, and poked his head out. With no one in sight, he emerged from the manhole, quickly slid the heavy cover back into place, and retreated toward the narrow treeline where he removed the waders and stuffed them back into the rucksack.

Ken was some distance from his family but the preplanned route that returned him to the picnic area was more direct than the one that brought him to the target. After walking into the park with no evidence of surveillance, he found another secluded spot, changed out of the street disguise, and returned to his wife and children looking the same as when he left a little more than five hours before.

Sharon, although relieved to see Ken, understood the operation was not yet a success. If surveillance had paid attention to her and the kids and noticed Ken's absence, that would raise questions. American men did not normally leave their families alone in a park for hours on end.

Wasting no time, they gathered up the children along with the picnic gear and headed back to the VW bus. Neither Ken nor Sharon relaxed on the drive home with the invaluable tape and tapping equipment in their possession. They drove carefully and continued observing cars behind and in front of their van, since no American was immune from accidental or KGB-staged automobile mishaps that could result in questioning or search.

No celebration awaited Ken, although the chief anxiously anticipated a reason to return to his office that evening. While he had received no emergency signal from Sharon, he went to the office, opened the door, flipped on the light, and saw a sheet of paper torn from a government-issued notebook taped to the wall. Scrawled in pencil on the otherwise blank page was "#1." It was Ken's simple signal announcing the success of a critical phase in the CIA's most advanced technical operation in the USSR.

Based on the analysis of Ken's sampling, CIA identified the primary target cable and authorized installation of a permanent tap. For several years CKTAW successfully recorded communications between Krasnaya Pakhra and the Soviet Ministry of Defense. Then, in the spring of 1985, something went wrong. An officer sent to recover the tapes from the recording device aborted his mission when remote interrogation of the unit returned a "tamper indicated" signal.[25] After a few weeks, a second trip to the site did recover the tapes, but the system had ceased functioning and the entire operation was shut down.[26]

What had gone wrong? For years, the virtually trouble-free American penetration of top secret Soviet data streams had gone undetected before suddenly coming to a halt. CIA counterintelligence officers speculated the reason behind CKTAW's loss was more sinister than just simple "bad luck."[27] Had the KGB discovered the tap because of an error with a case officer's tradecraft during infrequent trips to recover tapes? Could the KGB have a spy who was privy to one of the CIA's most tightly held operations? As with the roll-up of an agent, the failure of the device left a host of questions, though for CIA counterintelligence the answers were not long in coming.

On August 1, 1985, Commander Vitaly Sergeyevich Yurchenko,[28] deputy chief of the First Department of the KGB's First Chief Directorate (foreign intelligence), called the U.S. Embassy in Rome and offered to defect.[29]

Later that afternoon Yurchenko provided the chief, Alan D. Wolfe, with information about two penetrations of U.S. intelligence, one at the National Security Agency and a second, code-named *ROBERT*, within the CIA itself.[30]

Yurchenko claimed not to know *ROBERT*'s true name, but offered two important clues: *ROBERT* had sold classified information to the KGB in Vienna in the fall of 1984 and trained for Moscow operations, but was taken off the assignment just prior to departure.[31] Two days after receiving Yurchenko's information, the CIA's Office of Security informed the FBI that *ROBERT* was almost certainly the former CIA officer Edward Lee Howard.[32]

Howard, who joined the CIA in 1981,[33] had been selected in 1982 to become a case officer in Moscow. Since he had always been under cover as a CIA employee and had not been posted abroad, he was considered clean and less likely be identified as an American intelligence officer.[34] In preparation for the Moscow assignment, Howard and his wife, Mary, received six weeks of intensive training on the clandestine tradecraft necessary for operating against the KGB in their own backyard.[35] Training in detecting and evading surveillance was particularly rigorous with field exercises conducted against FBI surveillance teams.[36] Then, during a polygraph examination before his scheduled departure for Moscow, Howard admitted to drug and alcohol abuse, petty theft, and cheating during training.[37] The CIA fired Howard in early May of 1983.[38]

Bitter and angry at the Agency, Howard moved to Santa Fe, New Mexico, where he took a job as an economic analyst with the state legislature. But his troubles with alcohol and debts continued.[39] A series of bizarre telephone calls Howard made, including one to the U.S. Embassy in Moscow, were so troubling that on September 24, 1984, a former supervisor and a psychologist were dispatched to interview him at his home.[40] Howard revealed the unsettling news that in October of 1983 he loitered outside the Soviet Consulate in Washington, D.C., while considering whether to volunteer to the Soviets.[41]

Following receipt on August 3, 1985, of the CIA information identifying Howard as *ROBERT*, the FBI began surveilling Howard, but by the end of the month, Howard, sensitized by his training, detected the surveillance.[42] In early September the FBI began monitoring his telephone conversations.[43] Howard was under twenty-four-hour physical and technical surveillance.[44]

A little more than two weeks later, on September 19, the FBI confronted

Howard with information that he had been identified as a Soviet agent.[45] To protect the real source, Yurchenko, the FBI attributed the information to KGB defector Oleg Gordievsky.[46] Howard did not confess, but refused a polygraph test.[47] Interviewed again the next day, he stated his intent to engage a criminal attorney over the weekend and agreed to another interview on Monday, a meeting he never intended to attend.

During his CIA countersurveillance training, Howard had learned to use the JIB and how to roll out of a moving car while slowly rounding a corner.[48] Lacking an OTS-designed JIB, he created a field-expedient version with a toilet plunger, coat hanger, and Calvin Klein field jacket taped to the top. The dummy's head was a Styrofoam wig block and commercially available Jerome Alexander wig issued during his disguise training at "The Farm."[49]

Howard and his wife departed for dinner at a local Santa Fe restaurant on Saturday evening. While driving home they made a slow turn off Garcia Street onto Camino Corrales and Howard jumped from the car into the bushes.[50] Mary propped up the makeshift JIB in the passenger's seat, buckled it in place, and minutes later the FBI surveillance team clearly saw two "people" in Howard's 1979 Oldsmobile as it entered their garage.[51]

After spending the night at an airport hotel, the next morning Howard was on the first available flight from Albuquerque to Tucson, where he continued his secret journey to Moscow.[52] The FBI did not discover his escape until some twenty-five hours after he jumped from the car.[53]

Howard's defection had a catastrophic impact on Moscow operations. His devastating betrayal exposed to the KGB collection systems, tradecraft techniques, covert equipment, and agent-handling methods inside the Soviet Union and specifically CKTAW.[54] Howard had been one of the handful of officers to participate in the CKTAW operation and, in fact, during the polygraph examination that led to his dismissal, admitted cheating during an exercise in the mock-up of the manhole by replacing the weights in his backpack with cardboard to make it easier to get into the small opening.[55]

Almost a year later, on August 7, 1986, TASS announced that Howard had been granted political asylum in the USSR.[56] Reportedly, Howard continued drinking heavily and died at age fifty, in July 2002. Russian news reported the cause of death as a broken neck from a fall. Neither the Russian nor American intelligence services mourned his death.[57]

The significance of the American spy technology the KGB recovered from the manhole and the nearby cache was not lost on Soviet leadership. In a 1990 article that described and decried U.S. technical espionage against the Soviet Union, Lieutenant General Nikolai Brusnitsyn, deputy of the State Technical Commission of the USSR, complained that collection systems like CKTAW were interfering with arms control and reduction efforts between the two superpowers.[58] Brusnitsyn concluded by admitting to Soviet "anxiety over the ever-growing capabilities and scope of intelligence gathering and spying technology."[59]

Almost ten years later, in 1999, veteran KGB counterintelligence officer Rem Krassilnikov, author of *The Phantoms of Tchaikovsky Street,* provided a Soviet version of the CKTAW operation, which the KBG had code-named BILLIARD BALL.[60] Krassilnikov described an "inductance data sensor" in the manhole as being connected to a metal box containing a tape recorder, control system to turn the recorder on and off during conversations, transceiver, and an internal power supply capable of operating the device for four to six months.[61] According to Krassilnikov's account, the box was buried a half meter deep and located not far from the manhole cover. In this KGB version, the box was painted bright red with the Cyrillic inscription DANGER! HIGH VOLTAGE, protected by rodent repellent, and connected to a nearby ultra-shortwave antenna.

Krassilnikov, citing the KGB's technical analysis, reported the transceiver could be remotely interrogated up to 2.5 kilometers away. It responded with a coded signal indicating whether the unit had been disturbed, needed to have its tape changed, or required a replacement power supply.[62] The analysis estimated that it was necessary for the CIA to service the unit only every four to six months.

CKTAW's compromise did not diminish its achievement. For the CIA, the operation represented another remarkable fusion of technical reach and operational tradecraft, revealing vulnerabilities in the Soviet security apparatus. The imagination that conceived the operation, the engineering talent that built the system, and the operational execution represented a new kind of American technical collection capability. In the coming decades, the technologies developed during the CKTAW operation would be repeatedly applied to other equally critical targets.

LET THE WALLS HAVE EARS

Crest of OTS audio operations officers, 1970s.

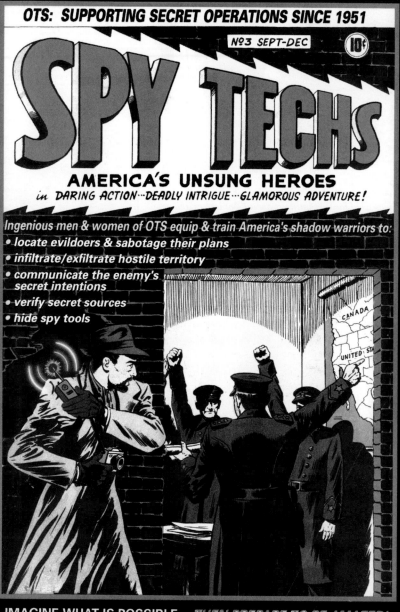

OTS fiftieth-anniversary poster, 2001.

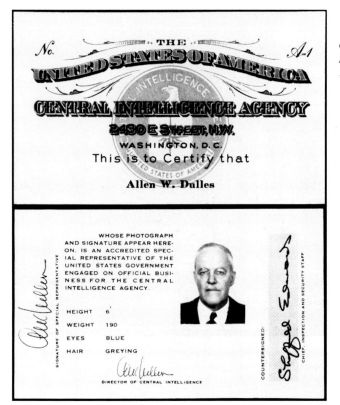

CIA credentials of DCI Allen Dulles, 1950s.

KGB photo of Oleg Penkovsky demonstrating use of Minox Model III camera during his 1963 trial.

KGB surveillance photo of CIA officer Richard Jacob preparing to clear the final Penkovsky dead drop in 1962.

KGB surveillance photograph, taken from across the Moscow River, showing Oleg Penkovsky inside his apartment copying CIA OWVL transmissions, 1962.

An OTS-designed rollover camera was an effort to enhance an agent's ability to copy documents secretly by moving the device across the paper. Normally three passes were required to capture a standard-size page of text. After techs had developed the film, the three-part image was reassembled in the darkroom and printed, circa 1968.

Seiko wristwatch concealment for T-100 camera shown in three positions: (a) Watch closed and functioning as a normal timepiece, (b) face plate rotated with lens open for operating the camera, (c) face plate removed showing camera assembly and film cassette, circa 1975.

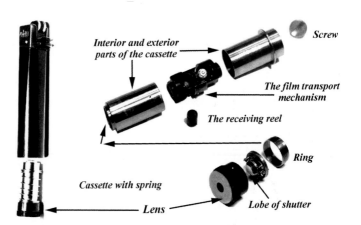

Components of the OTS T-100 camera as concealed in a Cricket cigarette lighter, 1970s.

Commercially available Minox III subminiature cameras were issued to CIA agents during the early part of the Cold War and were widely used by the KGB, HVA, MI6, and other intelligence services.

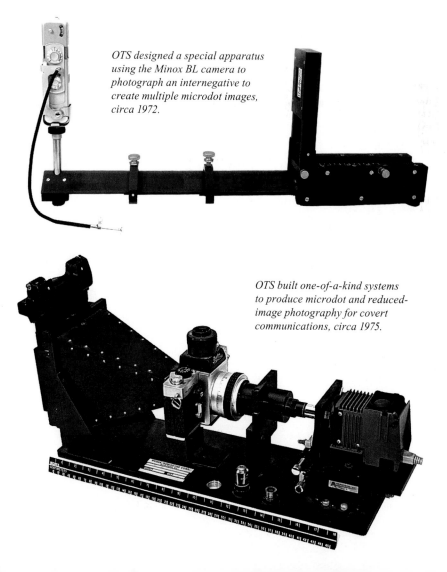

OTS designed a special apparatus using the Minox BL camera to photograph an internegative to create multiple microdot images, circa 1972.

OTS built one-of-a-kind systems to produce microdot and reduced-image photography for covert communications, circa 1975.

The tiny "bullet lens" reader was only slightly larger than a grain of rice but provided sufficient magnification for reading a microdot. Aspirin-size "scorch" tablets were dissolved to produce an invisible ink for secret writing.

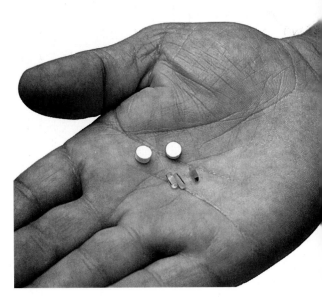

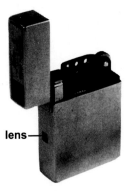

lens—

The Echo 8 cigarette-lighter camera was the smallest camera commercially produced during the Cold War. It was highly effective for surveillance photography in both business and social settings, but was not designed to photograph documents.

Microdot

CIA Mk. IV
Microdot Camera

"114" Microdot
Reader

From the left: Model "114" microdot reader; unbleached microdot (actual size less than one millimeter); CIA Mark IV microdot camera, 1965.

CHAPTER 12

Cold Beer, Cheap Hotels,
and a Voltmeter

If it ain't audio, it ain't shit. —Unofficial motto of audio techs

They were called audio techs and for more than two decades—from the 1960s until the 1980s—they were at the top of an unofficial OTS caste system. Aggressively mounting hundreds of audio surveillance operations around the world, these few hundred officers worked on the frontline of Cold War espionage. Their missions took them from European capitals to the most remote destinations on the globe, responding to stations that saw an opportunity to "collect audio."

Although CIA audio operations inside the borders of the primary Cold War "hard targets" (the USSR, China, Cuba, North Korea, and North Vietnam) were severely limited by internal security services, conditions in Europe and Third World countries were far different. In these countries, clandestine audio operations could be launched against the residential quarters, official buildings, and embassy compounds of the hard-target countries. This was particularly true in the Third World where security was often less stringent than in major European cities.

When operations levied a "requirement," it was expected a tech would catch the next plane for Paris, Rome, West Berlin, or a Third World country in turmoil whose new name had yet to be printed on maps or the new leader's visage on currency. The techs worked under aliases as businessmen, military personnel, or adventure-seeking vacationers, becoming whatever allowed them to blend in with other travelers and provided the most effective cover.

A tech might get off a plane in a South Asian country, spend a week planting a bug, then fly home for the weekend, change identities, and find himself in Africa by Monday evening.

Their exploits and sophisticated gadgetry rivaled the popular fictions of globetrotting intelligence officers. OTS audio techs slipped into basements and sewers to tap phone and communication lines, moved silently over rooftops, walked narrow beams through the crawl spaces of attics, and scaled the walls of fashionable villas in the dead of night. They were sent into diplomatic compounds, business offices, residences, hotel rooms, limousines, airplanes, and boats to plant audio devices. Accurate and sometimes embellished accounts of their operational adventures established what would become known as the tech culture, eventually defining, for better or for worse, all OTS techs regardless of their clandestine discipline. Among the most enduring of these professional and personal attributes was the techs' seemingly unwavering fondness for cheap hotels and cold beer.[1]

There were practical reasons for what could be perceived as "less than sophisticated" tastes among these world travelers who commonly logged more than 100,000 air miles annually and were away from home 150 nights a year. Although government employees on official travel were not permitted to accumulate frequent flier miles, they were allowed to keep any funds left over from the daily government food and lodging allowance.[2] The resourceful techs found they could supplement their modest salaries by staying at inexpensive hotels—a type of lodging that became known throughout the Agency as "tech hotels."

More important, the techs' choices when it came to accommodations and beverage made sound operational sense.[3] Whereas case officers traveling abroad needed to match their selection of accommodations with their cover position and operational role, techs usually wanted to get in and out of town while keeping the lowest possible profile. As a rule, they maintained erratic schedules, working when the target was accessible, regardless of time, weather, or holidays. If that meant access to a consulate or embassy could be obtained only after it closed for business, then the tech's workday would likely begin at nightfall and end before dawn. Whatever bargain lodgings may have lacked in comfort or basic amenities, such as heat, hot water, or a private bathroom, they made up for by way of discretion.

Luxury or business traveler hotels tended to boast large, attentive staffs that would notice an American businessman leaving late in the evening and returning in the early hours of the morning with traces of plaster dust on his shoes or paint flecks in his hair. On the other hand, tech hotel staff conveniently overlooked the comings and goings of their guests. If an American departed in the early evening and returned just before dawn with his hair matted with sweat and in need of a bath, it would not be the first time such a thing happened.

One technical operations officer realized he had ventured too far down the hostelry pecking order when the desk clerk asked incredulously, "You mean you want the room for the entire night?" Another tech found his own limits of frugality tested when he and his partner checked into a hotel offering a double bed for $3.50 a night. "You know, if we shared a room, we could cut the cost in half," the aggressively thrifty tech suggested.

As for their reputed penchant for beer—well, beer just tasted good and made other deficiencies seem less important. It was also safe. In countries where drinking the local water was a health risk, beer served double duty as thirst-quenching beverage and "water" for brushing teeth. During one operation that confined six techs in a target facility for several days, the team member in charge of providing food provisioned generous quantities of beer, canned tuna—and nothing else. In the sweltering heat of the empty warehouse, the odor produced by a sustained diet of tuna and beer led to a unanimous vote to prohibit that particular tech from outfitting the team with provisions again. Ever.

Despite living and working in these austere conditions, techs brought "technical magic" to operations. Their sophisticated gear tapped into secrets hidden behind the high walls and closed doors of secure installations. Their equipment combined with expert installation captured the private conversations of people to whom the Agency had no other access. The moods, personal doubts, and aspirations of potential recruitment targets held on reels of audiotape offered case officers a unique peek into their professional and personal world. Recordings of official discussions of policy and diplomatic strategies provided CIA analysts and policy makers with well-sourced, hard strategic intelligence as well as tactical counterintelligence that supported law enforcement.

None of this was lost on the techs themselves—neither the excitement of the operations nor the value of the intelligence take. "If it ain't audio, it ain't shit," became their unofficial motto and was proudly displayed on the desk of the audio chief during the 1970s. They were the "top guns" of the OTS and none of them doubted it for a second. However, the colorful tech culture, the technology, and the intelligence that flowed from it did not materialize overnight. The advanced technology, like the tech culture, evolved over two decades. The devices as well as the expertise to make them work were built from the ground up.

The Technical Services Staff was not a year old in 1952 when the CIA received information about an alarming audio discovery in Moscow. During an electronic sweep, the countermeasures team discovered a device secreted in the wooden replica of the Great Seal of the United States that hung behind Ambassador George Kennan's desk in his residence at Spaso House, barely a mile from the Kremlin, at No. 10 Spasopeskovskaya Square.[4] The seal had been hanging there seven years, after a group of Soviet Young Pioneers presented it as a token of friendship on July 4, 1945, to then U.S. Ambassador, W. Averell Harriman. The gift, presented by smiling children in neatly pressed uniforms, concealed a listening device that would baffle and frustrate the Agency for years. "The Englishmen will die of envy," Valentin Berezhkov, Stalin's personal translator, whispered to Ambassador Harriman during presentation.[5]

That the Soviets bugged the U.S. Ambassador's office should have come as no surprise to anyone. When American diplomats first arrived in Moscow in 1934, they discovered that listening devices inside their offices and residences would be the way of life within the Soviet Union.[6] U.S. Ambassador Llewellyn Thompson, posted to Moscow between 1957 and 1962, and again between 1966 and 1969, would walk in Red Square for private conversations.[7] As a young diplomat in the 1930s, Kennan personally played amateur spy hunter in Spaso House by hiding in the billiard room overnight—with a partner in the attic—in an unsophisticated attempt to catch the Soviet techs planting devices.[8]

Further evidence of the Soviet's use of eavesdropping was demonstrated by the NKVD's (forerunner to the KGB) aggressive operations during World War II. In the late fall of 1941, as German forces approached Moscow, the Soviet government ordered foreign diplomats out of the city to the relative safety of Kuybyshev (Samara). Then, with foreign embassies standing empty or hosting only skeleton staffs, the NKVD hardwired virtually every Western embassy in the city with embedded microphones. After the German army's advance halted just twenty miles outside Moscow in early 1942, the diplomats were allowed to return to the Soviet capital and their bugged embassies.[9]

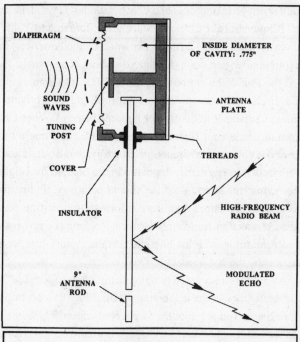

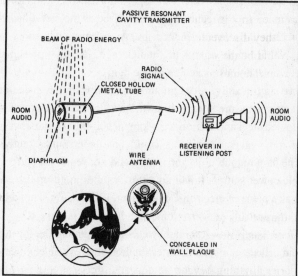

The bugged Great Seal of the United States that hung over the U.S. Ambassador's desk in Moscow contained a passive cavity resonator and went undetected for seven years, from 1945 to 1952. The CIA had no comparable audio technology at the time.

The NKVD bugged the rooms and conversations of President Roosevelt and British Prime Minister Churchill during the Tehran and Yalta summit meetings.[10] In Tehran, in 1943, Stalin maneuvered the American President into staying in the Russian compound amid rumors of a German assassination plot. At the 1945 Yalta conference, the NKVD techs employed an early version of a directional microphone to capture Roosevelt's private outdoor conversations at a distance of fifty to one hundred meters.[11] When Churchill's daughter, Sarah, made an off-hand comment that lemons might be just the thing with caviar (in another version of the story Sarah suggested lemons with tea), a lemon tree reportedly appeared in the orchard overnight.

After receiving the audio take from these operations, Stalin pushed his intelligence service for more and more information, insisting they report even the tone of voice in transcripts. After devouring the raw intelligence delivered each morning with his breakfast, Stalin would then act taciturn, even bored, during the actual meetings.[12]

However, what was discovered hanging behind the Ambassador's desk in 1952 was revolutionary in the technology of listening devices. Implanted in the middle of the carved wood of the Great Seal, cleverly hidden behind an air passage formed by the American eagle's nostril, was a device that was alarming as much for the technology it employed as the fact it had been active for more than half a decade. Indeed, four American ambassadors—Averell Harriman, Walter Smith, Alan Kirk, and George Kennan—presumably had their secret conversations picked up by the bug.

Differing significantly in design and function from any piece of covert listening equipment previously known, the device was constructed of precision-tooled steel and comprised a long pencil-thin antenna with a short cylindrical top. Agency engineers could not understand exactly how it worked. The stand-alone unit, apparently, did not require a battery or any other visible power source. It had no wires or tubes, nothing that identified the device as a piece of electronic equipment. If the oddly shaped length of metal was transmitting conversations, then how was it doing it?

"*The Thing*," as it was soon dubbed, bounced among the Agency's lab, the FBI, and private contractors for evaluation and reverse engineering. No one could offer anything beyond an educated guess as to how *The Thing* worked, and somewhere in its travels from lab to lab it was damaged from either improper handling or shipping.[13]

The Thing was eventually sent to Peter Wright, the principal scientist for MI5, the British intelligence service responsible for counterintelligence operations. Wright worked for more than two months to solve the mystery before eventually coaxing it into operation. He dubbed it a "passive cavity resonator."[14] *The Thing*, as Wright discovered, worked by reflecting radio waves, then picking those echoes up with a radio receiver.

To operate the device, the NKVD aimed a continuous 800 MHz radio signal at the seal from a listening post in the building across from Spaso House.[15] *The Thing*'s thin diaphragm at the top, which Wright had repaired, vibrated with the sound of a voice. Those vibrations created modulations in the reflected radio signal that bounced back to the listening post.[16] *The Thing* did not require internal power in the same way a mirror does not require power to reflect light. The radio transmitter and receiver, code named LOSS (or REINDEER by the Russian techs), were a marvel of signal processing, considering the technology available at the time.

According to Wright's own account, once he understood the principle and made the device work, he took another eighteen months to create a similar system for British intelligence. Called *Satyr*, his device featured aerials—transmitter and receiver—disguised as two proper British umbrellas.[17] *Satyr* proved to be a great success and Wright called it "black magic."[18] Then, as he observed, "the Americans promptly ordered twelve sets and rather cheekily copied the drawings and made twenty more."[19] The American version of the device, according to Wright, was called *Easy Chair* (also *Mark$_2$* and *Mark$_3$*).[20]

The concept behind *The Thing* represented a remarkable leap forward in eavesdropping technology. Viewed as a wake-up call at CIA for audio operations, it was the clandestine equivalent to Sputnik. The Agency had to play technological catch-up in RF audio as well as in space.

Ironically, although they did not know it at the time, many of the scientists who had worked and failed to decipher *The Thing*'s mystery would probably have recognized the name of its creator, Lev Sergeyevich Theremin.[21] A musical prodigy and physicist, Theremin was born in czarist Russia in the late 1800s, excelled at some of Russia's finest universities, and his genius was recognized by the Soviet regime. He lived in New York City during the 1920s and 1930s where he demonstrated the electronic instrument that bore his name. The theremin was played without touching it and enjoyed a brief life as a serious instrument. In the 1930s it captured the imagination of the avant-garde of the jazz age, though eventually it was relegated to creating the spooky sounds heard in horror and science-fiction films.

Those who crowded ballrooms and concert halls to hear Theremin's thoroughly modern instrument, or caught glimpses of him amid Manhattan's high society, never suspected that the socially sophisticated "Russian genius" was also covertly working for Soviet intelligence. Just before the Soviet Union's entry into World War II, Stalin summoned Theremin back to Moscow where he was summarily sent to a Siberian prison camp. Later he was transferred into one of the KGB's secret laboratories—a *sharashka*—to work on creating spy gear. After developing *The Thing,* Theremin was awarded one of the Soviet Union's highest honors, the Stalin Prize First Class, along with the equivalent of $20,000, a fortune in Soviet society, but remained imprisoned.[22]

At its formation in September of 1951, the CIA Technical Services Staff employed fewer than fifty people. Its scientists and engineers were housed in a makeshift lab at Indian Head, Maryland, thirty miles south of Washington, D.C. The facility, a one-time gun test facility for the Navy, sat on a large peninsula that jutted out into the Potomac.

"Somehow the Agency begged, borrowed or stole some space down there, but conditions were bad. We had to do our own janitorial service. We complained, but we did it," remembered Kurt Beck, an engineer who worked in that first lab. "When we started, there were a dozen people overall and maybe six to eight people working in the audio program."

The Thing arrived at the TSS lab in 1952, but the audio engineers had no point of reference against which to assess the device or understand how it worked. Data on covert listening and transmitting devices, comparable to research on military weapons systems or private sector consumer products, simply did not exist. Cataloging data on technical collection devices was a new concept for American intelligence and the Soviet passive resonator technology exceeded anything TSS could even attempt.

However, like all "finds" (discoveries of clandestine devices), *The Thing* revealed certain key capabilities of its host service.[23] The first was the cleverness of design. *The Thing* demonstrated that Soviet intelligence possessed a technical expertise not matched by the CIA. Second, the construction suggested that the Soviet engineering approach relied on creating one-off custom-made pieces specifically produced for each operation. The device was not mass-produced or made in a serial production of fifty or a hundred identical units with interchangeable components. To the TSS engineers, *The Thing* looked to be expertly crafted and built entirely by hand.

The Thing underscored deficiencies in U.S. technical capabilities compared to the Soviets as well as highlighting some Soviet weak points, such as a lack of technical innovations spurred by a consumer-based economy. Without private industries to offset some of the costs of building laboratories or production facilities, the Soviets' intelligence service had to underwrite all the costs for developing spy gear. This contrasted sharply to the OSS tradition of recruiting private industry to support intelligence. There could also be little doubt that *The Thing* was the product of Russia's long involvement with espionage dating back to the time of the czars while America's espionage capability, particularly in the technological sense, was still in its infancy.

The engineers at Indian Head recognized the huge gap as well as the larger, more basic and immediate question facing them: Precisely, how do you go about building a better bug?

The state of available technology in the 1950s limited the CIA's audio operations to little more than running a microphone wire or a telephone line to a set of headphones. Even among the war surplus equipment there was little of real value. The OSS technical staff had concentrated its efforts on creating special weapons and explosive devices for equipping partisan forces and operatives behind enemy lines. Few of the wartime agency's resources were devoted to technical aids for agent operations and even less to developing eavesdropping equipment.[24] The U.S. Army Signal Corps had produced a bugging kit but, at more than a decade old, it was sorely outdated.

While some equipment such as microphones and amplifiers was available on the commercial market, these devices were bulky and generally not suitable for covert duty. As a result, the Agency's supply of devices was limited to obsolete military surplus and whatever Kurt and his colleagues could scrounge, assemble, or modify from the telephone company or recording industry.

There was also little experience among the TSS staff. Most of the engineers, like Kurt, were just out of college, and in their mid-twenties or a little older if they had served in the war. The older TSS engineers' wartime experience tended toward equipment like radar, sonar, and large shipboard systems rather than clandestine audio surveillance. It would take the young engineers time, and trial and error to learn the new business of covert audio operations. And the world was on the verge of a technological revolution that put them in the right place at precisely the right time.

In 1947, scientists at Bell Labs demonstrated the germanium transistor and patented it the next year. By 1952, transistors were being used in specialty devices, such as hearing aids and military systems. Two years later, Texas Instruments, along with a small Indianapolis company, I.D.E.A, introduced the first transistorized "pocket radio." Called the Regency TR-1, it sold for $49.95 and appeared in stores just in time for Christmas.[25] The age of the transistor had arrived.

Through these electronic miracles, everyday consumer electronics could now be made smaller, less power hungry, and more reliable. For clandestine operations, the transistor's impact would prove equally profound. The same transistor—"solid state"—technology that allowed teenagers to listen to Elvis at the beach would provide the foundation for new capabilities in audio surveillance operations. Yet the transistor solved only half the problem for the CIA. TSS still lacked the necessary in-house laboratory facilities and engineering expertise to conduct basic development work, and the funding to pay for it. In response, TSS revived the OSS model of forming classified partnerships with industry and academia to develop technology, then contracting with companies to design and produce the systems.

In the booming postwar economy, competition was keen and companies were adding R&D engineers as fast as they could in an effort to feed industrial and consumer markets hungry for innovation. The decision to use classified contracts with private industry seemed a practical way to get the maximum "bang for their buck," or in this case, the smallest, most efficient audio devices for the dollar. TSS eagerly adapted new technologies from commercial research as well as spin-offs from the larger government programs to meet its operational requirements.

Because of the premium placed on reliability for clandestine devices, TSS bet that conventional and proven equipment, like transistors, microphones, and recorders, which worked well in the consumer and commercial markets, could be repackaged, reduced in size, and adapted to covert use. The same assembly processes used for making consumer products could also be redirected toward the production of spy gear.[26]

Companies that lent staff to the OSS during World War II were now asked by TSS for access to their proprietary research and their best engineering minds. At times it was a tough sell, just as it had been for Stanley Lovell. TSS was not as lucrative as the "big ticket" items like satellites, radar systems, missiles, submarines, and airplanes. The needs of TSS were relatively small—decimal dust—compared to billion-dollar military and satellite procurement programs. The CIA did not need production runs of 10,000 or 50,000 units. Fifty or a hundred small, reliable devices were enough and the companies could not publicly acknowledge or promote what had been produced. For large companies, in particular, these contracts and their limited production runs were marginally profitable in the short term and offered little potential commercial payoff over the long term.

Budgetary troubles severely limited TSS's audio program in the early years. One tech remembers the entire budget for worldwide audio equipment in 1956 was under $200,000, enough to buy a few tape recorders, microphones, and some other commercial equipment, but not nearly enough to mount either an effective engineering effort or aggressive operations. Many of the new prototype devices developed by private contractors and research projects never left the labs because TSS simply did not have the funds to order even a small production run.

Then in 1957, a CIA Inspector General (IG) report addressed the issue of whether audio surveillance would be a core CIA mission and a primary future method of collecting intelligence. The report urged the DDP and TSS to make audio operations a "top priority" and forever changed TSS's financial picture.[27]

A post–World War II engineer working in the audio equipment branch at the time, Tom Grant recalled that soon after the IG's report was issued, he received an unannounced visit from a senior CIA official. Asked about the state of audio surveillance equipment, Grant replied that both he and the research staff were frustrated and then outlined the problems in detail. For instance, when lab equipment broke down, Grant explained, he could not order a new unit because the limited funds available were needed for gear to support ongoing operations.

The visitor listened for some time, and then abruptly concluded the conversation by saying, "I'm willing to authorize ten times your current budget." Grant immediately renamed his guest "Mister Moneybags from Headquarters."

The additional funding, for a new program code named EARWORT, arrived in increments as TSS spent it.[28] To the techs, it seemed as if the financial floodgates had opened. Grant's job turned from scrounging the cheapest equipment possible to talking with development engineers about new mechanical and electronic devices to aid clandestine operations and ordering the most promising. Not every item was successful when put to operational testing, but at long last TSS was seriously in the audio surveillance game.

With new equipment coming online, the need for trained engineers to deploy and service the systems around the world was evident. Prior to 1958, TSS had only a handful of technical operations officers capable of planting bugs and setting up listening posts where the audio take was initially processed.[29] Tech equipment, cameras, microphones, and recorders were typically issued to case officers in the same way as concealments and one-time pads. TSS received an "equipment order," and then delivered the devices to the case officer along with some basic instruction.

However, the new, more complex—and delicate—devices brought with them the need for an increased expertise. For example, the first over-the-air audio transmitters incorporated glass vacuum tubes in their amplifiers, requiring special care in handling. Something as simple as a bumpy ride in the trunk of a car could play havoc with tubes mounted on a galvanized chassis amid the internal bird's nest of wiring. If a case officer opened a crate containing a large reel-to-reel tape recorder and found it did not work "out of the box," the entire system was shipped back to headquarters or a tech dispatched to troubleshoot the problem.

No body of knowledge, experience, manuals or testing protocols, either in-house or with the contractors, existed for those early systems. TSS had no method of documenting or benchmarking electronics destined for clandestine operations. The development and deployment processes were ones of trial and error—with a host both of trials and unforeseen errors. The most frustrating challenges were with technical devices that worked perfectly on the test bench, but then failed when installed for an operation.

In one early mishap, modified reel-to-reel tape recorders intended for telephone taps were designed to accommodate the waiting period between calls by turning the machine off when the line was not in use. In concept, this innovation saved recording tape, limited the need for constant monitoring at a listening post, and improved the efficiency of transcription. The recorders would start when the call began and switch off when the call ended. In testing, the system met the engineering requirements. It was only after deployment was a serious problem recognized. Neither the manufacturer nor the Agency anticipated the long wait times between calls. When a call ended, the recorders turned off the motors that advanced the tape, but did not disengage the magnetic heads. The recording heads, it was discovered, heated up after long periods in the pause mode and eventually melted the motionless recording tape pressed against them.

For an engineer, this was a small problem easily solved with a screwdriver and a quick change of relays to retract the head from the tape when not recording. However, for a nontechnical case officer in South America, the only solution was to call for a traveling tech to change the relays of each recorder at each station.

The installation of concealed listening devices sometimes unintentionally pitted the case officers against the techs. No technical skills were required for a case officer to pass a one-time pad and shortwave radio along with instructions for its use to an agent. However, the most basic mic and wire operation, which generally included concealing a microphone, tapping a power line, and running the wire back to the listening post, required a level of technical expertise beyond the typical case officer.

While the case officers designed the nontechnical elements of every operation, their limited understanding of the technical requirements for a successful installation did not pass without notice. "What are the two biggest lies techs tell?" went the question, followed by the punch line, "The time of my commute to work and the amount of time we tell the case officer the installation will take."

Most problems boiled down to understanding the capabilities and limits of the technology. Nontechnical case officers, seeing new technologies emerging in the 1950s and 1960s, were prone to either underestimating capabilities or overestimating benefit. Impractical operations proposed by case officers soon became part of the unofficial lore of techs who diplomatically described the most fanciful of these operations as "high-risk options." One courageous case officer suggested a bugging operation that would have had him lowered by helicopter onto the balcony of the target's apartment in a major European city. Although dramatic—even movie worthy—it failed the basic audio operational standard that surreptitious entry be clandestine. A helicopter hovering in central Madrid with a case officer dangling from a rope ladder would likely be noticed. Another proposal involved a tunneling operation in which the managing case officer suggested that the tons of dirt and debris generated could be flushed down toilets and carried away by the city's sewer system.

The growing awareness that technical devices offered enormous potential for intelligence collection demanded a new type of case officer–technical officer interface. If case officers were not qualified for the technical phase of most operations, engineers in the lab were equally untrained and unsuited in the basic tradecraft required for field operations. Laboratory engineers had neither the training nor experience to plan surveillance detection routes or select usable dead drop sites.

As more audio operations were conducted in the 1960s, case officers and audio techs found themselves dependent on each other's skills. Audio techs planting bugs were often in apartments without the knowledge of the resident or on the property of another government. They usually worked under time constraints and the stress of possible discovery. Mistakes and miscalculations were costly because second opportunities for an entry were rare and no traces of their work could be left behind.

Installations of audio devices also frequently called for technical improvisation when the operation was not accurately described in the planning phase or if equipment simply failed.[30] Troubleshooting a short-circuit in the bug with a voltmeter at 0200 hours in the basement of a foreign consulate, then cobbling together a fix from a limited selection of spare parts required a special personality. What the Agency needed was a hybrid of the technically oriented, creative, street smart, and pragmatic individual who thrived on adventure.

TSS cast its net wide, seeking out likely recruits among telephone companies, military bases, technical schools, and commercial radio stations along with the emerging television industry. They placed ads in popular technical and science publications without identifying the prospective employer.[31] While these first tech recruits may not have known how to design a new audio device from scratch, they knew how to read a schematic and their way around a soldering iron. They were natural tinkerers who grew up taking radios and cars apart, and then putting them back together. As kids, they played with erector sets and read magazines like *Popular Electronics*. Several were amateur radio operators who understood the basics of transmitting, antennas, and frequencies.

Grant's entrance into spying was typical of that first class. In 1952, he graduated from an electronics school in Kansas City and found a good job as a video engineer at the New Orleans television station WDSU. Then, a few days after receiving his draft notice, whether coincidentally or with other knowledge, a CIA recruiter called with an offer. Although he had turned down the recruiter once before, Grant now responded enthusiastically, took the job, and went on to serve in TSS, TSD, and OTS for more than thirty years.

The intelligence contribtution of techs like Grant, who thrived on fixing and experimenting with electronics, was imperfectly understood by the Agency's senior management. Dr. Herbert Scoville, the Deputy Director for Research in 1962 and 63, reportedly referred to TSD as "those tinkerers" at a staff meeting. The patronizing label received wide circulation within TSD and served to reinforce the techs' identification with the Directorate of Plans rather than with the new Directorate of Research.[32]

There was virtually no legacy of experience and TSS engineers and technicians received little or no formal training in audio operations. Whatever clandestine expertise they acquired occurred on the job. Many of these techs were technically intuitive and given to modifying consumer electronics for their clandestine work.

On an assignment overseas in the early 1960s, at a time when public address systems were often built into classrooms, offices, and hotels, a newly deployed tech realized the job of bugging every room in a large hotel was beyond the scope of a single person. However, within hours he had jerry-rigged a small circuit that turned the rooms' wall-mounted speakers into passable microphones that picked up conversations at the flip of a switch.[33] The innovation had security advantages beyond efficiency since the room speakers (usually used for announcements and fire alarms) were part of the woodwork and ignored by patrons.

Wall-mounted speakers were not the only type of in-place speakers whose wiring could be reversed. Television sets and table radios could be wired so that their speakers, when not in use, were turned into microphones. The techs could use televisions and radios as concealments for small microphones hidden behind the cloth or grill in front of the speaker. With the transmitter's circuitry masked by existing wiring, the audio device drew power whenever the set was plugged into an electrical outlet.

When a tech was directed to bug an apartment but could not enter it nor drill through the wall or ceiling to create an air passage, the solution came in the form of a contact microphone. Working on the principle that all hard surfaces in a room vibrate in the presence of voices and noise, placing a sensitive contact microphone on the reverse side of a wall or floor would pick up these vibrations and feed them to an amplifier and tape recorder. The design of the mic was similar to a traditional phonograph cartridge with the needle removed. The tech could hollow out a small furrow in the floor of the apartment above the target, place the microphone in the groove, and cover it with putty to eliminate ambient vibrations.

"Audio techs had to have a little larceny in their hearts to master skills

MAKING THE KEY IMPRESSION

What you need:

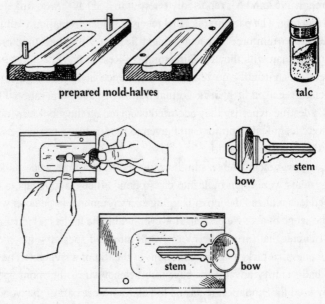

prepared mold-halves

talc

bow   stem

stem   bow

Use of a Key Impressioning Kit was often the first step in copying original or master keys for surreptitious entry into buildings and rooms.

needed in the field, including picking locks, duplicating keys, tapping phones, and surreptitious entry, to gain access to the target sites. Did we do anything illegal?" Grant recalled. "I guess, most of it."

Arriving in South America in 1958, an audio tech prepared for an operation to bug the trade office of an Eastern Bloc country. These government-controlled trade offices, functioning primarily outside the diplomatic community, posed as Western-style commercial companies and provided convenient cover for the intelligence officers.

The Soviets used both Amtorg, a trading organization, and Aeroflot, their national airline, as covers for espionage. From Amtorg offices, the Soviets conducted industrial as well as political espionage. Throughout the Cold War, the Soviets sought America's advanced technology by targeting individuals in private companies. In one year alone, the Soviets' New York residency supplied more than 180,000 pages of classified and unclassified documents along with samples of state-of-the-art technology.

It was a smart way to operate in the West since a business cover of "furthering peaceful trade and commerce" offered access to the latest manufacturing equipment and technical expertise, proximity to scientists and engineers who might be recruited, and a platform from which to recruit in nearby countries. In developing countries where business and politics frequently intertwined these front companies proved the ideal conduit through which Western economic and scientific secrets could flow back to Moscow.[34]

Not surprisingly, these commercial offices were priority targets for Agency audio operations. A microphone installation, taps on phone lines, or a telex intercept could yield high-value intelligence about local operations, sources, and recruited agents. Making commercial offices even more attractive was the fact that they were generally less well protected against intrusion than diplomatic facilities. To maintain the façade of a legitimate business, the trading companies leased office space in commercial buildings with security that usually conformed to local business practices. In some instances, the CIA officers would lease space directly above, below, or to the side of these Soviet offices, giving techs a common wall for audio ops.

In the South American country, the target trade office was located in a six-story commercial building in the heart of a major city. Erected before World War II on a main thoroughfare, the building was otherwise unremarkable. Arrangements were made by the local station to rent a small office in the same building to serve as a front.[35] The tech, posing as an employee of the front company, began his technical survey of the building's structure, architecture, layout, and entrances in preparation for an installation.

The tech found the stairway door leading to the roof and attic when, unexpectedly, the building's porter appeared. "In my best broken Spanish, with a big smile and plenty of hand waving, I said, 'Hey, how can I get up to the attic—there are many windows up there that overlook the city,'" the tech recalled. "'I've noticed how many beautiful churches you have and this would be a wonderful place to take pictures. I want to photograph as many of your churches as I can.'"

The porter, growing misty-eyed, gestured for the tech to follow him. With his new escort and communicating as best he could on their trek to the roof, the tech encouraged the porter to show off every possible aspect of the building, stairways, entrances, offices, and crawl spaces. When they finally reached the top floor, the tech found the attic appeared permanently unfinished, consisting largely of unused space with the partially completed floor offering clear access to the ceilings of the target suite. In short, the vacant attic was the ideal point to plant bugs exactly above the target and an easy way to run concealed wires back down to the listening post in the Agency's rented office several floors below. The tech took a few pictures from the attic window, thanked the porter (again in his best broken Spanish), and left.

Returning the next day with a high quality 8 × 10-inch photograph of the church across the street, the tech presented the image to the porter along with a small gratuity. The porter was so pleased he invited the tech back to take more pictures. With the porter an unwitting source for authorized access, the tech now had plausible reason to be in the unoccupied spaces of the building for extended periods. On the third day, the tech returned with his bag of tools, microphones, and wire, along with a more sophisticated photo system that included a large Speed Graphic camera and tripod.

Positioning himself in the attic directly above the ceiling of the target office, the tech set up his tripod, mounted the camera, and pointed it out the window in the general direction of a church. On the floor was one of the microphones the tech camouflaged to look like a foot switch with a cord running to the camera. As the tech was about to begin drilling through the floor for the microphone implant, he heard footsteps on the stairs. Quickly repacking his audio equipment, he found himself face to face with the porter, accompanied by his entire family, who came in to watch the American take pictures of the city's churches.

Clicking the camera, the tech explained that today he would be doing "time exposure" photography that required several hours to complete. After politely watching the camera and the tech for a while, the family became bored and left. With the porter and his family gone, the tech positioned the mic above the ceiling of the target's office and ran the wire from the microphone through the building's ductwork to the listening post where recorders would be installed. "It worked like gangbusters and, for all I know, it might still be there," the tech recalled proudly decades later.

Before 1960, bugging operations were primarily "mic and wire jobs" with techs running wires from the implanted microphone to the nearby listening post. In cases where wire could not be run out of a target facility, a concealed recorder was installed and tapes were changed regularly by a support asset, such as a recruited secretary or custodian with access to the target. What the Agency needed was a reliable radio signal transmitter from the target to a listening post.

There were advantages to each system. The clandestine transmitter eliminated the need to run wires, but usually required regular battery replacement, and the radio signal was vulnerable to sweep teams. The mic and wire jobs were difficult to detect since they did not emit a radio signal, but the wires took longer to install, and were susceptible to discovery during a physical inspection of the premises.

In 1963, a case officer arranged for the techs to have access to a Soviet Bloc residence while the diplomat was out of town. The operational plan was to embed mics in the concrete wall of the house, but when the techs began drilling they discovered their hardware store drills were ill suited for the task. They made too much noise for middle of the night work. The techs determined that liberal lubrication at the drilling point would dampen some of the sound and the next day a support officer was sent to the U.S. commissary to buy several gallons of Crisco cooking oil.

After dark, the techs returned to the job, and repeatedly dipped the drills into the Crisco to keep the hole lubricated. They were rewarded with much quieter drills, but as the drill bits heated in the deepening hole, so did the oil that coated them. Soon the poorly ventilated house was filled with the odor of fried food, posing a serious operational compromise should the residents unexpectedly return home.

While the drill team worked on the wall, two other techs were running the microphone wires across the diplomat's immaculately tended garden and manicured lawn to the nearby listening post. Working only with the available light provided by the stars and moon, the techs dug a shallow trench by hand using bayonets. It was tedious work. Not only did they have to stay on a course plotted across the wide expanse of lawn and flowerbeds in the dark, they also needed to carefully pick up any loose dirt on the grass and replace each piece of sod or flower as they moved along. Progress moved at a snail's pace as the techs spent the last hour of every night erasing any sign of their presence on the lawn.

"There was a very conscientious gardener who took care of the landscaping," one of the techs recalled. "We installed an observation post where we could watch the house during the day in case the diplomat returned early or visitors showed up. We began noticing that every morning, when the gardener came to work he'd walk over to the flowerbed where we'd been working, look down, shake his head, and go on about his business."

Panic began to spread among the techs and case officer. Had the gardener noticed the trench and was biding his time until the diplomat returned to tip him off? The restoration of the lawn looked flawless to the techs, but perhaps the professional gardener noted a disturbance or saw traces of the narrow channel they dug for the wires. Each day the gardener continued to inspect the flowerbeds with a worried look, but refrained from poking around in the grass. Finally, the case officer decided the only choice was to attempt to recruit the gardener since he clearly had noticed something amiss.

The plan worked. After the gardener agreed to a covert relationship with the CIA, he heaved a great sigh of relief. "Every morning for the last week," he said, "I come to work and the red flower would be where the yellow one used to be, the blue one was over here." Because of darkness, the techs did not always recognize the color of the flowers they replanted after digging, and inadvertently rearranged some of the plants. Reassured that he was not losing his mind, the gardener kept the secret and the installation operation came to a successful end.

CHAPTER 13

Progress in a New Era

A device called a transistor, which has several applications in radio
where a vacuum tube ordinarily is employed, was demonstrated
for the first time yesterday at Bell Telephone Laboratories at
463 West Street, where it was invented . . .

—*The New York Times*, July 1, 1948, last item featured in "The News of Radio"

The first new audio transmitter developed in partnership between TSS
and a private contractor arrived in the late 1950s. Designed specifically
for clandestine audio operations, the device was dubbed the SRT for Surveillance Radio Transmitter.[1] Composed of a hybrid mix of microtubes,
sometimes called "peanut tubes," that dated to World War II, and recently
introduced transistors, the SRT-1 was far from ideal, but a significant leap
forward in technology.

What operations needed was a concealable, reliable transmitter for audio
installations that eliminated the wiring of mic and wire jobs. While the
SRT-1 was functional and no longer required wires to connect the microphones to a listening post recorder, it was also the size of a shoebox and required so much power that batteries were impractical in most situations. To
make a battery-powered version, the techs tried to modify some of the circuitry, and built in a power converter to turn direct current from batteries
into alternating current. However, the converters proved to be even more
inefficient in terms of power consumption. The result was, in the words of
one tech, "a power hog to begin with combined with an inefficient interface

in the middle with a foot locker–sized thing on the back end of it. No wonder the operations guys didn't want to use it." With battery power impractical, techs needed to wire the device directly into the target's power lines to operate for any extended length of time.

The greatest value of the SRT-1 was that it benchmarked many characteristics not desirable in a piece of clandestine equipment. Its dimensions were too large for easy concealment and the "in the clear" signal allowed anyone tuning across the same frequency to pick up the transmission. It offered no "remote on/off" and broadcast continuously, draining power and making the signal relatively easy to find by hostile technical surveillance countermeasures (TSCM) sweep teams.[2]

However, despite the SRT-1's shortcomings, the engineers had created a new capability for remote collection against targets some distance away from the listening post. When the first SRTs were replaced with all-transistor units in 1960, covert audio operations multiplied around the world along with the expertise of the techs.

Regardless of whether the audio operation was "hardwired" or used a radio transmitter, each installation required a carefully planned entry into the target and the right tools to the job. Secure, temporary control of the target site was also required to provide time for installation, including running and hiding wires, constructing antennas, testing the system, and restoring any damage done to the surroundings. These exacting tasks could not be rushed. Frequently the work was done at night and in semidarkness. Excessive noise by the techs or the equipment could draw attention and lead to compromise. No debris or tools could be left at the site for later discovery and wires run through public areas were in danger of discovery. The job was akin to wiring a house for cable television without the occupant or his neighbors ever noticing.

To minimize installation time, TSD developed a portable installation device with a razor's edge that allowed the tech to slice open the wall, "bury" two wires only slightly larger than a human hair, and then close the opening. The compact "Fine-Wire Kit" could be operated with one hand. In addition to efficiency, the device provided a means to place wires in sections of open walls where no better concealment options were available.[3]

Given a choice, the techs preferred hiding wires behind wooden baseboards or chair-rail molding where they were less likely to be discovered and fewer restoration problems arose. For that purpose techs were issued a lightweight, aluminum "baseboard puller." L-shaped like a small pry-bar,

Potential locations for concealing an eavesdropping device exist in every office or conference room. 1. Picture frame. 2. Behind speaker grille of TV. 3. Table lamp. 4. Brace below table leg. 5. Cigarette lighter. 6. Ashtray. 7. Telephone. 8. Overhead light.

the baseboard puller came in two sizes, the smaller version less than a foot long. It could provide sufficient leverage to create a gap between the wooden molding on a wall large enough to slip the thin wires behind without damaging or leaving marks on either the wall or baseboard.

For audio techs, operations in the early 1960s became more numerous, complex, and increasingly audacious. Unstable Third World governments, especially in Africa where colonial governments transferred power to local authorities, added personal risk to clandestine activity. The techs, like other American visitors, were frequently viewed suspiciously and considered to be in league with the "colonialists."

Audio operations came to the personal attention of President Eisenhower following the downing of CIA pilot Francis Gary Powers's U-2 aircraft over Svedlovsk in the Soviet Union on May 1, 1960. Before the incident, the techs planned audio operations against the Soviet officials who would be accompanying Premier Khrushchev to the scheduled May 16 European summit meeting with President Eisenhower. The techs bugged several hotel rooms assigned to the Soviet attendees with mic and wire devices. When the summit collapsed after Khrushchev's public denunciation of U.S. spying, the techs received word that the President himself had a "collection requirement."

Eisenhower, the techs were told, wanted information about what the Soviet reporters from the TASS news agency knew about the cancellation and when they knew it. The President wanted specific information when he met the following morning at breakfast with his security advisors.

In fact, among the bugged targets was the room of the chief TASS correspondent. The "audio take" from conversations in the room revealed that the correspondent had telephoned Moscow after the cancellation to refile the story about the summit that he had written before departing Moscow. The audio operation left no doubt in the minds of the techs that the TASS official did not know in advance that Khrushchev would call off the summit.

Only infrequently did techs receive feedback on the value or use of the "take." Strict standards of "need-to-know" and compartmentation were accepted as part of the profession by both engineers and tech ops officers.

About the time of the scuttled May 1960 summit, the chief of TSD stopped by the bench of one of the lab engineers. "What are you working on?" the chief asked.

"A new concealment device," came the reply.

"What's it for?" the chief continued.

"I'm afraid I can't answer that," said the tech, "I don't know. I have a requirement from operations and I'm just making what they want."

Another tech who served during the same time agreed. "We had an operational culture that emphasized the need not to know about what our equipment might be used for and what results were obtained, and that's the way we did our jobs."

Failure rates in the field of early audio installations were unacceptably high, sometimes reaching 50 percent. Two primary reasons became apparent. First, on the production end, there were no protocols for testing and certifying performance of components or the integrated systems in place. Manufacturer certification of component performance was accepted as "final" and considered sufficient. Second, differing field conditions, particularly the temperature and humidity extremes of desert, subarctic, and tropical regions played havoc with electronic components.

Testing was an ad hoc affair in TSS's early years. Informal and unofficial systems developed. Engineers in the lab or the contracting company performed what they believed were good tests and then unofficially shipped a new device to a tech in the field. "They'd say, 'Don't tell anybody, but try this out,'" recalled one engineer. "'If it works we'll tell everybody. If it doesn't work, just tell me.'"

Typically, field techs received individual components from headquarters such as batteries, transmitters, microphones, and recorders, then covertly assembled them into complete systems in locations that could range from a government storeroom or hotel room to the office above a target's conference room. In many instances, the assembly at the target site was the first time all of the components of a device operated together. Too often, the techs discovered the system didn't work.

Changes were implemented within TSD to ensure that the design and packaging of clandestine audio equipment would operate under widely varying conditions. However, field conditions were neither stable nor consistent. Engineers in the lab needed to imagine a hot, cold, rainy, dry, humid, dusty, pristine, muddy place in which their device would be plastered, glued, screwed, or bolted into position after being dropped, kicked, crushed, and adjusted by a hammer.

Kurt, by then a lead engineer, recalled the early equipment problems. "Equipment failure in the field usually pointed back to design or testing failure in the lab. We had to learn how to test our stuff. There were no manuals that said, 'Follow these test procedures for your new, improved bug.' Nobody else in government was building these bugs. We had to think it out for ourselves. And it took us a few years to set up test and evaluation procedures; then we could give equipment a stamp of approval. It is one thing for an engineer in the Washington lab to say 'it works.' But that wasn't enough. How do we know it is going to work in Ouagadougou?"

The test protocols themselves were not without problems. On an African operation, a tech needed batteries for recorders in the listening post monitoring a bugged embassy. The tech requested and received six batteries from headquarters, each about the size of a car battery and weighing forty pounds apiece. They promised enough power to run the post for years. The tech hooked up the first battery but nothing happened. Nothing happened with the second, third, fourth, or fifth. And the sixth, too, was apparently dead. The operation had to be put on hold.

Still angry when he returned to Headquarters a week later, the tech made a beeline for the TSD warehouse determined to get to the cause of an embarrassing, time-wasting incident. Why would six batteries all fail at once? What were the odds? When confronted, the warehouse clerk was just as baffled by the batteries' failure. The clerk described how he tested each battery and recorded that indeed they reached peak and sustained power through the stated cycle time. With their performance "certified," the batteries were shipped to the field, the clerk unaware that he had drained all the power.

Given the uncertainty of equipment performance, it became common practice for techs to rebuild or retrofit devices in the field. Techs who knew something about electronics could look at circuit boards and see where, with just a little redesign, they could be made smaller, more reliable. Once a tech in Mexico City, discovering a circuit layout in an electronics hobbyist magazine, reconstructed a newly arrived audio transmitter in his home shop. "It was spy gear," he boasted, "untouched by Headquarters' hands."

Normally reliable commercial equipment, such as the microphones used for mic and wire operations, sometimes presented unanticipated problems. Top-of-the-line carbon microphones used by the stars of the American recording industry were so sensitive that, when first installed, they captured voice audio anywhere in a room. However, when the mic remained stationery for an extended time, the carbon granules settled and compacted, like cereal settling to the bottom of the box, dramatically reducing its sensitivity. In a recording studio or concert hall, this was not a problem because the mic moved with the artist, but hidden in a wall undisturbed, the mic's performance deteriorated over months or years.

To remedy these types of problems, TSD set up an equipment-testing division in 1964 to perform independent quality assurance evaluations of spy gear. The unit conducted independent tests and certified all of TSD's equipment whether it was created in Agency labs, built by outside contractors, or acquired commercially. Anything a tech eventually used in the field was tested for performance in heat, cold, wet, and dry conditions along with an array of punishing tests where the device was bent, dropped, abused, and vibrated. The techs welcomed the tests, jokingly saying they needed all the equipment to be "case officer proof and agent proof." Rough field treatment could be expected. DCI Richard Helms himself observed that case officers had to learn "not to fling these [devices] into the back seat of an automobile, but to treat them with the delicate hand they deserved."[4]

By the late 1960s the techs saw audio equipment reliability jump to better than 95 percent. Standardized testing eliminated most of problems before the equipment reached the field. Nevertheless, operational realities could still trump well-designed and thoroughly tested equipment. After one tech successfully installed a wireless bug in a European city, reception from the transmitter was punctuated by intermittent static at the listening post. The breaks in transmission appeared random. The techs were stumped when even at 0200 hours, with the audio working perfectly, static would suddenly obliterate the voices, then, a few moments later, die down and the clear audio resume. Troubleshooting of the equipment did nothing to eliminate the interference. Then, looking out the window during one of the static-filled transmissions, the tech noticed motorcycles on the street. The next time static occurred, he could see another motorcycle and realized motorcycles passing between the transmitter in the target building and the receiver created static. The bug was picking up ignition interference from the motorcycles on the street.

Another lesson learned was adapting technology to covert operating requirements. Normally, trade-offs between technology and environment are relatively easy. In the 1960s, if a new stereophonic hi-fi system did not fit on the bookcase, purchasing a larger bookcase with wider shelves solved the problem. In 2008, rearranging a roomful of furniture to accommodate installation of a large plasma screen TV solves a similar problem. However, such straightforward trade-offs are rarely an option for a clandestine operation in which technology has to be covertly introduced into a setting, and then operate at peak efficiency without visible changes.

The physical environment of a target represented a constant in the operational equation. When any physical features of a facility had to be altered, it was done temporarily and only if it could be meticulously reconstructed. Scratches, dents, chips, holes, odors, debris, sawdust, mismatched paint, wet varnish, rearranged furniture, cabinet doors ajar, foot tracks on the carpet, or tools left behind—any of these things could compromise the operation.

OTS techs carried paint-matching kits and used odorless, quick-drying paints and varnishes so that neither the look nor smell of a wall restoration would leave clues behind. The number of items in the tech's installation kit was memorized and each item counted before leaving a job. During the work tools were laid out on a cloth or rubber pad to avoid the possibility of leaving telltale oil or stains on floors and carpets and to keep all gear in a single location should an emergency bug-out (rapid departure from the site) be required.

Africa often offered the opportunity for creative solutions to problems encountered in the field. During one bugging job in a West African capital, the installation required surreptitious entry at night into the vacant building and extensive drilling for the mics and wires. The case officer and the techs faced the practical problem that sounds of drilling in an empty building at night would likely be considered unusual by neighbors. A means of masking the drill noise for an hour or more was required.

"How about bullfrogs?" the chief suggested.

The techs were confused. What possibly could bullfrogs have to do with the operation?

"I think that'll work," the chief continued. "The bullfrogs around here make an awful racket at night. We'll have our staff collect several sacks of bullfrogs and when you guys go in to do the job, we'll release them around the building. Their croaking will drown out your drills."

A dramatic technical breakthrough for audio problems appeared in 1960 with a new generation of radio transmitters. The SRT-3 addressed almost every operational deficiency of the SRT-1 and its rarely deployed cousin, the SRT-2.[5] For a clandestine transmitter, the SRT-1 had been a bulky, unstable, and power-hungry affair that comprised a hybrid stew of transistors and peanut vacuum tubes.[6]

Now came the model 3. About the size of a pack of cigarettes, with an all-transistor design, it transmitted on a hard-to-detect frequency above that of television station transmissions and was powerful enough to reach several hundred meters to an unobstructed line-of-sight listening post.

On a visit to Headquarters in 1961, a field tech noticed a little black box sitting on the desk of one of the audio program officers. Curious, he asked one of the women working nearby, "What's that?" With surprise, she replied with her own question, "You're putting in audio and don't know about this?" The tech admitted he did not and spent the next hour learning about TSD's new SRT-3, the first all-transistor transmitter, battery powered with a five-milliwatt output to the antenna. For the tech, it was love at first sight.

The system was not perfect, the SRT-3 had limitations, such as the amount of power consumed, the size of the battery pack for extended operation, and the fact that its unmasked signal once activated could not be remotely switched off. It transmitted a clear continuous signal until the battery died. However, the overall impact of SRT-3's reliability and performance revolutionized the CIA's audio surveillance program. For techs making audio installations, the SRT-3 was a thing of technological beauty and an operational joy. Housed in a plain, black metallic case, there were screws on the top to access the circuitry by sliding the top or bottom off and inputs for the mic, battery, and antenna.

Because the SRT-3's small size, battery power, and wireless transmission, targets of opportunity—or more accurately, opportunities to plant bugs—multiplied. The techs were delighted with the SRT-3's acceptance by the DDP stations. Never before had the CIA been able to field a battery-operated listening device small enough to be covertly planted in almost any wall, ceiling, or door and at the same time obtain reasonable performance for an extended operation.

Like drivers who must test the limits of a new car, the techs took the SRT-3 where audio devices had never been before. TSD's new device worked best when concealed within a hollow wall or wooden floor. Implanted in the floorboards or behind the wood of a wall with a mic pressed against a nearly impossible to detect pinhole-sized air passage that opened into the target room, the SRT-3 could provide high-quality audio for the life of the battery.

Since the first SRT-3s were not hermetically sealed and were susceptible to humidity, temperature, and other environmental hazards, makeshift fixes, such as wrapping the device in plastic or duct tape, were employed with varying degrees of success. It was all too common for both the techs and case officers to become frustrated when, after a well executed entry and installation, signals could not be sustained. The only means of troubleshooting the system was to make another covert entry, extract the mic and transmitter, replace the device, and send the damaged unit to the lab for evaluation. Time after time, the problem was found to be moisture caused by dampness inside walls. "Climate controlled" buildings were rarely found in the locations TSD worked. Without an adequate field solution, the problem persisted until engineers developed hermetic sealing techniques for components of the systems at the factory.

In Asia, techs received a sobering lesson in chemistry and civil engineering. The operation was to bug the new embassy building of an Eastern

Bloc country as it was being built. When reviewing the proposal, Seymour Russell, the TSD chief, expressed his "gut feeling" the operation was not going to be very successful and probably not worth doing. His senior technical ops advisor argued that TSD field techs as well as the DDP officers on the ground thought the target was both worthwhile and vulnerable to an audio attack. Russell allowed his operational inclination to outweigh his doubts and approved.

From beginning to end, it was the perfect operation. The case officer spent months recruiting construction workers who embedded dozens of audio devices into the wet cement at key positions throughout the building. Pinhole openings provided sound channels to the mics. The bugs were tested and planted without security leaks. As the embassy construction neared completion, the time came to switch on the audio. Nothing was ever heard.

New to the game of installing electronics in wet concrete, the techs had not considered that cement dries differently than clay or mud. The moisture did not evaporate from concrete. In fact, when water is added, concrete undergoes complex molecular changes called hydration, a process that produces an exothermic reaction essential in the hardening process. In other words, drying concrete gets hot. In fact, it gets *very* hot. Within an ordinary sidewalk a few inches thick, temperatures can reach over 100 degrees Fahrenheit during the hydration process. In a wall a foot or more thick, the heat becomes even more intense. Unknowingly, the techs had planted the devices in an oven.

"It's unbelievable how hot that gets—the trunk of an automobile is nothing compared to what happened inside concrete," said the tech who made the installation. "Our devices at the time couldn't withstand the heat." It was another step in the evolution of TSD's spy gear. Future packaging of audio devices would give rigorous attention to the physical environment of the operation, such as heat and humidity, comparable to the attention paid to transmitter performance.

"When the battery dies, the operation dies," became a mantra of OTS. As transistors delivered improved performance, batteries technology lagged behind, becoming the weak link in audio operations. "My wife often said I mumble in my sleep, but that I never said anything clearly," remembered a senior TSD manager from the early years. "Except one night, apparently, I sat up and shouted, 'Those fucking batteries!'"

The lack of small, long-life batteries constrained full operational use of the transmitters. After all, why launch an operation to bug a room if the bug transmitted for only a few days? In most cases, replacing the batteries was either impossible or added considerable risk of compromising the operation.

"Electronics and solid-state hardware rapidly moved ahead of battery chemistry," explained an OTS chemist. "Before transistors the majority of the power going into a piece of electrical equipment was to keep the filaments hot in the tubes. That's where power consumption was, burning watts in there. When the first transistor radios arrived, even the crummy batteries available gave acceptable life for the consumers."

But technology acceptable to consumers was not necessarily acceptable for clandestine operations. Standard U.S. consumer batteries were too large, provided inconsistent performance, and offered life cycles either too short or too unpredictable. For an extended-life (several months or longer) audio operation, dozens of batteries were sometimes required—each one adding weight and volume to a concealment or installation.

Commercial battery options in the 1960s did not resemble those nearly half a century later. A consumer going into the local drugstore could select from only a limited number of battery types. There was the D-cell to power flashlights and the rectangular 9-volt for the transistor radio. There were also large cylindrical dry cells sold in hardware stores for special uses, such as powering camp lights.

Compared to transistors and integrated circuits, batteries possessed little sex appeal as a technology. Research in private sector companies producing batteries was directed toward lowering manufacturing costs as opposed to developing and improving the science of the battery itself. Published manufacturer ratings that estimated power output of individual batteries were often imprecise or significantly flawed. Battery makers invested little effort studying or improving long-term performance of their products because few—if any—customers cared whether a cheap battery lasted one month or six weeks. The typical consumer did not require significant improvements in performance or reduction in size. With their low prices, consumer batteries were disposable and replaceable.

The techs found ways to work with the commercially available batteries, despite their limitations. They assessed the amount of space available to conceal the bug and put the device in with whatever number of batteries could fit inside. Wiring the batteries in "parallel," in contrast to "serial" wiring, did not alter the net voltage powering the device, but extended its operational life significantly.

"With commercial batteries we never knew for sure what their lifetime would be," explained Kurt. "I would make my best guess and tell the case officer it would run this many hours and that's it. Based on that, he would make the decision to do an installation or not. Sometimes, when we were lucky, the batteries ran 15 percent more than my estimate. But sometimes they ran 15 percent less and then there was a problem. If the audio was valuable and more was needed, the ops guys were faced with a decision of whether it was worth the risk to send us in again to replace the batteries. We had to learn to manage the expectations of the case officers. That became tricky when we weren't sure how a piece of equipment would perform."

This situation led a small cadre of TSD battery scientists to focus on mercury cell technology as offering the greatest potential for long power-life in a small package. In the mid-1960s, TSD established an extensive battery test program that produced more and better data on mercury cell performance than anywhere else in government or industry. These testing results led TSD to focus attention on a cell named the RM-1, made by the P.R. Mallory Company, and to create a specialized power sources unit for evaluating commercial batteries and developing smaller, longer-life cells for clandestine applications.[7]

Mallory first built its reputation during World War II with a mercury cell developed by the company's cofounder, Samuel Ruben. Not only did the mercury cell pack more energy capacity into the volume of the battery case than other chemistries used during the war, it also operated well even when deployed in tropical temperatures. After the war, the mercury battery design fell into obscurity. It was still manufactured, but just barely, and only as a highly specialized item for a limited number of industrial devices. "The RM-1 was in the marketplace for use in medical applications and as a reference cell for testing other equipment because of its consistent, precise voltage," explained Tom Linn, who headed the Agency's battery program, "but they were not widely used at all."

Still, it was considered the most viable cell available and TSD "forced" its use for extended applications. Although available in the right voltage and size for TSD's needs, the battery was not designed for use extending over months or years. It exhibited a tendency to fail prematurely, and like virtually all commercial batteries at the time, the RM-1 hated audio operations. The constant slow and steady current drain required by the SRT-3 fostered internal crystalline growth that could eventually short out the cell.

TSD studied the failure mechanisms of the RM-1 cell and through a process of identifying failure modes, correcting each one, retesting, and further correction, the RM-1 evolved into a deployable component. Eventually a series called the "Certified Line" was produced. "That was our certification," noted Linn. "It was certified for sure, certified for CIA's clandestine audio operations."

Later, when the heart pacemaker industry emerged during the 1970s, the manufacturer applied the knowledge learned in building the TSD battery. "I think it's fair to say the first pacemaker battery—a mercury battery—was a TSD special design cell," said Linn.

The requirements for power cells used in pacemakers and audio bugs were remarkably similar. Power must be sustained, reliable, and produced at a predictable, consistent level. Extended lifetime and small size are required since cells are not easily accessible after they are implanted, so replacing a cell is done as infrequently as possible.

The early SRT-3 battery–powered installations were configured with multiple standard D-cells linked in parallel. Eventually this would change when OTS began building cells in custom-made "cans" or containers. Not all of the specialty cans were metal containers; some of the housing materials could be molded to fit unusual concealment shapes or conform to human movement. Thin, flat, flexible, elongated, and curved shapes of metallic and nonmetallic containers for various power cell chemistries were developed and tested, all for increasing the options for concealment to enhance their operational use.[8]

TSD chemists also explored the possibility of using alternative substances to build what, in effect, would be a super-battery. "We did the calculations to try to find, of all the known chemical materials, what were the most energy-dense," said OTS scientist Stan Parker, who spent a lifetime working on battery chemistries. "We got some remarkable results from the calculations. But when we looked at the toxicity of some of those materials, we said, 'My god, I wouldn't want to be in the room if they were going to actually try to use these. I wouldn't want to even be in the same county.'"

It became clear that the volume of a battery could be shrunk only so much. A variety of exotic elements that yielded longer life in a more compact size were tested, but, inevitably, the laws of physics prevailed. "You can do a lot with chemistry, but then Mother Nature limits you," concluded Parker.

TSD's research chemists had run into one of Faraday's Laws of Electrolysis. Paraphrased, the law states: The amount of total power in any given substance is proportional to the quantity of the substance. Even OTS scientists, as clever as they were, could find no loopholes in Faraday's Law. When they wanted twice as much power, they needed twice as much of the substance. They could reduce the amount of the substance by half, and then get half as much power. Period.

One power source engineer explained the science to nontechnical case officers by saying, "You understand how electronics is shrinking the transmitter," he said. "Well, when the transmitter is reduced one more magnitude, we'll be able to disguise it as the label on this D-cell battery."

Faced with Faraday's Law, the designers of audio surveillance gear and covert communications became obsessed with power reduction. Instead of trying to pack more power into a smaller "can," the engineers began looking for ways to minimize power consumption in the listening and covcom devices themselves. Less power consumption could be traded off against longer life or a smaller "can" or both. Reaching for the goal of a five-year operational life for battery-powered installations meant they needed to squeeze all the power possible out of each cell.

First among the breakthroughs was a series of switch receivers that allowed techs to turn a transmitter off and on remotely. Operationally, this enhancement ended the inefficiency of power wasted by transmitting signals from empty rooms. The remote on/off did not change the total hours batteries would power a transmitter, but it maximized the effective life by limiting transmission to those periods when conversations were held. "At three o'clock in the morning, we didn't want that thing running, draining the battery life to send a signal that says, 'There's nothing going on,'" said Parker. "So, the listening post keeper could hit a button to turn on the audio in the room and listen. If something interesting was going on, if there was conversation, she'd turn on the tape recorder and continue monitoring. But if there's silence, or she heard someone snoring, she'd turn the transmitter off."

Remote on/off held another benefit. When a transmitter was switched off, sweep teams searching the radio spectrum for a bug would not be able to detect its presence. Equipment used by sweep teams to detect clandestine over-the-air transmitters in the 1960s looked for hidden metal objects and unauthorized radio frequency transmissions. Someone hunting for a bug could search the walls with a metal detector or modified radio receiver that automatically moved up and down the radio frequency spectrum seeking unknown or unidentified transmissions.

Because the on/off switch itself required some power to operate, it, too, became a candidate for additional power savings. One brilliant idea, the power saver circuit, came from an industrial contractor. The device was an extremely low power timer that turned the switch receiver to the on position for one second out of every twenty. If it did not receive a signal to start up the transmitter, the switch receiver obligingly went back to sleep for the next nineteen seconds. The savings in power was dramatic. Rather than running the switch receiver for twenty-four hours (1,440 minutes), the total time spent draining precious battery life could be reduced by more than 90 percent to just seventy-two minutes a day. Later, as portable, battery-powered devices began to multiply in the consumer marketplace, the same power saver circuit found a place in everyday products, such as pagers and cell phones.

The packaging of batteries presented another challenge for the techs. In some cases the battery's gases and corrosive chemicals tended to leak. While corroded batteries in a flashlight or camera create an inconvenience for consumers, chemical reactions could prove disastrous for clandestine operations. Not only did the audio system stop working, gas or liquid seepage that discolored areas around the installation could lead to detection, compromising the operation.

"A water-based system, like the mercury cell, can give off hydrogen or hydrogen and oxygen gas. The hydrogen is the bad actor," explained Linn. "If you get into a mode with gas escaping, leakage of the electrolyte results. This is corrosive and can change the paint color on a wall. We needed to be able to package every cell so it didn't leak liquid or emit gas." Balancing the laws of physics, audio engineering became a game of technological and operational horse trading, often conducted under the urgency of the "crisis de jour."

The technical trade-offs with operational requirements were seemingly endless. How long does the device need to remain operational? If only for a few hours in a hotel room, commercial D-cells could work, but bugging a foreign embassy's conference room for five years required a completely different technology. How large can the device be, including antenna? That answer was never the same depending on whether the antenna needed to be installed in a less-efficient horizontal position as opposed to the preferred vertical configuration to radiate the signal. What is the window of opportunity for the operation? If it must be done in the next five days, the tech ops officers had to use whatever equipment was available. Given an operational window of six months or a year, however, TSD engineers could redesign or adapt equipment and technology for a specific application.

"You'd get a call, 'Hey, listen, we're doing something, can you come to a meeting at three o'clock?'" remembered Parker. "So I'd say, 'What are we going to talk about? I'll bring the data I have on that.' And we'd all show up, sit down, and try to figure it out. Some of the ops people were very well versed and others wouldn't know an electron from a trumpet. They came in all flavors. But most of them you wouldn't want any other way. The guys who didn't have the technical knowledge sometimes had a whole lot of tradecraft knowledge they could bring to the party. One of the things I always used to say: There were enough problems to solve, so when somebody presented you with a solution, you left your pride at the door and said, Thank you."

CHAPTER 14

The Age of Bond Arrives

If you give me a target, I'll get audio in it.   —OTS audio tech, 1970s

The 1970s were heady years for audio techs in the field and scientists in the lab. The demand for audio ops in every part of the world accompanied the introduction of integrated circuit technology. With the new generation of miniaturized components capable of transmitting greater distances for longer durations, audio concealments seemed limited only by the tech's imagination. Installing audio in walls or wood blocks represented a "passive" concealment operation. The techs, however, also recognized that miniaturization and tiny electronic components offered opportunities to embed audio devices or cameras in hosts that continued functioning as they were designed. The devices were now small enough to implant into electronic concealments such as clocks, calculators, and radios. Expertise that OTS craftsmen applied to dead drops was now put to use in creating audio concealments. Watches and cigarette lighters were candidates for "active" concealments. With these concealments, the techs hid the spy gear in everyday objects an agent could carry, wear, or use for their intended purposes.

Audio devices were concealed inside furniture, books, cans of shaving cream, clothing, and in one case, a construction worker's hard hat. Maids or visitors leaving a gift or exchanging the desk lamp for a modified duplicate, could introduce audio bugs into rooms. CIA defector Phillip Agee featured on the cover of his autobiography a photograph of the lid of his typewriter case filled with sixty "poker chip" batteries, alleging they were part of a CIA operation to bug him as he fled around the world.[1]

Concealments were a mainstay of the OTS laboratory. On his initial visit to the lab, a newly appointed office director commented on the variety of high-quality woods that were in the inventory.

An OTS craftsman pointed to a piece of lumber and asked, "What do you see?"

"Cabinet-grade walnut," replied the director with pride in his knowledge of woods.

"No, sir," corrected the concealment specialist, "that is volume in a cellulose wrapping. And we can put anything we choose underneath the wrapping as long as it doesn't exceed the volume."

Once the "volume" of the bug package, consisting of microphone, transmitter, switch receiver, power cells, and antenna were reduced to six cubic inches or less, relatively small blocks of wood could encase all the system's components. The wood block became the workhorse for "quick plant" audio operations. "RF transparent" wood could be cut into almost any configuration with a milling machine or hand tools and then screwed, bolted, glued, or wedged into place. Small blocks could be fashioned to blend in with furniture, the molding in an office, or a picture frame by matching wood types, grains, and finishes.

For twenty years after the introduction of the SRT-3, each successive SRT model saw either the size of the transmitters decrease or improvements in performance or security.[2] Transmitter models in the mid-1960s also marked the introduction of signal masking systems to defeat audio countermeasures. Without masking, a technical sweep team could inspect a facility with electronic and magnetic equipment that scanned the RF signal spectrum and detected foreign objects to locate, lock onto, and expose the secret audio transmissions. Masking reduced the vulnerability of bugs by making their signal harder to isolate and identify as clandestine transmissions.

One masking technique commonly used by both the United States and Soviets buried the transmission in the signal's subcarrier. RF transmissions were designed to broadcast in two parts, much like stereo. The first part, a clear signal resembling white noise, was passed over as benign by someone scanning the radio spectrum. Then, just to the left or to the right on the dial—up or down the spectrum—was the subcarrier with the clandestine message. By tuning to the right frequency and tuning out the white noise, it was possible to hear the covert transmission. In principle, the use of subcarriers worked like hiding a piece of clear glass in a container of water. The glass remains invisible until the water is drained.

Other techniques for using subcarriers sent the audio signal along the existing AC power lines where it was collected and retransmitted to a listening post. Signals could be encrypted, masked, or both.

True to the nature of espionage, each technological advance was inevitably met by an effective countermeasure. In time, KGB counterintelligence teams began tuning in on the white noise in search of subcarrier transmissions. OTS responded and advanced the technology to the next level. "Concealing signals was an area I felt very strongly about," said an OTS manager who oversaw the program. "I wanted to come up with new modulation schemes, every year I wanted at least four or five brand-new ones to hide our transmissions. For a while we got into a pattern of using certain types of subcarriers almost exclusively and, unfortunately, the Russians knew what to look for in our 'offsets.'"

Eventually these techniques of hiding transmissions came to include a frequency-hopping technique in which short transmission bursts bounced up and down the radio spectrum in no apparent order. Without a receiver coordinated to the changes in transmissions, these frequency hops proved particularly difficult to identify and intercept since it was nearly impossible for sweep teams to anticipate the signal's pattern.

The complexities and opportunities presented by clandestine audio seemed endless. Installing an audio bug always put the techs at personal risk of discovery and arrest when entering, leaving, or working at a target. Building reliable miniature components for covert systems that could withstand extreme environments challenged the best engineering minds. Configuring the system to operate within the available concealment space required mastery of craftsmanship and design, but no matter how sure the tradecraft and skilled the engineering, none of that mattered without access, and some targets were virtually inaccessible.

The problem of access led TSD and its partner, the Office of Research and Development (ORD), to experiment with an array of exotic audio surveillance delivery systems.[3] In the early 1960s, Soviet diplomats in one Central American capital city often conferred in their embassy's courtyard on matters they believed were too sensitive to risk discussing in their offices, which they believed were possibly bugged. The courtyard, while surrounded by a security fence, was not walled and CIA officers observed that one bench seemed to be a favored gathering place for Soviets of particular interest. Adjacent to the bench was a large shade tree. Station officers had no means to gain access to the bench inside the embassy compound so the DDP turned to TSD to devise a means to bug the conversations that occurred around the bench. The open security fence surrounding the embassy led to the idea of shooting a bullet containing the microphone and transmitter into the tree just above where the diplomats usually conferred.

For the concept of the "bullet bug" to work, TSD would need an audio device small enough to fit into a projectile, a means to clandestinely shoot the package into the tree, and components that would tolerate the velocity and impact necessary to bury a projectile far enough into the tree to escape notice.

A TSD engineer took the concept to the president and chief scientist of America's leading hearing aid company, asking that they build a microphone small enough to fit into a .45 caliber bullet and rugged enough to function after hitting a tree. The problem of small size appeared solvable, but nothing in the company's inventory would tolerate the shock. As the technical discussion progressed, problem after problem arose. It seemed apparent the idea had no future, until the president suddenly interjected. "Well, it's a really good challenge, yes, let's do it." A team of engineers was formed to create a one-of-a-kind microphone with no manufacturing markings or signature.

After obtaining a similar commitment from a company specializing in small transmitters, TSD began testing and evaluation. Within three months, a 400 MHz transmitter, battery, and microphone small enough to fit into the projectile, somewhat larger than a .45 caliber bullet, were delivered. Battery life was limited to less than a day due to size limitations.

The antenna was a simple wire that trailed behind the projectile after it left the barrel of the gun, but presented a problem since it caused the projectile to wobble in flight and hit the target broadside. Over time, the techs found that by adjusting the antenna length the projectile would fly true, embed itself at the proper angle, and maintain the audio link to the listening post.

A vintage World War I rifle became the test weapon. Its long-rifled barrel enhanced accuracy by building up projectile speed and stabilizing the bullet and antenna before leaving the barrel. Test firings into three one-inch plywood targets clamped together were conducted at an abandoned rock quarry near Baltimore, Maryland.

Opting for safety, as they were using an old gun and unconventional ammo, the techs mounted the rifle to a table, placed sandbags around it, and attached a cord to the trigger for firing. After a few test shots, when the rifle did not fly apart, the more courageous techs fired it from a shoulder position. Repeated firings determined the correct amount of powder needed to limit the projectile penetration to no more than two inches, the maximum depth from which the microphones and transmitter could operate.

The techs found it impossible to use a standard silencer on the weapon so, to quiet the report, they jerry-rigged a fifty-gallon steel drum filled with acoustic baffles. Both ends of the drum were cut away and a center area of free space was created through which the weapon could be sighted. When fired from within the makeshift acoustic chamber, the sharp firing noise was reduced to a bass boom. Because the weapon was still too loud for operational use and they were without a technical solution, the planners envisioned a scenario in which two loud motorcycles would start at precisely the time the weapon was fired, masking the gunshot for anyone who might be within hearing distance.

From the first testing, the transmitter and battery components proved both reliable and functional. The microphones required several adjustments, in part because prior to the requirement, magnetic microphones were designed to withstand drop impacts only, a significantly lower stress than a bullet impact. Eventually, the microphones and other components proved consistently durable when the projectile traveled at approximately 500 miles per hour over distances up to 50 yards. The microphones picked up sounds of a portable radio sitting next to the plywood target and transmitted quality audio up to 250 feet.

In the next round of field tests, the "audio bullets" were fired into live trees to simulate an operational scenario. Once fired into the tree, two people seated nearby carried on a conversation at normal voice levels. Surprisingly, audio quality was poor compared to the plywood tests. Analysis showed no damage to the device, but the live tree wood proved different from plywood. Tree fibers when hit by the projectiles formed cones similar to the design of an echo-free anechoic chamber and swallowed up the audio.

Additional analysis determined that if the transmitter's size was increased, the necessary audio amplification could be attained. This would require, however, a larger bullet, increased firing noise, and a redesign of the weapon itself. The hole in the tree would also become larger and more noticeable. In the end, the DDP judged the value of the potential information insufficient to justify TSD's cost in time and dollars for additional development and the bullet bug was filed away.

While the project did not result in an audio "silver bullet" to bug the Soviets, technologies for high-reliability miniature microphones did emerge. Based on data obtained from the tests, TSD produced a series of very small microphones that could withstand high impact and high heat stress. This new generation of rugged microphones could endure rough handling, be installed in almost any wet or dry material, and perform at near zero failure rates regardless of where they were buried. Commercially, the research and design effort by the contractor produced shockproof microphones that enabled the size of hearing aids to shrink along with improving the microphones' performance in varied temperature and high moisture environments.

Animals as well as technology played starring roles in the quest to prove that any target could be "hit." When CIA operatives sought a means to penetrate the private meetings of an Asian head of state, reports reached Headquarters that during the target's long strategy sessions with his aides, cats wandered in and out of the meeting area. Feral cats were common to the region and generally ignored. Whether the concept of an "Acoustic kitty" came from a case officer or a tech is lost to memory, but the idea launched a research project that generated unwarranted ridicule and accusations after public disclosure.[4]

In fact, absent from the Acoustic kitty project were both cruelty and mutated, grotesque creatures from horror movies.[5] From the beginning, the techs recognized that the concept, undertaken jointly between OTS and the Office of Research and Development, fell into the high-risk category. At the time, embedding electronics inside animals or people was not a routine medical procedure.

The implant could not affect any of the natural movements of the cat nor could the cat experience any sense of irritation or the presence of the device lest it induce rubbing or clawing to dislodge components or disturb performance. The audio system components would include a power source, transmitter, microphone, and antenna.

Working with their prime audio equipment contractor, the techs produced a three-quarter-inch transmitter for embedding at the base of the cat's skull where loose skin and flesh provided a natural pocket. Implanting the transmitter proved viable, once a device was packaged to withstand the temperature, fluids, chemistry, and humidity of the body. Microphone placement presented a more difficult problem since flesh is a poor conductor. Eventually, the ear canal became the preferred location. An antenna of very fine wire was attached to the transmitter and woven into the cat's long fur. The cat's size permitted only the smallest batteries, a factor that restricted the amount of hours the audio could transmit.

Research to determine the performance of the various individual components and the most effective placement areas was conducted first on dummies and then live cats. Documentation of reactions of the cats to the "foreign materials" and to nerve stimulation refined the research and eventually produced an integrated audio system suitable for dress rehearsal. Agency officials reviewed questions of humane treatment of the animals and the potential negative publicity should the activity become publicly known. After those factors were weighed against the operational value of the project's success, the techs received authorization to go forward.

A small crowd stood behind the vet who conducted an hour-long procedure on a full-grown, anesthetized gray-and-white female cat in a clean, brightly lit animal hospital. The TSD chief audio engineer, seeing the first incision and a trace of blood, asked to sit down. No other complications arose, and after the cat awakened, she was put into a recovery area for further testing. Technically the audio system worked, generating a viable audio signal. However, control of the cat's movements, despite earlier training, proved so inconsistent that the operational utility became questionable. Over the next few weeks, Acoustic kitty was exercised against various operational scenarios, but the results failed to improve.

Acoustic kitty demonstrated that transmitters could be embedded in animals without damage or discomfort. The experimental animals could be directed to move short distances to target locations and people in a known environment. However, outside the experimental laboratory, Acoustic kitty had a mind of its own. Eventually, deployment of Acoustic kitty in a foreign environment over which the "handler" would have no assured control was judged impractical and the project was closed.[6]

Exotic detours aside, OTS's most productive audio ops followed a disciplined formula. Identify an operational requirement, select a target, survey the target, assemble the right equipment, establish a listening post, make the entry, install the device, test the system, restore any damaged area to original condition, dispose of any evidence of being there, and get out without

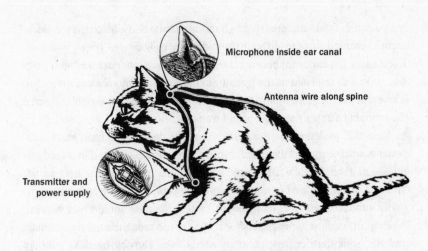

Microphone inside ear canal

Antenna wire along spine

Transmitter and
power supply

*"Acoustic kitty" was TSD's attempt to implant a
clandestine listening device in a cat, mid-1960s.*

getting caught. Audio techs improvised when it came to tools, combining an assortment of commercially available hardware store implements with specially fashioned gear made either in the lab or of their own design.

In one operation, a tech used standard well-drilling equipment—configured to operate horizontally—to drill more than a hundred feet from the listening post to the target site on the opposite side of a major thoroughfare. "We bugged every room in that building, and then brought all the wire leads down to the basement," the tech who led the operation remembered. "I surveyed in on where I wanted to enter the basement, then drilled a hole using just the azimuth and elevation data from the post to the target. I came out a foot away from target hole. The case officer said, 'You missed.' I said 'Shit, a hundred and nine feet underground, in a foreign city, I think I did pretty good,' and we figured out a way to compensate for the other twelve inches."

Some techs excelled at solving problems with their own inventions to meet specific operational needs. Although of little use beyond covert operations, the devices were invaluable for making installations. The *Nail Pusher,* or *Silent Hammer,* was used for restoration work on baseboards and molding. Essentially the device was a hollow tube with a plunger-type mechanism to reinsert nails silently without leaving traces of a hammer mark.[7]

One innovation that earned its inventor a unique, if dubious, reputation among his fellow techs was a new microphone housing. Techs had long been beleaguered by the challenge of securing a mic into position within the hole drilled to reach the target's wall. Too often after the tech carefully positioned a microphone in the hole against the pinhole, it slipped slightly away from the tiny air passage before being firmly anchored. If unnoticed at the time, the smallest misalignment produced a degraded sound. The tech's clever solution encased the mics in a sheath of pilable latex that fit snugly into the three-eighths-inch-diameter hole leading to the pinhole. Because of the phallic appearance, techs named it the *Peter Mic*.

As intelligence flowed through audio's reliable equipment, so did the audio techs' confidence in their tradecraft skills. Among the techs, and even case officers, the thinking became, "If access could be obtained, almost any target was vulnerable." In some respects, this was "spiral development." Hard targets required greater tradecraft skills and, as those skills were acquired, they were applied to even harder targets.

The increased sophistication of bugs and a willingness to take on the tough operations required better equipment. For instance, drilling holes represented a core skill for audio techs. Holes for bugs were drilled down from ceilings, up from underneath floors, and horizontally in walls. When the techs could not physically get inside a room to install a bug, they drilled through a common wall. The danger of such an operation lay in the fact that the techs were literally blind to what or who was on the target side of the wall.

These drilling operations had two major security risks: noise and unintended breakthrough. Electric drills were fast, but so noisy they were not an option for use in the middle of the night or with the target room occupied. Hand-turned drills were slow and difficult with harder construction materials. To drill quietly usually meant drilling so slowly an installation could require days, especially if multiple bugs were being installed.

In a typical operation, techs preferred to start with a three-eighths-inch drill bit (the hole had to be large enough for the circumference of the microphone) until reaching the final half-inch of material in the target's wall. At that point, depending on the size of the microphone head, drill bits of less than .050 inches were used to drill a breakthrough pinhole. The tiny hole created enough of an air passage for clear audio pickup while virtually invisible to normal observation.

With blind drilling techs never knew how close they were to the breakthrough point. Even for the best drill tech, it was a matter of guess, estimate, feel, and experience. Wrongly judged, the drill's breakthrough would leave a noticeable hole in the target's wall and debris on the floor. "If we don't know the thickness of the wall, then we don't know for sure when we are close to the other side," explained one tech. "So if we punch through a wall with a three-eighths-inch hole, somebody is going to notice. Sometimes when we did inadvertently drill through, we joked that our audio operation became a video operation."

Over the years, more than one tech accidentally broke through a wall, then looked into the hole only to see a curious eye peering back. During one operation, a tech drilled a larger than intended hole into a Soviet apartment. Several minutes passed, then without knocking, the Russian diplomat burst into their room and furiously began berating his "neighbors" for their carelessness in punching a hole through his wall while hanging pictures. The case officer apologized and assured the diplomat that his workers would be more careful in the future.

Targets were not the only ones displeased with tech mistakes. On a seemingly routine operation, the techs made an uneventful entry into a commercial building and began drilling the starter hole into the common wall with a Soviet trade mission. Suddenly, the drill bit broke through, creating a gaping hole in the adjoining room. With no way to repair the damage, the best that the techs could manage was to patch their side of the wall and retreat to the local chief of station's office to report their problem.

"We have a nice hole in the wall for audio," the head tech reported.

"Good," the COS replied, "very, very good."

"Well, what I mean," explained the tech, "is that we broke through with a really nice big hole."

The chief went ballistic. "Get out of my country and never show your face here again!" He spoke the order loudly enough that the techs heard each word and didn't even think of putting up an argument.

"I guess that tells us something about the lack of a sense of humor some of these guys have," whispered one of the team members.

A few weeks later, after the chief calmed down, another visiting OTS officer suggested that, given the importance of the target, another attempt was merited. The chief demanded and received assurances a similar mistake would not be made. The installation went off without a problem.

Listening to the live audio a few days later, the techs heard sounds of a work crew coming into the room. Not clear from the conversation was whether this was a sweep team or construction crew, but they were obviously looking carefully at the walls. The techs held their breath, waiting for what would happen next.

"Look at this," one guy said.

"Damn, what's that hole? That shouldn't be here," came the reply. "Well, we better get rid of it."

With relief, the techs listened as the conscientious construction crew repaired the wall with the three-eighths-inch hole, never noticing the pinhole a few feet away. This time even the chief of station was amused.

To avoid the disasters of noise or breakthrough, techs would often drill just a few revolutions per minute. To create the pinhole, they would slowly twist a six-inch cylinder shaft that held the tiny bit with thumb and forefinger, applying little, if any, pressure, and letting the bit pull itself through the final fraction of an inch.

The problem of measuring the thickness of the remaining wall between the end of a drill bit and "breakthrough" was partially solved by one of the OTS's cleverest tools, the *Backscatter Gauge*.[8] While the basic technology employed was not new, its covert application was a model of ingenuity. The principle behind backscatter technology is nuclear science. A tiny radioactive source emits a steady pulse of gamma rays, bouncing them off an object while a reader contained in the unit measures the number of pulses that bounce back. A thick material will repel more gamma rays than thin material. The gauge calculates the percentage of the returning rays against the number emitted. For example, an object that bounces back 50 percent of the pulses is twice as thick as one that returns only 25 percent. A more sophisticated version of the technology later evolved into security devices used to scan baggage and individuals at airport checkpoints. In this configuration, with advanced signal processing, the returned gamma rays actually paint a picture, similar to that of an x-ray.

When OTS adapted the technology for clandestine purposes in the 1970s, backscatter was widely used in industrial applications, primarily for quality control. "It was being used in paper mills to keep the thickness or consistency of paper constant," explained Martin Lambreth, the engineer who helped design the system. "By using the backscatter technique, the thickness could be measured continuously as it moved through production. As long as the radiation remained the same, the product was good. If it changed, they'd stop production. We wanted to use the same principle for measuring the distance from our drill bit to the surface of the wall we couldn't see."

With the OTS version of the technology, engineers developed a unit attached to a probe that fit into the three-eighths-inch drill hole. Techs would drill a short distance into a wall, then withdraw the drill and insert the probe to take a measurement. This drill-and-probe process continued as the electromechanical counter mounted on the unit recorded a wall's thickness at the deepest point of the drilled hole. In time, techs became so proficient that some abandoned consulting the small mechanical readout altogether, preferring to judge depth by the clicks of the counter. The faster the clicking sounds, the thicker the wall. Since drilling sometimes occurred in near darkness, this also reduced the need for illumination and added a measure of security to nighttime jobs.

"I just listened for the clicks. I'd be on a ladder, the gauge is on the floor attached by a long cable going click-click-click," said Martin, an experienced tech. "Then, after awhile, I'd hear click . . . click . . . click . . . and say, something's about to happen. I better be careful from now on." The reading, while not precise, was close enough to eliminate virtually all breakthroughs and earned praise from audio techs around the world.

A second drilling innovation adapted from industrial technology was the *Grit Drill*. Smooth plaster walls presented a particularly difficult problem for the audio techs. It seemed that every diplomat who happened to be an operational target also had an office or home with plastered walls. To cut through the plaster, some pressure was needed on the drill, but no matter how careful the tech tried to be, that pressure was just enough to spall the other side of the wall. Small chips of plaster from spalling were a dead giveaway to anyone doing a security inspection.

OTS management dispatched an engineer, working under an alias as well as commercial cover to obscure CIA interest, on a nationwide tour to find a solution. Lugging dozens of circular samples of brick, concrete, ceramic tiles, terrazzo, and other materials neatly packaged in plastic sleeves, he began a cross-country journey in search of a better drill.

He visited more than a dozen companies, large and small. Anyone who knew anything about drilling through hard materials was fair game. Those who agreed to meet with the engineer were told that his company was looking for a drill so fine and so tough it could penetrate every one of the circular samples with a clean one-millimeter hole.

He visited precision drilling companies that cut holes in circuit boards and scientists in labs working with microwave energy. In upstate New York, he paid a call on a company that excavated concrete and was anxious to help. "I think we can do it," said one of the mining engineers. "We would use a very small controlled explosive with great care." The tech found the concept intriguing but the idea of using explosives, no matter how small, was not something he could likely sell to OTS.

The engineer eventually found a research company in the South that said it had a scientist with a reputation for innovative engineering. So far, the search had been futile, but he went through the requirements yet again. There was little reaction except the scientist asked that some of the material samples be left behind. Because it was the last stop on the circuit, the tech was happy to discard the weight from his luggage. Back at OTS, he reported little progress on the problem.

Nothing happened for a few weeks, and then, unexpectedly, the OTS engineer received a call from the scientist who had kept the samples. "I've got the solution," offered the caller, and the engineer was on the next plane going south.

At the lab, the scientist rigged up an old thermal drill, a type used for a brief time by dentists. The drill utilized a very fine nozzle and air pressure to shoot a thin, high-velocity stream of extremely small aluminum oxide particles, essentially eroding the tooth's enamel to create a hole rather than drilling it out. While erosion eliminated some of the discomfort of pressure during drilling, patient complaints about the taste of the particles made the technology unacceptable.

The OTS engineer pulled out samples of every possible material and the two went to work. They drilled holes in glass, concrete, plaster, stucco, and ceramic tile. No material was spared from testing and a clean hole appeared in each sample. Both engineer and scientist admired what the drill had accomplished.

"Works great, you solved the spall problem. And it's no good to us," the engineer told the stunned scientist.

A lengthy conversation followed as the engineer pointed out that despite the precise, clean holes, the drill would not be usable in a clandestine operation since it sent a fine spray of particles through the hole at breakthrough. If a target room was on the other side of the hole, the drill would deposit a fine coating of dust on carpet, furniture, and files that would surely alert the room's occupant.

The scientist listened intently and asked questions about the nature of operations and the special tools that were required. "Can I keep working on this?" he asked.

The OTS engineer readily agreed. Here was a scientist with demonstrated creativity and who was clearly hooked by the challenge of clandestine requirements. Had he been a case officer, the engineer would have claimed a "recruitment."

A few days later, an OTS secretary took a cryptic phone message from the scientist. "Tell my friend to come on down, he'll know what it's about."

The following day the OTS engineer watched as the scientist attached a sleeve to the device that was inserted into the hole being drilled. The drill went into the sleeve, which was sealed around the outside of the hole. Extending from the collar was a hose that ran from a filtering system to a clear Plexiglas tube. "He blew the grit back into part of a vacuum cleaner bag, so he could filter the stuff out. Then coming out of the vacuum bag was a clear tube and in that tube was a Ping-Pong ball and a photoelectric cell on the side. That was the on/off switch," the engineer explained. "So, when the gas came on for drilling, it created air pressure in the tube and lifted the Ping-Pong ball. While drilling he had positive gas flow and ball stayed up. As soon as the drill tip broke through, the pressure in the tube dropped, the Ping-Pong ball fell, triggering the photoelectric cell, and the damned thing turned off."

The operational dust problem was solved along with the problem of cutting a clean pinhole through smooth plaster. OTS engineers reconfigured the device for portability and named it the Grit Drill. Helium was substituted for compressed air for its higher exit velocity. The entire Grit Drill and its accessories could be squeezed into a standard briefcase.

Once the *Grit Drill* kit was certified for deployment, the well-traveled engineer received orders to demonstrate the system to the overseas OTS tech bases. There, skeptical audio techs assembled for a demonstration of the "latest solution from Headquarters." For field types, the Headquarters' show-and-tells had a reputation for bringing more hype than practical value.

The engineer described the system, explained why it worked and demonstrated how to set it up. He then began punching tiny holes through materials the most susceptible to spalling. Each tiny hole was perfect, with no spall appearing on the target side of the sample materials. But before the demonstration was finished the chief of audio operations interrupted. "I've seen enough," he declared. "I'm doing an operation tomorrow; bring that thing along because you're going with me." The engineer was stunned. Not only had he never been on a clandestine operation, the *Grit Drill* had never been used operationally.

Three days later the engineer's anxiety had turned to excitement. The operation went smoothly with the engineer personally operating the *Grit Drill* during its first use in an audio installation. Now with an operational success as part of his briefing, he visited other techs in the region demonstrating the drill to interested audiences. But it was the thrill of the clandestine work that stayed with him. Like his scientist friend, he had been hooked by the excitement of clandestine operations. Back at Headquarters, after completing the show-and-tell TDY, the engineer transferred from his lab assignment to the cadre of audio tech officers.

As the audio surveillance equipment improved and inventories of transmitters, power sources, microphones, and installation tools grew, the techs themselves required additional training to configure and test equipment. The era of the "tinkerer" had ended. The days of on-the-job, trial-and-error audio training gave way to more rigorous and formal instruction aimed at greater professionalism and broader knowledge of increasingly complex equipment.

"If we weren't doing this, we'd be robbing banks," said Antonio J. "Tony" Mendez, one of the three OTS technical officers honored in 1997 as a CIA "Trailblazer."[9] No element of an audio operation created a more intense adrenaline rush than a surreptitious entry into a secure and guarded target. Surreptitious entry, defined as "an entry by stealth," accompanied almost every clandestine audio installation.[10] In most instances, the techs entered the premises or property of the target and then made additional entries into luggage, mail, or vehicles. Due to the risk involved, surreptitious entries were thoroughly scripted and rehearsed prior to the operation.

During World War II, the OSS created a surreptitious entry capability by enlisting former second-story men with prior experience in burglary, lock picking, and safe cracking. OSS created and issued a small "lock picking knife" containing "picks" instead of blades, that could be conveniently carried in the operative's pocket for quick access when needed. The original OSS surreptitious-entry manual cited the purpose for their work as helping the agent solve his problem:

> *He wishes to obtain access to secret documents, copy or memorize their contents, and leave the premises in the same condition as he found them. To arouse suspicion that the entry had been made, would in many cases be as fatal as being caught in the act of rifling the safe. The agent*

*should therefore learn thoroughly the technique of surreptitious entry,
so as to adapt it, as the occasion requires, to a similar job in enemy
territory.*[11]

The reasons for conducting entry operations did not change after World
War II, only now the "enemy territory" became the guarded official missions of America's Cold War adversaries scattered in countries throughout
the world. While all the audio techs were trained in the basics of surreptitious entry, a handful of officers specialized in the work.[12] These techs were
skilled at climbing ladders, bypassing alarm systems, picking locks, cracking safes, and performing room searches as well as installing devices. These
entry specialists demonstrated that, if given sufficient time and resources,
virtually any lock could be opened and any alarm system bypassed, though
there were always limits on time and the amount of equipment that could be
deployed at the job site.[13]

Regardless of how easy lock picking was represented to be on television
and in the movies, the techniques of manipulating locks required skill and
practice. At best, picking remained more art than science. One TSD tech remembered sitting at home in a European capital in the late 1960s working
through the weekend to "get the feel" for opening a new brand of foreign
lock. The first successful attempt required more than twelve hours, but, once
he had acquired the feel, he could open the lock in less than five minutes.

Target sites were usually well protected, and the more valuable the information inside, the more layers of protection surrounded it—locks, secured
doors, gates, windows, file cabinets, vaults, safes, and even the alarm systems. A lock specialist had to be proficient in dozens of mechanisms since
locks in different parts of the world varied in type and method of operation.
The techs discovered that German locks were particularly difficult compared to those in South Asia.

The grab bag of different locks and security architecture the techs found
in countries from Ghana to Paraguay ranged from early colonial to
state-of-the-art. With narrow time windows for covert installations, the
techs had to know how many minutes were required to break through locks
and security barriers and then restore and rearm the security systems. Information on all of these factors was obtained in a detailed preinstallation
clandestine survey of the target conducted by techs and included in the operational proposal. Only after headquarters approval of the survey could an
installation operation commence.

Under the best of circumstances, the CIA would obtain advance information that a Soviet intelligence officer planned to move into a new apartment or that the Chinese government was renting an office suite for a new trade mission. If possible, local support agents were recruited to rent or even buy office space, apartments, houses, or property adjacent to target buildings.

Techs infrequently picked a lock, preferring to find other methods to gain entry. Picking took too long, the results were unpredictable, scratches from the picking on the mechanism or housing were detectable, and once a lock was picked open, some had to be picked closed at the end of the operation. Sometimes, when the location was known far enough in advance, the techs could wire the vacant property before its occupants moved in. Architectural spaces, such as the attics of row house buildings, were particularly inviting, since their design offered a contiguous common space over each unit. Once the tech gained access into the attic, he had unobstructed movement to any top-floor unit in the row. Buildings might also provide common basement areas with several outside entrances that enabled the tech team to avoid being seen coming and going through the front door. Given complete access and unlimited time to do the installation, the techs planted multiple microphones and transmitters as well as ran wires, precluding the need to pick any locks.

Depending on the relationship with security services of the host country, known collectively as "liaison," the CIA could assign the task of performing an entry to liaison. In many countries, the internal security service already had duplicate keys to all rooms in every major hotel, and master keys for apartment houses and important commercial buildings.

The techs also made duplicate keys. By the early 1960s, master keys for every popular hotel in more than one major European city hung neatly on a rack in the techs' shop in local stations. When possible, case officers would "beg, borrow, bribe, or steal" master or original keys. OTS had key-cutting machines at its bases and the traveling techs carried portable key-making devices that fit inside a briefcase. From the properly sized and configured blank, the techs could cut a duplicate within a few minutes.

For keys that were briefly available, OTS developed a portable key-impressioning kit. The kit consisted of a small mold with two halves filled with plastic modeling putty.[14] The purloined key was placed on the putty and the two halves of the mold pressed together to capture a three-dimensional model of the key. Later, the tech could pour Wood's Metal into the mold and create an exact copy of the key.[15]

The last choice was to attempt to pick an unknown lock. For this contingency, OTS issued leather lock pick kits that were small enough to fit in a jacket pocket, but provided the necessary tools for picking and raking the tumblers of many commonly found locks.[16] These portable kits were most useful for opening luggage, desk drawers, and other smaller locks.

In the late 1960s, a TSD engineer developed a concept that mechanisms within key-operated locks could be measured and characterized remotely by marrying emerging ultrasonic measurement technology with an oscilloscope. Portable oscilloscopes had just been introduced and when combined with a small ultrasonic device, the techs would have a tool they could carry easily to the target, use to measure the lengths of the pins in a lock, and thereby acquire the precise data to make a key.

Once the engineer produced a prototype device that produced accurate calculations, OTS contractors refined the design for a field deployable unit. A year later, after the device proved itself by enabling several surreptitious entries into previously inaccessible targets, Cord Meyer, the Associate Deputy Director for Plans, recognized the engineer with a special award that included a $5,000 stipend. In his presentation, Meyer said he could not mention what was acquired from the entries, but added, "This is the largest stipend the DDP has ever awarded for a technical development. This gadget is right out of the James Bond movies."

Not all entry operations involved breaking into rooms or safes. HTLINGUAL, for example, was the CIA's controversial Cold War program that intercepted and examined U.S. mail to and from the Soviet Union.[17] Covert mail intercept required skills in "flaps and seals" to open and reseal envelopes, cartons, and packages thought to contain intelligence. "Surreptitious opening" and surreptitious entry shared the common objectives: to get inside a protected area and copy or steal the contents.[18] The two primary methods for opening a sealed gummed flap, such as on an envelope, were "dry openings" and "steaming."[19]

From 1940 to 1973 the FBI, and later the CIA, conducted covert activities to open and photograph suspect mail in the United States. The earliest techniques of chamfering (mail opening) were taught to the FBI by a friendly Allied intelligence service during World War II. Information obtained from these programs was sanitized to protect against revealing sources and was disseminated to the intelligence agencies, the Attorney General, and the President of the United States.[20]

As the Cold War intensified, the CIA initiated its mail-opening project in New York to target mail from the Soviet Union. The HTLINGUAL operation was conducted by the Counterintelligence Staff and the Office of Security with TSD's assistance. Over a twenty-year period, more than 215,000 letters to and from the Soviet Union were opened and photographed in New York.

The New York mail project originated in 1952 with a proposal to scan exteriors of all letters to the Soviet Union and record the names and addresses of the correspondents as a means of identifying U.S. contacts of Soviet intelligence. The project expanded when James Angleton, chief of the Counterintelligence Staff, advised Richard Helms, then Acting Deputy Director for Plans, of a need to open and examine a selected portion of the letters. He advised Helms that there was no capability for "searching for secret writing and/or microdots, or to determine whether items have been previously opened, and to open items sealed with the more difficult and sophisticated adhesives."[21]

TSD set up a lab in New York in 1961 to test letters for secret-writing chemicals and to study Soviet censorship techniques. Because technical examination was time consuming, only a small percentage of the letters opened and photographed were actually tested. The Manhattan field office of the CIA's Office of Security handled most of the opening and photographing. Those who opened the mail attended a one-week course called "flaps and seals" conducted by TSD at CIA Headquarters.

The basic method of opening the mail was simple. First, the glue on the envelopes was softened by steam from a kettle, and with the aid of a narrow stick, the flap was pried open and the letter removed. One of the agents who opened the mail testified "you could do it with your own teapot at home." It took approximately five to fifteen seconds to open a letter. At one point, the CIA developed a type of "steam oven" capable of handling about one hundred letters simultaneously, but its performance was judged inadequate and the agents soon returned to the kettle-and-stick method.

The original letters, which had been opened, photographed, and possibly subjected to TSD examination, were resealed and returned the next morning to the airport for reinsertion into the mail stream. Translations and summaries of the letters' contents were disseminated within the Agency and to the FBI.[22]

As the sophistication of the technology and number of audio installations expanded, OTS developed an intense, year-long training program for the audio techs designed around the lessons and mistakes from the early years. In addition to learning the ins and outs of the technology itself, novice techs were taught the basics of building walls, mixing plaster, matching paint, restoring wallpaper, and making repairs after a device was implanted. They learned how to pick locks, make key impressions, cut keys from blanks, manipulate combination locks, and do electrical and telephone wiring.

"There was every kind of reconstruction. One of our instructors was a master plasterer who, before retiring, worked in the White House and the Capitol," recalled one tech. "We had a dedicated facility, an old food warehouse in Alexandria, Virginia, where we'd learn how to mix mortar and lay brick. It didn't matter if you had a college degree or not. If you wanted to be a tech, you took that training."

For a typical lesson, the master plasterer assigned trainees to build a wall, plaster it over, then knock holes in it to simulate burying audio equipment, and replaster it. Then came the tough part. Not the least impressed by the tech's CIA affiliation, the plasterer would shine his flashlight on the gleaming wall, silently studying the work, then invite the tech to join him as he pointed out a ripple here and another there. No ripple was acceptable. "Nope, not good enough" were dreaded words. With those, the tech knocked down his wall and started over.

The course work on walls and plastering alone lasted a month, and then came paint matching, which included training with special paints that OTS formulated to be fast-drying and odorless.

Specialized soldering courses followed along with instruction about glues, adhesives, tapes, and fasteners that hold things together. The techs had to learn how to open and close all types of materials—fabrics, leather, wood, concrete, and masonry—in preparation for burying bugs in any concealment that might be in a target environment.

The techs received training to operate laser surveillance systems that, by projecting a laser through a window, could pick up audio from minute vibrations of the windowpane in the target room. Although these officers were audio specialists, overseas stations would not hesitate to ask their help in all other disciplines of OTS, so familiarization was provided in the full range of agent communication including microdot, secret writing, photography, and short-range electronic systems.

"They didn't skimp on training at Headquarters. It was thorough and hands-on. And in the field, every new officer got a mentor—a more senior audio officer—for their first tour," said the tech. "You didn't do anything on your own. You traveled with somebody else and they showed you the ropes. You drank beer together, stayed in the tech hotels, and, if you wanted to succeed, you listened to the lore, no matter how long that lasted late at night."

These field mentors also provided valuable unofficial training. Junior techs learned how to economize on space while taking the necessary tools for jobs that were never completely predictable. One tech always carried four types of tape wrapped around a No. 2 lead pencil. Individual rolls of tape added weight, required space, and contained far more tape than was ever needed on most jobs. Duct tape, double-sided tape, electrical tape, and copper foil tape were standards. Duct tape held devices in place while the epoxy dried, double-sided tape was used to stick components to walls or ceilings in temporary configurations for testing, electrical tape insulated and repaired wiring, and the copper foil with sticky backing made good practice or emergency test antennas. There was always room on the pencil to wind several loops of solder wire and utility wire as well.

Epoxy was the tech's best friend. The strength of small amounts and its brief curing time were important to completing a permanent installation quickly. It could repair broken housings, fill cracks and holes, and hold equipment suspended at awkward angles in almost any location. Epoxy could also teach a lifelong lesson. A tech on his first job and anxious to please his mentor, was asked to quickly prepare a batch of epoxy. Without seeing a container to hold the mixture, the tech squirted the two components into his palm and stirred them together. His palm became warm, then very warm, then very hot. Yet, the tech's professional pride and urgency to get the installation finished overrode the burning pain. He said nothing until the team returned to base and the medics treated him for a first-degree burn that permanently scarred his palm.

While techs were officially a "service" that responded to DO operational requirements, they frequently became involved in defining the requirement and developing the operational proposal for Headquarters. When operational proposals required technical detail, the writing responsibility fell to the tech. All technical operations required formal approval from both OTS, weighing in on the technical feasibility, and the operational division, which evaluated intelligence value and counterintelligence risk. Therefore, the field proposal issued under the local chief's signature became all-important. The COS always had the final word on the proposal, but the techs established informal codes to communicate differing opinions to Headquarters without crossing the chief.

One effective method for informing Headquarters of what the tech really thought involved the length of the proposal. In drafting the cable, concise and clear language signaled the tech's confidence in the operation, while a lengthy, excessively detailed proposal, with pros and cons, conveyed the tech's doubts and gave Headquarters plenty of information to "pick at" and challenge. In this way, when an operation was turned down, the chief directed his displeasure at Headquarters, not the tech.

Interactions between the audio techs and station management mirrored family relationships more than commercial customer-supplier exchanges. This was because OTS had no competitors for the services that the stations needed, but, more significantly, the case officers and techs shared a commitment to the common mission. Even so, disagreements between case officer and tech were part of the everyday picture.

"A regular philosophical battle was the 'time on target' discussion for an operation," explained a senior audio tech. "There's a school that said, when you go into a target site, stay as long as you have to and be as quiet as you can. You might be inside for three hours or four days, but take as long as needed to do the operation 'silently.' Another school said, you get in there, minimize the noise, but do it as fast as possible and get out in a few hours. I had one of these battles with a very tough chief of station. I lost, but he heard me out. He said go in, don't make noise, and stay as long as you need. It took five solid days to get through thirty inches of concrete drilling by hand. We smelled pretty ripe. But there wasn't any question about the chain of command. When the ops people made the operational call and Headquarters concurred, we saluted."[23]

Another tech disagreed with his case officer over his choice of cars for his cover status. "I had a 1957 Chevy, it was a wild-looking thing and the case officer was perturbed and thought it was too flamboyant," the tech remembered. "He wanted me to get something black and spooky. I said no, my cover is commercial and this red Chevy with fins is expected of a successful businessman. Plus, you can see that the fins aren't as big as those on some other models." The tech's argument prevailed.

An operation to bug a Czech intelligence officer in Europe almost never got to proposal stage. Headquarters rejected the station plan to send a tech on an after-dark walk around the target property to survey windows and doors, assess security, and observe activity in a neighborhood not frequented by Americans. The tech did not, Headquarters pointed out, have sufficient cover and plausible reason to be in that part of the city at night.

The tech and case officer agonized for a couple of days then sent a brief cable, "If caught the tech will admit to being a thief. Then we will go to the local service and get him out of jail." Headquarters reversed itself and approved the operation.

Techs, with their predilection for improvisation to get a job done found the glamorous locales often offered the least amount of operational freedom. "In Europe it seemed you had levels and levels of Agency management all wanting to review and second-guess every piece of a plan, and there was always concern about diplomatic niceties," remembered one tech. "In Africa, we shot a little more from the hip—the case officers did, too, and I think we got a lot more done. We liked working down there and in South America. In Europe we sometimes felt smothered by a lot of tradition and scrutiny."

A tech stationed in Central America during the Cuban missile crisis in 1962 undertook an ambitious operation to penetrate a Soviet embassy. After learning that the Soviets used a particular shop for typewriter repair, a case officer recruited the shop's owner. The next Soviet typewriter that came in was "lent" to the tech who disassembled the machine and installed a transmitter into the platen. Once back in the embassy, it was hoped the device would pick up sensitive conversations near the machine while audio analysis of the striking keys could potentially reveal individual letters or words.

Surveillance confirmed that the Soviet diplomat picked up the typewriter and took it back into the embassy. The plan was working, and after enough time elapsed for the typewriter's return to service, the tech sent a "turn on" signal to the device and heard . . . nothing.

The following morning, a CIA asset observing the embassy's entrance watched a Soviet emerge carrying a typewriter high above his head. With a theatrically extravagant flourish, he heaved the offending machine into the trash. For some unknown reason the bug had been detected and the audio operation thwarted, but wariness of the Americans' capability was raised to a new level. After the incident, Soviet case officers were not allowed to type their reports; all had to be handwritten.

Another operation made possible by miniaturization of audio devices assisted a foreign government in catching a Soviet spy. John Kennedy was President when a Northern European security service called on TSD with an unresolved problem. A Soviet diplomat, a suspected KGB officer, had begun meeting regularly with a senior government minister. While the meetings had legitimacy based the official duties of the two men, the security service suspected the minister of also spying for the Soviets. Yet, their investigations had turned up no hard evidence of espionage.

The alleged KGB officer was cagey and professional. He met the minister openly at expensive, well-known restaurants, ostensibly to discuss legitimate diplomatic matters. Several times counterintelligence officers had notice of a planned meeting and plotted with the restaurant's manager to install monitoring equipment at the table where the Soviet would be seated, though the ploy never seemed to work. In some instances, the Soviet canceled reservations at the last moment in favor of an alternative location. In other cases, the KGB officer and the minister objected to the table offered and insisted on another.

Tom Grant had been in the country several times providing technical support to joint operations. On one of his trips, Grant's contacts confided, "We're certain the minister is dirty, but we can't get the goods on the guy. We've failed every single time to record the meal conversation and haven't been able to identify the covert communications being used." Grant said he would think about the problem.

Grant was living in another part of Europe with his family. One day his wife discovered a local shop that featured Scandinavian goods, including some attractive pepper mills. "Buy every one they have in the store," he instructed his wife. "Tell the owner we have lots of friends back home who'll be getting these as Christmas gifts."

None of the pepper mills ever made the trip to America. Grant went directly to the audio shop where he had several cylindrical transmitters that would fit snugly inside the pepper mills. With the help of concealment techs, he disassembled the pepper mills and created a cavity sufficient for the transmitter, microphone, and batteries. By modifying the grinding and dispensing mechanisms, a small pepper reservoir was retained and the mill still functioned, providing the bugs with an active concealment.

When shown the pepper mill bug, the local security service agreed that it might work and sought the assistance of managers at three restaurants where the KGB officer and his minister frequently met. Rather than trying to place transmitters on a specific table, the service asked the managers to remove all pepper mills from the tables on the day of the next meeting and bring one to the table after patrons were seated.

"It worked like a charm. The Soviet made his reservation, then switched restaurants but used a restaurant we had planned for," said Grant. "When he arrived he asked for a table different than the one offered. Then the minister joined him. At that point, the headwaiter put our pepper mill right between the two of them. The guys in the surveillance van outside could hear everything."

The suspected KGB officer and the minister ordered a meal and engaged in small talk for over an hour. At the end of the meal, they ordered coffee. The conversation had centered on official government-to-government topics, boring the counterintelligence "ears" in the van. However, as the business lunch finally wound down, the Russian paid the bill, then, as he took a final sip of coffee, leaned across the table, right over the pepper mill, and gave the minister precise directions for a dead drop.

"Well, that woke up the 'ears,'" said Grant. "The guy with the headphones in the truck went bananas. Their service was elated. Even our case officer thanked me. I said 'I'm happy to help but you understand I have to get back my pepper mills for the Agency and besides, one of them has to go to my wife as a souvenir—she dragged me into that shop.'"

Later the government minister was arrested as a spy and the KGB officer expelled from the country.

In every part of the world, similar imagination and creativity were required from techs and case officers for the Agency to attack the wary and protected Soviet Bloc targets. An Eastern European diplomat posted to South America in the mid-1970s seemed virtually untouchable. Surveillance determined that his embassy office was well secured and his home always occupied by family, caretakers, and service staff. However, surveillance did identify an interesting pattern in his wife's regular Tuesday shopping excursions. As the case officer and the tech discussed the situation, the tech mentioned that concealment specialists had begun embedding a new generation of audio transmitters in desk and table lamps. The lamps functioned normally, the tech explained, and the transmitters operated without batteries by drawing power from the lamp's electric current. Not long after, a plan emerged.

A CIA officer residing in the country began a side business selling lamps, loading his merchandise into a van every day to establish a credible pattern of activity that made him a recognized figure in the area. On the appointed Tuesday, the officer drove to the shopping district and waited for the diplomat's wife. Arriving on schedule, she parked and went into the store. A few minutes later, the officer pulled his van into the adjacent parking place, accidentally dinging the fender of her vehicle, then waited until she returned.

Just as the target's wife saw the damage to her car, the lamp salesman quickly approached her offering profuse apologies. He explained that he waited for fear that somebody might have seen the accident and taken down his license number. Then, building a tale of woe, he confessed that if he reported another accident to his insurance company they would surely cancel his policy. Pressing far more money than required to fix the damage into the wife's hand, he implored her to accept the settlement and not report the incident.

The woman, sensing a good deal, accepted the money and, in a final gesture of gratitude, the lamp salesman urged her to select any lamp from the stock in his van. The diplomat's wife chose what appeared to be the most expensive lamp and the officer carefully loaded the gift in her car, making sure to activate the audio "on" switch.

Later that night, the tech and the case officer met at the listening post and with amusement heard the diplomat's wife relate the story of how she had "screwed over this poor American." The audio stayed on the air for about two months until the couple decided they liked the lamp so much that it should grace their second residence in the mountains, far out of the transmitter's range.

In another operation, two techs disguised themselves as local painters to gain entry to a diplomatic facility being readied for new occupants. Their job was to find places to install bugs. As the techs studied areas where audio might be best installed, their attention fell on two large ornately carved wooden doors piled among other materials. The doors, the techs learned, would eventually hang between two of the mission's conference rooms. The techs reasoned that by placing mics on either side of the doors a single transmitter could pick up and relay conversations from both rooms. As an added operational bonus, the huge doors could accommodate dozens of batteries, extending the life of the operation far into the future.

The techs had only a few days to work and the presence of construction crews limited their access to the facility. The techs needed to get the doors to their shop to plant the devices. Noticing the entire construction site was a mess, with scraps of wood, fixtures, cement blocks, flooring, and other debris laying around, one of the techs said, "Let's clean this place up. They'll be happy that we got rid of the trash."

The techs went to work, piling whatever construction trash they could find on the doors and carried them out like stretchers. Several stretcher loads went out and other materials were carried in. Eventually the doors went into the techs' waiting truck and returned a couple of days later loaded with mics, batteries, and transmitters. "Everything just went fine," remembered one of the trash haulers. "We got the audio in and no one ever suspected a thing."

However, sound plans did not always go smoothly and some were just victims of bad luck, falling into the category of "technical success, operational failure." In one memorable instance, the Soviet Ambassador in a European capital ordered a custom table for his home. The CIA got wind of the order and recruited the furniture maker, who agreed that the techs could put an audio device in the piece.

The operation had every earmark of success as the techs, observing from a safe house, watched the table as it was carefully carried up the steps to the Ambassador's residence. The tech and the case officer exchanged smiles and shook hands, but before they could pour the victory toast, the deliverymen appeared again. This time they carried the table out of the residence, back down the steps and into the truck. Later that evening, as the two looked over the returned table with the furniture maker, they learned the rest of the story. The deliverymen reported that when the Soviet Ambassador saw the table he was surprised to see the top was made of Formica. He uttered two words, "not cultured," refused delivery, and ordered the table out of his house.

At about the same time, the Agency learned of another Soviet official, a newcomer to the city, who lived in an apartment-hotel complex. The local station tracked the Soviet's pattern of movements for a few weeks, then the chief decided to bug the official's residence using a recently developed audio transmitter concealed in a standard three-way electrical plug. When the modified plug was inserted into a wall outlet, it drew power from the household circuit.

A week passed and Headquarters had not responded to the operational proposal. The station chief became anxious. He had an OTS technical team in the city ready to act. A detailed ops plan had been laid out. Choreographed in concert with the target's comings and goings, the installation opportunity window fell on a specific date and time, but approvals were not in hand. "If we don't hear by eight o'clock tonight, we're going to go ahead and do this," the chief told the senior tech.

The plan left little margin for error. A team of two techs would enter the apartment-hotel complex. One would take temporary control of the elevator, holding it at the floor of the Soviet's apartment and act as outside countersurveillance. The other tech would make entry with a duplicate key and insert the modified three-way plug in an outlet under the bed. He would then exit, lock the apartment, join his colleague in the elevator, and depart the building. The entire operation would require no more than five minutes.

At eight o'clock, without a Headquarters response, the operation commenced. The techs were well into the job when, at 8:15, the communications officer brought a cable marked IMMEDIATE to the chief. The message, although apologetic about the delayed response, left no ambiguity: PROPOSED OPERATION IS NOT APPROVED.

As the techs departed the Soviet's apartment complex, they received a signal to contact the chief immediately. He related the Headquarters disapproval direction.

"Well, it's too late," replied the senior tech. "In fact, we're already listening to him. He came back home just behind us. He brushed his teeth and went to bed."

The next morning the device captured the personal routine of the Soviet as he prepared for the day, then went off the air at noon.

The station chief was in a quandary. He had acknowledged receipt of the Headquarters order, but had not told his superiors what had already occurred. He was applying the operational maxim "What happens in the field, stays in the field."

The chief ordered the techs to go back to the apartment and retrieve the plug. A second entry, this one without Headquarters' knowledge or approval, was planned and executed. The tech crawled under the bed but there was no plug in the outlet. He quickly surveyed other electrical outlets in the apartment but saw no plugs.

For the chief, as well as for the techs, the situation was about as bad as it could get. Not only had they conducted two unauthorized entry operations, but a piece of the CIA's newest clandestine audio equipment and its concealment were lost, very likely compromised to the Soviets.

The next day, the techs met at the listening post with the Russian-language transcriber sent to TDY from Headquarters to translate and process the audio take. He had set up the post in a small room in the same building as the target apartment. The post would now have to be quietly closed down and the transcriber sent home. As the tech related the saga of the lost transmitter and the pickle the chief found himself in, he noticed that one of the post's tape recorders was connected to building power by a familiar-looking three-way plug. "Where did you get that plug?" the tech asked.

"I asked the maid for one and she got it," the transcriber responded, a little puzzled. The maid who serviced the listening post apartment also cleaned the Soviet's room. A couple of days earlier she had pulled the bed away from the wall to vacuum and saw the three-way plug. Concluding it was not being used, she put it in her pocket. Later, when the post's keeper asked for a plug for his tape recorders, receiver, and other equipment, she just happened to have one handy.

The tech's next call went to the chief who suggested that everyone meet for a three-hour, three-martini lunch. As the first round of drink arrived, the chief offered a toast: "Remember, 'Ask and ye shall receive. Seek and you shall find.' Look it up. Matthew 7:7."

A legendary audio semi-success occurred when a South American dictator discovered a transmitter in a wood block attached to a piece of furniture in his office. Those in the listening post recorded the dictator's outrage. Then, in dramatic fashion, he drew his pistol and fired several rounds into the device while denouncing the CIA and America to his staff. Satisfied his marksmanship had killed the device, he tossed the bullet-riddled trophy carelessly on top of a file cabinet.

Back at the post, the recorders continued to roll while the device kept transmitting. The bullets had shattered the block and struck a battery, but only wounded, not killed the device. Within the heart of the woodblock enough power still flowed to the undamaged transmitter that every word uttered in the office was heard for several more weeks until the remaining batteries eventually died.

Retrieving bugs could be as hazardous as installing them and just as critical to an operation's success. Spy gear abandoned in place poses the risk of later detection by the local service or, depending on future occupants, by another foreign government. Any piece of equipment discovered, even years after the operation, could reveal technology and tradecraft to the opposition. It is one of espionage's ironies that the very equipment used to acquire intelligence, once discovered by an adversary, becomes a valuable source of intelligence. Surveillance gear in the hands of a hostile security service could yield vital information for creating countermeasures, point to an agent, or expose concealment methods.

Lady Luck did not smile on a retrieval operation in Western Europe in the late 1970s. A long-running successful audio operation concluded when the targets moved out of their residence and the techs received orders to return to the now empty apartment and retrieve the four bugs installed in the attic. It was a typical nighttime operation that demanded stealth and sure-footedness for the techs to make their away across the building's narrow rafters. The summer night was hot and the attic increasingly uncomfortable for the techs who, after finding the first three devices, were having difficulty locating the fourth. "My partner was swearing like crazy and I'm tiptoeing across these little rafters looking for the fourth when one the rafters breaks," said one of the techs, remembering the incident. "The next thing I know, I'm hanging by one arm, looking down at a very expensive terrazzo floor."

With the crash echoing in the middle of the night and the dust settling, the tech's radio came to life. The lookout had heard the noise and anxiously asked what was happening in the house. "I said, 'As you can probably surmise, we've got a little problem in here. That was me, going through the ceiling.'" Then the fourth audio bug fell to the floor.

Rather than retrace their steps over the now suspect rafters, the two techs dropped down through the hole. Fortunately, since the apartment had been vacated, they had time to clean up the mess and repair the ceiling. "Afterward, we just told everybody we had set the Guinness record for the world's biggest pinhole," the tech joked years later. "A six foot by six foot pinhole will give you the best audio you ever heard. God, that made a lot of noise, I can't believe we didn't get caught."

A dramatic breakthrough in audio hardware occurred in the 1970s. The SRT family of transmitters, which was steadily improving, made an impressive leap forward with the "Century Series" that carried three number designators. Nothing like it had previously existed in the OTS arsenal of covert audio equipment. Known particularly for the tiny size as well as performance, the Century Series devices were microphones and transmitters made with integrated circuits jammed into packages of less than a cubic inch of space. OTS called it *fractional cubic inch technology.*

"To achieve the fractional cubic inch volumetric size, the whole package contained integrated circuits, very special integrated circuits," said Kurt Beck, who worked on the project. "Our contractors used custom technology and processes. There wasn't anything like it on the commercial market. It was a team effort. It was the contractor and the ops guys both asking, How do you get this stuff to be this small? How do you engineer the device into the size of this package?"

The volume of the new audio packages was comparable to that of six U.S. quarters stacked one on top of another. The housing around the bug's hermetically sealed components had slots on the side that allowed techs to plug in an external power source, antenna, and microphone as needed. In less than twenty years, OTS had gone from an unreliable vacuum tube SRT-1 to a stable but power hungry SRT-3 to a family of transmitters whose reliability, size, and functionality could be adapted to virtually any covert audio requirement.

"This was not your incremental, tiny improvement—this was a quantum step," said Linn, who built power cells for the Century Series transmitters. In terms of reliability and sophistication, it was the difference between a 1970s citizens band radio and a twenty-first-century cell phone. The James Bond gadgetry imagined in Q's fictional laboratory had arrived in Langley. Fractional cubic inch technology brought not only the ability to build audio into smaller concealments, but also a major reduction in power required to transmit. It allowed for simultaneous transmission from two or more microphones positioned within three feet of each other. Essentially working like human ears, the listening post could "steer" audio, filtering out background noise in the room to focus in on conversations that were of particular interest.[24]

"It wasn't just the electronics, but the power consumption. The power consumption was always the problem that would knock you in the head," explained Kurt. "So if we could achieve an order-of-magnitude reduction in the power consumption, we could make a corresponding reduction in the size of the battery. That to me is the breakthrough. The low-power technology. Every 10 percent savings in power drain translated into a big lifetime improvement for the size of the battery. It didn't make much difference if you halved the size of the transmitter, from a half a cubic inch to a fractional cubic inch, if the battery pack had to stay at ten cubic inches."

When OTS first envisioned the fractional cubic inch package, integrated circuits were in their infancy. A little more than a decade earlier, in 1958, Jack Kilby, working at Texas Instruments, and Robert Noyce, at Fairchild Semiconductor, independently came up with the idea of the integrated circuit. Kilby beat Noyce to the patent by less than a year and later won the Nobel Prize, but Noyce, who later cofounded Intel, came up with several technical solutions, such as how to connect the tiny components on the chip, which made production practical.

"We talked to these designers and the engineers and we found out there were a lot of trade-offs you could make in all this stuff," Kurt explained.

"When we started to push on them to get the power down, ideas began to crop up. The problem was getting these analog circuits to be efficient with the power supplied. Instead of having two percent efficiency, could we get twenty-five percent efficiency?[25] It makes a difference with the amount of power you need to run them. We pushed smaller and smaller. Our approach was to find the right designers. Give them some leeway. Don't stand over their shoulder. We gave them money and said, 'Go try it. If you have failure, do it again. Just don't give up.'"

In the end, the circuits designed for the new Century Series were both small and energy efficient. Techs called them "flea powered." The units drew only microamps from the batteries and signals were transmitted at the lowest possible power setting for reception by specialized antennas at the listening posts. There was almost no end to where the new devices could be hidden. Combined with new battery configurations, the Century Series could be hidden in books, wooden coat hangers, and even within the circuitry of other electronic devices, such as televisions or portable radios. Wood blocks, a longtime favorite among techs, could be made smaller as well.

Armed with new technology, techs along with the staff at Langley became more emboldened. Operations that would have been at best risky or impossible in the 1960s were now launched regularly. "Show me a target and I can get to it," one tech was noted for saying. Within the tech culture, this was more statement of fact than bravado.

Perhaps no operation better illustrated the techs' derring-do than one that took place in the 1970s against an implacable U.S. adversary. After years of futile negotiations to resolve an international dispute with the other nation, the President ordered his closest advisor to initiate secret talks at the highest levels of the foreign government with the objective of ending the conflict. Special assistance from the audio techs was requested for a risky and dangerous operation to acquire information on the intentions and strategy of the foreign negotiators. The techs were chosen for their unusual skills and proven courage in combat. One was an experienced mountain climber. The other, a combat-hardened former Marine, trained for the mission by climbing over slate roofs in his hometown.

Working in the early-morning hours of a moonless night, a tech, dressed in black, carrying mountain-climbing gear, crawled through a window of a safe house onto the steep slate roof of an adjoining building. A few floors below, the other tech waited anxiously with the newly designed audio equipment. When they could move unobserved, the techs skirted several roofs of adjacent houses leading to the residence of the chief of the foreign delegation and crawled silently over its slate tiles. Their targets were three chimneys positioned along the length of the roof's ridge. As they moved from chimney to chimney, into each they dropped a small device, called a "pinger," to measure the length of the fireplace flues that would eventually conceal an audio device. Resembling an oversized pistol, when the pinger reached the upper edge of the fireplace flue, the tech pulled the trigger. A small burst of radio wave energy—like radar—shot down the chimney and bounced back, instantly calculating the distance between the top of the chimney to the desired fireplace location for the bug. With data in hand, the techs retraced their climb and began planning the equally dangerous and risky installation.

"We returned a few weeks later, with mics and transmitters the lab had developed," recalled one of the techs. "They were encased in an asphalt bulb, maybe two inches in diameter, so when there was fire in the fireplaces, they wouldn't burn up." Wires, with precise distances based on "pinger" data, let the techs lower the devices to the proper length in each chimney before securing them to the top.

The audio collection produced transcripts of private strategy discussions held by the target negotiators that were immediately translated and hand-carried to the President's representative to prepare him for the next formal meeting. The techs never saw the transcripts and never expected to. That is the profession. Build the gear, put it in, make sure it works, and get out of the way. Let others use and benefit from the take. Besides, there was no reason to hang around; other stations had audio ops to be done—now.

The newly availabile fractional cubic inch transmitters encouraged planning for aggressive audio operations inside the USSR. Once the CIA had demonstrated that its officers could free themselves from surveillance in Moscow, technical audio operations followed. "In the late 1970s we were doing things in Moscow that were intentionally below the Soviet radar," remembers a tech. "We were trying to find the balance between high-, medium-, and low-technical operational acts on the street—the capability of the agent, capability for the case officer to meet the agent, capability to dead drop certain-sized things. All of that technology we provided the agent. Now we asked, can we bring audio into the mix?"

In one of the first operations of its kind in Moscow, a plan was formed to bug one of the police shelters located throughout the city. The small shacks, also set up at strategic points in the foreign embassy district, provided shade and warmth for the police, militia, and KGB surveillance teams as well. The young officers who manned the shacks had duties other than traffic control and maintaining civil order. Their presence was a deterrent to Soviet citizens from contacting foreign officials, because any Russian wandering in the area could be stopped, asked for identification, and questioned about the reason for being there. Equally important for the KGB were the reports from officers in the shacks who relayed the comings and goings of foreign officials to the KGB from these excellent observation posts.

As the CIA increased its clandestine contacts inside the USSR through the 1970s, the chief decided to bug one of the shacks. The secret audio could potentially provide valuable intelligence from an officer "calling out" the movement of diplomats to the KGB surveillance teams and from capturing other security instructions he received.

The target shack, which was manned twenty-four hours a day, seven days a week, was classically basic. Constructed of wood and roughly twice the size of an old-fashioned phone booth, it contained a small table built into the wall, a telephone, and a heater that offered minimal comfort for two men against the cold of the Russian winter.

Over several months, CIA observers noted that the officer in the target shack was often away from his post, across the street talking with a friend. They were able to estimate the size of the small table in the booth and predict the times when the officer took a break for periodic gossip.

The operational requirements for the audio device were specific. It needed to be small enough to hide under the table in the shack, large enough to hold enough batteries for extended transmission life, and capable of being installed in less than a minute while the shack was vacant.

The techs created a woodblock audio concealment matching the faded color of the wooden table. They set spring-wound screws into one side of the wood block with enough torque from the spring to secure the block to the table's underside. When the wood block was placed firmly underneath the tabletop's bottom, the protruding screw heads were depressed, which released springs to turn the screws.

Because the bug required so many batteries for power, the wood block was too long to fit into the briefcase the chief normally carried. This required the techs to create a sling for cradling the device that could be worn under a topcoat. Every day, regardless of weather, the chief wore the topcoat and walked by the shelter, awaiting a time when he could enter unseen, open the coat, kneel down, pull the woodblock out of its sling, put it under the table, and activate the screws. It could all be done in less than thirty seconds.

Several weeks passed with the chief carrying the device each time he walked by the shack. The opportunity finally arose one day as the chief was walking his dog. At a distance, the chief noticed the officer leaving the shelter and crossing the street to talk with a friend, exactly as the operational plan had envisioned. The chief stopped briefly, adjusted the dog's collar, ducked into the empty booth, planted the device, and continued walking. Later that evening, at the nearby listening post, the techs heard and recorded clear audio that was immediately forwarded to Langley for analysis of KGB surveillance codes and techniques.

Possibly more important to CIA operations than the intelligence collected from the little shack's tapes was the act itself. The CIA had successfully implanted an audio device that clandestinely collected KGB tactical conversations. The small breach of the KGB's internal security wall demonstrated that sound tradecraft combined with applied technology could compromise KGB communications. The little audio device became an early indicator of the possibility of future high payoff from technical collection operations inside the USSR.[26]

OTS officers who cataloged and analyzed foreign spy gear began to sense a peculiar pattern when it came to Soviet electronic gadgets in the late 1970s. Soviet technology seemed stalled. OTS testing repeatedly showed that the components and performance fell short of the kind of progress seen in Western spy gear.

The analysis proved correct. In a 1994 memoir, *The First Directorate*, the KGB's former counterintelligence chief, Oleg Kalugin, recounted a scene in which Nikolei Yemokhonov, the deputy for scientific and technical research, was "called on the carpet" by then KGB chief and future Premier, Yuri Andropov. Andropov reprimanded Yemokhonov for lagging behind American espionage technological developments and asked about an OTS transmitter obtained by the KGB.

"Well," replied Yemokhonov, "we don't have devices this size."

"What size have we got?" asked Andropov.

"Ours weighs about a kilogram," said Yemokhonov.

The American device weighed only a few ounces: everyone in the room knew that the bulky two-pound Soviet transmitters and receivers were barely suitable for clandestine work.[27]

Victor Cherkashin, a senior KGB officer, offered a similar take regarding OTS audio technology in his memoir published in 2005. Cherkashin recalled that information about U.S. eavesdropping operations inside the USSR "simply astounded the KGB" when provided by American traitor Aldrich Ames. Cherkashin recounted that at the time of Ames's initial betrayal in 1985, the CIA "was juggling several highly complex, technically advanced, ingenious operations inside the USSR without the KGB's knowledge including eavesdropping devices disguised as tree branches near research installations."[28]

On the defensive side, by the mid-1970s, the KGB had developed a significant countermeasures tool (code name MAGIC) to detect embedded audio eavesdropping devices. The KGB's first experiment with it inside their embassy in an Asian country found more than two dozen listening devices, some more than twenty years old with corroded batteries, hidden throughout the large complex. Called the *Nonlinear Junction Detector* (NLJD), the device could detect a transistor or integrated circuit inside a clandestine listening device even when it was not turned on.

The *Nonlinear Junction Detector* worked by setting up a field of energy—radio waves—that read reflected energy. Any circuitry containing a diode present in the field was read as a disruption. Unlike metal detectors, which searched for metallic objects by means of electromagnetic induction, the NLJD was more selective in its search, noticing only the junctions of diodes found in transistors and integrated circuits.

"It was the beginning of the end for classical embedded audio devices," said Sasha, a former member of the KGB counterintelligence, who claimed the KGB removed "hundreds and hundreds of listening devices from each continent. Europe, Canada, Great Britain, United States." Sasha asserted that the KGB "presented this nonlinear detector to Cuba, then Warsaw Pact countries, followed by Third World, so-called friends, such as Iraq, North Korea, and Vietnam."[29]

Once the NLJD technology was known, the United States countered with techniques to neutralize its effectiveness. The KGB found that bugs planted near naturally occurring junctions such as electrical sockets, rusty nails, or sections of walls containing pieces of dissimilar metals touching each other were "a nightmare for detection.[30] Certain components were covered with a variety of hybrid coatings to mask the circuitry within. Improved shielding techniques for audio packages rendered the devices invisible to KGB countermeasures, including the NLJD.

"We worried a little bit, and put additional filters in the circuits to keep the radio frequencies out," said Kurt. "We were always trying to shield them anyway, so the extra filtering became an incremental improvement. I don't think we lost many of our devices to nonlinear detection."[31]

Once the SRT audio systems were widely deployed, the CIA's capacity to process the "take" from the hundreds of installations worldwide became a continuing problem. Every audio op required listening post keepers and translators. Although much of the audio contained nothing of intelligence value, someone had to listen to the tape to make that determination. The promise of good intelligence from an audio installation often exceeded the results.

"I recall, over a ten-year period, fifty percent of the audio operations were terminated every year, and probably half of those shouldn't have gone forward in the first place," said one senior manager. "For a few years people were getting a feather in their cap because they were involved in an audio op. A case officer felt he really could not be promoted until he had run an audio op. That was part of the case officer checklist."

A study done by the same manager concluded that 5 percent of all audio ops produced 95 percent of all valuable information. But even that number was tricky. Some compared audio operations to diamond mining, the invaluable gems are found only after sifting tons of dirt.

For most of the last twenty-five years of the Cold War, audio dominated OTS operations. However, the emergence of computer-based information systems and cellular technologies in the 1980s and early 1990s created new target opportunities and eventually lessened the dependence on traditional audio for obtaining private conversations or communications. The target's technology, as well as the person, became an object of recruitment.

CHAPTER 15

Genius Is Where You Find It

The world has arrived at an age of cheap complex devices
of great reliability; and something is bound to come of it.

—Dr. Vannevar Bush, "As We May Think," *The Atlantic Monthly*, July 1945

E spionage novels and movies devote few pages or minutes of screen time
to the scientists and engineers who create spy gear. The notable exceptions are James Bond movies and the British gadget-master Q. Acting as the
proper British foil to Bond's more colorful persona, Q invariably anticipated
Bond's technical needs for each mission even as he fussed and fretted over
each piece of equipment that left his lab.

Contrary to Q's uncanny ability to provide Bond with just the right
gadgetry no matter how vague the mission, specific operational requirements preceded the design and deployment of OTS devices. In fact, operational requirements drove much of the innovation in the same way
competition in consumer products pushes companies to the next level of
technological sophistication with their products. Noteworthy is the fact that
innovation in clandestine gear is motivated not by market share or quarterly
profits, but by the need to ensure the survival of agents and officers. This remains as true today as it was for Lovell and the OSS during World War II.

Through the decades, the Agency had remarkable success in consistently
acquiring the required technologies and expertise. By necessity and tradition, OTS sought its devices from a surprisingly wide range of suppliers.
Over the years spy gadgetry has been produced by high-profile business

leaders and academics as well as obscure inventors. CEOs, attracted by a technical challenge and eager to serve their country, set aside manpower and facilities to establish covert technical units. Nobel Prize–winning scientists and internationally recognized engineers have volunteered to work on OTS projects in their off hours.

However, big ideas were often the products of the smallest companies with highly specialized expertise. A firm with only a handful of employees was just as likely to turn out an amazing piece of hardware as a multinational with nearly unlimited resources. "OTS had long-standing relationships with real garage-shop companies. Sometimes they were no more than ten people. That was the whole company, soup to nuts, including the accountants," said Gene Nehring, an OTS manager. "We always kidded about some of our suppliers. Some Agency managers would say, 'You guys deal with every garage shop around.' And yes, we do, and each one did one little thing better than anyone else, anywhere."

Perhaps nowhere was this truer than in the case of the T-100 subminiature camera, arguably the most productive piece of Cold War spy gear. Developed and manufactured by a tiny company housed in a nondescript industrial park on the Eastern seaboard, the film-based T-100 was the ultimate spy camera. Unlike the Minox, which was originally designed and marketed as a commercial product, the T-100's sophisticated optical and mechanical design was so highly specialized and technologically unique there were virtually no uses for the device outside of espionage. It operated like a point-and-shoot camera, but had no viewfinder and required a painstakingly precise process to hand-load the customized film on its miniature cassette. From design to operation, the T-100 had one function: enabling an agent to take a covert, clear picture of the writing, printing, or diagrams on a piece of paper directly in front of him.

"Think about it. That camera, as marvelous as it proved itself, was utterly useless as a commercial product," said Gene. "It could take a wonderful picture of a single sheet of paper at eleven inches. But it has a depth of field of about one inch, and no other applications."

The T-100's assembly was closer to watchmaking than any commercial manufacturing process. The owner of the company fabricated each camera himself under a large magnifying glass and halo light using a device he built specifically for the task. "He had all kinds of things that held the different components in place," explained Gene, who once witnessed the assembly process. "It was a real Rube Goldberg apparatus, but it allowed him to take these little tiny things and put them together. Imagine tying a trout fly and performing ten steps at the same time in three-dimensional space."

Because the camera was such a singular device, it offered a high level of operational security. Counterintelligence organizations, after all, cannot guard against a device they do not know exists. However, that same singularity and craftsmanship eventually became a cause for concern. By the late 1970s, with the T-100 proving itself such a valued piece of Cold War spy gear, operational managers grew concerned about future supplies of the camera. With the small company the sole source for the device and a single individual the only person able to assemble the tiny components, supplies could be jeopardized by something as ordinary as the owner of the company developing a twitch or injuring his hand.

The owner, recognizing the vulnerability of a national asset, provided the camera's specifications and engineering drawings to the Agency. The complex lens assembly, made up of more than half a dozen elements layered one on top of another, seemed a logical component for second sourcing. One of the premier optical houses in the country seemed the reasonable place to start.

"We said, 'Here's a design, what do you think? Can you make this?'" recalled Gene. "Well, they did their computer analysis of the lens and came back to us and said, 'Nope, it won't work. The light won't focus properly. You'll never get a picture out of this thing.' Naturally, we didn't tell them we already had fifty in stock and they were all working just fine, thank you very much."

The potential for another source arose after the Agency allowed a friendly intelligence service to borrow some of the prized cameras. Not long afterward, the service requested permission to build its own version. The CIA agreed to share the specs with the understanding that this overseas production run would become the needed second source. After a few months, the friendly intelligence service returned with news that they too had failed to duplicate the camera.

The inventors themselves could be as unique as the devices they created. One of the stranger meetings Gene remembered was in tracking down an inventor of a new type of long-lasting battery at his upstate New York home. "On a February day, I go flying up there," recalled Gene. "I'm picked up at the airport and as we're driving out to the house, my colleague says, 'This isn't your ordinary contractor. He's a little eccentric.'"

Not knowing what to expect, Gene arrived at the suburban home of the inventor to find him in the backyard digging a trench with a backhoe. After he completed his digging, the inventor jumped on a small Bobcat bulldozer and filled the trench in before beginning work on another. The trench, as it turned out, had no specific purpose. Digging holes and refilling them was his hobby.

After initial introductions and pleasantries, the inventor invited the two officers to his workshop for a tour. "We went into the cellar, and he had welders, drill presses, all of these tools, and everything said CRAFTSMAN on it. It was like walking into a Sears' tool department. He had one of everything," Gene said. "And that's where he'd assemble these little batteries by hand, in his basement with all of these Craftsman tools. But they were one-of-a-kind and met our needs."

Eccentricity was not limited to outside contractors. One of OTS's legendary engineers, Brian Holmes, is remembered as much for his personal style as his remarkable brilliance and creativity. Although Holmes's engineering work was unsurpassed, what drove managers and colleagues to distraction was Brian himself. Every week seemed to bring Holmes another security violation for leaving classified papers in the open or misplacing materials. Invariably these lapses triggered a broader review of security practices that disrupted the entire division.

"Brian was a nightmare, a wreck. He barely got his clothes on right, except the sonofabitch got medal after medal for coming up with things that nobody else could make," said Greg Ford, an OTS senior manager.

To make matters worse, Holmes's immediate supervisor, a by-the-book administrator, was nearly the exact opposite of the brilliant but disorderly engineer. That the bureaucratic fates had placed this odd couple in such close proximity was either funny, tragic, or both. Finally, at the end of his rope, the supervisor appealed to Ford in the plainest possible terms. He just could not take it anymore.

"I had to tell him, 'You're the best division chief I've got, but if I lose you tomorrow morning, I can replace you by the afternoon. If I lose Holmes, I can't replace him,'" said Ford. "'So we have to find a way to deal with this.'"

After giving the problem more thought, Ford hit on a solution. In another part of the OTS complex were several ultrasecure room-sized vaults built to hold equipment too large for the Agency's standard three-drawer office safes. These windowless rooms featured secure steel doors along with good lighting and ventilation. Ford moved Holmes's desk and equipment into one of the cavelike rooms, making it his new office and laboratory.

"He loved it, absolutely loved it," said Ford. "He had all his shit laying around on tables and everywhere else. He knew where everything was. It suited him. And at the end of the day, he didn't need to put anything in a safe, all he had to do was secure the vault door before he left. Naturally, I'd always have someone else check on it."

For Ford, the untidy Holmes fell into that rare and precious category of engineers he labeled "inventive bastards." A valued asset, they were aggressively recruited and then given enough freedom to work their magic. "Let me put it that way, if I have a hundred thousand Chinese, a hundred thousand Russian, and a hundred thousand American engineers, there's going to be about a hundred and fifty of these creative types in each group," said Ford. "One of my jobs was to find and convince them to work for OTS, and then protect them."

Finding, retaining, and protecting these engineers and scientists became an obsession for Ford and the Agency. After World War II and throughout the Cold War, the value of technology to intelligence operations steadily increased as devices grew smaller, more portable and concealable. "Science as a vital arm of intelligence is here to stay. We are in a critical and competitive race with the scientific development of the Soviet Bloc, particularly that of the Soviet Union, and we must see to it that we remain in a position of leadership," wrote Allen Dulles in the early 1960s. "Some day [sic] this may be as vital to us as radar was to Britain in 1940."[1]

Others shared Dulles's assessment of technology's importance to espionage and warfare, including MIT professor Dr. Vannevar Bush. During World War II, Bush served as chairman of the National Defense Research Committee (NDRC), the organization into which Lovell was recruited and from which OTS would eventually emerge.[2]

Even as the war was winding down, Bush was thinking ahead. Looking toward the future, he authored a seminal essay on science and engineering, "As We May Think," which appeared in the July 1945 issue of *The Atlantic Monthly*. His insights would prove prophetically accurate. "The world has arrived at an age of cheap complex devices of great reliability; and something is bound to come of it," Bush wrote.

> *Consider a future device for individual use, which is a sort of mechanized private file and library. It needs a name, and, to coin one at random, "memex" will do. A memex is a device in which an individual stores all his books, records, and communications, and which is mecha-*

*nized so that it may be consulted with exceeding speed and flexibility. It
is an enlarged intimate supplement to his memory.*

In the first decade of the twenty-first century, Bush's memex could be
called the personal computer, though elements of his predictions would
eventually turn up in cell phones, PDAs, notebook computers, and even the
Internet, all of which serve as supplements to our memories.

In a second paper, this one written for President Roosevelt that same
year, titled "Science the Endless Frontier: A Report to the President," Bush
argued that science is a vital resource of the United States, in peacetime and
war:

> *It has been basic United States policy that Government should foster
> the opening of new frontiers. It opened the seas to clipper ships and
> furnished land for pioneers. . . . Although these frontiers have more or
> less disappeared, the frontier of science remains. It is in keeping with
> the American tradition—one which has made the United States
> great—that new frontiers shall be made accessible for development by
> all American citizens.*

The quandary the Agency faced from the 1950s onward was in identifying applicable new technologies and recruiting the right engineers and scientists. This was no easy task. Men and women with technical skills were
becoming highly valued, emerging as the superstars of the post–World War
II generation. At Bell Labs, they designed transistors and then integrated
circuits. Xerox revolutionized computing by transforming an obscure
government-funded project into the first computer with a mouse and graph-
ical interface. Plastics and synthetic materials, jet engines, and televisions
were making industrial engineers wealthy and changing the way Americans
lived.

Even when the modest starting salary was not an obstacle for prospective hires, OTS faced other special problems in recruiting. Because of the
classified nature of the work, CIA employees were prohibited from publishing papers or obtaining patents. By working for the CIA, they could be assured of earning less than in the private sector and receiving no professional
prestige that would otherwise accompany publication or publicity of a technical breakthrough. The necessities of security demanded that their hard
work, though frequently invaluable, would remain secret. Finally, they

might never know how, where, or if their labors had paid off in field operations.

Nevertheless, the Agency found ways to tap America's engineering and scientific talent. The OSS model of collaborating with private companies that served America's intelligence effort well during World War II continued to provide TSS and OTS a window into leading-edge research. Eventually, this partnership model provided a decisive advantage over the centralized Soviet system, a fact not lost on some Soviet leaders. "We lack R-and-D and a manufacturing base," said Lavrenti Beria, head of the NKVD. "Everything relies on a single supplier, Elektrosyla. The Americans have hundreds of companies with large manufacturing facilities."[3]

The Soviet Union, by contrast, handled its need for engineering talent decidedly differently. Its engineers, scientists, and mathematicians who showed particular brilliance or promise were singled out and channeled into advanced studies. If they measured up, they were put to work in intelligence, the most talented sometimes held as virtual captives in KGB-controlled *sharashka* (prison labs).[4] From such facilities *The Thing,* along with some of the Soviet Union's most advanced weaponry, aircraft, and rocket technology, including early nuclear devices, emerged.[5] Russian aircraft designer A. N. Tupolev was held in one such prison in Bolshevo outside Moscow,[6] as was the physicist P. L. Kapitza. Aleksandr Solzhenitsyn, in his book *The First Circle,*[7] immortalized his own experiences in the *sheraska* known as the 01 Institute, which coincidentally also held Leon Theremin, inventor of *The Thing.*[8]

The Soviet scientists were left with little choice as to where to apply their talents. "Leave them in peace," Stalin was reputed to have said of the imprisoned scientists. "We can always shoot them later."

However, if Beria imagined all of America's industrial technology focused on defense or intelligence, he was mistaken. Post–World War II industry was largely geared for profit in consumer or industrial markets, and the trick, OTS discovered, was in adapting the innovative commercial and military technologies to clandestine use.

TSD's inventiveness encompassed aircraft as well as listening devices. The North Korean seizure of the USS *Pueblo* in January of 1968 became the backdrop to one of its most ambitious aviation projects. One of the frustrations facing both the Johnson and Nixon administrations was the seemingly limited options available to avenge such incidents short of declaring war.

Responding to the White House, in the spring of 1970, TSD was tasked to develop a means to infiltrate intelligence or paramilitary teams into hostile and otherwise inaccessible areas. "The project got started because of comments attributed to Nixon's national security advisor, Henry Kissinger," recalled one of the principal officers. "We understood he wanted a covert capability to access strategic North Korean targets if we ever decided to attack and destroy them."

Because the likely military or economic targets would be accessible only by air, a "silent" aircraft operating at night could potentially reach the targets covertly, thereby hiding the U.S. government's hand in the operation. From an intelligence perspective, such an aircraft could have the additional capabilities for deploying covert sensors for intelligence collection and conducting hostage rescue missions.

The initial requirement called for an aircraft that could fly 1,000 miles without refueling and carry a two-man team along with a modest 150-pound payload. Primarily because of its range, the Hughes OH-6 helicopter was identified as the platform for the project. OTS acquired an "off the shelf" OH-6 and went to work reducing its operating noise.

"First we slowed the tip speed down on the main rotor," said Jack Knight, the TSD officer who headed the project. "That required we change the rotor, so we made a five-blade version rather than four to get the same lift pattern. We could move the same amount of air at a lower RPM and maintain the same lift. We also changed out the tail rotor, going from two blades to four."

The engine noise presented a different set of problems. An initial attempt with a muffler failed when a contractor designed one weighing nearly 400 pounds, far too heavy for the OH-6. However, Knight had heard that a commercial aircraft manufacturer was running a program to quiet its jet aircraft and paid the company a visit. "There was a guy working on a 'quieting program' for a long haul passenger plane and we went to talk to him," said Knight. "We asked to borrow him for a few months, but the company had other plans for him. So that was a disappointment. But on the way out the door, he handed me a business card with his home number written on the back. I called that night and he said he'd work on the project during his off hours. He eventually designed an engine quieting system that weighed about thirty pounds, a perfect fit for us."

What the engineer did, explained Knight, was identify the sound frequencies coming from the engine making the most noise and attacked them by creating a series of sophisticated acoustic chambers. Just as high-end audio speakers are designed internally to acoustically enhance certain frequencies, the engineer's design performed the opposite function, trapping the sound waves in a carefully constructed series of baffles.

With the rotors and engine silenced—or at least quieted—Knight and his team next targeted the noises coming from the chopper's other moving parts. First, the transmission was quieted, and then they turned their attention to the converters (small generators) that provided auxiliary power, which were found to be extremely noisy. Their studies eventually led to converting the OH-6 over to solid-state electronics that required less power and smaller, quieter generators. "Then all of a sudden we found a real noisy valve in the fuel control system," remembered Knight. "You never hear it in a normal helicopter, but it was a screamer. I took it back to the manufacturer and told them to quiet it down. They looked at me like I was a nut case. But they did it using silicone insets to cushion the moving parts."

From start to finish, the project was completed and delivered in less than two months. The result was an OH-6, operating in "quiet mode," that could not be heard on the ground as it passed over at 500 feet. Flying at the optimum "quiet" speed of 85 knots, the helicopter was less fuel-efficient, while higher speeds increased the noise but improved fuel efficiency.

Richard Helms, the DCI at the time, followed the progress of the silenced helicopter with great interest. He called Knight in for personal briefings on the project as it moved through various stages, the conversations frequently focusing on the difference between "quiet" and "silent." One day, Lawrence Houston, Helm's senior lawyer, called Knight. "I want to go to California and hear this thing," Houston stated.

Knight obliged, taking Houston out to the Culver City Airport late at night to stand in the center of a darkened runway. Knight ordered a fly-by of a standard OH-6, which was audible from one end of the runway to the other. Houston and Knight stood on the runway tarmac as the sound faded and then vanished altogether. After a few minutes, Houston asked when the quiet helicopter would be coming. "It just did," Knight replied, and then radioed the pilot to make another pass and illuminate the helicopter when overhead.

"That sonofabitch *is* quiet!" Houston exclaimed. His report to Helms settled the question of "quiet" versus "silent."

The second major requirement for the quiet helicopter was a capability to "see into the night." Knight and his TSD team needed a Forward-Looking

Infrared (FLIR) system that would allow night flights at low altitude. Knight discovered that the smallest system available weighed several hundred pounds and produced poorly defined images that often resembled blobs.

In principle, infrared "sees" not the gradations of light like a video or still camera does, but differences in temperature. It picks up heat emanating from an object, much like a camera records light reflected from an object. At the time, FLIR was a new technology, somewhat comparable to early "tintype" photography during the Civil War.

When Knight asked a military components company to help with the FLIR problem, two recently graduated electrical engineers, both in their mid-twenties, were identified. Knight, listening to their ideas, did not leave the meeting until long after dark. With more enthusiasm than funding, the engineers saw in Knight and the Agency the opportunity to put their theories into practice.

The next morning, when Knight returned to the company, the FLIR manager was decidedly unfriendly. The manager sensed that the young engineers had committed the company to something that could not be delivered and he did not want the corporate reputation riding on an "impossible" project. Knight countered by writing and signing a letter on the spot absolving the company and the manager of any responsibility for the project's outcome. "I just wanted those kids, because I was convinced they could do something no one had done before," Knight recalled. "Those kids were going to run my program without interference from experienced nay-sayers."

Within sixty days, the two engineers had an operating prototype of their system. What they had done was to rethink the way IR receivers processed signals. Typical IR systems processed long, mechanically scanned linear arrays that had wide variations in line-to-line sensitivity. The young engineers reconfigured existing technology to create a single array of fifteen elements stacked together, which constituted a single-point detector with the capability to scan in both the horizontal and vertical planes. The additional elements allowed the system to take in more information, which was then processed into a more detailed image. The result was sensitivity so high that the FLIR scanned at TV rates.

"We told the engineers it couldn't weigh over eighty-five pounds and they gave us one that weighed fifteen. We were getting recognizable images—not TV quality—but dang near," Knight said. "People couldn't believe the world they were seeing was through the eyes of a thermometer.

It was so good you could pick faces out just from the sensitivity that registered vein systems close to the surface of the skin. It was so startling, I think it killed every other FLIR program going on in the country at that point."

Technology and the human agent were becoming interdependent as each gave the other capabilities and security that had previously not existed. Tiny, reliable, long-life audio devices could supplement an agent's information by remaining in a room after the agent departed. Small, concealable, low-light cameras enabled agents to clandestinely copy documents in supposedly secured areas. Low-power transmitters provided agents with a communications link to a handler he might never meet.

As the complexity of the technology increased, so did the intricacies of hunting it out. Many of the companies that were once little more than Gene's garage-shop contractors in the fifties and sixties had grown significantly by the 1970s, a few to multinational status. With their growth, some were no longer able or willing to accommodate the small production runs typical of clandestine equipment. The same problem Lovell faced in recruiting businesses into the specialized and marginally profitable field of intelligence thirty years earlier was now confronted by a new generation of Agency managers. However, these managers ran into an obstacle not encountered by Lovell.

The Cold War lacked the immediate urgency of World War II. Convincing a CEO to commit resources and manpower to clandestine endeavors, with the inherent risk of exposure and adverse publicity, became a tough sale. Although research funded by the Agency sometimes gave companies a temporary lead in the marketplace, such as it had done with battery power-saver technology, this ancillary benefit was never assured. In most instances, work for CIA had limited practical application beyond espionage.

In the early 1970s, OTS, in search of a digital device that pushed the limits of memory capacity, assessed the technology used for satellite-based reconnaissance that was moving toward digital imaging. The technology appeared to have clandestine applications. After receiving word that James Early was doing interesting work in the field at Fairchild Semiconductor, OTS sent Ford to investigate. Early, a member of the team that worked for Nobel Prize winner William Shockley on the transistor at Bell Labs, is frequently credited with pioneering efforts in moving the technology into commercial and industrial applications.[9]

By the time Ford stepped into Early's lab at Fairchild, the invention of the transistor was two decades in the past and Early, a senior researcher, revered within the engineering and scientific communities. However, Ford found a scientist unwilling to rest on his laurels and showing unrestrained enthusiasm for pushing the limits of digital technology. "I watched him work two blackboards on one side of the room for forty-five damned minutes with more formulas than it took to build an A-bomb," said Ford. "Finally, I said, 'What's it going to take to build this thing?'"

The problem Ford faced was that OTS did not have a budget for theoretical research. Whatever funds were spent had to be committed to a specific device, so Ford instructed Early to build a camera. Early put the price at $25,000 with a completion date of three months. Ford gave him $50,000 and made a mental note the completion date would likely be closer to nine years, rather than ninety days.

Three months later Early was in Ford's office setting up a contraption consisting of not much more than a small box with a 16mm lens mounted on one side and some wires trailing out from another to a picture tube and power supply. Ford watched as Early switched the device on, and saw one of the first digital images captured by a Charge-Coupled Device (CCD). "The thing worked perfectly. I called a friend of mine over at the Advanced Research Projects Agency (ARPA) and said, 'I don't know or care who you have in your office, clear them out, now!'" Ford recalled.[10]

Packing the device up and with his guest in tow, Ford set up a demonstration in an ARPA office a few miles away. "The ARPA engineer recognized exactly what the impact of this was," Ford said. "All he asked was, 'How much money can they sensibly absorb?'"

CCD technology would, in fact, revolutionize traditional tradecraft, make real-time images from space platforms possible, and transform the camera business in the consumer market. "An executive once asked me what was in the presentation that made me believe Early could pull this off," Ford remembered. "I told him, 'Nothing.' The guy lost me after the first foot of formulas on the board. But I'm looking at this sixty-something-year-old engineer, one of the coinventers of the transistor, and he's jumping around like a twenty-five-year-old kid. I'll give money to people like that."

Another problem facing the Agency was the nature of technological advancement itself. The speed at which technology progressed in the three decades after World War II placed OTS engineers in constant competition with consumer and industrial markets. "It's a race to get my device into the field before every other intelligence service has a countermeasure. Technology is my edge, so I have to get it into my clandestine product quickly," said one senior OTS scientist. "For instance, until the mid-1980s there were no cell phones and you couldn't buy a walkie-talkie small enough to use covertly. So we had to build special stuff. Now anyone can buy most of the devices that we had to invent during the Cold War."

The race against the consumer and industrial markets was one that OTS did not always win. Sometimes, developments in the private sector either overtook Agency engineers or shortened the operational life of a device to a surprising degree. In one notable case during the 1970s, OTS needed a better, more compact recording medium and contracted to have the standard-sized cassette shrunk down to allow for a smaller recorder. The contractor successfully delivered the device, but the effort was largely wasted when the first commercial and equally capable microcassette recorders appeared on the market a few months later.

However, the overall trend of technological proliferation in the consumer marketplace also brought operational benefits. With the spread of small, affordable portable devices, technology was becoming ubiquitous and transparent for people in every part of the world. For example, as Walkman headphones, along with low-priced pocket calculators, pagers, and digital watches became common in the 1980s, these everyday products were adapted or disguised for clandestine use. Audio receivers once hidden beneath lifelike molds of the user's ear could now be disguised as headsets for music or a cell phone.

Sometimes even standard commercial devices could be pressed into clandestine duty without modification. In an unwitting doctor's office in a European city during the 1980s, an answering machine, a new technology at the time, picked up calls in the middle of the night. Once or twice a month, a case officer would call the office, leave a brief message, and hang up. A short time later, an agent would call the office and tap in the code to access the messages on the answering machine. After retrieving instructions for a dead drop, he would then erase the secret message, leaving no trail back to his handler or even telephone records.

One OTS scientist recalled a conversation with a case officer returning from Europe in the mid-1980s. The case officer offered details of something called a "cellular phone." "I want to use this. You figure out how I can make covert calls," he told the scientist.

As a result of the conversation, the scientist linked up with an operations officer with a technical bent and a senior engineer to figure out how to make early cell phone technology an operational tool. The inspiration the three-man team needed came from the criminal world. At the time, drug dealers in major cities were monitoring cell phone calls and hijacking the phones to ply their illegal trade. The team developed similar technology to snatch random caller codes out of the air in selected foreign countries, creating a covert phone system dubbed the "portable pay phone." Short-duration calls could be placed using a random number to sever any connection between the case officer and the agent. The borrowed number added only a few barely noticeable pennies to the phone bill of its unwitting owner.

"What's happened is that as technology kept getting more automated, it became smarter and adaptive," said the scientist. "The more we could do, the more we're asked to do. New technologies let you do so many things you couldn't do before but they also made our older equipment obsolete more quickly."

Circuit boards and computer chips offered OTS miniaturization and flexibility for building equipment. Digital memory, a common component of modern electronic devices, became a blank slate upon which nearly anything could be written. Increasingly, even under close examination, spy gear was becoming indistinguishable from everyday objects.

Digital tradecraft also advanced the concept of the cloaking function in electronic form—as generations of spies had done with concealments and dead drops—by creating spyware buried deep within lines of software code. The process known as convergence in the consumer marketplace, in which a cell phone stores music along with a datebook or text messaging function, would become the twenty-first-century technical challenge for OTS.

SECTION V

PRISON, BULLET, PASSPORT, BOMB

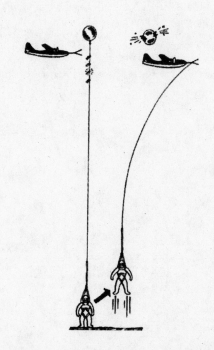

CHAPTER 16

Conspicuous Fortitude,
Exemplary Courage in a Cuban Jail

You are not expected to take anything with you in the field
that would reveal your identity or in any way show that you
are an agent of the government . . .

—U.S. Army order to an intelligence officer posted to Latin America in 1905[1]

On September 8, 1960, three American businessmen stepped off a plane in Havana. The passports and tourist visas they presented to Cuban officials identified them as Daniel Carswell, age forty-two, an electrical engineer from Eastchester, New York; Eustace Van Brunt, thirty-four, a mechanical engineer from Baltimore, Maryland; and Edmund Taransky, a thirty-year-old electrical engineer from New York City.[2]

Along with their official travel documents, the three carried credit cards, driver's licenses, and other pieces of identification confirming their identities. However, their names and all the material that supported their identities were fictions. TSD artists skilled in document fabrication and re-production had created all the mundane contents of their wallets—"pocket litter," in CIA parlance.[3]

Eustace Van Brunt was TSD engineer Thornton "Andy" Anderson, while Edmund Taransky was really Walter "Wally" Szuminski, an audio tech. The third tourist, traveling as Daniel Carswell, was Dave Christ (pro-nounced "Crist"). The most senior of the three, Christ had recently become head of TSD's audio operations. In that capacity, Christ carried in his head worldwide knowledge of the CIA's bugging capabilities, equipment, targets, and current installations.[4]

Cloaked in their false identities and a cover story, the trio entered Cuba on a weeklong mission to install clandestine listening devices. The target for the operation was not Cuban, but rather, the future embassy of a critical hard-target country. This rare opportunity was the result of Cuba's decision to embrace diplomatically America's adversaries.

The CIA learned where the embassy would be located and reached an agreement with the owner to allow Agency techs access to plant the bugs. Not only was the chance to install listening devices in a major target a golden opportunity, the plan was virtually risk-free. The owner could authorize access to his building to anyone, even three American tourists, at any time. No one would ask questions.

With open access to the building, the Christ-led team planned to conduct a thorough preinstallation survey and then work without fear of interruption. The team's single concern lay with the Cuban government's growing antagonism and suspicion toward the United States. Since American tourism to the once popular Caribbean island was becoming increasingly rare, the arrival of three Yankee engineers in search of tropical fun could very well attract the attention of Castro's immigration or counterintelligence officials.

While Cuba was still presenting a welcoming façade in the summer of 1960, unsettling changes were occurring under Castro's new government. During the eighteen months following the revolution, Cuba's reputation as a Caribbean playground was in rapid decline. Refugees were streaming into Florida while Havana increasingly became a city of civil unrest. Protests, which Castro aggressively countered with mass arrests, were becoming more common. Businessmen, who had supported the deposed dictator, General Fulgencio Batista, were branded as potential counter-revolutionaries and growing increasingly fearful of their new government.

Despite these troubling developments, Castro's true political orientation was still uncertain. In power less than two years, after seizing control on New Year's Day in 1959, he continued to deny communist leanings, although his 1960 decision to embrace the Communist Chinese at the expense of the Taiwan government should have been taken as a good indication of where he was headed.[5]

America still maintained diplomatic ties with Castro's government, but relations were strained and the situation between the two countries was clearly deteriorating. A conflict surrounding sugar imports to the United States along with American condemnation of tightening government controls on Cuba's press, trade unions, and universities angered Castro. Cuba had also resumed diplomatic relations with the Soviet Union, welcoming the new Russian Ambassador Sergei Kudryatsev. Fifteen years earlier, Kudryatsev had been declared persona non grata by the Canadian government after being linked to an extensive network of Soviet spies in that country.[6] The Soviet Union now had a toehold in the Western Hemisphere, a "friend" in Castro, and a presence just ninety miles from the United States.

From the U.S. perspective, the signals Castro sent out to the world in speeches and interviews were mixed. Human rights abuses were reported by Cuban refugees in Florida and the Cuban leader's pledges not to nationalize businesses seemed hollow. In one of his odder pronouncements, Castro banned Santa Claus along with the importation of Christmas trees in December 1959, bizarrely labeling both St. Nick and spruce trees as "imperialistic."[7]

As formal rupture of relations between the two countries became more likely, the CIA began making "stay behind" arrangements for intelligence activities on the island. Stocks of technical gear and espionage equipment were cached in the countryside. Agents who could no longer safely be met in person were given alternate communications plans and covcom systems, such as secret-writing materials and maps to the cache sites.

During the summer of 1960, CIA officers had spotted suspected surveillance by Cuban security officers, prompting the office in Havana to organize a small countersurveillance team of recruited Cubans to protect its operations.[8] At the same time, other elements of the CIA were already planning White House–approved covert operations against Cuba, including an armed invasion of the island by a refugee counterrevolutionary force and a possible assassination of the Cuban leader.[9]

The September audio operation took on urgency with the possibility that all official Americans could be ordered out of Cuba, rendering support for any technical attack virtually impossible. Already, a similar audio operation a month earlier against another target failed because of logistical problems. Adding to this growing list of concerns, September was midway through the hurricane season, and Hurricane Donna, forming in the Caribbean, threatened to disrupt travel to Cuba and upset operational timing.[10] Given the circumstances, a delay of even a few days could see this opportunity slip away.

A TSD team was needed to take advantage of a rapidly closing window of opportunity, but audio techs were in short supply. It was late summer and some were on annual leave, others scattered in transit to new assignments, and ongoing operations consumed the remainder.

Wally, an experienced field tech on home leave after a tour in Asia, was redirected to the operation, and postponed a visit to his parents. Andy, an engineer who developed audio gear, saw the operation as a chance to get his feet wet by helping in a routine installation. The firsthand experience, he believed, would help him to understand and anticipate operational problems and design better equipment for the techs in the field. Dave, as head of the personnel-strapped audio unit, also stepped up to the requirement. No one had any reason to believe the three would not be home within a few days.[11]

After making contact with their case officer in Havana, the techs unpacked the tools and audio equipment. Everything was in order, when fate threw the team a curve. Unexpectedly the building owner got cold feet and withdrew the offer for access. It was a disappointing development, but the case officer conveying the news to the techs assured them their trip would not be in vain. The local office had received authorization from Headquarters to bug an alternative target, the New China News Agency, located in the Seguro Medico Building, a new high-rise in the heart of Havana.[12]

This alternate operation seemed routine enough. An apartment above the news agency offices had been rented by a CIA contact, a Cuban dance instructor named Mario, and if any problems arose, the techs had a "bug out" plan to regroup in the apartment of an American secretary also living in the building.

The next day, the techs met with the case officer in a downtown sandwich shop to await a "go signal." When Mario showed up, drank a cup of coffee, and left without acknowledging the presence of the four Americans, the empty coffee cup signaled an all-clear for the techs to proceed to the Seguro Medico Building and survey the target. Because the local office feared an informant had penetrated its surveillance team, they would conduct the operation without countersurveillance. All agreed that since the techs would be working from the agent's apartment, little risk existed. They could control the apartment for as long as they needed to finish the job.

Arriving at the Seguro Medico Building for the survey phase, Andy and the others noticed that no concierge was on duty. That was the first good sign. "It was a Sunday and nobody was there. We got on the elevator and went one or two stories higher and walked down the stairs to the agent's apartment," remembered Andy. "We cased the apartment to determine the construction materials, where load-bearing walls were, and the location of power. We figured out what equipment we'd need to drill and make repairs and how we'd divide up the work. Basically we created the 'plan of attack' to minimize the amount of time to do the job. Then we went back to the safe house."[13]

The next day, after satisfying themselves and the local chief the operational plan was sound, the techs returned to the apartment to do the job. They would drill down through the floor and into the ceiling of the New China News Agency, opening pinholes of less than one millimeter to provide an airway for conversations to reach microphones placed snugly against the minute openings. "This wasn't exactly a blind drill," Andy explained, "because we knew the apartments were mirrors of each other. We were going into an area we believed they'd be talking in, like a bedroom office. You're never sure of how the audio will work, but in those days, when we had a good pinhole, we could usually get good reception and cover a couple of adjacent rooms."

The backbone of the equipment used for the mission was the SRT-3.[14] The all-transistor transmitter was about the size of a pack of cigarettes and broadcast an unencrypted clear signal.[15] A small switch receiver connected to the SRT-3 would allow the listening post keeper to turn the device on and off remotely to elude electronic sweeps,[16] and since the system would tap power from the building's electrical line, it could run indefinitely.

However, almost from the start, the job did not go as anticipated. The unair-conditioned apartment was stifling and soon the three stripped down to their shorts and tennis shoes. The apartment's thick concrete floors made the work slow going, even with the heavy-duty quarter-inch diamond drill bits.

For two days, the techs worked at the job, supported with food and supplies brought in by Mario and the case officer. With luck, they would finish up in three days and immediately leave the island.

On Wednesday, the operation began to go bad. First, a meter reader with the local utility company knocked on the door and was turned away without incident. Then a loud knocking summoned Dave to the apartment's front door. Opening the door, he found himself facing down the barrel of a large handgun in the hand of an unshaven young Cuban in olive green fatigues accompanied by four other armed young men, all dressed in the same style of fatigues. The five Cubans entered the room and silenced Dave so quickly that he had no opportunity to sound a warning.

Andy and Wally were in the bathroom. Andy had removed the two fluorescent lights on the top and side of the medicine cabinet and dug out a cavity in the plaster for the power line tap, while Wally was working on the antenna. "We just about had the thing pretty well finished, just the final touch, putting the tile back and plastering in the bathroom to cover the wires running from transmitter to the AC power line buried in the wall," said Andy. "We hadn't drilled the pinhole through the ceiling, but had everything else done. And that's when it all fell apart."

Andy and Wally had continued working away in the bathroom unaware of what was taking place just a few feet down the hall. Focused on the task, they assumed that either Mario or the case officer had come by, but when Dave did not return, Wally went to investigate. He too found himself staring down a gun positioned so close to his face that he could see traces of rust on the inside of the barrel. Then, when Andy turned around, he faced yet another armed Cuban. Hustled out of the bathroom, he joined Dave and Wally against the wall of the apartment's kitchen dining-room area.[17] To the three prisoners, the armed Cubans seemed indecisive about what to do next, which only added to their fear and anxiety.

"We stood against the wall for some hours. They ripped through the place, I don't think they knew what they were doing," said Andy. "First thing they took was all our money, then they took all my good cigars, and put us in the bedroom on a bed with the lights on."

Dave and Andy began communicating with each other by tracing letters on the bed. The Americans remained awake through the night as the Cubans waited to see who else might show up at the apartment. At one point, there was a gunshot from the living room. Had the Cubans caught and executed either Mario or the case officer? The techs' fears were unfounded. One of the guards had accidentally shot himself in the hand.

The next morning the prisoners were moved to the living room and faced the first of many interrogations. All of the equipment and tools they had brought in to do the job were arranged neatly on the floor. Repeatedly questioned, the techs maintained their cover story that they were Americans on vacation and saw an opportunity to make a little extra money doing some electrical work.

They were given coffee and then photographed with the equipment. Within days, the pictures appeared in the local papers alleging that the three were American spies.

Since the operation was put together quickly, the techs had little to back up their story, except the few documents they carried and their ability to brazen it out under interrogation. None of the three had received "risk of capture briefings" provided to military or intelligence officers who undertake dangerous missions. If one of the three alias identities or cover stories fell apart or if one of them broke under questioning, all three would be exposed as criminals or spies. The consequences were grave. Castro's government had already established a tradition of executions for political as well as social crimes.

Eventually the Cubans moved the trio from the high-rise to a military intelligence installation, only a few blocks from the U.S. Embassy. There they were fingerprinted and photographed. By early evening, the techs were separated, had their belts, shoelaces, and watches removed, and locked into different holding cells. Not much more than thirty square feet, with a shower and toilet combination at one end, each cell was packed with prisoners and triple-decker, GI-style bunk beds.

That night began the first of many interrogations. Escorted from his sweltering cell to a small, cold, air-conditioned room, Wally faced three Cubans. The one whom he would dub "Bad Teeth" took the lead. "What are you doing here? Come on, Mr. Taransky, tell us," Bad Teeth asked in English. "Why, you work for the CIA, don't you?"

Throughout the initial round of questioning, Wally stuck to the cover story and on the second day was driven back to the apartment and ordered to identify the equipment. Once there, he explained how certain pieces of the equipment functioned in general and managed to "accidentally" break some of the circuit boards to deny exploitation of the technology by the Cubans. The next day, he was again taken to the high-rise where he faced a horde of photographers and television cameras at a press conference. When one of the reporters asked if he was there of his own free will, Wally replied, "No, he brought me," and pointed to Bad Teeth. "He told me I'd be shot if I didn't cooperate."

With the press conference ended, Wally was put in one section of the facility and Andy in another. Dave was sent to a military base called "Columbia." In all, the three men would spend twenty-nine days undergoing middle-of-the-night questioning, being shuttled between steaming holding cells and freezing interrogation rooms. Their stories did not change.

During the twelve weeks leading up to the mid-December trial, relations between Cuba and the United States continued to worsen. The U.S. Embassy advised all American nationals to leave the country and Castro was hosted by a Russian delegation while visiting New York. After delivering a speech of more than four hours at the UN—setting a new all-time record—Castro discovered his plane had been seized as collateral against Cuban debts. The Soviets, eager to solidify their relationship with the Cuban leader, obligingly provided a plane.[18]

At CIA Headquarters, the arrest of the three officers caused a major flap. According to one memo circulated at the time, the situation was not hopeful. "The tourist cover used by the technicians was very light," the memo read. "The cover [asserted by the techs] could not be expected to hold up if the Cubans conducted a thorough inquiry and intensive interrogation."[19] For example, a check of the New York City address Wally used for his cover story belonged to a woman he was dating. If questioned, she would not have known an "Edmund Taransky."

In late October, the three were transferred to La Cabana, an ancient Havana fortress converted into a prison. Issued prison garb with a large P (for prisoner) stenciled on shirts and pants, the techs were again fingerprinted and then escorted to separate cells filled with common criminals, anti-Castro elements, and American adventurers caught in the revolution.

"At La Cabana they got serious about the interrogations. They had Dave back in the same facility where we were, but in different quarters," said Andy. "I never saw him except sometimes during interrogations, he'd be coming out and I would be going in. They'd bring you out of a hot room, put you in a freezing cold room, and threaten to pistol-whip you. Then they'd say, 'We're going to shoot you.' They had me convinced they were going to shoot me. I really thought it was going to happen. They said you have to cooperate or this is it; we know all about it. Once when the interrogator told me about our being in that sandwich shop, I thought, holy cow, we were dead before we went into the apartment."

Sometimes the questions would vary, with the interrogators accusing them of working for the FBI.[20] Bad Teeth would often claim that the other two prisoners already confessed, so not telling the truth was pointless. During one session, a young guard incessantly played with his gun, flipping the cylinder open and then pulling the trigger. "Tell him that men don't play with guns," Wally ordered Bad Teeth. "Only kids do." Bad Teeth obliged and the guard looked suitably chastened.

"Our attitude was that we didn't know what our fate would be. I was convinced I was going to be shot. I figured I'm expendable, but I'd never do anything to disgrace my children or the Marine Corps," explained Andy, who had served in the Marines from 1944 to 1946 and again between 1950 and 1952. "I made my peace with God, but it never happened, thank God."

The possibility of execution was, as the three learned at La Cabana, not an idle threat. Firing squads were busy day and night as Castro consolidated power by eliminating political opponents and malcontents. Reliable estimates set the number of political executions at upward of 2,000 by 1961.[21] "They were shooting five, six, seven every night. Right outside our window at one or two o'clock in the morning. One of them I will never forget as long as I live," recalled Andy. "His name was Julio and he was a doctor. He was educated in Spain and ran some kind of anti-Castro political cell. To save the people in his group, he took all the blame. He slept right above me and we became friends for a couple days. Then they shot him."

Prisoners at La Cabana were executed along the outer wall bordering a moat, long since filled in with dirt, that surrounded the old fortress. Firing squads of six to eight guards used U.S.-made World War II M-1 Garand rifles taken from one of Batista's armories. "Leaving and returning from our trial we got a particularly good look at the execution wall," Andy said. "When the .30-06 caliber round hits you it takes flesh and embeds it in the wall. I looked at that wall and could see exactly where guys had been standing. They shot some the night before our trial. We crossed over the moat on a bridge adjacent to the wall and we could see the results of the day's executions. First time I had ever seen something like that."

In the months leading up to the December 17 trial, Cuban–American relations reached the breaking point. President Eisenhower announced a ban on all exports to Cuba, except for a few foodstuffs and medicine. On October 25, the Cuban government retaliated by nationalizing all major banks, private sugar mills, distilleries, and stores, including American multinationals such as Sears, Roebuck, General Electric, and Coca-Cola.[22] Within days, U.S. Ambassador Philip Bonsal left the country.[23]

The techs were formally charged as "Enemies of Cuba" and the prosecutor asked for thirty-year sentences. Represented by a Cuban lawyer hired by the American Embassy, the Americans sat through a four-day trial, which consisted of a three-judge military tribunal. After the trial, U.S. Consul Hugh Kessler approached the three defendants. Trying to be upbeat he said, "You guys are great. Man, you're famous." Kessler would be the last American government official they would see for two years.

The verdict, never officially announced to the three in the courtroom, was guilty. They were sentenced to ten years apiece, which at the time seemed like a good deal. "The thing was, when you come back from your trial you either went to the left or the right," Andy explained. "If you went to the right, you went into a *copiea*, a little chapel-like room, and you knew you were going to get shot the next morning. For most prisoners, if you went to the left, you got thirty years. In those days thirty years was considered a pretty favorable sentence. So, before we were officially told what the verdict and sentence was, I realized that by making the left turn we weren't going to be shot."

A parade in Havana on January 1, 1961, featured Soviet tanks along with other weaponry. Relations between the United States and Cuba were officially severed two days later, just two weeks after the trial. The three techs, with their tourist cover still holding and their true identities still concealed, remained at La Cabana. That month, they heard that Mario, the agent in whose apartment the techs had been discovered but for whom the Cubans apparently had insufficient evidence to convict, was deported and joined his wife in Florida.

On January 22, 1961, prison authorities made an announcement over the public address system; the three were among 250 prisoners to be transferred to the Isle of Pines prison. Called the Presidio Modelo (Model Prison), the facility was located on a small, lush, 850-square-mile island a few miles off Cuba's coast and was perhaps the most dreaded of all Cuban prisons. Castro himself had been a prisoner on the Isle of Pines for two years following his 1953 attack on Moncado Garrison.[24]

The Isle of Pines had been the inspiration for Robert Louis Stevenson's *Treasure Island* and in the early twentieth century was known for its luxury resorts and sugar cane plantations.[25] However, in 1925, Cuba's president, Gerardo Machado, endorsed the idea of building a modern prison on the island. This would not be an ordinary prison, but a state-of-the-art facility that employed the latest "scientific" theories of rehabilitation. A Cuban envoy, dispatched to the United States to study prisons, returned greatly impressed with the new prison in Joliet, Illinois.[26]

Loosely based on the concepts of eighteenth-century British philosopher Jeremy Bentham, Cuba's new prison would be a panopticon, a circular structure in which cells faced inward toward a central guard tower from which the guards look outward toward prisoners in their cells. The design of the panopticon was based on the idea that the guards could see the prisoners but the prisoners could not see the guards because of the shuttered windows of the central guard tower. The theory behind the design held that prisoners would "behave" if there was the *chance* they were under surveillance. Once they behaved, they could be rehabilitated.[27]

Between 1926 and 1931, the Cuban government built four such circular structures, each connected by underground tunnels, arranged around a massive center structure, also round, that served as dining hall and something of a community center. Ninety-three cells circled each of the four buildings' five tiers, with a sixth floor remaining largely open and filled with support beams. Each cell measured approximately six feet wide by twelve feet deep.

What made the prison unique was that, in accordance to Bentham's concept, none of the cells had doors. Prisoners were free to roam within the building and prepare themselves to reenter society as productive citizens. "There is great care taken to fit each man into his own line of work, and there are workshops in which the prisoner may learn tailoring or boot making or any other trade his chooses, as well as classes for the backward or illiterate," enthused the *Illustrated London News* in 1932, just after the prison was opened. "At the end of the day, chess, dominoes, and cards are allowed, as well as more active games—the floor space of each round-house giving ample room for exercise ... Moving pictures and wireless programs are given in a large hall."[28]

However, what the three CIA officers found on the Isle of Pines bore no resemblance to the "perfect prison" cheerfully described in the British magazine with its pictures of pristine cells. Whatever scientific notions of rehabilitation may have inspired the original design had long been abandoned and the prison itself had fallen into a state of abject disrepair.

The cells were still doorless, but the prison was packed far beyond its 4,500-inmate capacity with 6,000 men crowded into the four structures known as the "circulars." Every level of every circular was filled with trash and vermin of every variety, from rats and lice to bedbugs and roaches. Prisoners were on their own to cope with the lack of sanitation that held the potential for disease and infection.

Each cell had a sink and a toilet, but no running water. A tap on the ground floor provided water that, as the three American soon discovered, was undrinkable. Smelling of fish, it could be used for bathing, washing clothing, and flushing the toilets, but not much else. Prisoners hauled this water to their cells in five-gallon buckets. Drinking water was trucked into the facility, and emptied into a cistern from which prisoners carried a gallon or two at a time up to their cells. Conserving water became a fact of life, and prisoners acquired new skills, such as using a single cup of the precious liquid to shave.

Since the toilets did not function properly, the prisoners designated two cells on each tier as communal bathrooms. These toilets frequently clogged, spilling sewage over their rims that eventually worked its way down to the ground floor, creating a half-inch of slimy scum across the terrazzo floor.

"Talk about stink," said Anderson. "Once in a while they would get some 'volunteers' for clean-up duty. The 'president' of the circular's prisoners was an ex-motorcycle cop that worked for Batista. He was a tall mulatto guy whose job was to control the prison and he would get a couple guys to bring some water up and they'd get a stick, pour water in, stir, and keep repeating until they finally emptied the thing."

Castro's prison held an odd mix of inmates, including many of Cuba's prerevolutionary elite of doctors, lawyers, and businessmen along with Batista loyalists and an American soldier of fortune. These political prisoners were held in circulars different from those that housed common criminals. "Of course there was politics. All the counterrevolutionaries hated the Batista people and vice versa," said Andy. "But they were all in there together."

The inmates received meager rations of food and the bare necessities of life. The lucky ones received care packages supplied by relatives and friends or bought food in the prison commissary. A few who had the financial means bought food from outside restaurants that prison officials would dutifully deliver. However, for those who could not afford to supplement their prison lifestyle with outside resources, existence was often unbearable. The inadequate diet of rice and beans prompted some starving prisoners to sell their beds for additional food. Prisoners who had no paper ripped off pieces of their shirt to clean themselves after using the toilet. Those who lacked the discipline for personal hygiene lived in filth.

The three techs sent collect telegrams to Mario's maid—who had acted as their outside contact when they were in La Cabana—asking for supplies to be sent in. "As a result, we got packages once in a while," remembered Wally. "That's the only way you lived in a Cuban jail. They give you almost nothing. They gave us those Batista military khakis with big P's stenciled on the back, but if you couldn't come up with razor blades, soap, toilet paper, spoons, dishes, and the like, you did without and they couldn't care less."

Eventually the CIA arranged for a private attorney to hire a woman to bring the techs packages from the States. The techs were surprised and encouraged as much by the packaging as the contents, since the type of tape and wrapping used were unmistakably the type used at the TSD warehouse near Washington, D.C. It was a clear signal they had not been forgotten.

"During the time we were in prison, the outside attorney's assistant continued to arrange to get us occasional packages—and sometimes we'd get a big plastic bag of Mixture No. 79 [pipe tobacco]," explained Wally. "When we saw how the packages were wrapped and sealed, it didn't take a rocket scientist adding up two and two to make four to guess who was packaging this stuff and sending it in. After I was released, a friend told me he was nosing around the warehouse and watched the guys packing supplies like underwear and Tang orange drink for shipment to tech bases around the world. But there was one older man over in the corner working all by himself. He would go over to the line, take some stuff out and put it in a box. My friend asked the guy, 'How come you're not over working with the rest them?' He replied very quietly, 'Don't say anything, what I'm doing is for our boys in Cuba. The others don't know that.'"

Less frequently, the techs received letters, including some from family or TSD colleagues. One letter to Dave included a picture of Mia, a TSD secretary oddly identified as Sally Wilson. In the photo, a male Agency colleague was embracing the woman. The two, who had no known personal association, were pictured walking across a stream. Although Dave thought it a strange photo, he was still happy to see familiar faces. Only after their release did the techs learn *S*ally *W*ilson was a clue signaling that the photograph contained secret writing and should be soaked in water. "If we had put the picture in water the back would have come off and secret-writing instructions would have appeared," said Andy. "Unfortunately, we audio techs weren't briefed on this type of communication, and didn't pick up on the code. So the attempt to establish a covcom link into the prison using secret writing fizzled."

Searches of cells, called *requisa,* were frequent. Following an attempted escape by one of the prisoners, the authorities moved all 1,400 men to the ground floor to stand naked in a semicircle, while their cells were searched. "We were there all day and if you lifted your head up, they made you lie facedown in this gunk on the floor," said Andy. "And people were defecating, taking leaks because we were there for fourteen hours with necks bowed and that was very painful. It got dark and the center tower was lit with what looked like 3,200 bulbs, and that was the only light in the circular.

They had all the guards with Czech carbines, a box magazine, and a flip-out bayonet that fit along the stock on the side. The guards, young kids, were nervous, too. All of a sudden, we heard them chamber rounds. The prisoners were tired, and many started crouching. We said we aren't going to cower like that, we're Americans. We were the only three standing, but if they started shooting, we'd be the first to get hit."

Riots over food and fights were common. When one hunger strike produced better rations, a second was organized. The response to this second protest was swift. Military personnel with bayonets were brought in and a *requisa* included throwing the contents of cells down from the tiers to the floor below while trigger-happy guards sent shots ricocheting through the facility to intimidate and control the prisoners.[29]

Suicides, as the three would soon learn, were routine. Prisoners would climb over the railing on the fifth floor and jump to their death. One day, when a newspaper astrologer named Dr. Carbell, who was serving two years for predicting Castro's downfall, started to climb over the railing, Wally and Andy pulled him back to safety. "The Cubans all stepped back because they were afraid of being linked with him. Andy and I moved fast and managed to grab him," recalled Wally. "He was educated somewhere in Europe and spoke with a cultured accent. He was also overweight and filthy as hell."

An American soldier of fortune, Richard Allen Pecoraro, had been swept up with anti-Castro plotters. Prison life drove Pecoraro mad; living in filth he huddled alone in his cell. Occasionally Cuban prisoners would come by and poke him with sticks, eliciting an animal growl. The techs befriended Pecoraro, a fellow American, cleaned him up, and brought him into one of their cells. They found a Cuban psychiatrist among the prison population who agreed to analyze the American through an interpreter. Eventually, they were able to get a supply of Valium shipped in from the outside for Pecoraro.

The prisoners most acclimated to incarceration were the common thieves. One trick the thieves showed the political prisoners was how to smuggle contraband into the circular. Guards at the prison's front door often slept, and trash was piled up among the high weeds surrounding the structures. With contraband hidden by friends in the weeds just beyond the walls, two prisoners would sneak out and walk around the perimeter of the circular in opposite directions to conduct countersurveillance. From inside a cell, a matchbox propelled via a slingshot-type device fashioned by the prisoners would fall at the feet of one of the men on the outside. He would then pretend to bend down to tie his shoelace and tie the newspaper or other contraband to the line to be reeled in.

To pass the time, the techs made a Monopoly board and taught fellow prisoners the game. "Wally had a bunch of Cuban friends who regularly came in to use our Monopoly set," remembered Andy. "And one day all the pieces came flying out of the cell and there was a lot of shouting. What happened was, Ernesto, a civil engineer who built the tunnels in downtown Havana, landed where someone had about four hotels. He just exploded, refused to pay the rent, and almost destroyed the set."

Andy fabricated a slide rule from a discarded cigar box and worked on logarithms from an old engineering text he found scattered amid the trash. Then the techs created a radio. Someone in the prison had smuggled in an earpiece and, amid the garbage-strewn jail, they had managed to scrounge up a few Russian-made transistors along with pieces of medical tubing used for intravenous feeding that could be used as additional earpieces. The tuning coil was created by wrapping copper wire around the cardboard cylinder from an empty roll of toilet paper.

A battery to power the radio remained a problem. "A battery is two dissimilar metals and electrolyte. We had copper from wiring that we ripped out of the walls and tin from galvanized pails, but we needed an electrolyte," explained Andy. "So we sent a guy to the hospital claiming he was sick, and he came back with a bottle of copper sulfate to treat the alleged ailment. It's a good thing the guards didn't make him drink it." When the crudely assembled materials were combined with the copper sulfate, the battery produced enough current to power the radio.

Another problem was the lack of a soldering iron. All the wires in the makeshift radio had to be tightly twisted together for a low-resistance contact. The antenna was another challenge. Consisting of a length of wire several hundred feet long, prisoners managed to undo a section of corrugated roof on the top tier to string the antenna along the outside.

When finally assembled, the radio could pick up broadcasts from WKWF ("overlooking the beautiful Florida Keys") and a high-powered, 50,000-watt New Orleans station. Among the news items that the techs particularly remembered was Roger Maris hitting his sixty-first home run and John Glenn becoming the first American to orbit the earth in space. "I used to go up to the roof at night for best reception," said Andy. "With those four tubes coming out of the radio, I heard American music for the first time in months."

Because of the constant danger of *javios*—prison snitches—the radio remained a closely guarded secret. Rumors of a radio prompted a *requisa,* but the radio itself was never discovered. With communication to the outside world established, the prisoners started an underground prison newspaper. "One of the Cuban prisoners was a radio operator," recalled Wally. "He was skilled at tuning the radio to find just the right sensitive spots. Once he got a station, he and an assistant, a stenographer who had been a legal secretary, worked with him. They would plug in the earphones and take down the news in shorthand from whatever station could be heard. The next morning, a copy of the handwritten 'newspaper' would be circulated among the prisoners."

On April 14, 1961, the three imprisoned techs went to bed as usual only to be awakened before dawn by the sound of gunfire and tracer rounds from .50 caliber machine-gun fire lighting the building. The invasion by Cuban exiles at the Bay of Pigs had begun. Throughout the winter, rumors of a possible invasion had circulated and now it was happening. The prison burst into chaos as a B-26 from the CIA-trained anti-Castro invasion force flew overhead and, a few days later, on April 17, an invasion force landed on Cuba's western shore at the Bay of Pigs.

Conceived during the Eisenhower administration, the invasion by 1,400 Cuban nationals was launched with the approval of President Kennedy. Originally proposed for an area known as Trinidad, in the shadow of Cuba's Escambray Mountains, the plan called for a relatively small invasion force to spark an uprising among the Cuban population. If the revolt proved unsuccessful, the invading forces would then retreat into the mountains to wage guerilla warfare.

However, in March 1961, Kennedy called the plan too "spectacular" and changed the landing site several times before finally settling on the less than ideal Bay of Pigs, a location surrounded by swamps.[30] Then, as the ships carrying the members of the Cuban 2506 Assault Brigade approached the island, President Kennedy called off the scheduled second and third waves of air strikes that would destroy the remaining planes of Cuba's small air force.[31] Spared from those strikes, Castro's air force was able to sink the brigade's supply ship, the *Houston*. With the invaders unable to establish a beachhead and without resupply or air support, failure was inevitable.

On the second day of fighting, the techs began to notice Cuban militiamen loading boxes into a utility tunnel under circulars three and four. There was no explanation for the activity until the bottom of one of the boxes broke and the prisoners could see it was dynamite. Apparently fearful that a mass escape would liberate inmates to join the invasion, Castro ordered the prison booby-trapped. In the dictator's mind, it was better to bring down the structures, killing all 6,000 men inside, than risk a small army of prisoners marching on Havana.

By April 19, the invasion was defeated. Of the Cubans who landed, 1,189 fighters were captured and a small number escaped back to the sea. When news of the failed assault eventually reached the prison, no hope was held for a second attempt.[32] Nothing more was thought of the dynamite until Thanksgiving Day in 1961 when guards began drilling holes in circulars three and four with jackhammers. Andy and Dave, assigned to work details clearing the debris, watched as crews drilled into the support columns of the two buildings.

After three weeks, with the holes completed, trucks arrived and boxes labeled *Mecha Explosiva* (explosive fuse) were unloaded into the holes. More boxes followed, these with TNT stenciled across the side. Judging from the number of cartons, the three Americans estimated that five tons of explosives were now underneath the circulars. Ominously, as the explosives were unloaded, some prisoners received black plastic rings with their prison numbers on them; other inmates were tattooed. Reportedly, this was done to identify bodies if the prison was brought down. Apparently, Castro was still fearful of a prison revolt.

The idea of living in a mined building did not appeal to the techs. As word spread throughout the circulars that the Americans intended to do something, a Cuban prisoner named Miro soon joined them. "Miro recruited two of his buddies—one guy looked like the Michelin tire guy," remembers Wally. "We went to one of the cells on the first floor that was being used as a toilet. *Servicio* is what they call it in Spanish. You didn't linger there very long, the smell was something awful."

With the "Michelin tire man" blocking the view of the guards, two other Cubans worked for four days to enlarge the hole in the floor leading to a utility tunnel where the explosives were emplaced. The team then recruited the smallest prisoner they could find, a fair-skinned Cuban who went by the nickname Americano. Standing just five-foot-five and weighing no more than 120 pounds, Americano was persuaded to squeeze through the small hole for a reconnaissance mission.

Lookouts were posted as the young man slipped into the tunnel one afternoon, instructed to bring back samples of whatever he found. Inside the six-foot-high by eight-foot-wide tunnel, Americano discovered enough explosives to bring the buildings down and two detonation systems, one electrical and the other a long length of primer cord. If one failed, the other could be put into play. The young Cuban also brought out a fifteen-pound block of TNT, which the three Americans told him to return, lest the guards find it missing.

A lieutenant in the Cuban army had headed the installation team and the techs now understood the job had been done well. "He was no dummy. He knew explosives and he knew what to do," said Wally. "They ran the lines from an outbuilding into the circulars with the primer cord encased in plastic tubing and the electrical line through separate tubing. When we understood what he'd done, we were left there scratching our heads. We're sitting on this thing and if it goes bang, we're dead."

Clearly, something had to be done, but it was not a simple matter of cutting both lines. Cutting the primer cord line would likely tip off the guards who would notice the slack at the detonation station. Severing the electrical line could also alert guards if they ran a test current through the system. The trick was to disable both systems without leaving any trace of sabotage. Technically the operation was not difficult. Under normal circumstances with a TSD tool kit, disabling the bomb would have taken minutes, but the techs only had a few simple knives, sewing kits, and razor blades.

"The electrical line looked like European cable, something like zip cord, AC cord. But it was built differently, the two conductors were a little bit separated," said Wally. "We came up with the idea to cut the plastic and then twist it. And that creates a short—and when you've got a short, it won't go. You try to energize the blasting cap and it won't fire."

The solution the techs devised was the electrical equivalent of putting a very tight knot in a length of garden hose. However, if interrupting the electrical circuit was relatively easy, the primer cord was a much more difficult matter. The Cubans were certain to notice if tension in the cord was released, so severing the line was not an option. The trick would be to create a gap in the primer cord while maintaining tension along its entire length. For this, the techs fashioned a special gadget that comprised a spool from a sewing kit and pins. By first cutting the cord, then inserting each end into the center of the spool, they could hold both ends in place with needles and pins. This would create a gap while maintaining tension along the line.

Because none of the techs was small enough to fit through the hole and any effort to enlarge it further would attract notice of the guards, Americano was again recruited to go back into the utility tunnel for the sabotage mission. Over four nights, working in the techs' cells in semidarkness, Americano trained for the mission. Using a sharp knife, he practiced exposing the wires and shorting out the line before slipping the insulation back over the exposed wires. Then he practiced with the makeshift thread spool and pins. Once in the tunnel, he would have only one chance to perform these acts perfectly and under time restrictions—beginning in the afternoon until just before the evening head count.

When the three Americans felt confident in his ability, Americano slipped into the hole. After a few tense hours, he reemerged and reported to Miro that the mission had been a success but, to the dismay of the techs, word about the operation spread. One prisoner ran up to the Americans thanking them for what they had done. Fortunately, the guards never discovered the sabotage plan or, if they suspected something, did not report it to their superiors. There was no *requisa,* the three techs were never questioned, and the sabotaged system remained in place.

"If and when they pulled the switch, we felt that would give us about twenty minutes before the guards realized what was happening," said Wally. "Then it became a case of breaking out—how the hell are we going to get out of there? Well, some of the Cubans had bars cut and whatever. But how many guys are you going to get out through a small window? Not too many. We'd have to go out through the front door."

Emboldened, the three Americans began thinking of weaponry they could have on hand if things came to a head between the prisoners and the guards. The first idea involved a flamethrower. Wally disassembled an old kerosene stove and spent four days grinding the brass valves using marble dust and toothpaste. Ultimately, his efforts came to naught when the stove consistently lost the pressure necessary for the flamethrower to operate.

Undeterred, one prisoner came up with the idea of making alcohol for Molotov cocktails.[33] Fruit was collected—oranges, grapefruits, mangos, and watermelons—and put into glass jars with water and sugar to distill. Then the extract was run off and cooked in a pressure cooker, the vapors run through a length of plastic tubing. The distillate was passed through the homemade still two or three times.

"We got the chief chemist from Bacardi, who was in the lockup with us, and we got some of the white lightning, handed it to him, and asked his professional opinion," said Wally. "He said, 'Yeah, man, this is good 95 proof.' He took the gallon of alcohol and disappeared. Three days later, we're all standing around, leaning over the edge, smoking cigarettes and talking about this and that, and he comes over and hands me a cup. I look at it, nice appearance. I taste it, and goddamn, it's good Courvoisier. I asked him how he got the color. 'Shoe polish,' he said."

The techs also improvised hand grenades. Americano was sent back down into the utility tunnel for some blasting caps and a small quantity of TNT. Melting down the TNT in a double boiler, they poured the liquefied explosive into condensed milk cans filled with nails, glass, and anything else that would serve as shrapnel. Blasting caps were attached to the top along with a length of homemade fuse.

Fuses for the grenades were created out of match heads ground into powder and impregnated into cloth. Andy, the true engineer among the three, set up a test program. A tech stood on the fifth floor, lit a fuse, and threw it over the side. A cooperating prisoner below would pick it up without attracting the guards' attention and report how much of the fuse was burned. Eventually, the techs determined that three inches of fuse would burn in about twenty seconds before igniting the blasting cap.

Although day-to-day life in prison did not improve significantly, these acts of defiance—building the radio, defusing the explosives, creating a small arsenal—encouraged and boosted morale among the techs and like-minded prisoners.

The TSD techs were helpless observers in the fall of 1962 when tensions between Cuba, the United States, and the Soviet Union escalated into an international nuclear crisis. An overflight of Cuba by an Agency U-2 in June indicated the Cubans were preparing for installation of surface-to-air missiles, although no missiles were seen.[34] Subsequently, U.S. intelligence observed both military advisors and equipment arriving in Cuba at unprecedented rates. U-2 overhead photography continued to confirm activity during September, including evidence that Soviet short- and intermediate-range missiles were about to be introduced on the island.

President Kennedy issued a national military alert on October 19 and addressed the world on October 22, explaining that the USSR and Cuba had conspired to install missile bases with the purpose "to provide a nuclear strike capability against the Western Hemisphere."[35] The threat of an international nuclear confrontation continued until October 28 when the Soviets agreed to remove their missiles from Cuba.

The world breathed a sigh of relief that nuclear war had been avoided, and on Christmas Day 1962 word began to circulate that a prisoner exchange was in the works. On March 16, 1963, James B. Donovan visited the prison. A New York–based lawyer specializing in insurance, Donovan (no relation to OSS General William Donovan) had served in the OSS as general counsel and then as a member of the U.S. prosecution team during the Nuremberg trials of Nazi war criminals.[36] In the years that followed, Donovan kept a hand in the intelligence business and at the request of the New York Bar Association, defended Soviet spy Rudolph Abel, then, several years later, negotiated Abel's exchange for U-2 pilot Francis Gary Powers.[37]

By the time Donovan arrived at the prison, he had already bartered the release of the Cuban members of the 2506 Assault Brigade and was optimistic about his chances for negotiating the release of Wally, Dave, and Andy. "Donovan came down to see us and brought his son," Andy recalled. "We understood he was someone Castro apparently trusted not to attempt to undermine the Cuban government. So the Cubans let him in and he insisted on seeing all of the Americans. I don't know if the place was bugged where we talked, but we acted like it was and were very discreet in what we said. I got the signals from him that they were working on it and 'Don't worry, we'll take care of you.' We took his actions as meaning that something was going on. But you also heard so many rumors and gossip you didn't take anything at face value."

A little more than a month later, on April 21, 1963, Wally, Dave, and Andy, along with eighteen other prisoners, were told to gather up their belongings. They were transported back to La Cabana and released in exchange for four Cuban nationals held in New York on charges of sabotage conspiracy.[38]

Taking off from Havana for Florida's Homestead Air Force Base, they were well into the air when a CIA medical officer told Wally that his mother had died. The news touched the deepest emotions of the techs who endured two and a half years of depravation and uncertainty. They cried together.

When the plane's hatch opened to a media pack on Homestead's tarmac, the American soldier of fortune, Pecoraro, was the first to step off the plane to freedom. The three techs, to avoid the cameras, lingered behind the others, and were then hustled away to a nearby safe house to see their families, receive medical attention, and undergo the obligatory debriefings.

The techs had been in captivity for 949 days, and for the entire time their cover and aliases held. In prison, they had refrained from discussing the operation or personal reminiscences about home and family lest other inmates overhear the conversation and their cover stories erode. Yet, within days of their return, someone whispered to the press that the three American tourists held in Castro's prison were, in fact, CIA officers.

The CIA debriefings lasted for about a week. They were interviewed by psychologists, counterintelligence officers, debriefers, and subjected to polygraph examinations. Personal security became a concern after their identities were leaked to the press. During the summer of 1963, the techs waited for the phone call that would return them to duty. Wally went north to be near his father, while Andy was sent to a fishing camp in Florida owned by Agency retirees. Dave remained in the Washington area.

Eventually certified fit for duty, the three returned to new assignments in the fall of 1963. Dave Christ, against his preference, was transferred to the Office of Research and Development in the newly formed Directorate of Science and Technology. Andy and Wally continued working in TSD. Andy became head of an equipment testing and certification unit at the OTS laboratory while Wally remained in audio operations.

TSD Chief Seymour Russell told Andy that he should not expect to be treated any differently from other techs—he would be judged on the quality of his future work, not the past. Initially, even within TSD, the returnees were avoided by some of their colleagues and business was conducted around them. The only senior Agency official who formally acknowledged what the techs had endured was Executive Director Lyman Kirkpatrick, during a brief meeting with them in his office. Quietly, each received a one-grade promotion but, otherwise, for the official bureaucracy, their nearly three years spent in a Cuban prison never happened.

Christ, seeking no personal recognition, submitted a lengthy and comprehensive recommendation to CIA management in late 1964 that Andy and Wally be given "the highest possible" Agency award for the courage, imagination, and fortitude they exhibited during the ordeal.[39] The recommendation was ignored and Christ retired in 1970.

When Andy announced his intention to retire from the Agency in 1979, David S. Brandwein, then director of OTS, conducted a routine review of his personnel file to determine what retirement award might be appropriate. Included in Andy's file was a copy of Christ's 1964 memo. The graphic description of the conditions in the prison and professionalism shown by the techs under the horrific circumstances so impressed Brandwein that he immediately brought the matter to the attention of the CIA's senior awards panel.

At Brandwein's urging, the panel conducted a full review and recommended that all three techs be awarded the Agency's highest medal for bravery. DCI Stansfield Turner accepted the recommendation and personally presented David Christ, Thornton Anderson, and Walter Szuminski with the Distinguished Intelligence Cross in May 1979, sixteen years after their return home.

At the time, only seven others had received the DIC in the CIA's thirty-year history. The citation for each of the techs read:

> The DISTINGUISHED INTELLIGENCE CROSS is awarded in recognition of exceptional heroism from September 1960 to April 1963. During this period [the recipient] endured hardships and deprivations with unquestioned loyalty, great personal courage and conspicuous fortitude. [His] exemplary conduct as a professional intelligence officer was highlighted by his unswerving devotion to the Agency and by his disregard for his own personal safety in order to assist others. [The recipient's] performance in this instance reflects the highest credit on him and the Federal service.[40]

While the three techs were finding ways to survive in prison, the CIA and TSD were planning to eliminate the Castro government. Both the Eisenhower and Kennedy administrations pushed the Agency to develop new capabilities for dealing with what was seen as an intolerable political problem in Cuba. Rather than test the international consequences of a military invasion, both Presidents turned to the CIA for secret and covert means to accomplish a policy objective.

The CIA's Directorate of Plans developed two parallel paths to solving the Cuba problem. Support from the TSD was sought for both plans. Beginning in March 1960, the United States began equipping and training an indigenous "secret army" composed of Cuban exiles and former Batista supporters to invade the island. The second path, direct action against Castro himself, was aimed at incapacitating or killing the Cuban revolutionary.

TSD specialists trained the "Cuban exile army" in clandestine skills needed for a sustained guerilla war. The Cubans were taught clandestine photography and film processing, secret writing, signaling, and use of cover and alias documentation. TSD issued a numbered identity card to each of the trainees and indirectly created the exile army's name. By selecting the

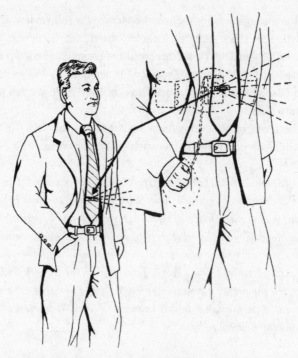

*To acquire close-up, clandestine photos of people or objects during the
Cold War, small Robot cameras, camouflaged by clothing, were designed to
shoot through tiny openings in buttons or tie tacks, 1960s.*

number "2500" for the first card with all others following in numerical order, a TSD tech reasoned that Castro's intelligence service would think the
force was larger than it actually was.[41] Then, when recruit number 2506,
Jose Santiago, died a few weeks later in a training accident, the Cubans adopted the name "Brigada 2506," or later, more formally, the "Brigada de
Asalto 2506," in honor of their fallen colleague.

The only TSS-TSD-OTS officer killed in the line of duty between 1947
and 2008 was a casualty of anti-Castro operations. Four days before the Bay
of Pigs invasion, TSD explosives experts were training members of the force
of Cuban nationals in constructing and arming small charges for harassment and sabotage operations. As Nels "Benny" Benson, a forty-five-year-
old native of Eagle Bend, Minnesota, and one of TSD's experienced explosives
officers, demonstrated how to mold a charge composed of thermite and C-4

into a form that resembled a soap dish, an errant spark ignited the materials. The resulting fire threatened to spread to adjacent explosives.[42]

Benson immediately picked up the flaming mixture and carried it away from the site. Critically burned, he died in a Miami hospital three weeks later. One of the nearly one hundred stars chiseled into a granite wall of the lobby of the Original CIA Headquarters Building commemorates the life and sacrifice of Benny Benson, who died doing what duty demanded.

The CIA's other solution to the Castro problem drew TSD into the ultimately objectionable policy of sanctioned assassination. Both TSD's chemistry branch and its explosive devices branch had the expertise to create lethal materials and delivery mechanisms. TSS had previously developed poisons as part of the U-2 program in the mid-1950s and impregnated needles for pilots to carry as an alternative to capture and torture. The so-called suicide needle was created by the devices branch of the Special Operations Division of the U.S. Army Chemical Corps at Fort Detrick Maryland, a military research organization with whom the CIA, through TSD, worked closely. The poison on the needle was saxitoxin, a naturally occurring toxin found in contaminated shellfish and one of the most lethal substances known.[43]

L-pills had been part of the OSS defensive inventory for issuance to agents on particularly dangerous and sensitive assignments. TSD continued to make those available for CIA operations. Little imagination was required to envision that the same or similar potions could be used offensively as well. Likewise, the guns, bullets, darts, and camouflaged explosives in the TSD inventory for covert paramilitary operations could be considered for use against a specific individual.

After the failed Bay of Pigs invasion, President Kennedy and Attorney General Robert Kennedy admonished senior CIA official Richard Bissell, the Deputy Director for Plans, for "not doing anything to get rid of Castro."[44] As a result, in November of 1961 Bissell instructed CIA officer William Harvey to direct a program called ZR/RIFLE, focused on the Cuban problem.[45] From the outset of the Kennedy administration, the CIA had been urged by the White House to create new assassination capabilities, referred to as "executive action," the "magic button," or even the "last resort beyond last resort and confession of weakness."[46]

The CIA attempted to assassinate Castro at the time of the Bay of Pigs invasion, but it disintegrated into a Keystone Kops comedy.[47] The plot involved the CIA's Office of Security engaging a former FBI special agent turned private detective, Robert Maheu, to recruit members of organized crime to carry out the assignment. Maheu contacted a former member of the Capone crime syndicate, who hired two Mafia members judged to "have experience."

The plotters faced numerous complications, and TSD's chemists struggled to find the correct weapon—a lethal but slow-acting pill that would give the agent time to slip away before taking effect. The first batches of poison capsules failed to dissolve adequately in water, but the second batch worked in trials on monkeys and was issued to a Cuban "official" for the attempt.[48] The official later returned the pills after getting "cold feet."[49]

From 1960 through the end of 1965,[50] various schemes, some whimsical and some serious, were discussed and evaluated to eliminate Castro. Of the conceived and planned attacks on Castro that encompassed public embarrassment as well as assassination, none came close to succeeding.

The variety of plans that were considered to undermine Castro's charismatic appeal by sabotaging his speeches was reminiscent of William Donovan's admonition to Stanley Lovell to "raise merry hell." These included:

Hallucinogenic Spray: One TSD scientist and bioorganic chemist proposed spraying Castro's broadcasting studio in Havana with a chemical to induce LSD-like hallucinations. TSD eventually found out, however, that the chemical was unreliable.[51]

Hallucinogenic Cigars: Since the method of introducing the spray into the room was not possible, Schieder suggested impregnating Castro's cigars with a special chemical to produce temporary disorientation during one of his long, rambling speeches, which were broadcast live to the Cuban people.[52]

Contaminated Shoes: A plan was evaluated to attack Castro's famous beard when he was traveling abroad by contaminating his shoes when they were placed outside his hotel room door at night to be shined. The idea was to "dust" the inside of the shoes with thallium salts, a strong depilatory which, when absorbed into the body, would cause Castro's beard to fall out. TSD procured the chemical and tested it successfully on animals before the DDP scrapped the plan when Castro cancelled his trip.[53]

Depilatory Cigars: Similar in concept to the failed shoe attack, under this scheme, Castro's cigars were to be treated with a powerful depilatory that would cause his beard to fall out, leaving him hairless as a means of damaging his macho image. The special box of cigars was to be provided to Castro during his appearance on a television talk show hosted by David

Susskind. After CIA officer David Atlee Phillips questioned how the operation could ensure that only Castro, and not others (including Susskind) would smoke the cigars, the idea was abandoned.[54] Phillips made the point that assassination schemes required both effective technical substances and precise operational planning.

Poisoned Cigars: The CIA recruited a double agent to offer Castro a Cahiba cigar, his favorite brand, treated with botulin, a toxin so deadly that the target would die shortly after putting the cigar onto his mouth. CIA records indicate that the cigars were passed to the double agent in February of 1961, but he apparently decided against carrying out the plan.[55]

Exploding Cigars: During a Castro visit to the United Nations the CIA considered a plan to plant a box of exploding cigars at a place where he would smoke one "and blow his head off." The plan was not carried out.[56]

Exploding Seashells: In early 1963 TSD was asked to construct a seashell with explosives, to be planted in the ocean at a spot in which Castro commonly went skin diving. After a technical and operational review, CIA discarded the idea as impractical.[57]

Contaminated Diving Suit: A proposal was made for a U.S. lawyer involved with official negotiations over the release of prisoners captured at the Bay of Pigs to present Castro with a contaminated diving suit. TSD bought a diving suit, dusted it inside with a fungus, which would produce Madura foot, a chronic skin disease, and contaminated the breathing apparatus with a tubercle bacillus. The plan was abandoned when the lawyer decided to present Castro with a different diving suit.[58]

Poisoned Pen: On November 22, 1963—the day that President Kennedy was assassinated in Dallas—a CIA officer offered a poison pen to a Cuban agent, *AMLASH*, in Paris for use against Castro. TSD had modified the ballpoint pen with a hypodermic needle designed to be so fine that the target (Castro) would not sense its insertion and the agent would have time to escape before effects were noticed. *AMLASH* was instructed to use Blackleaf-40, a commercial poison, with the device, but in the aftermath of Kennedy's assassination, he decided against taking the pen back to Cuba.[59]

Suppressed Pistol and Rifle: The CIA subsequently provided *AMLASH* with a suppressed pistol and suppressed FAL rifle and scope, as well as highly concentrated explosives.[60] *AMLASH* took no action, and in June of 1965, the CIA terminated contact with him.[61]

The DDP's planning of assassination attacks on other foreign leaders, such as Patrice Lumumba of the Congo, also drew on TSD's research, production, and delivery capabilities.[62] The initial plan to eliminate Lumumba involved putting poison into his food or toothpaste. A syringe, surgical mask, rubber gloves, and a vial of toxin were sent to the Congo for the operation. However, the plot failed when moral objections to assassination were raised by senior officers of the Agency's DDP, as well as the difficulty in gaining access to Lumumba's entourage.[63] Ultimately, the plot became unnecessary when opposition Congo forces killed Lumumba in January 1961.

With the exception of the *AMLASH* operation, assassination of foreign leaders as a policy option for the United States ended on November 22, 1963, when President Kennedy was shot in Dallas, Texas. It would be more than a decade before the Rockefeller Commission (1975) and the Church Committee (1975–1976) provided the American public insight into the CIA's secret role in the assassination schemes that clustered in the 1959–1963 years. Obscured by the sensationalism and intrigue of the plots and technologies were the conclusions of both investigations. With respect to assassination planning, the Rockefeller and Church reports determined that CIA officers acted on accurately understood policy direction from the White House under both the Eisenhower and Kennedy administrations.[64]

In response to the two reports, President Ford issued Executive Order 11905 that contained the provision: "No employee of the United States Government shall engage in, or conspire to engage in, political assassination." Subsequently, a revised 1981 Executive Order 12333 governing intelligence activities reaffirmed the prohibition: "No person employed by or acting on behalf of the United States Government shall engage in, or conspire to engage in, assassination." The EO added that "no agency of the Intelligence Community shall participate in or request any person to undertake activities forbidden by this Order," language that explicitly prohibited "indirect participation" in assassination.

Serious public discussion of assassination as a U.S. policy option ended with these Executive Orders but it would be rekindled after the September 11, 2001, al-Qaeda terrorist attacks. A December 2001 *Newsweek* poll found that 65 percent of those surveyed supported assassination of al-Qaeda leaders. The dramatic change in public opinion likely reflects the contrast between the potential danger perceived from Castro and the reality that al-Qaeda, a non-state organization, carried out attacks on American cities, airlines, and civilians. Even in that environment, however, it remains unlikely the U.S. public would support authorized covert assassination operations against the head of a recognized foreign government.

CHAPTER 17

War by Any Other Name

We had a war going, but nobody knew.   —OTS officer in Vietnam, 1962

TSD officer Pat Jameson was sitting on a hard bench in Saigon's Tan Son Nhut Airport in 1962 studying the aircraft traffic as he waited for another Agency officer's flight to arrive. A Pan American plane landed for refueling and Jameson watched as a group of American tourists disembarked. Walking across the tarmac through the glare of the Southeast Asian sun, they made their way to the promising shade of the drab building, eager to see what exotic souvenirs the ramshackle terminal might hold.

Perhaps drawn by an American, or at least Western face, one of the tourists approached Jameson. "I hear there's a war going on down here, is that right?" the tourist asked Jameson casually, as if he were inquiring about the weather in some distant city.

Jameson nodded toward a corner of the tarmac. "Look out there. You see that? There's a bunch of people being taken from that plane to ambulances," he said. "And there's some new guys with fatigue creases in their pants getting on that same plane to go up-country to replace those dead and injured ones. That's the story that we're living with here."

"God, I never knew that!" the tourist exclaimed as he stared at the scene.

More than forty years later, Jameson reflected on a scene he remembered vividly, "We had a war going, but nobody knew."

That a tourist on a brief layover was unaware of the situation in Vietnam was not surprising. For most of the American public in 1962, Vietnam was an obscure and distant country of little consequence. Seemingly just another former European colony in turmoil, news of Vietnam's problems was usually confined to the back pages of the morning paper. The French Indochina War of the early 1950s had been largely forgotten or ignored outside of foreign affairs wonks at the CIA, State Department, and Pentagon.

For those concerned about Vietnam's history and future, 1954 was the year keenly remembered. That spring the French suffered a decisive defeat at Dien Bien Phu when 70,000 Vietnamese soldiers overwhelmed their 13,000-man outpost.[1] The Vietnamese had dragged howitzers and other heavy artillery along the ridgeline of the isolated valley north of Hanoi, and then fired directly down into the French garrison.[2] Led by General Giap, tens of thousands of troops endured not only the hard physical labor of moving artillery over jungle trails, but also repeated strafing by French aircraft.

DCI Allen Dulles directed the CIA's "front" airline in Southeast Asia, Civil Air Transport (CAT), to fly resupply missions during the siege using unarmed C-119 "Flying Boxcars" cargo aircraft while the U.S. military sent fifty B-26s for air support operations to aid the beleaguered garrison.[3] Despite this U.S. assistance, which did little to turn the tide, President Eisenhower thought the French government's decision to "make a stand" at Dien Bien Phu ill advised and its efforts to keep Vietnam under colonial rule an invitation for the communists to gain an advantage.[4]

Eisenhower's assessment had been correct. The commander of the French garrison, sensing defeat was at hand, committed suicide with a hand grenade.[5] The siege, which lasted from March until early May, effectively ended French colonial rule in Indochina, but brought no lasting peace. An international conference, convened in Geneva during July of 1954, offered a plan to create a unified Vietnamese government following democratic elections in 1956. However, the Geneva agreement was not endorsed by the United States and resulted in a negotiated standoff that included a temporary division between north and south along a Demilitarized Zone at the 17th parallel.[6]

Two countries emerged from the agreement, the communist-ruled Democratic Republic of Vietnam in the north and the Republic of Vietnam in the south. Almost immediately, Ho Chi Minh's regime embarked on a protracted campaign of guerilla warfare to unify Vietnam under communist rule. The United States, practicing a policy of Cold War "containment," was determined not to let that happen.[7]

When Jameson encountered the American tourist in Saigon, U.S. paramilitary support to the South Vietnamese consisted of advisors from both CIA and U.S. Army Special Forces. The Eisenhower administration committed American assistance to South Vietnam, but limited efforts to advising and assisting the South Vietnamese government in unconventional warfare, paramilitary operations, and political-psychological warfare.[8] This role expanded during the Kennedy administration to include CIA paramilitary support with substantial assistance from Special Forces to interdict material flowing from the north to the Vietcong in the south.

With limited news coverage and the relatively small commitment of U.S. forces, few Americans recognized Vietnam as a war zone. The fighting did not resemble the battlefields of Europe during World War II or those of the more recent Korean conflict. The Vietcong guerillas, with no means to mount large-scale military attacks, concentrated on building espionage networks within the South Vietnamese government and carrying out terrorist-like attacks on selected targets.

Jameson had been sent to Vietnam by TSD to support the Agency's covert action program.[9] His role, as an "authentication" officer, carried on a tradition that reached back two decades to similar work done by OSS. Just as the OSS had reproduced German and French documents for agents sent into occupied Europe, TSD was now outfitting South Vietnamese agents with documents and clothing for infiltration missions into the north to conduct intelligence gathering, sabotage, and harassment operations.

However, the situation Jameson found in Vietnam suggested that TSD could do more than just provide documentation. Another small TSD unit had experience in training and equipping paramilitary forces through its involvement with the ill-fated 1961 Bay of Pigs invasion. Now, a year later, TSD's paramilitary and "authentication" units were combined to form a covert action element. As the only member of the new group in Vietnam, Jameson assumed the paramilitary responsibility as well.

The primary problem that confronted the CIA and the South Vietnamese government in 1962 was halting the flow of munitions and personnel entering South Vietnam from the north. The principal infiltration routes were Highway 1, an intermittently paved road running along Vietnam's eastern coast and the better-known Ho Chi Minh Trail, an intricate 20,000-kilometer network of roads and jungle trails. The Ho Chi Minh Trail ran along Vietnam's western border, cutting southward through Laos and Cambodia.[10] For Jameson and other Agency personnel, shutting down the flow of weapons and personnel meant taking the fight to the enemy by destroying the infrastructure along both supply routes.

It was to become a counterinsurgency war, fought by small, fast-moving teams. Employing unconventional warfare tactics and clandestine weaponry similar to those used by the OSS, U.S. military advisors worked with special units of the South Vietnamese army and indigenous groups, such as the Montagnards and ethnic Chinese Nungs.[11] However, waging this type of war required training and detailed planning.

"When doing sabotage, folks tended to focus on the 'big bang,' the explosive charge in your hand," Jameson said. "Part of my job was to make sure all the other pieces were in place. Leave out one of those and you'll leave a bridge standing or lose your team."

Planning for sabotage, Jameson recalled, required exhaustive sessions, sometimes taking two or three days for a single mission. Every detail, from the daily rations to intelligence about the target's precise orientation, materials, and appearance had to be addressed. Destroying just one bridge required logistics, explosives, first aid, communications equipment, and a means to get the team in and out safely. The team had to be trained to handle explosives, set charges, and improvise in the field when necessary. All the intelligence about possible entry and exit routes had to be assembled and considered, since there would be only one chance to bring down the bridge.

"We had to diagnose the construction of that bridge, often with little data, and design explosive charges to do the job," said Jameson, "but not use P equals Plenty. If you want something destroyed, use the right amount of explosives at the weakest point. We taught the Vietnamese how to attach the explosives quickly and set the time delays that allowed the team to get out before the explosion."

Jameson devised an easy to remember acronym, CARVER, to guide preparation of the target package. "Criticality" assessed the importance or critical role of the target for the enemy. "Accessibility" asked if the team had a reasonable chance of getting to the target. "Recognizability" meant the team would know the target when they saw it. "Vulnerability" focused on a realistic appraisal of the degree of damage or destruction that could be done to the target. "Effect" addressed the impact destruction of the target would have on the enemy. "Recoverability" estimated the time and effort required for restoration or reconstruction of the target.

Taken together, the elements of CARVER provided both the planners in the field and those authorizing an operation at Headquarters with a risk-and-benefit analysis to make sound operational decisions. It was foolish, Jameson reasoned, to engage in high-risk operations unless the probability of success was also high. "Target analysis told us how to get the most 'bang for the buck' with limited assets," said one of Jameson's fellow officers. "It worked like a flow sheet that described how something worked, and then led your thinking to identify the weakest point to be attacked."

TSD became engaged in Vietnam as early as 1961, when a marine engineer was dispatched to Hong Kong to overhaul Agency-purchased Chinese junks. Although still conventional in appearance, the junks were far from ordinary by the time the engineer was done with them. The tech replaced the standard propulsion systems with Gray Marine 671 diesel engines that boosted speed from a modest three knots to impressive fifteen knots. He also added a pair of 55-gallon fuel drums lashed to the masts concealing .50 caliber machine guns and a battery of camouflaged 3.5-inch rockets on top of the wheelhouse rigged to a firing switch within the captain's easy reach. Finally, the engineer built a covered hiding place beneath the deck for a pair of crewmen armed with 9mm Swedish K submachine guns. The junks were deployed for covert patrol and infiltration operations along the Vietnamese coast north of the DMZ. If approached by a hostile patrol boat, the junk's reaction was both surprising and devastating.[12]

A seaworthy rubber raft known as the Zodiac, which grew out of a TSS project code-named RB-12 during the Korean Conflict, became the mainstay of amphibious infiltration operations into North Vietnam. These rafts carried landing teams launched by the modified junk mother ships to the insertion points along the coast. To track the rafts after launching, the techs adapted "cherry top" flashers similar to those used on early police cars, lining them with Kodak gelatin filters (numbers 87, 87C, 88A, or 89B) through which light only in the infrared spectrum passed. When lit from within, the covert "flashers" were invisible to the naked eye, but could be seen by using a T-7 metascope—a hand-held, battery-operated infrared optical device.[13]

Training of poorly motivated Vietnamese guerilla fighters proved problematic. "The Vietnamese government provided the requisite numbers of bodies for sabotage and harassment operations, but they were unqualified, and difficult to train," Jameson explained. "Many of the young men they supplied were just people the government wanted to get off the streets, city kids, and not country boys who understood how to hunt and shoot. We kept saying, without much response, 'give us some country boys, and don't give us these city thugs.'"

This became apparent during one covert mission that called for a CIA-trained four-man Vietnamese team to infiltrate the North and destroy a bridge on Highway 1, cutting off—at least temporarily—a major Vietcong supply route. As planned, the team would launch from the junk in a Zodiac raft, land on the beach, trek five miles inland to the bridge, plant the time-delayed charge, then hike back to the Zodiac hidden on the beach and return to the junk.

On the night of the operation, the team launched on schedule and began reporting their location by radio as the Zodiac approached the landing site. At the command post on the junk, Jameson noted the progress as the team reported reaching "point one," "point two," and "point three." The mission looked good. A few minutes later came the next report: "point three," then "point two." Clearly, the team was returning. When the Zodiac pulled along-side the junk, one of the Vietnamese team members pointed to water in the bottom of the raft from what seemed to be a leak.

Jameson, skeptical, inspected the raft and found multiple holes—all made by knife blades. When questioned, team members admitted that after launch they had become frightened and sabotaged the mission. "Those guys went to jail," Jameson recounted, "but we learned an important lesson. Find a higher caliber of agent. We recruited four Nungs and trained them for the operation."

Several weeks later the Nung team launched from the same junk and encountered no trouble reaching the beach, landing, or finding the right trail to the bridge. Following the operational plan, they attached the explosive charges to the bridge support structure. Once secured, they activated the time-delay devices and departed the area using a different trail than the one used to approach the target. Overhead photos the next day confirmed that the bridge "dropped" exactly as planned.

Other high-value targets called for a more creative approach. A North Vietnamese petroleum depot that provided fuel for equipment moving into South Vietnam was one such a target. Unlike the bridge, the heavily guarded facility was surrounded by chain-link fences and unapproachable by saboteurs using conventional explosive charges. Since air strikes were not authorized at that stage of the war, the most effective option was to attack from a "stand off" position. A rocket attack could potentially take out the fuel tanks, but only if a small team could carry enough firepower close to the target. Then they would have to set up, aim precisely, and fire all the rockets for a reasonable chance of hitting the fuel tanks and elude whatever response the attack elicited.

The solution devised by TSD would later be called the *Triple Tube Rocket Launcher.* "The genesis of the *Triple Tube Launcher,* what we called the TTL, started with a single antitank rocket fired from an improvised launcher that wasn't much more than a piece of angle iron," said one tech who followed the development of the device. "The original concept was to fire a 3.5-inch antitank rocket by stuffing a wad of match heads and time fuse in the back end. Crude but simple, it was used in urban guerilla warfare scenarios like the Hungarian uprising with a civilian population combating tanks. We thought that if one rocket was good, then three ought to do a better job with greater chance of hitting a target. So we set the rockets with a three-degree spread between tubes and added electrical firing for more precise command and control."

The three-tube launcher was mounted on a modified backpack frame that allowed team members to accurately sight and adjust inclination. "We tried to do all this so the saboteurs didn't have to do any thinking," Jameson explained. "They'd just go in there and go right to the place where they've been shown, aim it like this, raise it, hit two buttons, and go." For the rockets to penetrate the tanks' steel and ignite the fuel inside, the techs added incendiary adaptors, aluminum packages filled with magnesium, that would burn fiercely when exposed to oxygen after the initial explosion.

Time delays were attached to the launcher package, allowing the team to initiate the firing sequence and head out of the area before launch. "We didn't want to have our guys firing the rockets, then running like hell for nine miles to the boats," said Jameson. "So we adapted time-delay mechanisms that gave the team several hours to get back to the rubber boats and head downriver before the rockets went off."

To prevent spent rocket launchers from later being used against American troops, the techs added a self-destruct mechanism with half a pound of explosives to destroy the unit after firing. But that left another concern. If an enemy patrol discovered the launchers before they fired, the NVA would acquire an effective weapon. TSD engineers responded by incorporating an antidisturbance device. If tampering was detected after the safety was removed, the rockets launched automatically and 1.35 seconds later, the explosives detonated.[14]

Another TSD innovation, the firefight simulator, resembled a collection of fireworks and other explosives, set on a timer. The device mimicked the sound of automatic weapons fire, mortars, and grenades. U.S. military units infiltrated the simulators into enemy base camps to create diversion and confusion. In one instance, when the device went off in the middle of the night, the panicked and disoriented North Vietnamese began shooting each other.[15]

In 1962, the Kennedy administration initiated the transfer of covert Southeast Asian paramilitary programs from CIA to military control. The official date for the transfer was to be November 1, 1963. However, the timing was disrupted first because of the overthrow and murder of South Vietnamese President Diem on November 2, then, three weeks later, by the assassination of President Kennedy on November 22. In December, a plan was approved by Defense Secretary Robert McNamara to increase covert attacks into North Vietnam and in January 1964 the U.S. Military Assistance Command, Vietnam (MACV) organized a clandestine unit composed of Air Force Air Commandos, Army Special Forces, and Navy SEALs under the Special Operations Group (SOG).[16]

As the war effort expanded and troop numbers increased, the Agency continued to maintain an active presence in Vietnam with TSD playing a key role. Much of the early gear supplied by TSD was reminiscent of equipment issued by OSS during World War II, such as escape and evasion devices with concealed compasses and saws as well as radios and clothing. "We had a little package about the size of a pack of cigarettes that illuminated an internal map," said Bill Parr, a TSD engineer at the time. "By pushing down on the top of the package, the map inside would be dimly lit by a back light. We gave the maps to teams working in North Vietnam, so if a team had to move at night they could orient themselves."

The backlit maps eliminated the need for flashlight illumination that could give away the presence of a clandestine team and the small size replaced otherwise bulky sheets of large foldout terrain maps. Later, TSD lent some of the units to NASA for evaluation as a tool for astronauts to store navigational information when working in darkness during the early space flights.

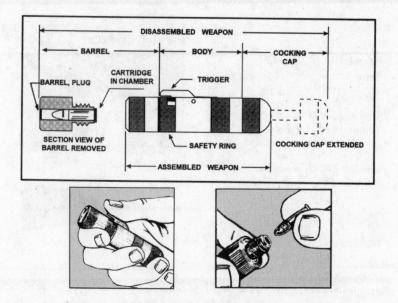

The CIA Stinger was an easily concealed .22 caliber single-shot pistol effective for targets at short range. It could be fired from the palm of the hand toward a target in the same room or passing in a crowd.

Other devices included concealments that would blend into the environment. "I got an idea one day going through one of the stores down in Naha, Okinawa," said Parr. "I noticed the way the thermoses were designed. You could buy a big thermos and a little thermos and switch out the glass insert from the little thermos into the larger body. That gave you a big whopping cavity for concealing papers and documents. It became a standard item for our agents in Vietnam. We put a left-hand thread opening on the bottom so that an unknowing person who attempted to open it would actually make it tighter." Techs then improved on the concept after realizing that dropping the thermos would break the inset and soak any papers inside. By using water-soluble paper, an agent would have a self-destruct mechanism built in his thermos concealment.

There was also a new clandestine radio. The RS-6, a portable shortwave radio station, was intended for use by an agent operating behind enemy lines. Transmitting in the 3–15 MHz frequency range at distances up to 3,000 miles [using continuous wave (CW) or Morse code], the RS-6 received both CW and AM signals. The radio was small enough to conceal

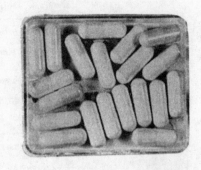

"Puppy Chow" tablets, tranquilizer capsules mixed with ground beef, silenced guard dogs. The average dog required four tablets or more if the animal's ferocity warranted it. The effects lasted up to four hours with no aftereffects beyond temporary loss of balance and lethargy. A Syrette filed with antidote could be injected to speed the animal's recovery if necessary.

inside a standard briefcase, or could be separated into four smaller components buried inside waterproof pouches. The versatile set could be powered using storage batteries, AC lines, or a hand-cranked generator.[17]

A harassment device known as B-3 *Dust Powder* consisted of finely powdered tear gas in a small plastic squeeze bottle. *Dust Powder* could harass and disperse groups of people or be used for personal protection. When the nonlethal powder came in contact with the moist tissues of the eyes, nasal passages, or throat, it caused coughing, tears, loss of breath, and nausea. Once the person left the contaminated area, the effects disappeared in a few minutes.[18]

A tranquilizer, *Puppy Chow*, was a plastic case filled with twenty tranquilizer capsules, and two Syrettes containing antidote. The kit was used to silence guard dogs by feeding them tranquilizer capsules mixed with ground beef. The recommended portion for the average dog was four capsules, but was to be increased if the animal was particularly ferocious. After ingesting the special mix, the dog became unconscious for up to four hours, but suffered no ill effects other than loss of balance and lethargy during the recovery period. If needed, a Syrette filled with antidote could be injected to speed up the animal's recovery.[19]

A covert *Document Copy Attaché Kit* that contained a complete photographic reproduction system concealed inside a standard briefcase performed the function of a portable photocopier. Once assembled, it offered a one-position copy device for photographing documents up to nine by fourteen inches. The system employed a fixed-focus Pentax camera modified for silent operation. An agent or case officer could use the system to produce properly exposed negatives consistently with no prior training.[20]

Early in the Vietnam War, the need to carry adequate food rations posed a persistent problem for covert infiltration teams whose operations could last ninety days or more. Packed in cans with liquid, C-rations, the typical military fare, were bulky and inconvenient. They were heavy and their metal and paper packaging waste had to be carried out since any trash would leave evidence of the mission.

"When I first got to Vietnam I found that the ninety-day supply of food, clothing, ammo, and other equipment for one infiltration team took up about four pallets, with food rations taking up most of the space," said Jameson. "We could airdrop the pallets into remote areas, but for the team to find it and unpack it without leaving a trail was nearly impossible. So I set out to reduce that bulk as much as I could."

Working with one of America's leading breakfast cereal companies, TSD engineers thought they found a solution with a product they called "CD rations."[21] Resembling today's energy bars, the CD rations contained concentrated servings of protein and other nutrients that were rehydrated in water and cooked in the field to provide all the nutrition of C-rations without the bulk. "I had TSD's Asia shop make special survival vests with a lot of pockets of the exact size to carry the new product," remembered Jameson. "Then I found a patrol that was going out for several days and was willing to taste test our new rations. They would be gone for a week and subsist only on those rations. We needed to learn how they would be accepted by soldiers under the stress of combat."

Jameson accompanied the patrol, which included U.S. Special Forces and eight Montagnards. "No one complained about the food, but the patrol hadn't been out for more than a couple days when some members started getting sick and throwing up the rations," said Jameson. "We didn't have any backup food, because I hadn't allowed us to take any. I knew that given a choice, they probably wouldn't eat what they had left."

The team found a village with an orange orchard and loaded up their backpacks. "Well, those oranges were extremely acidic," Jameson explained. "We ate them like apples, just spitting out the rinds. So, by the time we got back to the base camp, the acid was eating the lining of our mouths and we were all bleeding. We were a sight. It's safe to say future patrols didn't request rations from me."

Jameson filed a report on the debacle and the techs returned to the drawing board with the cereal people. "Eventually that work produced a much better bar," said Jameson. "But I could never introduce it. Not after the riding I took from those Special Forces guys."

Throughout the early part of the war, Stanley Lovell's .22 caliber silenced *Hi-Standard* pistol was a favorite among CIA and Special Forces.[22] Compact, accurate at close ranges, and reliable, the World War II gun was marking its third decade of service as OTS engineers worked to improve and adapt the weapon to new missions. Among the first enhancements added was an attachable shoulder stock that essentially turned the pistol into a rifle for increased accuracy.

Parr, who headed the project, added features to make the pistol suitable for the skies as well as the jungles. "After we put the shoulder stock on it, we fitted it with a holster that would survive an enormous amount of G force, so it became a survival weapon for U-2 pilots." said Parr.[23] "That was a cool project. I went to the California test area and sat in the cockpit of an SR-71. I needed to see what it was like, if the pilot had to bail out with a holster and weapon on his hip. Same way for the U-2 aircraft. We talked to the pilots, 'What are your druthers, guys? What makes you uncomfortable? What can you live with? What additional equipment can you handle, on top of all the other crap that you've got?' We went through all that."

Reviewing the *Hi-Standard* pistol design, TSD engineers discovered that if the sear—the part of the firing mechanism linking the trigger to the hammer—was trimmed down, the weapon functioned as a machine gun, emptying its ten-round magazine in about a second and a half. "You'd pull the trigger once and it would go 'burp.' You've put ten rounds into your target and multiple hitting is always more effective than a single shot," said Parr. "But with the silencer on there was a tendency for the muzzle to drop rather than rise. You have to compensate in that direction. With the shoulder stock, you had a degree of control. The whole trick was cutting the sear, and then you had to do a little work making sure the magazine spring was correct. Our goal was a silent machine gun and we made pretty good progress."

During their research, engineers found that silenced weapons frequently malfunctioned and jammed when firing standard military ammunition. For a weapon to be silent as well as lethal, powder loads needed to be precise. Too much powder and the weapon was noisy, too little powder and it suffered in terms of velocity and lethality. "That was always trouble with the

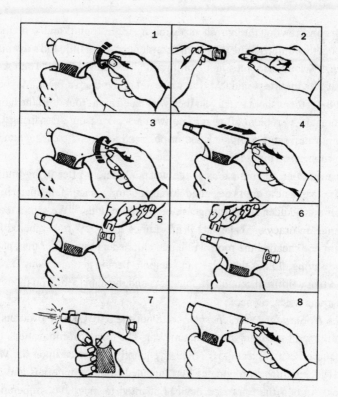

The 9mm CIA DEAR Gun *in the mid-1960s was a low-cost personal weapon accurate only at short distances and a successor to the OSS-designed* Woolworth *or* Liberator *pistol.*

silent-weapons program, using standard-issue ammunition. There was a wide variance in loading consistencies and performance," explained Parr. "It turned out the 'silenced' guns just didn't work with regular ammo."

TSD engineers also updated the OSS *Liberator* pistol, the single-shot .45 caliber handgun designed for large-scale distribution to partisan forces behind enemy lines during World War II.[24] The Vietnam edition, called the *DEAR Gun* (for DEnied ARea Weapon), was a small, inexpensive cast aluminum handgun with a blued steel barrel that fired a single 9mm parabellum round. As with the *Liberator*, it was intended for use by partisans to obtain another, more powerful, weapon from the enemy. Packed in a Styrofoam box with illustrated instructions, the plan called for an airdrop of the weapons behind enemy lines.[25]

"By unscrewing the barrel, inserting a single round, and screwing it back together and cocking it by pulling a plunger like a little kid's toy pistol, the *DEAR* would shoot a single 9mm round," explained Parr. "Then somebody at Headquarters said, 'Maybe we're not doing them a favor. Maybe we ought to let them shoot twice.' So then we loaded extra rounds in the handle. That created a problem of what to do with a stuck casing after firing. That led to our designing a simple stick, or push rod to eject the casing attached to the rubberized butt of the pistol."

The question then arose as to the weapon's safety after firing multiple rounds. Would the gun, designed for firing only once, fall apart in the shooter's hand after discharging five or twenty rounds? "We sent a tech to the range and he must have shot it fifty times in a row," remembered Parr. "But on one shot, he lost control of it and chipped a tooth. After that, people started saying, 'Oh, it's a dangerous weapon. People get hurt using it.' That spread like wildfire through Headquarters and the plan to airdrop thousands of the guns was scrapped."

The *Hi-Standard* and *Liberator* were not the only OSS innovations that saw a second life during the Vietnam War. The technical innovations from TSD and other CIA technical offices frequently improved upon the World War II technology with updates that included modern materials and electronics. Among the improved devices adapted to meet 1960s operational requirements was the *Stinger*, a .22 caliber weapon designed to be fired from the palm of the hand at a person sitting in the same room or passing closely in a crowd. The 1962 update of the original design improved the concealable, reloadable gun that featured a lightweight aluminum firing tube (four and a half inches long by three-quarters of an inch in diameter). Issued with a spare barrel, seven rounds of ammunition, and a pictorial instruction sheet, it could be concealed rectally, or camouflaged inside a lead-foil tube of mechanic's grease inside a tool kit.[26]

Another gun, code-named *Golden Rod*, concealed a 9mm machine pistol inside an ordinary looking flashlight. About a foot long and two inches in diameter, the weapon had an internal circular magazine feed. "You loaded all the rounds around the barrel, then pressed the firing button, and this thing would spit out these 9mm rounds at a faster rate than you could distinguish an individual shot," said Parr. "You just leveled it toward the target and it went 'burp, burp.' It stopped when you let your finger up and started again when you put your finger down."

One of the more colorful weapons of the Vietnam era was the *Gyrojet*.[27]

The pistol was actually a handheld rocket launcher designed by a California contractor. Constructed of stamped steel and plastic, it fired 13mm projectiles powered by solid rocket fuel that reached a speed of 1,250 feet per second within sixty feet after leaving the barrel. Because the fuel burned quickly, the gun had virtually no recoil and was nearly silent, except for a distinctive "whooshing" sound. Despite its low noise level, it packed a punch. During one test, the projectile penetrated the door of a three-quarter-ton truck and tore through a fifty-five-gallon drum filled with water before embedding itself in the opposite door of the vehicle.[28]

Unfortunately, the gun had two major flaws: inaccuracy and unreliability. The expended rocket fuel, venting through two holes at the base, often sent the projectile off target. "It stabilized itself by using miniature canted jets to introduce a spinning rotation along the axis of flight. It would almost 'spin up' in the barrel as it was getting ready to depart, and that was part of its problem," said Parr. "You'd pull the trigger and you'd hear the ignition cartridge fire and it would sizzle, and then a whoosh. It was like a bottle rocket—it took a little while to build up momentum. It was spinning as it was leaving the barrel. Compared to a firearm there was a delay."

Solving the problem required spending huge sums to machine the mini-jets on each round to precise tolerances. "I was managing the *Gyrojet* contract and the contractor was trying to defend its accuracy," recalled Parr. "So, he said, 'Come on out to our range and I'll prove it to you.'"

The contractor escorted Parr out to the company range, which consisted of a porchlike affair overlooking a dirt patch. Squeezing off several rounds from the *Gyrojet*, the contractor made shots that clearly were off target. "The rounds were all over the map. And the contractor says, 'Well a .45's no more accurate,'" remembered Parr. "He handed me a .45. I fired once and hit right in the middle of the target. Pure luck, but it made my point."

TSD discontinued the *Gyrojet* contract because of its accuracy problems, though the gun found limited deployment with SOG. First Lieutenant George "Ken" Sisler was armed with a *Gyrojet* when he single-handedly charged a North Vietnamese platoon to rescue injured members of his squad. After saving his fellow soldiers, he was shot a short time later by a sniper and awarded the Medal of Honor posthumously.[29]

Throughout the Vietnam War, Dien Bien Phu remained a potent reminder of the defeat of the French colonial power. With the North Vietnamese Army (NVA) continuing to use the garrison as a military base, the former

French stronghold presented an inviting psychological as well as a military target for some U.S. planners.

"Headquarters wanted desperately to send a message to the NVA and we thought we could put together a team with modified 2.75 air-to-ground rockets to attack the target," explained Parr.

The idea was to put Dien Bien Phu "under siege," at least temporarily, through a covert operation. The techs needed to develop a means to launch rockets into the NVA garrison in rapid fire to simulate artillery, and do it with a single team of a dozen Montagnards. However, to make the system work, the team would have to position the rockets properly and set coordinates because the angle and direction of flight had to be precise for the rockets to strike their targets. "We couldn't train Montagnards to use an optical system, it was too complex," said Parr. "We had to make it simple."

First, the techs calculated the firing angle from the designated launching site to the target, then produced an easy-to-use sighting system. The only requirement was to align a small free-moving arrow on the rocket casing with a black mark on the bracket that held the rocket in position. Once properly located and with the inclination set, the team could arm the rockets and clear out. Similar to the *Triple Tube Launcher*, the rockets had timers that allowed the twelve-man team to be far from the area when the attack began.

To avoid detection and to deflect attention away from the target, the insertion point was a three-day walk to Dien Bien Phu. Each of the twelve men carried one modified rocket and locating beacon so the techs could track their progress. After reaching the site, the rockets were positioned, armed, and the team extracted. The rockets fired as programmed and hit their targets inside the garrison. "We all considered the operation a roaring success," said Parr. "We picked up NVA communications stating that the attack came during one of their staff meetings. It was like sticking an elephant with a pin but remarkably satisfying."

By 1968, Laos had become a major battleground as the CIA fought to slow the flow of North Vietnamese troops and material along the Ho Chi Minh Trail that ran through northern and eastern Laos into South Vietnam. The North Vietnamese government protected these routes, in part, by supporting the communist Pathlet Lao insurgents who controlled the region.[30] From Udorn, Thailand, the CIA and U.S. military advisors conducted joint operations with Laotians under the leadership of General Vang Pao. The immediate objective was to interdict supplies and men moving on the Ho Chi Minh Trail and retake Pathlet Lao–held territory.

TSD sent a four-man survey team into Laos in 1968 to assess the technical requirements needed to support paramilitary operations. The base for CIA containment operations in Laos, near Udorn, was located on the Thai side of the Mekong River that divides Laos and Thailand. In addition to housing the CIA's Joint Liaison Detachment, the base was also home to Air America and a photo interpretation center.

The survey established a need for sustained technical support in the area. A technical shop that included an electronics bench, machine shop, photo darkroom, woodworking equipment, and briefing room was constructed. Prominently in the center of the floor sat a brown craft paper–covered worktable around which the techs met for operational planning and evaluations. The center table in a workshop without individual offices became a symbol of the cooperation and integration required from all parties in operations planning. TSD techs worked alongside other CIA officers in northern Thailand and Laos until the U.S. withdrawal from Vietnam.

Electronic monitoring of the Ho Chi Minh Trail demanded technical innovation from TSD as the Agency attempted to use seismic technology to differentiate the types of vehicles traveling along the trail. "To test the prototype devices, we went out to one of our stateside facilities, strung them out on the road, and ran trucks and bicycles by them," Jameson explained. "We tested the concept until we figured out how we could do this. I went back to Laos and started burying the detectors alongside the complex of roads and paths that made up the 'the trail.' We buried the transceiver off to the side of the road where they were well camouflaged." As convoys traveled south along the Ho Chi Minh Trail, the devices detected and registered seismic disturbances from traffic and sent the data to a recording unit buried off to the side. Planes then made daily circuits to "interrogate" the recorders by radio.

The Americans also trained indigenous forces to observe at key positions along the trail and report on North Vietnamese convoys. The observers had radios to "call in" when they saw traffic, but information had value only if the teams understood and reported precisely what they were seeing. "Some of them didn't know the difference between a truck or jeep or trailer," Jameson noted. "We needed a better way for them to communicate." To compensate for the language problem, the reconnaissance team units were issued the

Elephant Transmitter with buttons that featured icons depicting a man, bicycle, elephant, truck, troop transport, or tank. The observer punched the icon matching the vehicle's shape to record traffic types and numbers.

The techs also tapped North Vietnamese communications wherever lines or transmissions were accessible. One program tapped transmission wires, recorded conversations, and transmitted to a hidden "repeater" that was actually the first of a string of relays that stretched across Laos to a safe listening post in Thailand. In essence, the system functioned like a modern cell phone network, with the signal bouncing from one cell, or repeater, to another, until reaching the listening post. "The biggest problem with this operation wasn't the transmission," explained Jameson, "it was finding places to hide the relays."

The tapping equipment used a mixture of commercial and Agency audio devices and batteries reconfigured in the field for the operation. Some of these were concealed in wooden telephone poles. In a typical operation, a two-man team of Montagnards would be inserted by helicopter to within walking distance of the target. It was, at best, a difficult operation, the team carrying a pole filled with batteries and transmitters through miles of hostile territory. Once at the target, they climbed the pole, tied down the telephone line, substituted the original pole with the TSD replacement, and reattached the line. When an operation went smoothly, the listening post immediately received a stream of North Vietnamese conversation. "We were getting an awful lot of intelligence on the movement of supply trains coming down along the Ho Chi Minh Trail from North Vietnam to the south," said Jameson. "That information allowed us to target supply convoys and understand something of their order of battle plan in the South. Trucks loaded with 7.2 [rifle] ammo compared to truckloads of mortar rounds indicated preparations for different types of engagements."

The development of electronic and high-frequency signal and homing devices for clandestine operations may have been TSD's most significant contribution of the Vietnam and Laotian conflicts. Historically fires lit on top of high places and hilltops served as beacons for navigation and pathfinding and alerts of approaching enemy forces. Paul Revere's midnight ride that warned patriots of the advancing British army began with a beacon of light in a church steeple.

By the 1960s, ground-based radio direction-finding equipment became a primary navigation aid to "mark" a position for guiding aircraft or a ground party to a specific location. The techs installed hundreds of HRT-2c aircraft beacons at defensive and logistic sites throughout Laos to enable pilots to find locations for airdrops, landings, or close support.

Although the HRT-2c was compatible with automatic direction-finder systems already incorporated into all CIA, military, and commercial aircraft, it presented problems in placement and maintenance. Because each beacon required an antenna at least fourteen feet tall with a ground plane and grounding wires, finding a suitable location for placement could be difficult. The wires were vulnerable to people or animal traffic, and sections of the antenna's fiberglass pole were pilfered by villagers who discovered they made excellent pipes for opium smoking.

Despite these problems, ground navigation beacons guided pilots throughout Laos and enabled flights in most weather conditions. Agents and reconnaissance teams were equipped with hand-held receiver-transmitters for positioning, authentication, identifying resupply sites, marking targets, air strikes, and calling for extraction. The small cylindrical-shaped unit, resembling a swagger stick when extended, featured a collapsible antenna at one end and a push-to-transmit button on the other.

To support teams on extended missions in Pathlet Lao–controlled territory or North Vietnam, resupply pallets of food and equipment were air dropped, but the covert operations precluded radio communications with pilots. An alternate means of locating the pallets was required. The answer came in a new type of beacon in the form of portable commercial FM receivers and hand-held direction-finding units. The receivers would detect signals from high-frequency transmitters attached to the pallets that began signaling only after the pallet landed.

Some air strike operations involved concealing small transmitters in U.S. radios or rifle butts intentionally left at the scene of a firefight with the expectation they would be scavenged by enemy forces. When these were carried back to a base camp, signals from the bugged weapon would silently pinpoint the location for a precision air attack.

"Counting all the reconnaissance, resupply, and sabotage operations, I estimated TSD equipment was used in thirty to forty missions a day in Laos and Vietnam," said Jameson. "[The technology we used there] wasn't particularly advanced compared to what was available for top-of-the-line audio bugs; from our perspective, it seemed that Headquarters was more focused on building new audio devices. However, we had guys with enough technical talent that we could build our own 'bombing beacons.'"

One of these, called the HRT-10, was about the size of a transistor radio with only enough battery life for a few hours. The unit could be concealed in items such as backpacks or PRC-25 radios and issued to agents who infiltrated the Pathlet Lao or North Vietnamese base camps. "They'd get up close to the camp and stop to take a crap and hang the antenna on a tree," Jameson explained. "Then the agent would come back and tell us where he put it—for instance, it could be fifty yards north-northeast of a base camp in a certain area. Our aircraft could pick up the weak signal once they were in the general area and strafe or bomb the target."

Jameson never believed TSD invested sufficient money and effort in developing and upgrading combat support receivers and transmitters. "We ended up using commercial receivers and made the modifications ourselves," he recalled. "Eventually a receiver was modified for our aircraft and linked to direction-finding antennas on its belly.[31] It helped, but it was limited to 'left and right' indicators and was so complicated that it required a tech to operate it. We flew in unarmed Air America aircraft to guide the pilot to the target site. The CIA pilots would then call in the Air Force for a bombing raid based on the coordinates."

Because the planes could not fly directly over the target without alerting the enemy below, TSD developed a number of techniques for "off-set targeting." As the spotter plane flew along the side of the target, a tech took readings from the signal, then calculated the target's true coordinates. That information was relayed to the bombers already coming into the operational area. The spotters' planes then released a smoke grenade to provide a visual confirmation of the target for the attack pilots. In the next evolution of the system, a smoke grenade and time-delay mechanism were incorporated into the beacon itself. Since the delay did not ignite the smoke grenade until attack aircraft were on station, the procedure reduced the time enemy forces had to scatter or disable the smoke grenade.

The most successful operations were ones in which an agent carried a beacon into an enemy camp with the time delay. Walking sticks were common throughout the region and were large enough to conceal beacons and batteries. "We devised the means for a timer to switch on so that the asset could depart the area before the signal came on," said Parr. "The Air Force then did 'mini–arc light' strikes with Phantom jets coming in wingtip to wingtip at dawn flying directly in on one of our signals."

Exfiltration of downed pilots and imprisoned soldiers from behind enemy lines was a CIA and military priority throughout the Vietnam War. The

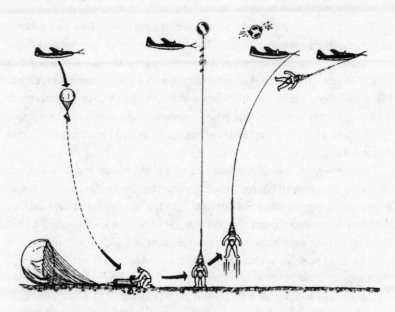

During a covert action in Indonesia in the late 1950s, Civil Air Transport pilot Allen Pope was shot down while delivering supplies. TSS worked on two plans to rescue Pope from a low-security jungle jail. One plan involved using the Skyhook air-ground extraction device, another a collapsible rubber aircraft. While the equipment functioned as designed, neither plan proved operationally practical and Pope was eventually rescued by other means.

captured and missing would not be forgotten or abandoned. In 1958, when CIA pilot Allen Pope was a prisoner in Indonesia, the Agency came up with an audacious plan for his rescue.[32]

Intelligence determined Pope was being held in a remote jungle region of Indonesia under house arrest. Although he had relative freedom of movement in the general area, there was little chance for him to survive an escape through the jungle. TSS hit upon the idea of a portable inflatable aircraft. With the help of the Goodyear Company, TSS designed a small rubber plane that could be bundled and airdropped into a jungle clearing.[33] All Pope would need to do was add water to special pellets inside the bundle and a chemical reaction would produce enough gas to inflate the plane. "We tested it and it worked out pretty good. But we got ready to run it to him and somebody, politics I suppose, canceled the operation," recalled Jameson. "As clever as it was, I don't believe the aircraft was ever used operationally."

The CIA eventually used a submarine to insert two of its paramilitary officers on the Indonesian coast. They went into the jungle, located Pope, and walked him out to safety, and the rubber airplane faded into Agency lore. "We put it in a warehouse where it stayed for years," Jameson recalled. "One day there was some operation and I said, 'I think we can solve this with that rubber airplane.' But when we went to the warehouse, we found the rubber was all dried up and cracked. Nobody had maintained it. So we threw it away."

Another extraction device considered for Pope was the *Skyhook*, an invention of Robert E. Fulton, who envisioned that an airplane outfitted with a hook dangling beneath it could safely snatch a suitably harnessed individual from the ground. Fulton's inspiration, the "All American System," traced its origins to a mail recovery technique from the 1920s in which pilots snagged mail pouches suspended between two poles and winched them into the airplane.

An attempt by the U.S. military during World War II to modify the technique was only partially successful. In July 1943 tests by the Army Air Force, instrumented containers were extracted from the ground by aircraft, which recorded accelerations following the pickup in excess of 17 gs, far more than the human body could tolerate. Changes in the pickup line and modifications in the parachute harness eventually brought this down to a more acceptable 7 gs. The first live test, with a sheep, failed when the harness twisted and strangled the animal. During subsequent tests, the sheep survived. Lieutenant Alex Doster, a paratrooper, volunteered for the first human pickup. On September 5, 1943, after a Stinson engaged the transfer rope at 125 mph, Doster was yanked vertically off the ground, and soared behind the aircraft.[34]

Later Fulton began experiments that would lead to the *Skyhook* in 1950 by devising a harness that could hold either a person or cargo attached to 500 feet of braided nylon rope. The rope was lifted into the air by a large balloon filled from a small bottle of helium. Once airborne, the rope could be snatched by an aircraft equipped with two steel horns that automatically locked it into place and released the balloon. As the line was pulled tight, it would glide past the side door of the fuselage. Crewmen, standing in an open hatch, could then attach the line to a winch—as the front lock was released—and pull in the rescued pilot or agent to safety.

Over the next few years, Fulton refined the system. Using a Navy P2V for the pickups, he gradually increased the weight of the pickup until the line began to break. A braided nylon line with test strength of 4,000 pounds solved the problem, but early experiments met with mixed results. In one instance, a test pig was picked up successfully. Flown through the air at 125 mph, the pig arrived inside the plane unharmed, but expressed displeasure with the experience by attacking crewmembers.[35] Human volunteers testing the system returned to the aircraft in considerably better spirits and John Wayne brought the device public fame in the movie *The Green Berets*.

The *Skyhook* became a favorite among Special Forces as a test of courage. One Special Forces officer, a parachute rigger who had made more than 5,000 jumps, was said to enjoy "flying" alongside the aircraft Superman-style for extended periods, before being reeled in. In one instance, now legendary among the Special Forces, the pilot of a plane about to pick up a high-ranking officer testing the system slowed his airspeed in an attempt to give the officer a gentler ride, but only succeeded in bouncing him along the ground, breaking numerous bones.[36]

An intelligence requirement for *Skyhook* came in 1961 after the Soviets were forced to abandon a suspected submarine-monitoring station in the Arctic Ocean because the facility's ice runway was collapsing. In one of the *Skyhook*'s few operational deployments, two officers, from the Navy and Air Force, were selected for the mission appropriately code-named Operation Cold Feet. They would be dropped by parachute, spend seventy-two hours at the station, floating in the Arctic, assessing and collecting items of intelligence value. *Skyhook* would then pluck them from the ice along with photographs, papers, and whatever else they discovered on the abandoned facility.

The officers landed as planned on the ice and collected a hundred pounds of intelligence materials from the station. With the collection mission completed, the first intelligence officer picked up by *Skyhook* was dragged 300 feet by wind blowing the balloon before the plane hooked onto it. When the second investigator inflated the pickup balloon, he held on tightly to a piece of equipment on the ground to avoid being dragged. Eventually both men were safely "*Skyhook*ed" into the airplane, bringing information that confirmed the facility was a submarine monitoring station.

Every day Brian Lipton arrived at 0800 hours for work at the Agency and departed at 1700 hours or earlier if he had a late-afternoon softball game. Lipton, a member of Gottlieb's cadre of university-trained chemists hired in the mid-1960s, seemed like a good fit with TSD. He shouldered a heavy workload, was amiable, and always put in a full day's work. Few of his colleagues suspected Lipton led a double life inside his primary CIA cover. At night after the parking lot and the offices of TSD's South Building headquarters emptied, Lipton returned to his office alone. There, behind a locked door, he worked on one of the most closely held projects of the Vietnam War era—creating covert-communication systems for U.S. prisoners of war in North Vietnam.

For the POWs who were isolated, inundated with propaganda, and subjected to continuous physical and psychological torture, these covert communications were often their only link to the outside. Confirmation to the beleaguered POWs that the U.S. government knew they were alive gave hope, and for many, enabled them to hang on. The communications were also a means for the U.S. government to identify who was alive, detail their living conditions and treatment, and possibly plan rescue missions.

"As a new TSD officer, I initially thought this was the kind of stuff that happened every day. A person is called in and gets assigned an important and super-secret job," said Lipton. "We understood there were many things going on up and down the hall that you didn't know about. It wasn't until later, when we started to see successes, that I appreciated how valuable the work was."

The project traced its origins to a single POW, Navy Captain James B. Stockdale. Shot down over North Vietnam on September 9, 1965, Stockdale at first used letters from prison to his wife, Sybil, to communicate the last names of fellow prisoners, employing a simple code, uncharacteristically referencing several "football mates" from his class at the U.S. Naval Academy. In fact, the three men had not played football with Stockdale, but their last names were the same as missing aviators from his squadron whose status was then unknown.[37]

U.S. Naval Intelligence was presented with the information and afterward asked for Sybil Stockdale's cooperation in sending a secret message to her husband using a "special picture."[38] Preparing the first images took weeks, but Sybil began by including a photograph in the one letter her husband was allowed to receive each month. The "photograph of the month" ploy was designed to create a routine pattern of the correspondence so the special photo would not be alerting to North Vietnamese censors.[39]

In the fall of 1966, the special photo was ready and Sybil devised text in the letter that secretly instructed her husband to submerge the picture in water. Once soaked, the photograph's layers would separate, allowing her husband to read an enclosed message.[40] Sybil's signal involved a photo of a stand-in for her mother-in-law shown enjoying herself in the ocean. Sybil knew that her husband would immediately recognize the woman was not his mother.[41]

Sybil's two letters reached Stockdale just before Christmas of 1966 and were handed to him during a filmed North Vietnamese propaganda session intended to show the humane treatment of American POWs. Stockdale instantly knew the photograph was not of his mother, but remained puzzled as to why his wife sent it. Several days after Christmas, in a moment of frustration, he decided to destroy the photo, then thought: "It's dumb to throw away something from the States without doing more with it; James Bond would soak it in piss and see if something came out." Stockdale's middle name actually was Bond.[42]

In the few minutes of privacy before the guards made their next rounds, Stockdale filled his drinking cup with urine and stuck the picture inside. After a half hour, nothing happened, other than the "cheesy Polaroid paper starting to fray at the edge." When he pulled at the edge and separated the two white pieces of the photo, he began to see "small specs" emerge. Suddenly, an entire paragraph appeared on a "decal-like thing" on the inside of the photo paper.[43] With only seconds before the guards arrived, he tried to memorize the message, which read:

> *The letter in the envelope with this picture is written on invisible carbon. . . . All future letters bearing an odd date will be in invisible carbon. . . . Use after you write a letter. . . . Put your letter on hard surface, carbon on top and copy paper on top of that. Write message on copy paper with firm pressure but not enough to indent papers below. . . . Best to write invisible messages in lines perpendicular to lines of plain language of letter home. . . . Use stylus directly in invisible carbon if copy paper not available. . . . A piece of invisible carbon can be used many times. . . . Begin each "carboned" letter with "Darling" and end with "Your adoring husband." . . . Be careful; being caught using carbon paper could lead to espionage charges. . . . Soak any picture with a rose on it. . . . Hang on.[44]*

Stockdale's first priority was to construct a list of the names of every POW known to be alive and send it back to the people "in intelligence." On January 2, 1967, he prepared his first letter containing forty names of confirmed POWs, using the carbon technique along with the "darling/adoring husband" phrases.[45] The message was then unknowingly returned to the United States by a leftist anti-Vietnam organization named Women Strike for Peace.

Two weeks later another opportunity to communicate arose when Reverend Muste, a Presbyterian minister and frequent Hanoi visitor, passed through and offered to take mail back with him. Stockdale used the carbons for a second secret message, adding more names of confirmed POWs along with information on the camp's location and a list of potential nearby targets.[46] Its secret writing, when developed, revealed ominous content: "Experts in torture . . . hand and leg irons—16 hours a day."[47]

Commander Robert Boroughs of Naval Intelligence, who worked with Sybil, attempted to expand communications with other POWs, but the project stalled. One Agency officer who became aware of the project later on said, "What I understood was that more senior Pentagon officers took the position, 'Absolutely not. These guys have got it tough enough right now, they're being tortured to death. They're in the most miserable situation in the world. The last thing we need to do is make them spies.'"

At that point, the Navy turned to the CIA. "It's my understanding that after the Pentagon decided not to continue communicating covertly with Stockdale, the Navy decided to do it 'under the table,' because they believed their guys needed it. They also had a senior pilot [Stockdale] who asked for it," said a TSD officer who served at the time. "The Navy request came in from outside the normal chain of command, but we had a Director or Deputy Director who accepted their request and TSD was directed to assist."

Unfortunately, there had also been private attempts by some of the families of POWs to communicate clandestinely with their loved ones. In one well-meaning but failed effort, the wife of a POW secreted a transistor radio in a jar of peanut butter without telling anyone. The North Vietnamese guards stole the peanut butter for their own use and discovered the radio, leading to thorough searches of all packaged material. For example, candy bars were crushed into small pieces and toothpaste tubes squeezed empty before being given to prisoners.

Then, in April of 1967, Commander Boroughs introduced Sybil to an "expert from Washington," Bruce Rounds, from the State Department. Sybil recalled the way Rounds laughed when he said he was a State employee—it made her think he really worked for the CIA.[48] Rounds suggested a new code that would allow Stockdale to communicate even if guards were dictating each word he wrote. Rounds taught Sybil the code, which she then used in a May 25, 1967, letter that included a photograph of the roses she received on Mother's Day. The picture with the roses was intended to be "soaked."[49]

Fourteen months passed before Stockdale was allowed to send another letter. He inserted a phrase of special interest, "I often think of Red and wanted to include a message, but it is impossible; tell him next time." His reference to Red alerted the intelligence community that he was unable to send a covert message. "Red" was Retired Vice Admiral William F. "Red" Rayborn who, until June of 1966, had been Director of Central Intelligence.

As more letters with coded information were received, the few people who knew about the project were hard pressed to explain how they were getting access to POW information, such as prisoner names and camp locations.

"Over the time that I worked at night on the project, I had the deeply satisfying personal pleasure of seeing how grateful the military was that they had this channel. For years, it had been unknown what happened to many of the guys, whether they were KIA or MIA or POWs," said Lipton. "After we had the communications link, not only did the military know, but a lot of these families also began to get reliable information about their sons, fathers, and husbands."

After Stockdale and the other POWs were released in 1973 and books began to be written about their experiences, concern mounted that the closely held secret methods of communication would be exposed. Word reached DCI William Casey that Stockdale intended to include a detailed account of the covert communications in his autobiography and a subsequent movie starring James Woods.

Stockdale, now retired from the Navy as a Rear Admiral and awarded the Medal of Honor, seemed determined to write about POW communications. He reasoned that since the war was over, secrecy surrounding the covert channels was no longer necessary. "The Agency told Stockdale, 'You can't do this,'" Lipton remembered. "And Stockdale said, 'The hell I can't. What are you going to do, court-martial me?'"

With negotiations mired, Lipton offered to visit Stockdale. Along with an Agency lawyer, he met the retired Admiral at his Southern California home in a face-to-face attempt to dissuade him from revealing the POW secrets. "It was a very hostile environment when we walked in—electric, fiery," Lipton recalled. "It took two hours, but we convinced him to remove the most sensitive references to our capabilities. Then he bought us lunch at the Hotel del Coronado." Back in Washington, Lipton and the lawyer went directly to Casey's office. "I don't know how you did it," the DCI said. "I couldn't, but you did. Congratulations and thanks."

For his work that made secret communications possible, an association of American POWs called Nam-POWs Inc. declared Lipton an honorary member and "prisoner of war in Vietnam." Lipton remembered his first POW reunion. "A heck of a lot of guys came up to me and said, 'I wouldn't be alive today if it wasn't for what the CIA did. That's what kept me going.' That's how I was able to go in and work all night long, then come back and work the next day. I knew that we were doing things that really made a difference; not only in military value, but for those warriors and their families."

The U.S. withdrawal from Vietnam did not end OTS's involvement in covert warfare. President Reagan and DCI William Casey believed that Central America was in danger of falling under the influence of Communist, pro-Soviet regimes in the early 1980s. The leftist Sandinista government that came to power in Nicaragua in July of 1979 bore troubling similarities to Castro's Cuba. The five-member junta ruling the country after overthrowing the former leader, Anastasio Somoza, almost immediately established close ties with the Soviet Union and began building up military forces.[50] The Sandinistas had an estimated 5,000 guerilla fighters in 1979, a number that quickly grew to an army of 70,000 troops by 1982.[51]

American satellite photography also showed an expanding military infrastructure with three dozen new bases constructed in rapid succession. Sandinista pilots were training to fly advanced Soviet MiG aircraft and the army was being equipped with Soviet tanks and artillery.[52]

The effect of another Marxist government in the Western Hemisphere suggested a breach in the U.S. policy of containment and the potential to destabilize pro-American governments in nearby countries such as El Salvador. U.S. intelligence reported weapons flowing to the El Salvadoran guerillas through Nicaragua. Reagan and Casey agreed that CIA covert operations should be part of the strategy to thwart Soviet-Marxist ambitions in Latin America.

Under Casey, reconstituting the CIA's covert action infrastructure, dismantled after Vietnam, became a priority. Casey decided to visit one of the

OTS's newly upgraded facilities for training of foreign counterinsurgency teams and target analysis. Jameson and Parr acted as hosts.

"Casey sat with several OTS special missions officers until three in the morning, drinking beer and planning the CIA's action in Central America," said Parr. "Jameson prepared an analysis of how he believed the Agency could effectively counter problems caused by the Sandinistas in countries around Nicaragua. It became a loud and raucous evening with Casey's encouragement and participation. He was interested in the special weapons, explosives expertise, and training OTS could bring to the problem. Impressed by Jameson's analysis, the DCI concluded the evening by saying he had the funds and directed OTS to expand its training capabilities.

In early 1984, the CIA, acting on a classified authorization from President Reagan, began mining selected Nicaraguan harbors in an attempt to disrupt the country's economy. The DO tasked OTS to develop, test, and produce special mines that could not be traced back to the CIA or U.S. military suppliers. OTS engineers used thirty-inch sections of ten-inch-diameter industrial sewer pipe and C-4 explosive was tamped into the pipes with Louisville Slugger baseball bats. The mines used the military Mark-36 impact fuse system triggered by hull pressure, magnetics, or the sound of a ship passing overhead. The techs also installed a self-destruct capability set for varying times depending on the anticipated length of the operation.[53]

An initial plan to airdrop the mines was abandoned when the operational managers decided that parachutes floating downward failed the covertness test. As an alternative, the CIA used high-speed boats launched from a mother ship positioned in international waters to lay the mines. The ocean-going racing vessels, called "Cigarette" boats, obtained courtesy of the U.S. Customs Service, had confiscated them in drug busts.[54]

OTS engineers modified the boats for paramilitary operations by adding superstructure to accommodate 25mm chain guns powerful enough to penetrate tank armor.[55] They also increased the boats' speed to greater than 60 knots, assuring a rapid withdrawal after launching an attack.[56]

Between February and April of 1984, as many as seventy mines were planted in the harbors on both the Atlantic and Pacific coasts of Nicaragua.[57] Several ships were hit, including Dutch, Japanese, and Russian vessels. The mines created a loud noise, caused some limited damage to the ships' structures, and killed no one. Nevertheless, after becoming aware of the mines, a number of merchant ship pilots refused to sail into Nicaraguan ports and secret intelligence reporting indicated the Sandinista government was preparing for negotiations to stop the mining.[58]

However, in early April, the mining operation became public knowledge followed by a political uproar. Senator Barry Goldwater, Chairman of the Senate Select Committee on Intelligence, sent DCI Casey a blistering letter on April 9 accusing the Agency of failing to inform him of the operation.[59] Goldwater's assertion was in dispute, but the public furor in Washington could not be contained. The intense reaction to the "secret mining" led to a U.S. Senate 84–12 vote passing a resolution that condemned the CIA's actions. While the resolution was not legally binding, the Reagan administration bowed to the will of Congress and ended the operation in Nicaragua.[60]

"Oh boy, there was this hue and cry about mining neutral waters and acts of war. The Navy and Congress both got their noses out of joint," remembered Parr. "They'd [Congressional leaders] all been briefed, but some 'conveniently forgot' that when the crap hit the fan."

Through its first thirty-five years, OTS officers, in response to policy directives of Republican and Democratic Presidents, supported covert military and paramilitary operations in Cuba, Vietnam, Laos, and Central America. Over the next two decades many of these same officers would be called again into harm's way in Iraq, Africa, and Afghanistan.

CHAPTER 18

Con Men, Fabricators, and Forgers

The Americans were rookies. They never learned to lie as well as we.

—Oleg Kalugin, retired KGB Major General

Jean-Bedel Bokassa, leader of the Central African Empire, was harboring a grudge against the United States.[1] It was 1972 and Bokassa had in his possession documents written on what he assumed was official U.S. government letterhead alleging a plot to take control of his country. Already distrustful of "colonialists" and "imperialists," Bokassa now had what he believed to be proof of his grim suspicions.

In reality, the United States had little interest in Bokassa or taking over the former French colony. Located between Chad and the Congo, the Central African Empire, later to be renamed the Central African Republic, was an uninviting piece of real estate, even for a developing country. Although containing some uranium deposits, the impoverished country consisted primarily of subsistence farmers, limited industry, and poor roads.[2]

Bokassa, although a particularly unpleasant head of state, was primarily a survivor. As a French Legionnaire, he had survived one of the bloodiest battles in Vietnam at Dien Bien Phu, earning himself France's Croix de Guerre and membership in the Légion d'Honneur.[3] Returning to Africa a war hero, Bokassa headed his young country's armed forces before staging the military coup that landed him in the presidential palace in 1965.[4] Once in power, he began bestowing on himself a series of increasingly grandiose titles, starting with President and Prime Minister, and then President for Life.[5]

Intent on holding control of the country, Bokassa showed little patience for dissidents, sometimes by throwing them to the crocodiles.[6] In 1979, he would outrage the world by imprisoning and executing schoolchildren who dared to protest the new mandatory school uniforms. Designed and sold by Bokassa's wife, the elaborate uniforms were both too hot for the region and priced far beyond the modest means of the average citizen.[7]

Now, in the summer of 1972, Bokassa turned his anger toward the United States.[8] The documents in his hands outlined America's intention to oust his government by force. The far-fetched proposition fell well within the worldview of an African despot whose own path to power included a violent overthrow of the government.

The U.S. Ambassador, in an effort to counter the false, but diplomatically troublesome accusation, turned to the CIA for assistance. Within TSD was a small team of forensic scientists who operated a Questioned Documents Laboratory (QDL). Primarily used to detect forgeries created by the KGB's "active measures" element of the Soviet propaganda machine, the QDL also examined questionable foreign documents to support counterintelligence operations. Although QDL experts received little prominence in a division known primarily for its gadgets, disguises, and concealments, their expertise could be of value in this case.

TSD dispatched QDL examiner Dr. David Crown to Bangui, the Central African Empire's capital city, in July 1972. The U.S. Ambassador knew the documents were bogus, but confided to Crown that he had been unable to persuade Bokassa. It would be up to the QDL examiner to convince the ego-driven and paranoid ruler that the documents were, in fact, forgeries.[9]

Pointing out the blatant inaccuracies would be a simple matter under more ordinary circumstances. Even to the untrained eye, the language and linguistic flaws in the documents were obvious. The letterhead was emblazoned in gaudy green and red, hardly an approved style of Washington's sedate bureaucracy. The name of the supposed agency that generated the plot—"Communications in Superior Science Activities N.S.C. Fairbanks: The White House, The State Dept.; The Defense Dept., Washington 25, D.C."—was a ridiculous jumble of terms and government entities and included an apparent reference to the CIA's highway signs along the George Washington Parkway northwest of Washington, D.C. There, the fictional agency was headed by a "Richard Brelland" whose supposed title was "Chairman of the Board of Directors." The presence of a two-digit postal zone, not used since 1961, and the fact that the writing on all the documents was from the same hand, marked them as sloppy, if not outright amateurish, forgeries.[10]

Still, the language was inflammatory and threatening. Chairman Brelland's supposed communication read: "The boss has put the finality to the future of CAR'S GBB, he believe that we should use agents stationed in both Kinshasa and . . . Monrovia. Col. No 7 cabnet [sic] believe that the best time would be between the 2nd and 6th, but Ext. 9 is insisting on the 10th to 15th of which the big has already approved of." In a later paragraph, the U.S. Ambassador was implicated: "The Ambassador proposed a new idea; he said if there isn't time for a Mil. Coupe [sic], we could hire an assassin in the amount to $1 million and make sure the head is out of the way."

Far too clumsy to have been produced by any Soviet Bloc intelligence service, the documents were nevertheless troubling for American diplomats. Rumors of coups and invasions could affect America's foreign policy in Africa no matter how amateurish the forgery, unreliable the source, or outrageous the claims. Like Bokassa, rulers who had come to power by force were intuitively, if not realistically, fearful of their own ouster by similar means. Rumors leaked to the press could take on a life of their own, and gain acceptance as fact throughout the African continent in a flood of news stories and angry editorials.

So, in utmost seriousness, Crown found himself seated before President for Life Bokassa in a former colonial administration building that served as the presidential palace.[11] With the U.S. Ambassador acting as interpreter, Crown, posing as a U.S. government expert, launched into a technical description of the documents' many flaws, presenting a detailed analysis of fonts, typewriters, and handwriting used in creating the forgeries. To debunk the document was simple enough, but with each defect Crown pointed out, Bokassa seemed to grow increasingly uninterested.

Feeling the situation slipping away and fearing the possibility of becoming the first forensic document examiner to bore a head of state to death, Crown knew he had to switch tactics. He reached into his pocket and pulled out a fifty-dollar bill. Bokassa, recognizing the U.S. currency, immediately perked up. The expert then offered the President for Life a friendly wager. Breaking diplomatic protocol, Crown pointed his finger at the African ruler and challenged him to have his secretary call the phone number on the letterhead: 1-8338-91-65886. If the secretary got an answer in the United States, then Crown would pay for the call with the fifty-dollar bill and Bokassa could keep the change. On the other hand, Crown explained, if no one answered, the secretary was to dial a different number in Fairfax, Virginia, Crown would speak with his wife, and Bokassa would pay for the call.[12]

Sensing a sucker's bet when presented with one, Bokassa smiled. "Not to worry," he said at last. "Tell Mr. Nixon I am not mad, all will be nice."[13] The documents crisis passed and the threat seemed forgotten. Bokassa proposed an official dinner that evening for his American guests, but then failed to show up for the fish-and-chicken spread. As toasts were exchanged with government dignitaries, Crown made note of the volatile and unpredictable personalities the Ambassador faced daily.

A few years later, on December 4, 1977, Bokassa declared himself Emperor and, enigmatically, Apostle.[14] The French government, still eager to maintain friendly ties to the uranium-endowed country, provided a golden throne, jeweled crown, and scepter for the ceremony that mirrored Napoleon's self-crowning as Emperor.[15] Within two years, the French tired of Bokassa's antics, and supported a 1979 coup that ousted the Emperor.[16] Forced into early retirement, Bokassa lived in exile in France and then Côte d'Ivoire for a few years before returning to his home country in 1987 to stand trial for torture, murder, and cannibalism.[17] Still the survivor, he was released after serving seven years in prison and eventually became something of a nationalist figure at the time of his death at age seventy-five in 1996.[18]

No sooner was the case of the fraudulent documents settled in the Central African Empire, than more documents, all bearing the same fictitious letterhead, began appearing throughout Africa, including Sierra Leone, Côte d'Ivoire, Ghana, Mali, Upper Volta, Niger, Senegal, Gabon, and Guinea. Each detailed some fiendish American plot, including invasion and assassinations.[19] Although laughable in retrospect, these false rumors and obvious forgeries were taken seriously by many Third World leaders. In Guinea, President Ahmed Sékou Touré announced he would cut the throats of all Americans in Guinea if America were to invade his country.[20]

Africa was rife with supposed plots of "imperialist"-sponsored invasions based on fabricated documents that found eager buyers among the unsophisticated intelligence services of newly formed African countries. Over the following months, Crown traveled to half a dozen African nations to discredit the forgeries through irrefutable scientific analysis and defuse the suspicions of the Third World leaders.

The end of this forgery scheme came in December 1972 when Côte d'Ivoire security officials apprehended a Liberian citizen, Lemuel Walker, in Abidjan.[21] His merchandise claimed to show an impending CIA invasion of the countries of Gabon and Mali based on fabricated letters, memos, letterhead stationery, blank documents, cachets, and identification cards from the nonexistent American group. After being arrested, Walker was unceremoniously rolled up in a rug to immobilize him and sent back to Liberia to face charges. There, under interrogation, he admitted to at least eighteen political misinformation operations targeting Guinea, Ghana, the Central African Empire, Egypt, Libya, Gabon, and Lebanon.[22]

Among Walker's possessions, authorities found a forged ID that identified him as Walter H. Clifford, "Assistant Director" of the "Central Security Commission—Special Interlligence [sic] Agency, National Security Counsel [sic]."[23] Walker had a dozen aliases along with a well-practiced sales pitch and rap sheet that dated back a decade. His approach to African intelligence services had been both clever and simple. Flashing forged identity papers produced by a local printer to establish his bona fides, Walker approached embassy officials by claiming to work for a Western intelligence agency. Once inside an embassy, he would spin a tale of intelligence intrigue for his diplomatic audience. His pitch was straightforward: As a "secret agent," he became recently blessed with a conscience that compelled him to turn against his imperialistic employers in favor of the greater good of African nationalism. His merchandise, though bogus, contained a few kernels of truth gleaned from reading *Time* and *Newsweek* magazines.[24]

The embassy officials, with little ability to verify Walker's identity or authenticate his materials, would accept and forward the documents to their governments. Playing on the paranoia of local leaders, Walker found a willing, if not somewhat frugal, market for his clumsy forgeries. During his trial, the profit from the scheme was estimated to total a modest $3,307, which he supplemented along the way by forging checks and skipping out on hotel bills.[25]

Common criminals, such as the entrepreneurial Walker, are not new to political forgeries. Unstable political environments and intelligence services with limited resources or experience have historically proven attractive targets for forgers and con men. Immediately following World War II, forgers flourished by selling documents to the West eager for information relating to the Soviet Union and Eastern Europe.[26] Working alone or in émigré groups, these hustlers sold bogus intelligence reports generated in so-called "paper mills" that sprang up throughout Europe.

Nearly always containing a few grains of truth culled from public sources, these counterfeits purported to provide intelligence on everything from Soviet troop strength to chemical weapons research. For the fledgling CIA, establishing or debunking the bona fides of the sellers and tracking down the source required considerable effort. At one point, 50 percent of the intelligence on file about the USSR was attributed to such "paper mills."[27] Eventually, as Western intelligence services built a capability to detect and catalog the forgers, they began circulating the names of known fabricators and con men in what became known as "burn lists." These lists, which often included details of how the fabricator operated, were akin to an intelligence agency's Better Business Bureau reports.[28]

More insidious than the criminals peddling false intelligence for a quick profit were the Soviet and Eastern Bloc intelligence services that produced a steady stream of forgeries in an effort to discredit American foreign policy and leadership. In contrast to Walker or the paper mills of post–World War II Europe, these schemes were not the product of intelligence outsiders motivated by money or a personal political agenda. These were professional enterprises backed by the resources of a sponsoring government's intelligence service. Timed to coincide with political events and executed with precision, these forgeries were specifically aimed at creating a political advantage for the Soviets.

Forgeries as a political weapon have a rich heritage. Author, printer, scientist, diplomat, and signatory to the Declaration of Independence Benjamin Franklin played the role of forger and fabricator during the Revolutionary War. Franklin skillfully created a fictional letter in 1777 from Germany's Frederick II of Hesse Kassel to King George III advocating more aggressive use of German mercenaries in combat against the colonists.

The Franklin forgery complained that not enough Germans were being killed to turn a decent profit, since the British paid bonuses to German royalty for each fatality. Perhaps, the spurious document suggested, it would be more humane to deny the wounded mercenaries medical attention and let them die, rather than live as invalids. Combined with offers of amnesty and farmland from the American side, Hessian soldiers deserted en masse. According to records, 5,000 of a reported 30,000 Germans put down their arms.[29]

As the Cold War's geopolitical battlefield widened beyond Europe during the early 1950s, the KGB turned to forgeries and fabrications as an intelligence and foreign policy tool. A well-placed forgery could effectively strain diplomatic relations between otherwise friendly nations.[30] When such documents appeared in the media, they could also weaken support for a government's policies among its citizens or turn the tide of public opinion.

The KGB inherited an appreciation and expertise for disinformation from its predecessor organization, the Okhrana, which in 1903 published *The Protocols of the Elders of Zion*. The ambitious fabrication claimed existence of a worldwide Jewish conspiracy to create an "intensified centralization of government" and monopolies, and "revealed" the practice of ritual sacrifice of Christian children in religious ceremonies.[31] A masterpiece of disinformation, the *Protocols* reportedly sparked anti-Jewish pogroms across Czarist Russia.

Although later analysis of the document placed it as a combination of an 1865 work opposing Napoleon II and a piece of fiction by a Prussian postal employee, the *Protocols* became an enduring "bestseller" of political forgeries.[32] Eventually spreading beyond Russia's borders to the West, it was later adopted by Hitler as a propaganda tool. Remarkably, *Protocols* continues to have credibility—more than a hundred years after its appearance—particularly in Middle Eastern countries as well as with a handful of extremist groups in Europe and the United States.[33]

Political forgeries intensified with Soviet rule. Beginning with a program of forgeries in 1923, these actions would eventually become known as "active measures" (*aktivnyye meropriyatiya*).[34] By the 1950s, Soviet forgeries were primarily crafted to discredit the West in general and the United States in particular as well as create divisions between Western Allies.[35]

In 1959, KGB forgers were consolidated into their own organization when the First Chief Directorate created Department D (for the Russian word *dezinformatsiya*) and staffed it with between forty and fifty specialists. When Western officials exposed the work of Department D, the KGB simply changed the name to Department A and continued turning out bogus documents.[36]

So intense was the Soviet forgery campaign, the U.S. Senate called for hearings. Testifying before a Senate Subcommittee in 1961 and using analysis supplied by Crown and other TSD document examiners, then Assistant Director of the Central Intelligence Agency Richard Helms presented 32 examples of forgeries or disinformation from the Soviet Bloc offensive.[37]

"Of the 32 documents packaged to look like communications to or from American officials, 22 were meant to demonstrate imperialistic American plans and ambitions," Helms testified. "Of these, 17 asserted U.S. interference in the affairs of Communist-selected free world countries. The charge of imperialism is the first of the two major canards spread by the Soviet bloc in Asia, the Middle East, Africa, Europe, and wherever else they command suitable outlets."

The second theme, Helms pointed out, was that the United States was a menace to "world peace."[38] Other documents charged everything from secret agreements to plots by private businesses to take over local industries.[39] In a world where new nations were emerging from former European colonies, the spread of Soviet lies could prove both destabilizing to fragile nations and devastating to U.S. foreign policy.

Technically and linguistically, the Soviet forgeries had little in common with the amateurish work of Walker and his documents filled with misspellings and convoluted syntax. Unlike the fiery editorials or self-serving news stories printed in *Pravda* and broadcast over state-run airwaves, the Soviets gave their forged documents credibility by eschewing the stultifying political rhetoric common to official Kremlin pronouncements. In fact, the forgeries often appeared first in newspapers or press reports outside the Soviet Bloc countries before turning up as "news" in the USSR's government-run press.

These government-manufactured forgeries followed a pattern that separated them from the work of a single con man. They employed talented graphic artists who produced nearly flawless reproductions of letterheads and official stamps, paid close attention to the smallest details, such as the quality of the paper, and appropriately used colloquial or official language.

After British MI6 officer Harold "Kim" Philby defected to the USSR, he found work in the early 1970s in the KGB's Active Measures Department churning out fabricated documents. Working from genuine unclassified and public CIA or U.S. State Department documents, Philby inserted "sinister" paragraphs regarding U.S. plans. The KGB would stamp the documents "top secret" and begin their circulation. For the Soviets, the Cambridge-educated Philby, a one-time journalist and senior British intelligence officer, was an invaluable asset, ensuring the correct use of idiomatic and diplomatic English phrases in their disinformation efforts.[40]

One favorite KGB tactic was a pattern of "surfacing" a grainy photocopy or reprint of a slightly out of focus photograph, rather than a sham original document. The fuzzy photograph or barely legible photocopy not only added clandestine credibility to the spurious document for the gullible, it also made detailed analysis all the more difficult for the expert. "Newsworthy" documents were often distributed to several recipients anonymously by a "concerned citizen" who asked for no payment in return.[41] The KGB developed distribution lists of sympathetic newspapers or well-known writers whose replay of the information would add even more credibility.

Unlike the work of amateurs, who often overreached by professing to uncover large and complex plots filled with international intrigue, the professionally produced forgeries were focused, well written, and subtle in their content. They offered a façade of plausibility by implying, rather than directly stating, the propagandist's lie and by the arrangement of selected verifiable facts or the exclusion of others.

Walker's productions, the post–World War II émigré "paper mills," Soviet "active measures," and even Ben Franklin's forgeries, all played on the preexisting fears and prejudices of the intended audience or buyers.[42] Just as important as the quality of the forgery, the Soviets well knew, was the ability of the forger to assess the emotional sensibilities of their targets. That is to say, a document confirming the fears of a target audience was likely to be believed, even if it was not perfect.[43] In the opinion of President Ahmed Sékou Touré of Guinea during the early 1970s, "Documents can be forged, but the information is true."[44]

For Crown and other techs working with questioned documents, the discrediting of quality forgeries involved precise procedures and laborious, complex processes. Questioned signatures could be compared with known exemplars of an alleged writer. Examiners made comparisons of inks using ultraviolet light, infrared radiation, and the microscopic examination of ink tracks. To assist its examiners, the QDL maintained a collection of envelopes, inks, and specimens from typewriter fonts that could reveal the make, model, and date of manufacture. QDL files were always searched for evidence of prior use of specific typewriters in previous forgeries. The older typewriters with "swinging key bars," IBM Selectrics, and daisy-wheel typewriters could sometimes be identified by specific wear or damage to characters.

Paper analysis under x-ray diffraction and microscopic examination of fibers could identify paper filler type and establish the source of paper pulp. Because countries used different inorganic chemicals as paper fillers, the origin of the paper could be compared to the alleged identity of the forger.

For example, German paper would be filled with barium sulfate; French paper utilized talc. To the QDL examiner a document on paper containing barium sulfate that was said to originate in southern France became suspect.

With the introduction of photocopy technology, the Soviets began creating forgeries in the form of multigenerational copies to counter scientific examination. The technique assumed that laboratory analysis of photocopies could not establish whether a document was authentic. TSD examiners, however, found traces of evidence on photocopy documents that revealed clues of fabrication and sometimes even the origin of a suspect document. Details such as minor differences between font balls used on IBM Selectrics sold in Europe and those sold in the States could be detected in photocopies.

Subtle linguistic differences could reveal a document as a forgery. In one instance, a purported official U.S. document was labeled RESTRICTED! but carried a date after the U.S. government had ceased using the designation.[45] Another telltale sign were formats used in official letters. A government organization might use the date format of "July 4, 1990" or "4 July 1990" but one form would be consistent on all official documents.[46]

While scientific analysis could confirm the authenticity of a document, determining the ultimate source of a forgery was more elusive. In this regard, the QDL examiners would offer their opinion based primarily on the forgery's delivery, audience, and public replay. To this end, the basic thrust of the document examination included consideration of the decidedly nontechnical question: Who would benefit if the documents were believed? Through the combined skills of QDL examiners and counterintelligence specialists, data such as when a document surfaced and how it became known were evaluated to match prior modus operandi. "We found during the 1960s and 1970s," Crown observed, "that the Soviets could produce excellent technical forgeries but were rarely ever able to disguise their motives."

As the Soviet forgery offensive continued through the early 1970s, the QDL was regularly called on to debunk professionally constructed documents appearing in the Middle East, South America, Africa, and even Europe. In one instance, the Soviets surfaced an Airgram (a form of telegram) aimed at destabilizing NATO. Dated December 3, 1974, the document outlined instructions for bribing foreign officials and engaging in espionage against friendly countries. The forgery was unmasked by the signature of a nonexistent official, Robert Pont, and several format mistakes, such as the use of slash marks in place of parentheses.[47]

The KGB was a longtime supporter of organizations with public-appeal names, such as the U.S. Peace Council, and cultivated members of the U.S. press.[48] They trained forgery departments in Eastern Bloc intelligence services for the specific task of targeting the West. In the mid-1970s, the Soviets refocused their "active measures" campaign to capitalize on the Watergate scandal, Congressional hearings into alleged CIA abuses, and the Vietnam War aftermath after perceiving a receptive world audience for propaganda that described U.S. policy "mistakes" and "abuses of power."[49]

In one particularly ugly case, a counterfeit document surfaced during the Carter administration intended to stir worldwide controversy. A small San Francisco newspaper called the *Sun-Reporter* published a 1980 forgery of a Presidential Review Memorandum on Africa. The headline the paper ran could not have been clearer: "Carter's Secret Plan to Keep Black Africans and Black Americans at Odds." As the White House issued angry denials, the Soviet news agency, TASS, made the story available in a variety of languages.[50]

This was not the first time the Soviets had "played the race card" in a disinformation campaign. In 1971, forged pamphlets supposedly from the Jewish Defense League were mailed to radical African-American groups. Yuri Andropov, then head of the KGB, approved the bogus pamphlets that called for a Jewish campaign against "black mongrels." Then, in 1984, the Soviets forged Ku Klux Klan material timed to coincide with the Olympic Games in Los Angeles. The taunting material, distributed to African and Asian countries, read, in part: "The Olympic Games For Whites Only! African Monkeys! A Grand reception awaits you in Los Angeles! We are preparing for the Olympic games by shooting at black moving targets . . ."[51]

Often believed by members of the press as well as heads of state, the Soviets' propaganda could not be ignored by U.S. diplomats and intelligence. However, even when programs such as QDL conclusively proved a document a forgery, it was only after the fact. When the documents were verified false and, in the rare instances, a retraction printed, the initial damage had already been done, making disinformation a particularly effective weapon in developing countries. The intensity of these campaigns continued nearly unabated throughout the Cold War, prompting hearings in 1961, 1980, and 1982.

So significant was the threat to U.S. policy and the need to combat disinformation that in September 1979, DCI Stansfield Turner asked Crown to brief President Carter on the extent of the Soviet efforts and the CIA's capability to detect and defeat the campaigns. Turner and Crown met in the historic Old Executive Office Building, then walked through an underground passage to the White House.

In the Oval Office were President Jimmy Carter and National Security Advisor Zbigniew Brzezinski. Using samples of forgeries to describe the OTS methodology for identification and debunking, Crown's presentation lasted several minutes beyond the allotted time. Fascinated by the Soviet offensive, Carter and his adviser recognized a powerful weapon that could strain diplomatic relations, generate headlines, and possibly determine elections both at home and abroad.

At a January 1984 conference of senior officers, the KGB's First Chief Directorate reaffirmed a priority to "work unweariedly [sic] at exposing the adversary's weak and vulnerable points."[52] In the context of the statement, "exposing" meant fabricating disinformation by the Active Measures Department of the First Chief Directorate.[53] Some attendees at the conference may have been aware that one of the most cruel Soviet disinformation campaigns had already begun in 1983, aimed at laying blame for the spread of the AIDS virus at the U.S. doorstep.[54]

The Soviets launched a story in the Indian newspaper *Patriot* based on out-of-context U.S. Congressional testimony and quotes by anonymous scientists. The story, which attributed the disease to biological weapons research, seemed to have no journalistic legs at the time and quickly faded from sight.[55] However, two years later, with AIDS spreading rapidly and public alarm growing, the Soviets replayed the allegations in a Soviet publication, *Literaturnaya Gazeta*. Specific details included assertions that American scientists at Fort Detrick, Maryland, created the disease for use as a biological weapon on specific population groups and were using military personnel to spread it worldwide.[56] Within days, newspapers in Europe, Latin America, and Asia picked up the story. Like a successful marketing campaign for a new brand of soap, as the lie sold, the Soviets increased their efforts.

By 1986 at a Nonaligned Movement summit meeting in Harare, Zimbabwe, a bogus scientific paper in English was released detailing "evidence" that the United States had created AIDS. The story spread throughout Africa's media while the KGB supplemented the campaign with rumors, posters, and flyers along with repetition on radio and television broadcasts.[57]

The United States responded in 1987 by increasing the distribution of

AIDS-related information worldwide and offering a stern warning from then Surgeon General, C. Everett Coop, that a joint AIDS research agreement would not progress until the disinformation campaign ceased. Shortly thereafter Soviet scientists held a press conference to confirm the natural origins of the disease.[58]

In fact, the fabricated AIDS story had historical precedent. It was essentially a rehash of a 1952 Soviet propaganda campaign claiming the United States used biological weapons in the Korean conflict. The new element was the hot-button mention of AIDS. What was so damaging in the AIDS disinformation campaign was that fears of the dreaded disease gave the story credibility in popular culture. Similar to *The Protocols of the Elders of Zion*, the AIDS disinformation campaign continues to live in urban myth and conspiracy theory. A 2005 study conducted by the U.S. National Institute of Child Health and Human Development showed that almost half of 500 African-Americans surveyed believed that the AIDS-causing HIV virus is man-made; more than one quarter believed that AIDS was produced in a government laboratory; and 12 percent believed it was created and spread by the CIA.[59]

In the Sudan in 1969, a group of goat herders found a cache of weapons in a drainpipe along a rural trail. Hidden for the purpose of discovery, the eclectic collection included a "pen gun," box of ammunition, and magnetic *Limpet* mine that could be attached to the side of a ship. There was also a note implicating a U.S. State Department official in an alleged political plot against the Sudanese government.

The timing of the goat herders' discovery was nearly flawless. Sudan had broken diplomatic ties with the United States following the 1967 Arab-Israeli War. Thereafter, each country maintained "official" relations through "interest sections" housed in an embassy that flew the flag of a different nation. Adding to an already complicated situation were the policies of Sudan's new government that seemingly leaned toward Moscow.

American diplomats in Khartoum needed to reveal the materials as fabrications and defuse the potential for political crisis certain to follow any public release about the find. A break came when a senior Sudanese official, Major Farouk Othman Hamdallah, Minister of Interior and State Security, privately let one of his American contacts know that he would make the goat herders' find available for examination if the examiner was not an "official U.S." representative.

Crown was briefed on the situation and instructed to go to Ethiopia. With the Ethiopian visa in his passport, he then sought a tourist visa from a Sudanese consular clerk in Washington. Since he would be in the neighboring country, Crown implored, he would like to take a few additional days as a tourist to visit the Sudan and experience its culture, history, people, and land. It was his lifelong dream to walk in the footsteps of the Great Mahdi who had slaughtered General Gordon in Khartoum in the 1880s. In return for the performance, Crown received a tourist visa and checked into Khartoum's Acropole Hotel a few days later.

Unnoticed by Sudanese officials, Crown was introduced to U.S. Chargé d'Affaires George Curtis Moore, who had a reputation for being less than friendly to CIA activities. However, Moore's apprehension evaporated as the two discovered a common sense of humor along with the realization that Crown's expertise might resolve a difficult problem. Moore arranged for Crown to examine the goat herders' find at Major Hamdallah's private residence.

There Crown found the *Limpet* bomb, a pen gun, ammunition, and a typewritten note set out on a small table. After greeting the examiner, Hamdallah gave instructions to his servant and left the house along with Moore. "Have the servant call me when you're finished," said Moore, "and I'll come back to pick you up."

"Great idea," Crown remembered thinking. "Here I am by myself, an American on an irregular tourist visa in a country the United States doesn't have diplomatic relations with, unable to speak the local language and sitting in the Minister of Interior's house handling deadly devices. If something happens, I don't even know how to use the phone. If a couple bad guys come through the door, I'm in deep trouble."

There was little to do except get to work. The servant offered the examiner bottles of Coke and Fanta. Based on briefings he received prior to his trip, Crown, although not an explosives expert, identified the *Limpet* mine as a U.S. device, probably recovered by the Cubans from the Bay of Pigs and subsequently given to the Soviets. The pen gun was of Pakistani manufacture and easily available in the Horn of Africa region.

Major Hamdallah's servant brought another round of Coke and Fanta.

Crown examined each letter on the typewritten note under a microscope and referred often to his typewriter key classification. He eventually concluded the note had been written on an Olympia typewriter of recent manufacture. While American government officials were not known to use that model of Olympia typewriter, it was widely available in both West Germany and East Germany. These findings, although limited, were based on solid data and professional examination.

That evening Moore, Crown, and Hamdallah sat on red vinyl cushions in the Minister's green curtain–draped living room. In presenting his findings and his conclusions, Crown described the typewriter involved, talked about how the U.S. government produced its official documents, and offered the determination that the circumstances of the find, the origin of the materials, and the content of the document itself pointed to a covert plot by the Soviets or their allies.

Hamdallah listened politely, said little, and offered no commitment to cooperate. He did consent to a request by Crown to pose for a photo, while Moore expressed appreciation for Crown's work, which, he believed, had been sufficient to instill doubt in Hamdallah. The case was closed.

Crown's visit had an unexpected impact on both Moore and Hamdallah. Within two months, Major Hamdallah requested to see Crown again. This time Sudan would welcome the examiner as an official visitor and Moore would insist that Crown stay at his residence. Hamdallah wanted more details about typewriter markings and other forgeries in Africa at the time. From Moore's perspective, the meetings between U.S. officials and a senior Sudanese minister represented otherwise unobtainable openings for diplomatic contacts.

Sudanese President Jaafar Numeiri gradually shifted away from his Soviet Bloc orientation in 1969 and 1970, prompting an unsuccessful leftist coup in 1971. Hamdallah, one of the coup supporters, was in London at the time, but decided to return to Sudan in an attempt to reestablish an anti-Numeri organization.[60] Arrested en route, he was returned to Sudan in custody, tried, and shot.

After Hamdallah's death, details of his secret life were revealed. The Sudanese minister had cooperated at various times with East German as well as U.S. intelligence services.[61] Markus Wolf, former head of East Germany's foreign intelligence arm, described Hamdallah as an "intimate contact" with whom he had "developed a close personal as well as professional friendship." [62]

With Numeri still in power and Hamdallah replaced, Moore arranged for Crown to brief Ziada Satti, a senior Sudanese police official, on the scope of Soviet and indigenous forgeries circulating in the region. Satti listened attentively and recommended Moore and Crown take their information to the new Interior Minister. The Minister expressed particular interest in examination techniques, and quizzed Crown on U.S. prison construction before introducing the team to Brigadier General Rashid al-Din, head of the Sudanese National Security Agency. Al-Din was likewise receptive to the presentation, even mentioning the East Germans by saying, "You won't see them around here anymore."[63] Then, unexpectedly, al-Din brought up a name from Crown's past. "Major Hamdallah had been a patriot, although now a dead one."

A month later, the climb up Sudan's diplomatic ladder continued with the CIA's senior document examiner meeting President Numeiri himself. Crown recited his well-practiced litany of the many Soviet-originated forgeries bouncing around Africa. Numeiri was attentive and raised the subject of the 1969 find and the .22 caliber pen gun. Crown retold the story and his conclusions that the cache likely originated with the Soviets or East Germans.

Still suspicious about the CIA and its possible connection to the failed 1971 coup, the African leader probed Crown for information about his background and professional credentials.[64] Crown responded that as a midlevel Pentagon forensic examiner, he cooperated with the Department of State and other U.S. government agencies when asked to do so.

Numeiri soon moved on to alleged Chinese support for Anyanya forces opposing his government, showing him photos and posters to analyze, then requested training for Sudanese examiners. The diplomat, Moore, quickly injected himself into the discussion, commenting that the U.S. government had "many ways" it could help and he would be pleased to work out the details.

As they left the president, Moore observed, "Dave, we make a great vaudeville team."

Crown replied, "Yes, and we played the Palace."

Both noted the irony in the fact that three years earlier a likely East German forgery intended to discredit the United States with the Sudanese had become the key to unlocking the door of normalized diplomatic relationships between the two countries.

An invitation to return to Khartoum arrived on Crown's desk in March 1972. Satti, who had been promoted to Director General of Sudan's Ministry of Interior and President Numeiri, were asking for another briefing. Crown reviewed his findings on the Chinese photographs, which were likely the work of Soviet propaganda. Numeiri expressed annoyance at policies of the

Soviet Bloc, Libya, and Egypt intended to influence the Sudan. The U.S.-Sudanese relationship warmed with Numeiri's changed policy, and career Foreign Service Officer Cleo Noel became the U.S. Ambassador to the Sudan. George Moore remained at his post in Khartoum, serving with Ambassador Noel until March of 1973 when both were taken hostage during a reception held at the Saudi Arabian Embassy and executed the next day by Black September terrorists.[65]

That event was one of many during the 1970s that would commit Crown, the QDL, and every other component of the TSD, to America's war on terrorism. Increasingly the document examiners would turn their attention to understanding, tracking, and exposing the travel and identification papers used by terrorists. Already it was evident that passports, visas, and other documentation essential for terrorism could be forged, purchased from a commercial vendor, created from stolen blanks, altered from valid passports, or procured with assistance from corrupt officials.[66] According to a CIA report, "Clandestine Travel Facilitators: Key Enablers of Terrorism," passports for terrorists were available for purchase from drug addicts in foreign countries or forgers based in Chad or Saudi Arabia. A genuine passport could be purchased in Pakistan with bogus visa stamps to age it.[67]

Putting their knowledge of forgeries to work, the QDL created a passport examination manual, known as the Redbook. The manual, with a distinctive fire engine-red cover, included samples of forged passports, stolen passports, and fabricated entry/exit stamps.[68] In all, the Redbook contained exemplars of thirty-five forged passports and cachets from forty-five different countries.[69] Each example was illustrated in color with its faults clearly marked. Printed in six languages, the Redbook was made available to U.S. Customs and Immigration officials and countries cooperating with U.S. counterterrorism and counternarcotics programs.

The 1986 edition of the Redbook reported that over fifty individuals carrying forged passports provided by terrorist organizations had been identified before they could carry out their terrorism assignments. The Redbook's authors also concluded a brief introduction to document examination with a cautiously optimistic yet ominous paragraph that read:

> *Terrorism is a plague that threatens all of us. It must be stopped. Use the REDBOOK! . . . whether at border control, police registration, or visa application. If we screen travelers and check their passports, as experience proves, terrorists will lose their ability to travel undetected*

and international terrorism will come one step closer to being stopped!
The threat is real![70]

The threat was indeed real. In 1986, as international terrorism continued to grow, OTS document specialists trained hundreds of immigration and border control officials to spot spurious passports, visas, travel cachets, and other forms of documentation. OTS supplemented the Redbook and training with a film titled *The Threat Is Real* that was translated and distributed among law enforcement personnel in any country willing to cooperate with the U.S. counterterrorism effort.[71]

By 1992 use of the Redbook and a companion passport-examination manual had been credited for the apprehension of more than 200 individuals carrying forged passports provided by terrorist groups. The manuals were annually updated as the quality and sophistication of the terrorist documents improved each year.[72]

Time showed that terrorists became better at forging passports, and rapidly adapted computer software to help them with forgeries. Instruction booklets began circulating among terrorists on how to "clean" visa cachets and alter passports. Genuine official documents came from Arab "volunteers" fighting in Afghanistan who were ordered to give their passports to commanders upon arrival at their unit. If killed, the "volunteer's" documents were altered, usually by photo substitution, and passed along to another operative.[73] Prior to 9/11, al-Qaeda mastermind Khalid Sheikh Mohammed traveled on a Saudi passport with bogus cachet stamps to "age" it.[74] Now, twenty-five years after Curt Moore's death, the war on terrorism would dominate U.S. intelligence.

CHAPTER 19

Tracking Terrorist Snakes

We have slain a large dragon but we now live in a jungle
filled with a bewildering variety of poisonous snakes . . .

—R. James Woolsey in Congressional testimony after the collapse of the USSR

The Agency's war against terrorism began decades before September 11, 2001. The CIA's chief in Athens was assassinated in December 1975. In October 1983, terrorists used a truck bomb to blow up the Marine barracks in Beirut, murdering 241 U.S. soldiers. That same year, terrorists bombed the U.S. Embassy in Beirut, killing 63, including CIA officers. Then in 1984, CIA's chief in Lebanon, William Buckley, was kidnapped, tortured, and later murdered in 1986. That same year, DCI William Casey created the DCI Counterterrorist Center staffed by representatives from the principal intelligence community agencies, with a mission to "preempt, disrupt, and defeat terrorists."

While radical groups with diverse agendas, such as the Baader-Meinhof Gang, the IRA, and the Weathermen, committed terrorist acts during the 1960s, America saw the face of terrorism up close during the 1972 Olympics in Munich. Broadcast live on international television, the dramatic scenes of masked Palestinian terrorists killing eleven Israeli athletes seemed as senseless as it was shocking.

To the CIA, the only new aspect of the appalling spectacle in Munich was its global broadcast. Acts of terror carried out by religious and political fanatics, either independently or with government sponsors, have influenced societies and destabilized governments for centuries.[1] In the eleventh century, a Muslim sect, the Order of the Assassins, conducted suicide missions with the promise of "paradise to follow."[2] English Catholics conspired to blow up Parliament in 1605 in hopes of creating an uprising against King James I.[3] The list of terrorist acts through history includes virtually every country and continent. However, in the last half of the twentieth century, the frequency of terrorist strikes accelerated dramatically. According to one account, 8,114 terrorist incidents occurred worldwide during the 1970s. In the 1980s, that number increased nearly 400 percent to more than 30,000.[4]

Advances in technology have aided terrorists in their efforts. New chemistries have reduced the amount of explosives needed to inflict significant damage, and television coverage typically allocates airtime relative to the number of people killed or injured. It can be argued that the net effect motivates increasingly more heinous acts in an attempt to attract the attention that drives public fear.

Targeting civilians is a fundamental terrorist tactic. In the nineteenth century, the radical German revolutionary Karl Heinzen envisioned a time when weapons of mass destruction, powerful enough to destroy entire cities, would be in the hands of terrorists.[5] Stopping that from happening became one of CIA's most important missions.

The first OTS officer assigned to Vietnam, Pat Jameson, may have been the first OTS tech drafted into the war on terrorism. During the early 1970s, terrorists were using tools already well known to Jameson, including false identities and documentation along with special weapons and improvised explosive devices. His experiences in Vietnam and Laos with target analysis and planning paramilitary operations allowed him to understand how terrorists identified target vulnerabilities and traveled undetected.

When the United States left Vietnam in 1973, OTS reassigned Jameson from Laos to one of its covert European bases. Now a seasoned tech with experience in both paramilitary and audio surveillance operations, he would become a primary Agency resource for creating counterintelligence programs in Europe and Middle Eastern countries. Early counterterrorism operations concentrated on putting audio devices into residences and offices of suspected terrorists. The strategy, Jameson recalled, "paid off handsomely," particularly in West European countries where members of terrorist cells were often longtime residents and believed their radical activities were protected. For the Agency, these operations not only assisted countries in preventing attacks, they also provided a means to solidify and expand cooperation with local security services.

Jameson found that many Middle Eastern countries were overly confident in their attitudes, thinking themselves immune from terrorist acts. Several countries, including friends of the United States, ignored or failed to take action against terrorist groups at home, as long as they conducted their attacks elsewhere. In late 1974, a CIA chief invited Jameson to assess one country's overall counterterrorism security picture. He confided that his contact, a general who headed the local intelligence and security service, seemed unconcerned about the potential for terrorist attacks on the government. In keeping with a tradition established by Donovan and Lovell, Jameson and his counterpart spent an evening drinking good brandy and devising a plan that would form the country's first counterterrorism program.

A few days later, Jameson secured a meeting with the general, intending to offer a special training program for a new counterterrorism team. However, upon entering the general's office, Jameson faced an unreceptive host despite the recent problems in the country from a foreign terrorist cell. The general dismissed the incidents as insignificant and isolated and remained adamant that the country had no terrorist problem that needed CIA assistance.

Jameson saw an opening. "Well, that's good," he responded, "but I have a bet for you. I believe I can walk down to your marketplace, make a few purchases, and within a few days create an explosive device that I can then plant in a public location to kill any important foreign guest that comes into this country."

The intelligence chief nearly exploded with anger. Jameson had challenged his fundamental responsibility, the protection of the country's leader, and his international guests. Several minutes passed and when the conversation calmed Jameson suggested that the general personally choose one of his officers to accompany him as a guide and translator for a few days to prove or disprove his bold wager.

During the next three days, Jameson and his guide wandered the markets of the capital city and purchased remote-controlled model airplanes, ammonium nitrate fertilizer, diesel fuel, and parts to make improvised detonators—everything necessary to make a powerful bomb.[6] Jameson then checked into the hotel directly across the street from the intelligence chief's office and spent two days observing and making notes about security at the hotels, on the streets, and at entrances to government buildings.

Returning to brief the general, Jameson was greeted with a slight smile. "So, you wasted your time, didn't you? You see, we're really secure here," the intelligence chief said. Jameson pulled a small U.S. government–issued green-covered notebook from his pocket and began reading from his notes.

"I bought radio control switches to activate a bomb. I bought ammonium nitrate for the explosive. I bought a tourist suitcase to hold the bomb," he recited. "I bought fuses, timers, and initiators. Here's a list of the prices I paid for each item. I can get as many as I want. I can make one or several bombs and I can make a big one or a small one. See, it's all here and your man knows that I have it all in my hotel room."

The general remained unconvinced. "That might be so, but you can never use it operationally here," he said. "You could never get close to my VIP guests."

Jameson was prepared. "Sir, look outside your window. There's a big sedan parked across the street right in front of the hotel where all your guests stay. I would load that car up with explosives, sit on my balcony, and wait for your guest to come out. I'd push the button and *boom*—no more VIP."

The general was stubborn. "No, that wouldn't work either. My security guys are well trained, well armed, and know how to protect VIPs. They would spot anything out of the ordinary like that car."

"We've checked," Jameson shot back. "That car hasn't been moved for three weeks."

The intelligence chief agreed to consider what Jameson had reported.

Included in Jameson's final security report was an item that would prove eerily prophetic. It warned of the dangers posed by a longstanding tradition in the country whereby petitioners could ask the head of state for assistance by pleading their case in person. Jameson assessed the vulnerabilities of the petitioners' ceremony and concluded that a high probability of assassination against the leader existed. He noted that petitioners approached the ruler one at a time without being screened, searched, or x-rayed. To emphasize the point, he added that close associates and extended family members had nearly unrestricted access to the head of state. Nothing would prevent one of these from carrying a concealed pistol and shooting at close range.

While that portion of Jameson's report was ignored, other pieces of a

U.S.-proposed counterterrorism program began falling into place. The general understood the vulnerability presented by the example of the unidentified automobile parked near his office and a few weeks later a commercial airplane with its windows covered with blackout paper landed at a covert CIA training site. The passengers were members of a special counterterrorism and VIP-protection team sent by the general for a month of intense training by OTS specialists. As the team disembarked, Jameson personally greeted each one. Now having proved himself a friend of the country, he began having regular meetings with the general to offer advice and planning assistance.

In the mid-1970s, after completing one of his now frequent meetings with the general, Jameson headed home. During a stopover, as he approached the boarding gate to catch a connecting flight, he was surprised to see another CIA officer waiting for him. The general was demanding that Jameson return immediately. The country's leader had been assassinated.

"Let me guess. He was holding an audience, the people were standing in line and he was sitting there talking to one of them and they shot him in the head," Jameson said.

"How do you know that?" the colleague asked, taken back by the specific details.

"I told the general six months ago that could happen. I didn't have a premonition. But I saw how the petitioners approached and how he was unprotected from his inner circle, the risk was obvious."

"You were right. The assassin was a relative."

Jameson knew he was in a delicate situation. Not only had the assassination occurred uncomfortably soon after his report, but he had just departed the country hours before the shooting. If the general believed he was somehow implicated in the murder, Jameson could be arrested the instant he stepped off the plane. On the other hand, if Jameson didn't honor the request to return, they might assume that he, along with the CIA, was guilty of something.

Jameson spent a sleepless night in limbo as cable traffic was exchanged with Headquarters. All eventually agreed that it would be best for the OTS officer to go directly to the general, tell what was known, and offer personal as well as Agency assistance.

It was not a comfortable meeting. The general launched a tirade about the ineffectiveness of CIA counterintelligence training while implying that

Jameson might have had some connection with the assassination. Letting the accusations pass without comment, Jameson replied, "I told you how I would go about doing this, I didn't say I was going to do it. I told you what was needed to correct the situation, but you didn't correct it. What has happened is past. Where do we go in the future?"

"You stay here for now," the chief ordered, concluding the meeting.

After several days of being under virtual house arrest, Jameson requested another meeting. The general had calmed down and conceded that no blame for the assassination lay with Jameson or the CIA. He was free to leave the country. However, for the remainder of his career, Jameson would be part of CIA's counterterrorism mission and within a few months tragedy struck first hand. In December 1975, Jameson's CIA chief in Athens, Richard Welch, died from a terrorist's bullet.

Among terrorists' standard weapons are small amounts of explosives fashioned into improvised devices and concealed in everyday items. Although small, these bombs can create physical damage as well as instill fear in a larger population. Less than an ounce of explosive is needed for a letter bomb capable of killing or maiming anyone around it. A few pounds of explosive hidden in a purse, briefcase, or suitcase can bring down an airliner, while a few hundred pounds in a car or van is capable of destroying an entire office building or embassy.

Bill Parr had a cramped office in Southern Europe during the late 1970s but that mattered little since most of his work was on the road. As one of the few OTS "bomb techs," Parr frequently received the first call after bombings in Africa or the Middle East. CIA stations found that after bombings foreign services were especially receptive to hosting techs like Parr who knew how to conduct post-blast investigations and analyze security weaknesses.

Parr was asleep at home when a phone call awakened him at 2 AM. "Come in immediately." The caller did not need to identify himself. Parr recognized the urgency in the voice of the CIA's senior communicator. Arriving at the office, Parr saw the message headed with the word IMMEDIATE and followed by NIACT, for "night action," which required an immediate response regardless of the time of day.

This message originated from a country friendly to the United States, and whose political leaders were frequent targets of terrorists. Reportedly, the country's intelligence chief had acquired what seemed to be a suitcase bomb from a terrorist cell. For reasons unknown, the intelligence chief had taken the suitcase to his office, opened it, and noticed wires in one corner along with some unidentified materials wrapped in black tape.

"Then he apparently suffered a sudden attack of brilliance and decided not to mess with it anymore," Parr recalled. "That's when he called the CIA for help."

Parr replied with his own IMMEDIATE-NIACT with questions about the appearance of the device and remained at the office for the rest of the night responding to additional messages before catching the morning flight out to see the device in person. Reaching his destination before noon, Parr was taken to an office where the opened suitcase sat on a desk. Almost certainly a bomb, the suitcase contained a messy collection of wires, small boxes wrapped in black tape, and miscellaneous packing materials. When an x-ray machine was located at a nearby prison, the suitcase was carefully transported there. The x-ray image revealed a coil pattern in one of the taped packages that Parr concluded was detonator cord wrapped around another unidentifiable substance, probably high explosive.

The trigger mechanism to control the detonation was circuit board wired to the coil. Parr calculated that he could pull the electronics away from the explosives without initiating a detonation. Taking the device to a remote area, he attached a hook on a long line to the electronics, and gave it a hard yank from a safe distance. The electronics, which had been attached to the blasting cap with black tape, pulled cleanly away from the explosives. After cutting off the blasting cap, he x-rayed the device again to recheck the circuitry.

With the device rendered safe, follow-up operational ideas began emerging. The bomb had been obtained through an agent who penetrated a terrorist cell. If a signaling device could be attached, and the suitcase reinserted into the cache, it could be tracked to determine its intended target. However, to do so, the bomb would have to be reassembled.

Parr knew that since he took it apart, he could put it back together, but the agent had only a few hours to return the device to the cache before someone discovered it missing. Not only did that rule out a tracking operation, Headquarters, quite sensibly, disapproved of returning a live bomb to a terrorist cache. The tech would have to reassemble a disabled bomb without leaving any traces of tampering. Using a layer of epoxy to short out the wiring from the switch to the blasting cap was relatively simple, but reassembling the bomb in its case without leaving signs of alteration required several hours. The sleep-deprived tech finally completed the reassembly and delivered the bomb to the local service for return to the cache. Sometime in the future at an unknown time and place, a terrorist's plot would fail.

In the course of disassembling and reassembling the components, Parr examined the electronic circuit board, took photos, and made sketches. The device contained components he had not seen in terrorist devices, revealing a new type of timing device that subsequently appeared in other bombs as a trademark for a particular PLO bomb maker.

"You deserve a medal," the chief told Parr as he departed.

"I don't think so," Parr replied, "I've spent most of my life in Vietnam and Laos making devices that help kill people. It seems a little ridiculous to give me a medal when I disarm a device." Nevertheless, a few months later, based on the recommendation of the chief, Parr did, in fact, receive the CIA's Intelligence Star for "a voluntary act of courage performed under hazardous conditions."

In December 1988, John Orkin was heading an OTS unit responsible for conducting technical design and performance assessments of spy equipment deployed against American targets by the Soviets or other adversaries. These devices were usually discovered either through technical surveillance countermeasures or acquired from a friendly liaison service.

Typically, Orkin's engineers analyzed each recovered device to determine its country of origin, function, materials, design, and capabilities. This was no easy task, since intelligence organizations routinely mask the country of origin of a device by "sanitizing" spy gear. Since a bug found in the wall of a diplomat's office was not usually stamped MADE IN THE SOVIET UNION or came with an instruction manual, Orkin's job was to figure out how a mysterious device worked, who made it, and how it might have been deployed.

With a well-equipped suite in the sprawling OTS covert laboratory outside Washington, Orkin joked that he was far enough from the Langley Headquarters to do "real engineering work" without interruption or micromanagement. The lab was largely immune from the crisis-to-crisis atmosphere that dominated many intelligence operations.

Orkin began his CIA career during the early 1970s evaluating OTS equipment. Responsible for testing every piece of spy gear produced by OTS before certifying it for deployment to the field, the unit functioned as the

Agency's in-house "Underwriters Laboratory." Far removed from the glamour associated with agents, spies, and back-alley intrigue, testing the frequencies of transmitters and battery life of communication devices in a government laboratory seemed a world away from the frontlines of espionage.

The certification of OTS devices and analysis of hostile gear involved both elements of reverse engineering and similar testing procedures. However, during the 1970s, analysis of foreign devices was an ad hoc affair. "As a device was recovered the lucky engineer-of-the-day got assigned the project to test and write a report on how the foreign equipment performed," explained Orkin. "We were dealing with mostly technical surveillance equipment—microphones, transmitters, communication and concealment devices. Eventually it became apparent that we were reinventing the wheel with each analysis. When a device came in, we often didn't remember if we'd seen the same thing three years before or if there was already a report in the file. So, one of the engineers finally said, 'Give them all to me. I'll do the work.' Over time he became the in-house repository of knowledge about foreign equipment."

Eventually, this led OTS to establish a unit with expertise in reverse-engineering foreign equipment and a proficiency in spotting patterns that accompanied technological evolution. Although the number of new devices might be fewer than ten a year, analysis was needed to establish continuity. "We needed to develop data to do a side-by-side comparison to look at the evolution of devices," he said. "We needed to know with certainty if two devices were identical and, if not, then document the changes and record the improvements."

The Soviet Union remained the focus for OTS foreign-equipment testing well into the 1980s. The work analyzing Soviet and East European spy devices offered invaluable counterintelligence data for the FBI, State Department, and U.S. military security components that were also developing countermeasures.

However, with the number of terrorist organizations multiplying in Africa, Europe, the Middle East, and Asia in the 1970s, the Agency began acquiring explosives and bomb fragments from terrorist caches and post-blast investigations. As most early terrorist bombs were individually fabricated from whatever parts and materials the bomb maker had at hand, these were collectively known as Improvised Explosive Devices or IEDs. The OTS expertise in reverse engineering and knowledge of foreign electronic circuits proved to be the perfect match for unraveling the firing, triggering, and timing mysteries of these weapons. It was in this way, almost by accident, that Orkin and his colleagues became the Agency's focal point for collecting, analyzing, and cataloging instruments of terror.

"The analysis of terrorist devices evolved slowly; it was an occasional item here and there. In fact we initiated some activity ourselves, going to the counterterrorism officers and asking them, 'What do you have?' And they'd say, 'Oh, here's a device that I picked up in Jordan in 1978.' And we'd offer to do a report," remembered Orkin. "Initially we were working backwards on stuff found a few years before. If anything looked like it had been made in more than one quantity or if it looked like something we might see again, we'd take it and do a report on it."

Orkin began detecting a disturbing pattern of state-of-the-art technology making its way into terrorist devices in the early 1980s. Terror organizations that relied on crude timers and other components for their bombs a few years earlier were now acquiring advanced technology that greatly increased the lethality of terrorist bombs.

"The reason that technology became available was because of state-supported terrorism. Terrorists could buy the needed technology because Syria, Libya, and a few other countries like Iran and Iraq were giving them money," said Orkin. "Initially they used very simple devices, like timers that you had in your kitchen, alarm clocks, and wristwatches with the hole drilled in the center—they really used the kind of basic stuff you see in the movies. I put 1984 as the date when they began to apply newer technologies to these devices. We began seeing multiple examples of the same device, indicating that they were fabricating them in small 'production runs.'"

This was a significant turn of events. Terrorists could now get electronics engineers to design the timing circuit for a bomb instead of relying on someone with little training working in his cellar with a soldering iron. Terrorists, who had traditionally been impassioned amateurs, were finding allies, tutors, and funding from the intelligence professionals of rogue countries. Established governments were now directly aiding terrorism by providing cash and enabling networks for procurement and shipment of components and devices.

"Throughout the eighties the terrorists progressed from devices that were fairly simple to those of professional quality," explained Orkin. "We saw the devices evolve. We would think, boy that's not a good way to do this, and when the next device came in, we'd see the problem fixed."

Orkin, an intelligence officer as well as engineer, pinpointed a potential vulnerability that might be exploited. The more advanced and technologically specialized each component became, the more likely it could be traced back through supply channels to specific manufacturers. Embedded in every chip and circuit board were technical and engineering details that, when pieced together, sometimes revealed the genealogy of the device, including its sponsor and even assembler.

For instance, lot numbers denoting manufacturing runs used for inventory control, coupled with standardized fabricating processes, gave virtually every component a personal history. Working out the "technological DNA" of terrorist devices, the foreign-finds experts uncovered yet another disturbing trend: terror organizations were no longer working in isolation. They were now forging ties with rogue nations and each other.

In one instance, when analyzing a device uncovered in the Middle East, he noted features of an elaborate, high-powered radio receiver with British markings. "We kept finding components that were marked from a British company. And when we showed it to the British, they identified it as PIRA [Provisional IRA] technology," recalled Orkin. "British analysis had concluded that PIRA traded technology for guns and explosives with the Libyans. Libya also trained some of the PIRA guys who handed them the technology. There was another incident in Peru. And when we looked at it, we said this looks like the other device from PIRA."

What eventually emerged was a complex, interlocking network of terrorist organizations. For example, PIRA technology could support the ETA (Euzkadi Ta Askatasuna) in Spain and ETA could hand off some of the equipment to Shining Path in Peru, and so on and so on. While these groups did not always share political or social agendas, they did share a common desire for bomb-making technology. "It was getting harder and harder to differentiate between groups by their technology," explained Orkin.

The ringing bells and songs of carolers on Christmas Day 1988 sounded with little joy or peace for families of the 259 passengers and crew of Pan Am Flight 103. Four days earlier, just after 7 PM on December 21, a bomb ripped a hole in the forward fuselage of the Boeing 747 as the plane reached its cruising altitude of 31,000 feet over Scotland. Stunned air traffic controllers at London's Heathrow Airport watched the aircraft vanish from their radar screens, replaced by small blips as nearly 700,000 pounds of airborne debris, including flaming jet fuel, rained down on the small town of Lockerbie.

Two hundred seventy perished that day, all those aboard the plane and eleven residents of Lockerbie. Death and destruction in the air and on the ground turned the anticipation and joy of the Christmas holidays that traditionally unites friends and coworkers in a spirit of joyous goodwill into a season of mourning. The next day, CIA casualty officers suspended holiday plans after learning that among the dead was a CIA officer. Another star would be chiseled into the white marble of the Agency's memorial wall of the Original Headquarters Building.

Even for those not directly touched by the tragedy, the grim images of the wreckage on covers of magazines carried a stark reminder that evil honored no holiday. The debris field, spread over more than 800 square miles, was a horrific scene of human remains, personal effects, and pieces of aircraft. The plane had not exploded, but broken apart as it plummeted to earth. Few news organizations captured or broadcast images of the most gruesome elements of the crash site; however, one photo that shocked and sickened the world eventually became a morbid visual shorthand for the tragedy. That was the image of the battered nose section of the 747 with the cheerful script MAID OF THE SEAS resting on its side in a muddy field. Nearly twenty years later, even those who can recall few of the details surrounding the tragedy of Flight 103 remember that image vividly.

Terrorism was not new to the Agency or unknown to viewers of the evening news. Nevertheless, Pan Am Flight 103 seemed different. The plane was not in a war-torn country. Those aboard were students, husbands, and mothers. Their faces, revealed to the world smiling in family photos, showed ordinary lives ended in a violent and grotesque manner. What could the perpetrators possibly hope to gain by murdering college students returning home for Christmas vacation? What message were they hoping to send? The senseless slaughter defied all reasonable motivation and strained even the tortured logic of terrorism.

CIA officials quickly concluded that this was a terrorist attack, even if the identity of the perpetrators and their motives were unknown. Reports from around the world identified groups eager to claim credit for the cruel spectacle. Critical details began emerging over the following weeks, as investigators assembled information from agent reports, signals intelligence, and fragments of the plane. The bomb had been concealed in a commercial Toshiba radio cassette player, the Bombeat (Model RT-SF 16), available in consumer electronics stores. Fragments of the radio, as well as the owner's manual, were recovered in the wreckage of the cargo container where the blast originated.

Traces of the plastic explosive Semtex were found on the radio's circuit board and experts calculated that no more than 400 grams (about 14 ounces) positioned close enough to an outer bulkhead was enough to blast an eighteen-inch a hole through the fuselage. Within seconds after the detonation, the plane decompressed, suffered structural failure, and tore itself apart in the sky. But more puzzling than the presence of Semtex or the cassette player was what remained of a brown Samsonite suitcase. Tests revealed traces of Semtex on the suitcase fragments, indicating that it more than likely held the bomb. Yet, no such suitcase was checked in as baggage at Malta, the departure point for the baggage container where the blast occurred and investigators were unable to match the brown, hard-shell piece of luggage to any passenger aboard the plane.

From the start, the investigators focused on the use of Semtex. The explosive of choice among many terrorists, it is difficult to detect, but relatively easy to obtain. Counterterrorism experts suspected Palestinian groups based in Syria, which had used the explosive in past attacks and had a history of relying on consumer electronics as carriers. Iranian terrorists also favored the explosive, a fact that focused attention on an Iranian national aboard the flight's Frankfurt to London leg, who disembarked before the plane's departure for New York.

Although investigators understood what had happened, their study of the debris field and wreckage yielded little about who did it and how the bomb had been placed on the airliner. The break did not come for eighteen months. Then, nearly eighty miles from the center of the debris field, a local man stumbled onto the remnant of a T-shirt bearing the label of Mary's House, a store in Malta's port town of Sliema. Embedded in the material of the garment was a thumbnail-sized fragment of circuit board, about 0.4 inches square. That tiny speck of forensic evidence would eventually lead to the unraveling of a terrorist plot and test the international justice system.

However, the trail to the Lockerbie bombers began not in a European capital or the Middle East, but in sub-Saharan Africa. The Republic of Chad, a former French colony, is known primarily for the exotic name of its capital city, N'Djamena. Despite decades of ethnic warfare and limited strategic significance in the Cold War, the United States maintained relationships with Chad's government. In the early 1980s, the local security service uncovered what they assumed to be a case of espionage and quickly moved in for an arrest. When the suspected spy was taken into custody, his suitcase surprisingly contained not a collection of standard spy gear, but a quantity of Semtex attached to a portable radio.

Without the capability to conduct an in-depth technical analysis, Chad's intelligence service passed the device on to the CIA. The device, Orkin noted, was unexpectedly complex. The sophisticated circuitry was controlled by a standard pager traced to a firm, Meister and Bollier AG (MEBO), based in Zurich, Switzerland. Known to have ties to both Libyan and East German intelligence organizations, the company's circuits were apparently now being used in a terrorist IED.

A few years later, in the autumn of 1986, local authorities uncovered a terrorist cache of weapons in Togo's capital, Lome, and U.S. authorities were invited to examine these materials. Among the hodgepodge of aging weaponry and munitions, two state-of-the-art timers stood out, one of which Orkin acquired for analysis. Like the find in Chad, the device was also traced back to Switzerland and MEBO.

Then, in February of 1988, two known Libyan intelligence operatives were arrested in Dakar, Senegal, as they disembarked from their flight. Items found in one of the passengers' briefcases included a silenced pistol, ammunition, four blocks of TNT, and two blocks of Semtex, along with a timer. Local authorities allowed CIA officers to photograph the timer. The photograph, like the timers from Chad and Togo, eventually landed at Orkin's lab. Again, MEBO appeared linked to the device.

"In retrospect, the devices from Chad were first-generation," Orkin explained. "Then, we observed that the device from Togo was a prototype. We said, 'If you cut the corners out here and change this and change that, you can put it in a box and make it look nice.' And that's exactly what we got in the Senegal device, it was a packaged unit." The terrorists had learned the engineering principles of form, fit, and function.

Orkin now had three increasingly sophisticated devices, all from Africa, all physically linked to the same manufacturer in Switzerland, and all with some Libyan connection. A report on each rested in the OTS archives as technological curiosities of no immediate intelligence value.

Then came Pan Am 103. By September of 1989, Scottish investigators converged on Mary's House clothing store in Malta, led there by the scrap of T-shirt. The shop's owner remembered the customer who bought the T-shirt, describing him as a Middle Eastern man who shopped indiscriminately, purchasing items without regard to size, as if he were simply trying to fill a suitcase, which, of course, is precisely what he was doing. The police produced a sketch of the buyer from the store owner's recollections, though there was as yet no name to go with the face.

The Scots were still having a difficult time identifying the tiny piece of circuit board found embedded in the shirt. Unable to match it up to any piece of the plane or known electronic component, they sent a photograph of it to the FBI with the understanding that it would not be released outside the Bureau. Analysis of the photograph yielded little, if anything, new. After six months, the FBI received permission to show the photo to other U.S. government agencies. That same day, Orkin received an unexpected call.

"The guy calls me from the FBI's explosives unit. He came out to my lab, pulled out the picture, plus a one-page report with a brief description of the fragment that says it's epoxy fiberglass and seven-ply board," said Orkin. "It also makes note of the fact there was green solder masking. Solder masking is used to prevent solder flowing to parts of the board where you don't want it."

Orkin had seen the same design before. The green solder masking, coupled with the curve of the board, matched previous reports done on the devices found in Togo and Senegal and associated with Libya. Then he discovered that the fabrication techniques, along with a modified connector, matched the device found in Chad. They all pointed to the Swiss firm MEBO. The connection, though still tenuous, represented a starting point. Beginning with this information about the components found on the African circuit boards, the FBI launched a global investigation.

The FBI traced the components, eventually tracking the timing crystal to a specific company. Asked about the component, the firm consulted its records, which showed that one hundred of them had been sold to MEBO, establishing a definite link back to Swiss firm. Next, the investigators went to MEBO. As Orkin later heard the story, the investigators were told, "Yes, we built ten with the plastic case and ten without the plastic case for the Libyans." Then, realizing what had been revealed, the manager decided not to talk with the investigators anymore.

Called the MST-13, the timer unit was made specifically for the Libyan government's Ministry of Defense. The precision device could be set to delay an explosion up to 10,000 hours or 10,000 minutes. "This was an elaborate timer with crystal timing controls that were very accurate," explained Orkin. "This timer had been in a suitcase placed on a plane going from

Malta to Frankfurt. In Germany, the luggage was changed to another flight going to London, and then put on Pan Am 103. It was set to wait many hours before triggering the explosion."

Based on the recovered evidence, the timer apparently had been set with the intention of sending the plane down into the Atlantic. However, an unexpected weather delay at Heathrow Airport and high winds diverted the flight path over Scotland. Had the plane taken off as scheduled, the bomb would have detonated over the ocean and all evidence likely destroyed or lost forever.

With these discoveries, suspicion shifted from the Syrians and Palestinians to the Libyan government. A defector named two Libyan intelligence officers whom he claimed were behind the Pan Am 103 bombing. The first, Abdel Basset Ali al-Megrahi, was identified as the Mary's House customer who purchased the T-shirt and other clothing used to fill the suitcase containing the Semtex. The second Libyan, Al Amin Khalifa Fhimah, had a cover job with Libyan Arab Airlines at Malta's airport. This put them in a perfect position to smuggle the brown Samsonite suitcase into the plane's cargo hold. The problem for investigators and prosecution was that both suspects were now back in their home country, where Libyan leader Muammar Qadhafi was steadfastly rejecting requests to hand them over for trial.

More pieces of the puzzle fell into place in March of 1990, following the collapse of the Soviet Union. Semtex, the plastic explosive used, was manufactured in Czechoslovakia. The name was derived from the first four letters of the company that first manufactured it in the 1960s, the Semtin Glassworks, plus the suffix "ex" denoting the English word for "explosive." During a well-publicized press conference in London, Czechoslovak President Vaclav Havel revealed that his country, while under Communist rule, sold an estimated 1,000 tons—2 million pounds—of Semtex to Libya. The Libyans, Havel added, had refused to return the explosives.

By 1992, with evidence pointing decidedly at Libya, the case came before the United Nations Security Council. Sanctions were imposed and negotiations with the Libyan government dragged on for six long years. Then, in 1998, as sanctions became an economic burden, the Libyans finally agreed to turn over the two accused men for trial. The trial was scheduled to take place at Camp Zeist, a former U.S. military base in the Netherlands that had been declared within Scottish jurisdiction for purposes of the trial.

As the prosecution began assembling the case, it became clear the timers would play the key role in linking the two Libyan intelligence officers to the flight. According to Scottish law, the prosecution was obliged to show the defense how the investigators were led to the timing device. This put Orkin, whose entire professional life had been conducted under a cover unassociated with the CIA, in an unusual position. He would appear in open court as an expert witness in an internationally watched trial.

The Agency offered unprecedented assistance in the trial. Not only would Orkin testify, but two former chiefs of station would also be made available, if necessary, to appear in open court. Contents of classified operational cables were provided as evidence.

Orkin's role was to deliver the critical testimony that linked the devices to Libya, illustrating a technological trail that started in Switzerland, progressed through Africa, and ended, tragically, in a field in the Scottish countryside. He was the single witness who could provide irrefutable scientific credibility to a detective story that began with a minuscule piece of plastic found by chance. But the trade-offs were stark. His presence at the trail could potentially blow his cover and, if he were identified, put his life and the lives of his family in jeopardy.

To protect their identities, the court agreed that Orkin, along with the chiefs, could testify using alias names and wearing tailored disguises. The alias, chosen sometime earlier for the tech who had a reputation for clever puns and elaborate word play—he was "Mr. Orkin," the bug killer. Whether the Libyans recognized the popular American extermination service was unknown, but as the trial date approached, the key witness became increasingly concerned with his choice of alias. What seemed amusing and fitting at the outset was now worrisome. Could the defense attorneys somehow use the pseudonym to discredit his testimony? Was he being too clever in such a serious matter?

To prepare the witnesses' disguises, OTS selected a senior disguise officer who combined a sense of artistry to fit the person to the materials as well as match the materials to the subject. It was an exacting process. Orkin needed to not only look natural, he had to feel genuine within the disguise and be relaxed on the witness stand. The obscure OTS engineer would be transformed into the internationally recognized electronics expert "Mr. Orkin."

The OTS disguise specialist eliminated Orkin's trademark facial hair, shaved here and added there where previously none existed. "I had a nerdy, old blue suit, white shirt. I almost put a piece of tape on my glasses to finish the effect," he remembered. "Then there was a stink raised by the Defense lawyers. They said, 'We don't want people coming in here in Shirley Bassey fright wigs.' When we finally sat down with the defense attorneys for a preliminary meeting, after twenty minutes or so of lawyer-type discussion, they asked, 'Are you in disguise now?' I was and they couldn't tell. So much for fright wigs."

Orkin flew to the Netherlands and met with the prosecutors for a briefing on courtroom procedure. Using technology as evidence and experts to explain it was tricky business. Detailed scientific mumbo jumbo could confuse those without a technical background, and engineers, trained to think and speak with precision, could be trapped by clever legal questioning. For example, Orkin was reminded, a lawyer might ask if a scientific result is "100 percent accurate," to which an engineer might answer in the negative, because the results were only 99.99 percent accurate. The possibility that the two suspects would escape justice through just such wordplay weighed heavily as the prosecutors prepped Orkin on how to make his data clearly understood in the courtroom.

"We're going to ask you this, this, this, this, and this, and we're going to establish your credentials," the prosecutor told Orkin. "Don't volunteer any more information. Just answer the question I ask you. I know what you know, and if I want more, I'll ask for it. If I don't, then let it go. But don't start expounding."

Prosecutors warned Orkin that the two defense attorneys had already established a "good cop/bad cop" routine during the previous days of testimony. "They told me that one of the attorneys stands up, is a real sweet guy, and tries to get you to say something you shouldn't say. Then the other guy gets up, questions your parentage and everything else to try to get you upset," he recalled.

Orkin arrived in court in his nerd disguise and spent forty-five minutes on the witness stand answering questions for the prosecution. He detailed the elements of reverse-engineering that linked the fragment to the timer and the timer back to MEBO. Whether the technical details confused the defense attorneys or they merely discounted the importance of the science was unclear. However, after Orkin's extensive prosecution testimony, the defense "good cop" attorney simply said, "No questions." The "bad cop" lawyer then rose and asked a question that revealed a naïveté of technical matters.

"Isn't it true that all electrical devices contain electronic components?"

"By definition," Orkin answered bluntly.

No more questions followed.

By the end of the proceedings, there was no doubt that the timers originating with MEBO ended up in the hands of the Libyan government. Added to the mountain of technical evidence was the testimony of two Toshiba Corporation employees who confirmed that in 1988 Libya purchased 20,000 cassette recorders of the same model that concealed the bomb.

In all, the defense attorneys for the Libyan operatives called only three witnesses during the eighty-four-day trial. On January 31, 2001—more than thirteen years after Pan Am Flight 103 was destroyed—the three-judge panel returned a verdict: one conviction and one acquittal. The judges wrote:

The evidence, which we have considered up to this stage, satisfies us beyond reasonable doubt that the cause of the disaster was the explosion of an improvised explosive device, that that device was contained within a Toshiba radio cassette player in a brown Samsonite suitcase along with various items of clothing, that that clothing had been purchased in Mary's House, Sliema, Malta, and that the initiation of the explosion was triggered by the use of an MST-13 timer.

The court found Abdel Basset Ali al-Megrahi, the mastermind of the plot, guilty of murder. His sentence was a minimum of twenty years' imprisonment. His accomplice, Al Amin Khalifa Fhimah, who, authorities alleged, supplied luggage tags and assisted in getting the brown suitcase placed on the flight, was found not guilty by virtue of lack of evidence.

Terrorism moved from Scotland to sub-Saharan Africa when al-Qaeda sponsored bombings of the U.S. embassies in Kenya and Tanzania on August 7, 1998. These were immediately followed by credible reports that other embassies in Africa, Europe, and Asia were also in danger of attack. A U.S. embassy in the Balkans was identified as a specific target. As a precaution, U.S. personnel were temporarily relocated to a compound outside the city. Shortly thereafter, the local security service captured one al-Qaeda operative believed to be involved in the plotting but a second suspected member of the cell remained at large.

The at-large terrorist suspect, identified as a primary al-Qaeda forger who specialized in altered travel documents, was married to a local woman who claimed to have no knowledge of her husband's whereabouts. His capture could produce a wealth of intelligence through the identification of the alias identities of other al-Qaeda operatives along with exemplars of their passports, driver's licenses, and other travel documents. The death and injury suffered by U.S. officials in the Nairobi bombing added special urgency to the search. Finding the other terrorist became an obsession for the handful of case officers and techs.

OTS became involved in the hunt when the techs received word that communications had been intercepted between a suspected al-Qaeda cell in Western Europe and the Balkan terrorist. One particular exchange suggested that the European cell was providing logistical support and funds to the Balkan cell through a female cutout.

An operation combining OTS tracking and audio devices to find the terrorist was proposed. The concept involved implanting a tracking device as well as an audio transmitter in a package sent to the cutout who could be expected to deliver it to the target. Although a solid plan from the intelligence-gathering standpoint, the technical aspect was problematic.

Placing an audio and tracking device in a small package would be difficult, as both would be transmitting for an indefinite time. In addition, after the modified package was inserted into the European postal system, it would be outside of CIA control. Finally, any suspicion of the package by someone in the supply chain would alert the terrorist to his vulnerability.

Brian Mint, assigned to head the OTS tech team, understood the complexity of the operation and voiced his skepticism to the operational team. "We know he needs money," the senior case officer insisted. "He's gotten funds from Western Europe before and he's looking for more. You put money in some package with tracking and audio devices and we can get him."

Although Mint's confidence that a technical package would work did not match that of the case officer, he assembled a team of engineers to tackle the problem. The tracking and audio devices would require a host that the target and cutout would accept without suspicion. Since money would be the bait, the package also needed to be large enough to accommodate several hundred dollars' worth of European currency in small bills along with the two devices. In that sense, the host needed to serve as concealment for spy gear as well as money—a gift holding two secrets.

A tech was dispatched to look for suitable concealment hosts. European tourist trinkets offered the best options, since an inexpensive "gift" would not be alerting to customs officials or others handling the item during transit. The terrorist would likely assume the gift was more than it appeared based on its point of origin. Further, he would see the souvenir as a clever piece of tradecraft by his contacts and not examine the object more thoroughly once he discovered the hidden money. The psychological aspect of the "double concealment" was critical. The terrorist must have the intuitive sense to open it, but not suspect its second covert purpose.

Since cavities could be made and restored most easily in wooden souvenirs, the techs settled on a 14 by 10 by 3/4-inch wooden wall plaque. Mounted on the plaque's face was a thin metallic plate engraved with the outline of an Italian cathedral and a script below reading SAINT SUSANNA.

The techs removed the metal faceplate and hollowed out a cavity in the center of the plaque large enough to hold small-denomination bills worth several thousand dollars. An Arabic linguist constructed a handwritten note: "Brother we are with you. Hopefully this will get you by until we're able to contact you again." The faceplate was refastened to the front of the plaque with an adhesive strong enough to hold it in place, but could be easily pried off. The techs then carved a second compartment at the edge of the plaque large enough for electronics and batteries for two weeks of continuous transmissions.

The package, addressed to the cutout, was shipped with labeling to give it the appearance of originating with the European terrorist cell. An OTS team began monitoring the tracking and audio transmissions from specially equipped vehicles.

"We didn't know for certain where the package was going. All we knew was the address of the person we thought was the cutout," said Mint. "If that assumption was wrong, the operation ended. Further, we didn't know if we would be able to keep our van with the audio receiver within the transmitting distance of the bug. We did what we could and hoped for the best."

The techs tracked the delivery to the female cutout in the Balkans. Immediately opening the package, she read the Arabic-language note and fifteen minutes later, with the package tucked under her arm, began walking across town. Tracking signals suggested she was performing a basic surveillance detection run by making several stops and doubling back on some streets. She boarded a bus that took her into one part of town, changed buses, and headed to a different section. The surveillance team, which was unobtrusively following, was eventually led to a neighborhood known for a militant Islamic presence. The cutout entered a two-story house with upper and lower apartments inside a walled compound. When she came out, she was not carrying the package.

Local security established a 360-degree perimeter around the building while the techs set up a listening post in a nearby house within range of the transmitter. An assault team assembled. The audio recorded talk among several unidentified people about the package and then sounds of the metal plate being removed from the front of the plaque. The techs and case officers wanted to cheer as the concealment passed its first test. The target had recognized that the souvenir was more than it appeared and found the cavity concealing the money.

The techs listened as an excited conversation between the terrorist and his wife was translated. The communication channel with the European cell was working. Throughout the afternoon, the techs continued to receive strong tracking signals from within the house as well as audio. In the early evening, the terrorist's voice suddenly became agitated and his wife sounded emotional. The techs speculated that perimeter surveillance might have been spotted as the terrorist had prepared to leave the house. Then the techs heard the distinctive sound of a weapon being cleared and a round chambered.

After night had fallen and no more conversation was heard, the local operational commander directed an assault be launched. The techs listened to the commotion as the team entered the house and rushed to the second floor where the terrorist and his wife were believed to be. After a quarter hour of searching, the team reported they had not found the target.

"I thought, How could this happen? We know he was in there. We had him," recalled Mint. "There were only two possibilities. Either he managed to slip through the security around the house or he was still in there, hiding somewhere."

Another hour of continued searching produced nothing. The wife claimed no knowledge of the suspect's whereabouts. Believing the operation had come to a dead end, the techs entered the house to retrieve the plaque, which they found opened underneath the bed of the second-floor apartment. Then, just as they turned toward the door to leave, pistol shots and bursts from automatic weapons fire came from the kitchen, followed by loud noises.

The assault commander had continued searching the house and entered the kitchen for a final look. Either curiosity or policeman's instinct prompted him to move a small washing machine from against the wall. As he struggled with the surprisingly heavy appliance, a cavity between the back of the machine and the wall was exposed. At that moment the armed terrorist, who had been hiding inside the washing machine, fired a single shot that hit the commander in the chest. Another nearby officer returned fire. The terrorist rolled out of the machine and continued to shoot until he was killed with a burst of automatic fire from the assault team.

Closer inspection revealed that the working elements of the washing machine had been removed to create a hiding place just large enough for one person. Access to the concealment was obtained by removing the loosely attached tin backing and crawling through to the cavity.

Press reports the following day made no mention of the Agency's operational or technical role in the action, though the assault team, along with the wounded commander, received deserved accolades. The OTS techs were satisfied to have played an unpublicized role in removing another terrorist from the seemingly endless war.

The six-member OTS special missions team would miss another New Year's Day with their families. Created in the mid-1980s to provide an immediate global response to acts of terrorism, the team was prepared to deploy within hours of being alerted. Team members were trained in post-blast investigation techniques and ordnance disposal and for nearly two decades had disarmed bombs, thwarted terrorist attacks, and led foreign authorities to terror suspects. In more instances than the officers cared to remember, their urgent deployments inevitably clustered around the holiday season. Terrorists, it seemed, favored December for their murderous acts.

In 1983, American and French embassies were bombed during December, then a few days later the Provisional Irish Republican Army planted an explosive device in London's most famous retail emporium, Harrod's. Peruvian revolutionaries under the banner of the Tupac Amaru Revolutionary Movement took several hundred hostages at the Japanese ambassador's Lima residence in December 1996, and the same month a suicide bomber struck during an election rally in Sri Lanka, injuring the president. Libyan terrorists bombed Pan Am Flight 103 over Lockerbie, Scotland, just before Christmas 1988. HAMAS blew up a West Bank restaurant in December 2000 and, at about the same time, a bomb in India's Parliament killed thirteen men and women.

However, December 2001 was not like previous years. The United States was at war with al-Qaeda and its protector, the Afghan Taliban government, and the mission would take the bomb techs directly into the combat zone. Some of the team's members came from Headquarters while others flew in from field locations. All were volunteers.

*OTS designed and printed propaganda leaflets to support
the 2001 war against the Taliban in Afghanistan.*

Little more than two months had passed from the first October air strikes
against the Taliban and al-Qaeda terrorist training camps to the liberation
of the Afghan capital, Kabul. The Taliban and its al-Qaeda cohorts were
clearly on the run. While some Taliban forces were surrendering en masse
in Kandahar, others had taken to the hills, literally high-tailing it on horses
and on foot to the White Mountain range and honeycombed cave complexes
of Tora Bora.

While the large-scale military operations were all but over and the rapidly
advancing forward elements of the CIA's paramilitary units now faced only
sporadic small arms fighting, the country was still far from stable. As news
broadcasts showed a jubilant population defiantly enjoying activities forbidden under Taliban rule, such as kite flying and men shaving off the formerly
mandatory beards, the political and security situations remained danger-
ously volatile.

The war itself was progressing with lightning speed and, within weeks, the Agency's Afghanistan mission shifted from tactical support of the Northern Alliance to assuring a safe transition to a new government. The request for the OTS officers came as the Taliban abandoned control of the southern third of Afghanistan and the key city of Kandahar. The city had been among the last of the Taliban's strongholds and concern over what unpleasant surprises might be left behind was fully justified. The need for expertise in handling explosives and skills to assess, identify, and clear buildings of IEDs and conventional munitions became imperative.

The team was given seventy-two hours to ready itself for deployment. The mission was as ambiguous as it was urgent. In reality, no one knew what the team would encounter on the ground or exactly what equipment was needed. Broadly defined, the team would provide ongoing explosive detection, assessment, and disarming capabilities for Agency and military personnel. They would be operating in areas the Taliban and al-Qaeda had controlled for years, some of it captured only days or hours earlier. They would also likely be called on to provide secondary functions, such as establishing emergency communications, field engineering, and photography.

Intelligence reports from the field described hundreds of discoveries of tons of ordnance either discarded or cached during the country's two decades of nearly continuous warfare. This meant the team had to prepare to work with explosives of Russian, Chinese, and Pakistani origin, much of it unstable. Most would have to be destroyed to prevent it from falling into the hands of local warlords or opponents of the new government. Ironically, some ordnance and weapons were quite familiar to the OTS officers, as it came from stocks the CIA provided the Afghanistan mujahedin during their 1980s war against Soviet occupiers.

The team's housing would be primitive and amenities such as electricity and running water scarce or nonexistent. They would live next to the native population, some of whom were deeply suspicious of the new U.S. presence or still loyal to the Taliban. Almost every Afghan man was armed. The team could expect sixteen-hour workdays that would leave them covered in dust from unpaved roads and bat guano from caves in a hot combat zone to handle unstable explosives. Nevertheless, it was the kind of job they trained and lived to do.

During an intense thirty-six hours, the team assembled enough gear, about 5,000 pounds, to sustain them for a series of operations whose length

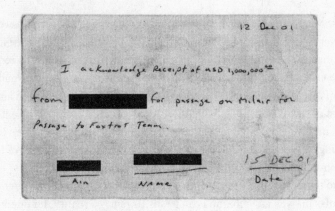

Handwritten receipt for one million dollars in cash carried in a duffel bag by an OTS officer to Afghanistan, December 2001.

and intensity could not be known in advance. The gear they loaded into the C-17 Globemaster at Andrews Air Force base outside Washington included everything from portable x-ray devices to explosives and ammunition.

As the plane's cargo doors swung shut and the four jet engines were about to rev up, an urgent message arrived that a final package was en route. With the deadline for departure perilously close, they waited. Then an un-marked truck raced across the runway and pulled up to the plane. Lugging an ordinary-sized duffel bag from the truck, the courier presented a three-by-five card that served as a receipt for the package. "Just sign," the courier in-structed the team leader, Mark Fairbain. "You have a million in U.S. hundred-dollar bills bound in $10,000 stacks. Trust me, you don't have time to count it." Looking inside the bag, Mark saw stacks of bills held together by small lengths of cotton string. That was enough and he signed the receipt.

With the gear and million-dollar duffel bag secured, the plane lifted off into the night sky over Washington. Then, as the heavy transport gained altitude, the smell of burning electrical insulation filled the cavernous cargo area. The team did not argue with the pilot's decision to divert to Charles-ton, South Carolina, rather than continue the transatlantic flight with a cargo that included a good deal of explosives.

After swapping out the two and a half tons of gear to the second plane, the team took off for another try to cross the Atlantic. "As soon as we were

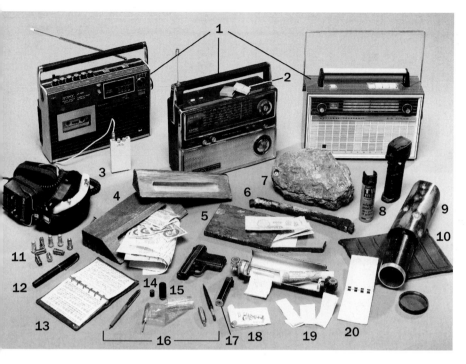

Annotated captured CIA technical equipment displayed by Russian counterintelligence, circa 1975–1986.

1. Shortwave radios to receive one-way voice-link messages

2. Frequency converter

3. SRR-100 scanning surveillance detection radio

4. Furniture concealment device

5. Dead drop container with money

6. Dead drop container

7. Dead drop rock

8. Containers

9. Dead drop canister

10. Waterproof dead drop pouch

11. T-100 camera bodies

12. T-100 fountain pen camera

13. Small binder with ciphers

14. Microdot viewer

15. T-100 keychain camera

16. Soft film with instructions inside ballpoint pen

17. T-100 lighter camera

18. Diagrams of drop sites and signal sites

19. Onetime pads

20. Note pad with special paper for secret notes

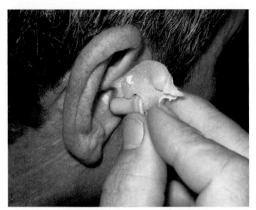

Proper placement of radio receiver and ear camouflage. Each ear camouflage was custom molded for the wearer, late 1970s.

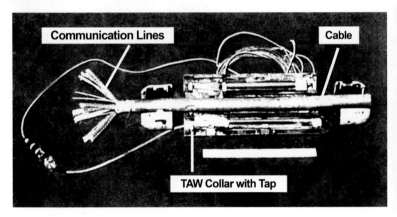

KGB photo of CKTAW cable tap found in Moscow, mid-1980s.

KGB photo of equipment used to record communications from the CKTAW cable tap, circa 1985.

"Stand By to Bug," an unofficial crest of the audio operations officers in the Technical Services Division, 1966.

OTS audio tech's traveling tool kit, circa 1982.

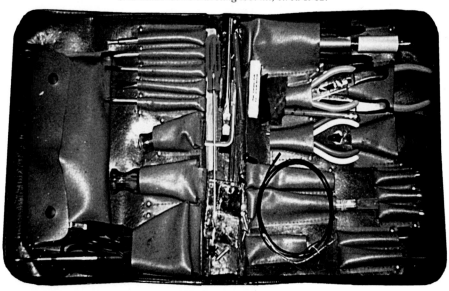

The Escape & Evasion Rectal Suppository was a concealable multipurpose toolkit packed within a smooth waterproof black plastic or aluminum shell, circa 1955.

The dart gun was designed by the Army Security Agency for tranquilizing guard dogs or other animals. When displayed by DCI William Colby before the Church Committee Senate hearings in 1975, many incorrectly assumed it had been an OTS-developed device.

Vials of CIA shellfish toxin shown during Congressional investigations, circa 1975.

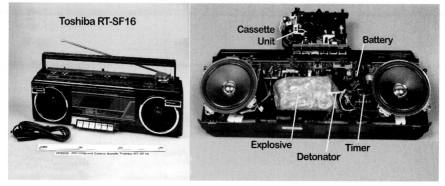

Left: Toshiba radio of type that contained the explosives that destroyed Pan Am Flight 103 in December 1989. Right: Reconstruction of how the bomb was concealed in a radio.

Evidence shown during the trial of two Libyans accused of the Pan Am Flight 103 bombing. Following discovery of the tiny fragment shown on the tip of a finger in the lower right, OTS analysis eventually led to the arrest and conviction of a perpetrator.

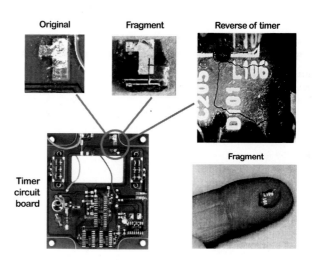

Cover and selected pages from the OTS-produced REDBOOK passport-examination manual designed to assist immigration and customs officials in detecting fraudulent passports, visas, and other travel documents used by terrorists, 1986.

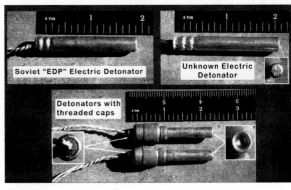

Detonators discovered by OTS officers in the booby-trapped Governor's Palace, Kandahar, Afghanistan, in December 2001.

Munitions recovered from the booby-trapped Governor's Palace, Kandahar, Afghanistan, December 2001.

Approximate placement of the primary explosive charges above the main meeting room of the Governor's Palace, Kandahar, Afghanistan, December 2001.

Post–9/11 CIA poster.

in the air we were off the jump seats to find a spot on the floor to get some rest," remembered one tech. "The only problem was we had packed away the sleeping bags and ground mats. So we slept on the metal floor with just our coats draped over us."

After an uncomfortable night, and already six hours behind schedule, they awoke in Ramstein, Germany, for a refueling stop. When airborne again, they continued on to Bahrain and from there to a secret airbase in Pakistan. Unable to make up for the six-hour delay, and with dawn breaking, the techs were informed the plane would attempt a dangerous daytime landing, the first U.S. aircraft to do so since the war began.

The southwest area of Pakistan, although not technically hostile territory, had yet to be declared completely secured. The airbase, which had first served as a launching point for search-and-rescue missions into Afghanistan, was now hosting large transport aircraft, such as C-130s and C-17s. The increasing Western presence ignited the ire of local militants whose protests included random ground fire ranging from small arms to antiaircraft guns. Pakistani armed forces found it difficult to suppress the dangerous but largely ineffective attacks.

On approach, the team strapped themselves into their seats just in case the aircraft needed to deploy countermeasures or take evasive action. The precaution, to the team's relief, proved unnecessary and after an uneventful landing, the techs set about off-loading the two pallets of gear and rested in tents that had sprung up in the "boom town" airbase. They would make the trip to Kandahar that night under the cover of darkness.

The final leg would be a 300-mile trip from Pakistan to Kandahar in Air Force MH-53J Pave-Low helicopters. Outfitted with advanced avionics that allow flight close to the terrain's contours, the Pave-Lows were also heavily armed with two mini-guns on the side doors and a .50 caliber machine gun mounted at the rear.

According to the plan, four Pave-Lows would fly in formation crossing into Afghanistan, then separate. Two would go to Kandahar and two to another base. Flight time was estimated at three hours, getting the team to Kandahar sometime after midnight.

Just before dark, the techs loaded the two helicopters with the two and a half tons of containers, boxes, and bags. An extensive preflight briefing covered topics ranging from landing positions to combat search and rescue (CSAR) procedures. Each team member was issued a Glock 9mm before boarding the craft.

Bad luck boarded as well. A warning light blinking in the cockpit of one Pave-Low a short time after take-off sent the OTS team back to Pakistan airbase. Once on the ground, the team reloaded their gear and million-dollar duffel bag onto a new Pave-Low, only to be told the replacement helicopter also suffered from mechanical problems. At the last possible minute for the night operation to continue, the ground crew isolated and fixed the problem.

As they lifted off from the base, the noise inside the Pave-Low drowned out conversation. Gunners positioned themselves at the opened side doors and the back ramp was opened for the .50 cal gunner. Crossing into Afghanistan, all three gunners opened up and test fired. Adding to the noise, the opened doors created an uncomfortable internal wind tunnel. "Only then," recalled Mark, "did we learn what a cold rough flight really was, when the helicopter began bouncing up and down from the wind of the southern mountain range."

Three hours later, the chopper hovered over bomb craters and debris before touching down on a dark runway of what was once a state-of-the-art airport. Built in the 1970s, Kandahar International Airport was at one time the largest and most modern in Central Asia, but it had been severely damaged in the Russian invasion in the early 1980s. After the Soviets left, the airport became a stronghold for Taliban and al-Qaeda forces, which turned its runways into minefields. Recent U.S. bombing raids had added to the destruction, pockmarking the tarmac with deep craters and littering it with the debris of war. Now, with the airport recaptured by U.S. Marines and Pashtun guerillas, control of the once modern structure—less than twenty miles outside Kandahar—signaled the end of the Taliban's fixed presence in the area.

With the chopper's blades still turning, the team rushed to off-load the gear in the fifteen minutes the pilot allowed before taking off for the return flight to Pakistan. No reception committee greeted the six as the Pave-Low vanished into the night, leaving the team alone on the darkened runway at four o'clock in the morning. The team stood beside thousands of pounds of high-tech equipment and the money bag they had hauled from plane to plane and chopper to chopper, used en route as a pillow, footrest, and bed. Mark noticed that the strings securing the stacks of hundred-dollar bills had come untied leaving a bag filled with 10,000 loose bills.

Armed with only Glock 9mm sidearms, the team had no choice but to wait in the open on the runway. "We had been told that the Marines moved onto the airfield a few hours before, but they were nowhere to be seen," recalled a team member. "We had no contact plan. Six sets of headlights suddenly popped up at about three hundred meters and headed our way. We hoped they were friendlies." The headlights were from vehicles belonging to two Marine units supported by Delta Force operators.

The Marines greeted the OTS team in short-bedded Toyota pickups woefully inadequate for hauling 5,000 pounds of equipment to the Governor's Palace in downtown Kandahar. Two trips would be necessary, the trip taking an hour each way.

The palace compound had until recently been home to the Taliban's reclusive one-eyed spiritual leader Mullah Omar. A chief architect of Taliban rule, Omar had issued the religious decrees that turned Afghanistan into a repressive regime that provided a sanctuary to Osama bin Laden and the al-Qaeda terrorists. The complex consisted of two large parallel buildings, one the residence and the other the meeting hall/auditorium, with a courtyard between them and narrow walls at each end to create an enclosed rectangle.

The team once more transferred the gear, this time into the palace. The nearly three-day trip had left them exhausted and sleep was a priority. However, not four hours later Mark was awakened with urgent news. A man claiming to have important information had come into the courtyard telling a story about explosives buried within the palace. Fighting grogginess and working through an interpreter, Mark engaged the slightly built man dressed in tunic and turban.

Speaking in a calm, deliberate voice the volunteer explained that the retreating Taliban hid explosives in the palace's earthen roof. The explosives, he said, were to be detonated just after sundown that day, at the start of the Muslim holiday of Eid al-Fitr, the three-day celebration marking the end of Ramadan.

The walk-in seemed credible—so credible that Mark could not suppress the suspicion that he had helped plant the explosives and was now having second thoughts. With the fall of the Taliban, loyalties in the country were shifting daily. The volunteer, who spoke matter-of-factly about explosives possibly only a few meters above their heads, would not have been the first Taliban loyalist to switch sides.

Inside the palace, everything seemed normal with the American and British troops and Afghanis, all involved in their own tasks. More than fifty people were already present, preparing for the Eid festivities, and within hours the assembly hall would be filled with the principal leaders of the southern third of Afghanistan, guests of the new governor, Gul Agha Sherzai, who was hosting the event.

For experienced OTS officers who had picked through many post-blast scenes, it was easy to imagine the sudden death and destruction from a rooftop blast. Destroying Afghanistan's southern leadership on a Muslim holy day would be a cruel, audacious act and a devastating blow to the new government and the United States. Delay could prove fatal to the local officials and guests, more than a hundred U.S., Australian, and British military personnel, as well as America's Afghan policy.

Within fifteen minutes of the conversation, Mark made the decision to send one member of the team to the roof. Putting an officer on the roof in daylight was risky. If the building was under surveillance, anyone on the roof was certain to be spotted and a terrorist could decide to detonate ahead of schedule. Mark calculated the risk, taking into account the approaching nightfall and the start of Eid. Both were less than four hours away.

Frank Shumway, a tech experienced in using thermal imaging equipment, was rousted from a sound sleep. After Mark explained the situation, Frank agreed to climb onto the roof. Strapping on a hundred pounds of equipment, including communications gear to keep him in constant contact with the team, Frank would report each movement and every observation. These transmissions would be recorded, and should a detonation occur, by accident or command, the recorded information could provide valuable data for post-blast investigators and future operations.

After navigating the narrow ledge of the perimeter wall that joined the two buildings, Frank climbed a ten-foot ladder to reach the earthen roof above the palace's assembly hall. Its smooth and hard packed surface showed no signs of recent disturbance, but as soon as he switched on the thermal imager, the results immediately contradicted everything Frank saw through his own eyes.

The small screen identified four distinct "hot spots," each giving a signature of recent excavation. What the human eye cannot see registered clearly on the screen. No matter how well a recently dug hole is refilled and disguised, it will absorb heat at a different rate than an undisturbed area. Foreign objects just below the surface—such as IEDs—can also enhance the thermal signature.

"Four holes are set in an L-shape; one hole is positioned directly above the reception area and the three others run along the lateral axis of the primary assembly hall," Frank reported. "The imaging also shows a narrow line of disturbed earth from hole to hole, then off toward the edge of the building."

Frank's information bolstered the credibility of the walk-in's report. Should the holes contain even a moderate amount of explosives, the blast would, at minimum, collapse the roof. While the imager could not provide clues to what, if anything, was beneath the surface, the pattern, size, and shape of the hot spots were consistent with what was known of how mines or munitions could be deployed. If the holes concealed an explosive array, the work had been extraordinarily well done.

Satisfied that the roof had been thoroughly imaged and convinced that explosives were likely buried there, Mark recommended to the U.S. officer in command that military personnel and the Afghanistan locals preparing the celebration be ordered to the far side of the compound. The soldiers complied, but the Afghanis declined to evacuate or even suspend their preparations. It was only two hours before dark and guests had already be-gun to arrive in the reception area.

With the most experience in explosive ordnance disposal of any member of the team, Mark made the next foray onto the roof to probe the hot spots. Information from the walk-in, combined with the hole patterns and a nar-row line of disturbed earth running from the holes to the edge of the roof all strongly suggested the explosives were configured in a command detonated array. With command detonation, the four connected charges would be set off with a single signal. Working with the command detonation hypothesis the team prepared a small electric charge to cut the command wire.

The potential for disaster was enormous. The team still had no idea what type of IEDs lay buried inches beneath the surface, whether there was a secondary detonation system, or if they were booby-trapped.

As the reception hall rapidly filled with Afghan dignitaries and nightfall approached, Mark took the narrow staircase to the roof and began gently probing around the areas shown on the imager. Several inches underground, his probe hit what felt like steel. He carefully swept the crumbled earth away to expose a small section of the buried object. Using the thermal im-ages as a road map, he continued working slowly and cautiously, and even-tually identified an eight-gauge detonator lead wire buried just beneath the surface that connected the four holes and led off to join a tangle of other communications and electrical wires that fed the building from the outside. He had seen enough. There were several explosives beneath the roof linked for command detonation with the signal wire trailing off to some remote place in the war-battered city beyond the compound. He attached the electric cutting charge, gave the signal, and the line was severed.

Mark left the roof just as the sun was setting. Afghan dignitaries were now crowded into the reception hall of the palace oblivious to what was likely thousands of pounds of live explosives only a few feet above their heads. The end of Ramadan was announced as Mark made his way down the narrow staircase and a liberated Kandahar erupted in celebratory gunfire. He could not suppress a smile at the thought that somewhere among the city's celebrants was a terrorist, his finger repeatedly pushing a button in vain, wondering why in the name of Allah his best efforts had come to naught.

There had not been time to consider removing whatever munitions lay buried beneath the surface. The palace compound was as secure as possible and Mark reasoned that leaving the munitions buried until the end of Eid carried little risk.

After Eid, an Afghan military de-mining team was brought onto the compound to dig the ordnance out of the roof. Mark estimated that it would take one day per hot spot—four days in all—to safely excavate the explosives. With the Afghan squad assembled, two OTS team members began a four-hour refresher course in explosive ordnance disposal (EOD) procedures and safety. The introductory remarks were never concluded. The senior Afghan officer interrupted to announce that his troops were well prepared for this type of work. No further training was necessary. It was the shortest refresher course in OTS history.

The OTS team watched at a distance as the Afghanis attacked the job with frightening enthusiasm. What Mark had estimated as a four-day job was finished in less than one. Before the day ended, the Afghanis had removed more than 2,200 pounds of hard case explosives, including fifty-five 122mm tank rounds and more than a hundred antitank mines from the four hot spots. A fifth hot spot turned out to be empty.

The rooftop IED would be classified as rudimentary, but what it lacked in sophistication, it compensated for in size. Detonation would have reduced the palace to rubble, killing or injuring everyone inside, and in all likelihood taking out the U.S.-occupied structure across the courtyard.

"We've heard a dozen times from our paramilitary colleagues and the special ops guys that our entire deployment was paid for in full that first day," noted a team member.

A few days later, with the components of the bomb piled up outside the palace as a photo op, the local commander called a press conference to announce the find and the successful defusing. A single bored reporter, along with a photographer, listened politely and took notes, but no story ever appeared.

The walk-in was invited back to the palace to receive a reward. Afghan and U.S. personnel staged a semiformal ceremony. After drinking his tea, he was given $1,000 in hundred-dollar bills. He showed little reaction to the reward money, which he accepted graciously, and then offered a short speech, declaring that his only motivation was to help his country.

The OTS team remained in Afghanistan for another six weeks, clearing hundreds of tons of ordnance from homes and remote hideaways. While hiking the rugged White Mountain range with Delta Force operators, they discovered caches of aging explosives crammed into the man-made caves. In one cave, stockpiles of mortars and mines were packed from floor to ceiling, extending far back into the mountains. A B-52 bombing strike was required to detonate the massive cache.

On the outskirts of Kandahar, in the rubble of an al-Qaeda training camp, the team discovered and then destroyed dozens of drums of chemicals used to produce the explosive known as triacetone triperoxide, or TATP. A favorite of suicide bombers, TATP was used by would-be shoe-bomber Richard Reid in his foiled attempt to bring down an airplane in December 2001, and again by terrorists in the July 2005 London bombings. The number of lives they saved may never be known.

Upon their return to the United States, members of the team were awarded the CIA's Intelligence Star for Valor in a ceremony presided over by DCI George Tenet. They became members of an elite cadre of some twenty techs whose courage and service has earned the honor of receiving the Intelligence Star during the fifty-year history of the Office of Technical Service.

FUNDAMENTALS OF TRADECRAFT

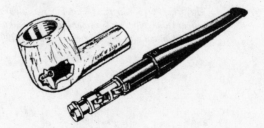

The prized OTS "Spyman" statuette was awarded to officers for honorable service while assigned to the OTS technical and engineering laboratory, 1991.

Author's Note by H. Keith Melton: As a young naval officer returning from service in Vietnam in the late 1960s, I continued my interest in the world of espionage. The exploits of famous spies were fascinating, but my engineer's curiosity focused on the more obscure topic of clandestine technology. Many books, almost all produced by nontechnical writers, chronicled noteworthy spy cases, but rarely could I locate details about the gadgets used to secretly photograph documents, plant listening devices, and accomplish other amazing feats. I watched the James Bond movies of the era and wondered whether there were such gadgets in the real world or was Q only part of movie magic. Finding the answers to these questions became a passion that has consumed the last forty years of my life and led me on treasure hunts around the world.

My quest began in Washington, D.C., and eventually required many trips to Russia, Germany, Poland, Czechoslovakia, Great Britain, France, Israel, and into Asia and South America. Repeatedly my travels took me to KGB headquarters in Moscow and to the Berlin study of Markus Wolf, legendary head of the East German Intelligence Service (HVA). I became a regular guest of Walter Pforzheimer, the late dean of intelligence bibliophiles and founder of CIA's Historical Intelligence Collection, and through him a lifelong friend of his eventual successor, Hayden Peake, the noted historian, author, and intelligence bibliographer.

Eventually I discovered a commonality among all the world's spy agencies, that each selects its intelligence officers for the ability to recruit and manage agents and not for their technological skills. Only rarely does an operations officer understand the technology inside the spy gear employed in secret operations. For this necessary technical assistance expertise and creativity, intelligence services created a cadre of specialists known as techs, to support and, at times, even conduct the operational activity.

Techs were usually recruited because they had preexisting knowledge in fields such as photography, radios, electronics, chemistry, woodworking, fabrics, or communications. Techs working for the KGB, HVA, CIA, MOSSAD, MI6, DGSE, or DGI shared a technical language. Each intelligence service had an internal component dedicated to examining espionage devices captured or recovered from its adversaries. Analysis of "foreign finds" could identify the originator of the gadget, provide new technology and techniques, and lead to countermeasures. Over time, many of the technical tools, regardless of the nationality of the service, began to look similar. Commonality of functions resulted in commonality of forms.

In compiling my research, I also discovered that the fundamental work of clandestine intelligence could be grouped into five general categories, and within each grouping, technical support was critical. Photos and illustrations of spy gadgets used by various services appear in my previous books, *Clandestine Warfare* (1988), *OSS Special Weapons and Equipment* (1992), *CIA Special Weapons and Equipment* (1993), *The Ultimate Spy Book* (1996), and *Ultimate Spy* (2003). Early in the preparation of this book, the authors faced the dilemma of using tradecraft terminology in the text without having space to provide a definition and explanation each time the term appeared. The solution has been for me to write a primer that draws together explanations of the essential technical terminology used throughout *Spycraft*.

In this six-chapter section, I have attempted to integrate the gadgets with the doctrine of intelligence that lies behind their development and use. Chapters 20 through 24 describe the five pillars of tradecraft common to all intelligence operations. When merged with the clever devices created by innovative engineers, these pillars distinguish the professional intelligence services from those operations that are performed by quickly apprehended amateur "spies." Chapter 25 summarizes the revolutionary changes that digital technology and a global Internet have brought to each of the pillars.

CHAPTER 20

Assessment

> If one attempt in fifty is successful [for recruitment],
> your efforts won't have been wasted.
>
> —British turncoat and KGB spy Harold "Kim" Philby, as quoted in *The Literary Spy*

Clandestine intelligence operations using human agents, whether conducted in the eighteenth century by America's Revolutionary War spymaster, General George Washington, or in the twenty-first century by Islamic terrorists, have common characteristics. Five categories of recruitment and agent handling are so universal and fundamental that they can be called the "pillars of tradecraft." These are:

- Assessment
- Cover and disguise
- Concealments
- Clandestine surveillance
- Covert communications

Depending on the stage of an operation, one of these disciplines will assume dominance, and every effort will be made to execute it flawlessly. For the Central Intelligence Agency, OTS had the responsibility to develop and support technical tools for each pillar that would provide U.S. officers and agents with a comparative advantage over their adversaries.

Assessment is the first step in recruiting a spy. Selecting the right target from among the thousands of individuals who could potentially help an intelligence service requires identifying the one or two with the motivation and ability to sustain the double life required by espionage. Sound tradecraft demands more than guesswork.[1] Based on the experience OSS had with assessment and testing procedures, the CIA employed a small group of professionally credentialed psychologists to assist operations officers in winnowing the prospective recruitment pool and identifying the most "vulnerable" targets. Like their OSS predecessors, the psychologists of OTS employed a variety of assessment techniques and tests to gain insight into a target's dominant personality traits and potential behavioral responses to specific situations.

Recruitment often encompassed months of patient cultivation by the case officer of a target before moving the contact into a clandestine "handler-agent" relationship. Infrequently, however, recruitment could occur during a five-minute pitch in which an unsuspecting foreign official would be asked, "Would you work with the CIA?" Operational circumstances determined whether an individual was the subject of extended development or a cold pitch, but in either case, the assessment conducted before the question was asked loaded the dice in favor of the case officer.[2]

Assessments provide the CIA with a good sense of the target's likely reaction to a pitch and their long-term value to the CIA. However, under the best of conditions, acceptance of a pitch can never be assumed and sound assessments will anticipate the possibility of an angry and hostile response. If the pitch goes well, an agent is recruited. If the recruitment offer is rejected, assessment will have provided information to minimize blowback and operational compromise.

Motivations to become a spy are as complex and varied as human nature itself. Because of unpredictable individual differences and cultural variations among foreign officials identified for recruitment, identifying a target's dominant motivator to conduct espionage became the primary function of the operational psychologists. One grouping of motivations became known as "the MICE model." MICE, the easy-to-remember acronym of money, ideology, coercion, and ego, describes crosscultural characteristics that often translate into vulnerabilities that become a basis for recruitment.

Money holds particular attraction to targets from countries whose culture places high social value on achievement, status, and material possessions.

Ideology becomes a powerful incentive for individuals who hate the political or economic system, which they cannot otherwise escape or oppose.

Coercion represents a negative motivator that could be effective only in selective circumstances with particular personalities.

Ego frequently motivates acts of espionage by individuals who believe their talents, capabilities, and importance go unrewarded by their employers or unrecognized by professional colleagues.

CIA psychologists found three of the most significant indicators of a willingness to spy were split loyalties (potentially evidenced by extramarital affairs or intense dislike of a supervisor), narcissism (when seen as excessively self-absorbed, arrogant, and vain), and dissidence in parental relationships. Added to these were contributing circumstances such as failed careers, marriage problems, infidelity, and substance abuse. Seldom was there a single motivating factor, and most recruitments were based upon a combination of vulnerabilities. Clandestine audio operations became one of the most useful ways to gather unfiltered information about a target's private motivations in unguarded conversations with family and friends. CIA psychologists concluded that for most agents the susceptibility to recruitment and the willingness to act is the highest between ages of thirty-five to forty-five, a time of personal reevaluation and mid-life crisis commonly experienced in many cultures.[3]

In addition to targets who were cold pitched and those who were recruited after development, volunteers constituted a third pool of potential agents. Some of history's best spies have been volunteers. These individuals, also known as walk-ins, sought out an intelligence service to which they could offer their information or services. Volunteers are treated with caution because many have an exaggerated sense of the value of their information or are seeking an emotional thrill of becoming part of the espionage game.[4] More significantly, volunteers could also be dangerous "dangles" or "plants" controlled and directed by a rival intelligence service. If the bait of a dangle is accepted, the hostile service is in a position to run a double-agent operation to either acquire information about the sources, operational methods, targets, and technology of its rival or feed false information to the enemy.

Regardless of how a potential spy came to the attention of the CIA, recruitment occurred only after a favorable judgment was made about an individual's access, motivation, and ability to lead a clandestine existence. The process of evaluation that precedes the decision of whether or not to attempt to recruit a target is called "assessment."

Two questions are paramount when assessing a prospective agent. The first is: What access to information of intelligence value does this person have now or will have in the future?

The level and value of an agent's access are determined through questioning, verification of whatever personal bona fides are presented, and evaluation of the initial information the source provides. An individual's official position, social or family contacts, career progression, skills, and the quality of information are all used to confirm the potential agent's level of access. When Aldrich Ames gave the Soviets the names of nearly a dozen active CIA agents in June 1985, his access was confirmed and a willingness to commit espionage demonstrated.[5] When the National Security Agency evaluated Victor Sheymov's initial reporting in 1980 on Soviet communications security, the quality of information immediately established Sheymov's access to exceptionally valuable intelligence. The Soviet and U.S. response to Ames and Sheymov demonstrated the willingness of intelligence agencies to move quickly to recruit a volunteer without lengthy assessment when access to critically important intelligence was demonstrated.

The second question considered by a recruiter is: Can a prospective agent live the life of a spy and do what is required by the task of espionage? This assessment requires insights to predict, with reasonable accuracy, the future behavior of the target. Like buying an automobile, expectations and desired outcomes at the time of the initial transaction can sometimes trump reality. If either the automobile or the agent turns sour, the frustration and expense of owning a lemon can turn into disaster. Bringing professional skill and the tools of modern psychology to the process of assessing the situational behavior and personality of would-be spies, foreign leaders, and current agents became the core work of the OTS operational psychologists.

From its beginning, OTS employed a staff of professional psychologists to conduct operational assessments of foreign targets. The assessments were based on the best psychological science available and used both commercial and specially designed psychological tests to evaluate a target's personality, motivation, and aptitude for clandestine work. Raw data for assessments was acquired from reports of operations officers who observed personal and

behavioral traits of targets. The OTS psychologists then applied their expertise to evaluating all of the information gathered on the individual.

The psychologists provided professional personality assessments of recruitment targets, individuals who volunteered to work with the CIA, and
defectors. Depending on the specifics of the case, the assessments were
used for guidance in building a relationship, refining a recruitment pitch,
addressing agent-handling problems, minimizing issues at agent termination, preparing for agent resettlement, and framing counterintelligence
judgments about assets. The assessments were frequently combined with
results of polygraph testing administered by the Office of Security for the
fullest possible understanding of the subject. Defectors, whose bona fides
were in question, such as the high-profile cases of Yuri Ivanovich Nosenko
and Anatoly Golitsyn, were assessed by the TSD psychologists both to support both counterintelligence analysis and to assist officers responsible for
resettlement.[6]

These assessments could be either direct or indirect depending on
whether or not the psychologist could personally interact with the subject.
When personal meetings were not possible, assessments relied on the psychologist's analysis and interpretation of credible, secondary data.

The most complete assessments included direct personal meetings between the OTS psychologist and the target. For security, these operational
meetings usually employed various elements of clandestine tradecraft, including disguise, alias identity, and surveillance detection runs. Under normal circumstances, such meetings with the target were conducted in a
manner that did not reveal either the psychologist's true profession or intended purpose.

The psychologists conducted the assessments in whatever venues could
be arranged for meeting with the target. In Germany during the mid-1980s,
a leader of a terrorist cell had been an intermittent contact of a case officer,
but little progress was made toward recruitment. The question of whether or
not to continue recruitment operations against the individual came to OTS.
Because the target frequented a nightclub that drew its patrons from the international community, an OTS psychologist was directed to make the
nightclub part of her weekend activities. For her disguise, the psychologist
chose a "blonde bimbo" look based on knowledge that the target's eye
gravitated to every blonde that entered the club.

On a particular Friday evening, the psychologist, with the assistance of
disguise specialists, selected a slinky dress, put on a curly blonde wig,

blue-tinted glasses, rosy pink lipstick, and blue eye shadow. As she walked out of her office, the psychologist passed the chief's secretary offering the standard "Have a good weekend" greeting. The secretary looked up with surprise to ask, "Who are you? Have you signed in?" After a moment of silence, both found amusement and appreciation for the superior work of the OTS disguise officers.

At the nightclub, the psychologist observed the movements and interaction of the target and put herself in a location to attract his notice. The ploy worked and the two engaged in a conversation that moved quickly from introductory chitchat to increasingly friendly banter. It was a good night for the psychologist, whose questions were so readily answered that she needed to periodically go to the powder room to jot notes and confirm that her disguise elements were in place. As the evening progressed, so did the personal level of their conversation. The terrorist, clearly enjoying the pursuit of his blonde prey, became increasingly familiar and uncomfortably suggestive with the psychologist. Seated in a dark corner of the nightclub, he leaned very close and whispered a well-practiced line, "I'd just like to run fingers through your blonde curly hair." The psychologist choked back the overwhelming urge to rip off the wig, hand it to the terrorist, and in her silkiest voice reply, "It's all yours if you will stop annoying me now."

A direct assessment might involve pretext testing or face-to-face interviews with targets. The target would be unaware of the true purpose of the interview since the psychologist would be introduced by the case officer as a friend, colleague, or knowledgeable specialist about a common area of interest. Thereafter, the psychologist would observe and record the target's verbal skills, interaction with the case officer, body language, temperament, and other personality and behavioral characteristics.

The "unexpected" usually became "expected" during the interviews. In support of an operational project with a cooperating service to build a new counterterrorism team, an OTS psychologist posed as the American official who would make the decision on members for the team. Over the course of several days, the psychologist administered assessment tests to several dozen candidates under the guise of "the final interview."

After the team members had been selected, several additional individuals were nominated to become office manager for the project. As the psychologist talked with a young woman about her interest in the office manager position, it became evident that the candidate had no applicable skills for the work. She could not type, claimed no previous work in an office environment, had never done filing, acted as a receptionist, or exhibited any knowledge of office procedures.

The perplexed psychologist finally blurted the question, "Well, what are you good at?"

"Hijacking airplanes," replied the applicant.

Inquiries about office skills ended, and further questioning by the psychologist confirmed that the woman had been part of the planning and execution of three hijackings. She was reclassified from potential office worker to possible field operative.

In situations that precluded personal interaction, OTS psychologists made discreet observations of targets at a distance. These could often be done during diplomatic receptions, social events, or while seated at an adjoining table in a restaurant. The evaluation of clandestine video or audio surveillance tapes of a target represented another quasidirect assessment technique. These clandestine observations supported both operations for assessment of recruitment targets and collection of personality information on foreign leaders.

A daring, but ultimately abandoned, plan for assessment by discreet observation occurred when Soviet Premier Nikita Khrushchev visited the United States in 1958. A TSS psychologist was directed to remain at his home on a specific day for a special assignment. The psychologist's residence adjoined an empty field, large enough for a helicopter landing area. When he saw a helicopter land in the field, he was to climb aboard for a ride to Camp David where Khrushchev was scheduled to confer with President Eisenhower. Once at Camp David, the psychologist would be slipped into a closet in the room where the two heads of state were meeting. From a nonobvious peephole in the closet door he was to observe the Soviet leader's demeanor, voice inflection, body language, and any other characteristic that might provide insight into his mental and psychological state. The psychologist waited throughout the day, but no helicopter appeared. In the tradition of "need to know," no reason was ever given for scrubbing the operation.

For personality and behavioral assessment, OTS selected psychological tests and procedures applicable to the target's position, nationality, prospective operational role, and relationship with the case officer. The tools OTS used for assessment testing fell into three classes: commercially available tests that measured intelligence, psychological characteristics, aptitude, interests, and personality traits; modified commercial tests that were adapted for particular operational purposes; and CIA proprietary in-house-developed test and evaluation procedures.

The CIA's primary direct assessment tool, a largely culturally neutral test, was developed by TSS psychologist John Gittinger in the 1950s. The test questions could be administered openly or covertly by a case officer or psychologist in any language and the responses fed an assessment method named the Personality Assessment System (PAS).[7] Gittinger, who joined the CIA in 1950, had developed his skills as the director of psychological services at the Oklahoma State Mental Hospital in Norman, Oklahoma. By interpreting data derived from patients tested against the Wechsler intelligence scale, Gittinger determined that he could make basic judgments about personality. He eventually collected Wechsler data on 29,000 subjects representing social groups ranging from hobos to fashion models, and businessmen to students. He was an early user of computers for compiling large quantities of test data to develop comparative relationships. At the CIA, he refined the methodology and built a mature PAS model. By the time he retired in 1979, Gittinger's conscious emphasis on cross-cultural orientation for assessments and a demand for systematic, rigorous research-based judgments had become the basis for the CIA's acceptance of operational psychology as a technical tool for agent operations.

While Gittinger's system had detractors ranging from those who thought all psychology was suspect to professional peers who questioned the methodology, the PAS proved valuable to case officers involved in operations where time for personal contact with their targets was limited. The PAS results were so strong that the test became the standard method for assessing and predicting agent motivation and situational behavior. The insight the OTS psychologists provided about foreign targets by interpreting data from the PAS earned them the nickname "the wizards."[8]

Indirect assessment involves reviewing all reporting from case officers about a target's personal history, behavior, demeanor, and reactions to his contact with the case officer. All information available from or about the target, including public and private speeches, publications, and letters, as well as news stories and commentary by associates or relatives, are factored in. Covert audio or video transcripts, when available, also become a valuable part of the assessment package. The psychologists apply accepted analytical tools to the material and conduct an internal peer review from staff colleagues with multicultural backgrounds and foreign language skills in addition to their professional credentials. Direct assessments yielded higher quality and more extensive data for analysis than did indirect assessments, but the latter were necessary when the subject proved to be inaccessible.

During the Cold War years, when many targets lived in countries with severe travel restrictions, OTS maintained a small staff of handwriting specialists called graphologists.[9] Graphology, a discipline more respected in Europe than in the United States, seeks to identify psychological characteristics of an individual based on measurable letter formations and line strokes in handwriting. The graphologists measured three dimensions (the vertical, horizontal, and depth of strokes or letters) for as many as twenty-one different characteristics of writing. Handwriting analysis has demonstrated the ability to distinguish between mentally healthy persons and those with mental illness. The OTS graphologists applied the same methodology to identify essential characteristics of persons who were unidentified, would not agree to a structured assessment (such as VIPs), were writers of anonymous letters, or were held in captivity.[10]

Advocates asserted that by analysis of handwriting, which in graphology is called "brain writing," psychological characteristics and personality traits important to the CIA on otherwise unknown people can be identified.[11] Although psychologists disagree on the value of graphology as a stand-alone tool, many Agency operational managers agreed that, as a supplement to direct assessment or in the absence of direct assessment opportunities, handwriting analysis done by trained graphologists contributes valuable insight into a target's mental state.[12]

The best graphological analysis required a page or more of current handwriting for comparison against a similar amount of writing from some years earlier. Rarely did the graphologist have the luxury of being in possession of that much information and at times had to lower his expectations of the science. When presented with a collection of Stalin's doodles after the dictator's meeting with U.S. diplomats in the early 1950s, one TSS graphologist declined to provide a current psychological assessment. The sketches were clearly depictions of wolves, the graphologist commented, but he could offer nothing more than conjecture as to how those reflected Stalin's mental state.

In another instance, during the summer of 1983 a graphologist was given the handwritten signature of Soviet Communist Party General Secretary (and former director of the KGB) Yuri Andropov. Comparing the recent signature with previous Andropov signatures, the graphologist concluded that the writer had an inflexible commitment to ideological ends and little interest in compromise. At a time when the U.S. government questioned whether Andropov represented a "new, more Western" type of Soviet leader and was uncertain whether his health would limit his tenure as the Soviet head of state, the graphologist added that the handwriting comparisons showed evidence of increasing stress and difficultly in controlling moods. Causes of the stress and the reaction could, she concluded, be related to physical health or pressure of the position or both. In fact, Andropov's subsequent policies did not demonstrate new flexibility and, less than six months later, he died.

More recently, in the early 1990s, a classic handwriting analysis occurred when a CIA officer unexpectedly received a folded piece of silk from a fellow parishioner at a Catholic mass in Rangoon, Burma. The message, whispered the parishioner, had been written by a political prisoner who arranged to have it smuggled out of the heavily guarded prison and intended it to be given to the U.S. government. When the message on silk arrived at Headquarters, an OTS graphologist was asked to assess the writing but was given no information about the author or the circumstances of its acquisition. She studied the writing for several days, applying the standard techniques of letter and stroke measurements, and reported: "The writer possesses the genuine humility of those who are truly at peace and genuinely altruistic. Independent and individualistic, the writer is a true visionary . . . extraordinarily idealistic but at the same time sophisticated, manipulative, savvy, and subtle. Peaceful conflict resolution is a forte."

What the graphologist did not know was that her work was playing a key role in a major foreign policy decision. The assessment request came from a presidential envoy who was considering whether to meet with Burmese prodemocracy leader Aung San Suu Kyi. The meeting eventually occurred and afterward the diplomat credited the analysis with preparing him for an encounter with "a skilled, dynamic leader with keen political instincts and a flair for the dramatic" and, who "through courage and determination had repeatedly faced down the Burmese military and endured." Aung San Suu Kyi received the Nobel Peace Prize in 1991 and the U.S. Presidential Medal of Freedom in 2000.

The introduction to a 1954 national security assessment prepared for President Eisenhower titled "Report on the Covert Activities of the Central

Intelligence Agency" asserted: "If the U.S. is to survive, long-standing American concepts of 'fair play' must be reconsidered. . . . We must learn to subvert, sabotage, and destroy our enemies by more clever, more sophisticated, and more effective methods than those used against us. It may become necessary that the American people will be made acquainted with, understand, and support this fundamentally repugnant philosophy." This stark perspective, articulated in a top-secret report prepared by a special study group headed by James H. Doolittle, reflected Washington's perceived danger from the Soviet Union in the mid-1950s.

Assessment programs developed by TSS and TSD in response would earn respect and commendation for their operational value from case officers to the most senior Agency officials.[13] Yet it was precisely in the effort to understand, predict, and control the response and behavior of operational targets that the CIA has also drawn some of its harshest criticism. In the mid-1970s a series of revelations about secret CIA programs from the 1950s and 1960s created a public image of an organization flooded with research programs on mind control, behavior modification, brainwashing, hypnosis, and out-of-control drug experimentation. For five years, from 1972 to 1977, CIA Directors Helms, Schlesinger, Colby, Bush, and Turner were compelled to explain and defend programs and activities that management had begun closing down more than ten years earlier.

In April 1953 DCI Allen Dulles and Richard Helms, Assistant Deputy Director for Plans, authorized the Technical Services Staff to conduct a supersecret behavioral research program under the code name MKULTRA. Because the research involved recently synthesized drugs and pharmaceuticals (including LSD), the program became the responsibility of TSS's Chemistry Division, headed by Dr. Sidney Gottlieb. In concept, MKULTRA descended from OSS's World War II research and subsequent authorized CIA drug-testing programs Project BLUEBIRD (1950) and Project ARTICHOKE (1951).[14]

As chief of the OSS R&D organization Stanley Lovell had worked on chemical and biological weapons. After the war, the Army Chemical Corps investigated the effect of drugs on interrogation for both offensive and defensive use. At the same time, the CIA was receiving reports that the Soviets were experimenting with so-called mind-control techniques and drugs with some success. Fear that brainwashing techniques had been perfected by the Communist Chinese and the North Koreans added impetus to the mission to understand better the science of human behavior. Among the hallucinogenic drugs, LSD held a particular fascination, in part because of reports that the Soviets had shown an interest.

CIA Director Dulles had voiced public alarm over America's limited understanding of how people's thinking could be influenced. Speaking from a prepared text to alumni of his alma mater, Princeton University, at Hot Springs, Virginia, on April 10, 1953, Dulles asserted that the U.S. government had been "driven [by the tensions of the Cold War] to take positive steps to recognize psychological warfare and to play an active role in it." Dulles described a "sinister battle for men's minds" being waged by the Soviets and questioned whether America recognized the magnitude of the problem. He suggested the ongoing conflict be called "brain warfare."

The speech accused the Soviets of attempts at mass indoctrination of the population of countries they attempted to control and the perversion of minds of selected individuals. Under the latter circumstances, Dulles commented, a person's brain "becomes a phonograph playing a disc put on its spindle by an outside genius over which it has no control."[15] Before the month ended, the DCI had followed up on his public description of the threat by approving the ultrasecret MKULTRA research program. Its intent would be to understand the human mind in order to counter Soviet capabilities for mind control and create tools that could be used by U.S. intelligence officers for agent recruitment and handling. The project would sponsor research and experimentation with any available chemical and biological materials and tap into expertise across the disciplines of psychology, psychiatry, pharmaceuticals, and hypnosis.

During its eleven-year existence (1953–1964), MKULTRA remained a tightly compartmentalized Agency program that eventually involved 149 individual subprojects.[16] The Technical Services Staff was the logical organizational location for the activity because, before 1962, TSS had the scientific research responsibility for CIA (the letters MK denoted a TSS-managed program). The initial program was aimed primarily at creating new operational defensive capabilities to protect American assets from Soviet psychological or psychopharmaceutical manipulation. Understanding the effects of drugs and alcohol on human behavior would be a major focus. MKULTRA research and development would also produce a handful of new offensive capabilities involving incapacitating and lethal toxins, which would eventually draw intense unfavorable attention and prove to be of little operational value.

After MKULTRA was approved by Dulles, Dr. Sidney Gottlieb reasoned that any drugs or chemicals developed would be of limited value without a means of covertly administering them. Gottlieb contacted John Mulholland, one of America's best-known and most respected magicians and an expert in sleight of hand, or "close-up" magic, for advice.[17] His goal was to engage Mulholland to teach the techniques of magic, especially sleight of hand and misdirection, to case officers for delivering the MKUL-TRA "potions" to their targets.[18]

Mulholland agreed to Gottlieb's request, and proposed an outline for a training manual that would include[19]:

- Background facts to correct erroneous facts about magic and enable a complete novice to "learn to do those things which are required."
- Descriptions of the covert techniques necessary to "deliver" material [chemicals] in a solid, liquid, or gaseous form. Included would be explanations of the necessary skills and instruction on how to learn them.
- Examples and studies to explain how the techniques and mechanical aids could be employed in various operational circumstances.

Mulholland put the cost of the manual at $3,000 and agreed to write it in a manner to provide total secrecy.[20] To protect against the manual falling into the wrong hands, no references were made to "agents" or "operatives"; the intelligence officers were to be called "performers" and covert actions would be referred to as "tricks."[21] His early draft of the manual contained five sections: (1) Underlying basis for the successful performance of tricks, (2) Background of the psychological principles by which they operate, (3) Tricks with loose solids, (4) Tricks with liquids, (5) Tricks by which small objects may be obtained secretly.

Mulholland noted: "As sections 2, 3, 4, and 5 were written solely for use by men working alone, the manual needed two further sections. One section would give modified tricks and techniques of performance to be performed by women and the other would describe tricks suitable for two or more people working in collaboration."[22]

By the winter of 1954, the manuscript, titled "Some Operational Applications of the Art of Deception," was complete.[23] Mulholland wrote in the introduction: "The purpose of this paper is to instruct the reader so he may learn to perform a variety of acts secretly and undetectably. In short, here are instructions in deception."[24]

With the first 100-page manual completed, Gottlieb invited Mulholland to work on a new project "on the application of the magician's art to the covert communication of information."[25] The work "would involve the application of techniques and principles employed by magicians, mind readers, etc., to communicate information, and the development of new [non-electrical] techniques."[26]

In 1956 Gottlieb proposed expanding the scope of Mulholland's work "to make Mr. Mulholland available as a consultant on various problems, [for] TSS and otherwise, as they evolve. These problems concern the application of the magician's technique to clandestine operations, such techniques to include surreptitious delivery of materials, deceptive movements and actions to cover normally prohibited activities, influencing choices and perceptions of other persons, various forms of disguise, covert signaling systems, etc."[27]

Mulholland's TSS work continued until 1958, when his failing health limited his ability to travel and work.[28] The CIA's interest in solving intelligence problems using the skills of the magician, however, continued. In 1959 TSS considered revising and adapting Mulholland's work on "deception techniques (magic, sleight of hand, signals) and on psychic phenomena."[29]

By 1962 it had become evident to CIA managers that MKULTRA had produced few operationally usable products or new capabilities. A critical 1963 Inspector General report on the value and administration of MKULTRA, combined with little support for the projects from the chiefs of the operational divisions, led to the decision to terminate the program. Before the end of the decade, all questionable subprojects had been closed, leaving only a residue of noncontroversial research contracts in place.[30]

At the ending of MKULTRA, Gottlieb wrote:

It has become increasingly obvious over the last several years that the general area [of biological and chemical control of human behavior] had less and less relevance to current complex operations. . . . On the scientific side . . . these materials and techniques are too unpredictable in their effect on individual human beings . . . to be operationally useful. [Operationally] the emerging group of new senior operations officers has shown a discerning and perhaps commendable distaste for utilizing these materials and techniques. They seem to realize that, in addition to moral and ethical considerations, the extreme sensitivity and security constraints of such operations effectively rule them out.[31]

MKULTRA had encompassed a research area that used new, untested drugs to produce unanticipated effects on humans. It had been launched in the interest of national security by a DCI with the assistance of a senior officer, Richard Helms, who would eventually become DCI. However, in the 1960s, at a time when priorities for national security began to shift and standards for conducting experiments involving human subjects were evolving, controls over the MKULTRA experiments that might have seemed appropriate in 1953 were judged inadequate.

Ultimately, the CIA was cited for a failure of "command and control" for only two MKULTRA drug experimentation projects, but both were dramatic and tainted every other activity associated with the project.[32] For several years TSD retained eleven grams of shellfish toxin in CIA-classified storage despite a presidential order that all material of this type be destroyed. While the retention represented the inaction of a single officer to comply with the order rather than an organizational effort to defy policy, and although no harm to any individual occurred nor was any use ever made of the toxin, experimental or operational, the fact of its existence several years after the presidential directive reflected poorly on the CIA. In a second area of drug testing on unwitting human subjects, however, TSS's failure to obtain required official approval before conducting an LSD experiment that went horribly bad resulted in decades of personal tragedy, legal entanglements, and official inquiries.

Dr. Frank Olson, a biochemist and researcher in biological warfare at the U.S. Army facility in Fort Detrick, Maryland, who worked on a MKULTRA subproject, died in New York City on November 25, 1953. He fell to his death from a hotel room window more than ten stories above the street below. Dr. Olson was likely suffering from delayed reactions caused by ingesting LSD several days earlier. The previous week, at a TSS-organized retreat at the Deep Creek Lodge in western Maryland, Olson and several other "researchers" had shared a bottle of Cointreau. The liqueur had been laced with 70 micrograms of LSD without their knowledge.

Due to the political and operational sensitivities of the MKULTRA program, the CIA withheld details of the circumstances surrounding Dr. Olson's death from Olson's family until they partially surfaced in the 1975 Rockefeller Commission investigation of CIA activities. Subsequently, the 1976 U.S. Senate Church Commission report added substantial additional details on the MKULTRA program to the public record.[33]

Following the CIA Inspector General's internal report in 1963 and the

1976 Church Report, a third airing of the MKULTRA saga occurred in 1977. A few months after the Church Committee closed its investigation, some 8,000 pages of previously unidentified MKULTRA financial records were discovered in response to a Freedom of Information Act (FOIA) inquiry on the project. The newly found documents had been filed with contract and financial records at the CIA's Records Center rather than under the MKULTRA project title.

These records had escaped the shredder in 1973 following DCI Helms's directive to Gottlieb to destroy all MKULTRA research and operational files, and then were inadvertently missed during the records search in response to the Church Committee.[34] Helms described his thinking on ordering destruction of the MKULTRA records in a taped interview with journalist David Frost in May 1978:

> It was a conscious decision [to destroy the records] that there were a whole series of things that involved Americans who had helped us with the various aspects of this testing, with whom we had a fiduciary relationship and whose participation we had agreed to keep secret. Since this was a time when both I and the fellow [presumably a reference to Dr. Gottlieb] who had been in charge of the program were going to retire there was no reason to have the stuff around anymore. We kept faith with the people who had helped us and I see nothing wrong with that.[35]

The 1977 find was reported immediately to the White House and the Senate Select Committee on Intelligence (SSCI) and congressional interest was rekindled. That year, the SSCI convened a joint hearing with Senator Edward Kennedy's Subcommittee on Health and Science and called DCI Stansfield Turner as the primary witness. Appearing before the committees, Turner testified that the documents added little information to what was already known about MKULTRA's methods, experiments, operations, and the breadth of the program. The SSCI agreed, and the joint hearings were concluded after one session.[36] However, the redacted materials, subsequently released under FOIA, became the basis for John Marks's *The Search for the Manchurian Candidate,* a bestselling account of CIA research in the 1950s and 1960s into human behavior.[37]

The negative publicity surrounding MKULTRA far exceeded its modest contribution to intelligence and the negative aspects of the program acquired undeserved legendary status in the mind of the public as well as conspiracy theorists. Secret government-sponsored mind-control research, dangerous experiments on unwitting people, covert assassination tools, and white-coated chemists mixing unknown concoctions in hidden laboratories produced vivid images in the public's imagination. Virtually none of this was a reality, but more than five decades after Allen Dulles and Richard Helms initiated the ultrasecret program to counter what they believed to be a grave threat to free thought, MKULTRA continues to generate public intrigue and controversy. The officer chosen to carry out the program, Sidney Gottlieb, did what he understood duty demanded, and paid a heavy personal price.

The breadth of Gottlieb's life as a scientist, CIA official, builder of enduring intelligence capabilities, humanitarian, respected office director, and patriot was obscured even at his death on March 7, 1999. *The Washington Post*'s headline on Gottlieb's obituary read CIA OFFICIAL SIDNEY GOTTLIEB, 80, DIES; DIRECTED TESTS WITH LSD IN '50S, '60S.[38] Like the headline, the first sentence mischaracterized Gottlieb and his work by focusing exclusively on "mind control experiments and administration of drugs and LSD to unwitting human subjects." In fact, LSD, drug testing, and the procedures adopted had been a small part of the authorized MKULTRA research program in the fifteen-person chemistry branch Gottlieb headed. Like many of their scientific contemporaries of the 1950s, TSS engineers applied their talents to national security work consistent with Gottlieb's policy: "If it is technically possible, do it and put it on the shelf. The policy maker will decide whether it is ever used or not."

The obituary ignored Gottlieb's remarkable contribution to America's security during his eleven years as Deputy and Director of the Technical Services Division. Under Gottlieb's leadership, TSD built worldwide clandestine technical capabilities critical to virtually all significant U.S. clandestine operations in the last third of the twentieth century. Eleven of the obituary's twelve paragraphs focused on drug, poisons, and mind-control themes while ignoring the fact that with Gottlieb at its head, TSD conceived and built the technical devices that enabled the CIA to break the back of KGB counterintelligence inside the Soviet Union. Only the obituary's closing paragraph alluded to Gottlieb's humanitarian activities—that after retirement he worked in a leprosy hospital in India for eighteen months.

Another *Washington Post* story published two years after Gottlieb's death and three months after the 9/11 terrorist attacks on America more accurately captured his life and work. The author observed that Gottlieb, the longest-serving Chief of the CIA's Technical Services, had served his country as "the coldest warrior" while also living as a "humble and compassionate [man], an altruist eager to ease the miseries of the weak and sick."[39]

Yet, regardless of Gottlieb's public service and personal charity his name will be inextricably linked to the ten-year MKULTRA program and the sinister implications of associated words such as drugs, LSD, assassination, and mind control. No consensus is likely to emerge on how well MKULTRA and Gottlieb's role served the national interest at a time when America's leaders sensed a "clear and present danger" from the Soviet threat. However, whenever the question is debated, an indisputable fact, articulated in the final report of the 1976 Church Commission, remains: Under the administrations of four Presidents—Eisenhower, Kennedy, Johnson, and Nixon, "[the] CIA has been responsive to the presidency throughout. No rogue elephant."[40]

Cover and Disguise

They must lead a double life . . . it is a vexing existence.

—David Atlee Phillips, as quoted in *The Literary Spy*

False or assumed identities are a way of life for intelligence officers. While conducting their work, case officers and technical officers alike have learned to live "normally" with alias names by combining a con man's verbal skills to spin a plausible cover story with unassailable identification papers. The intelligence officer must convincingly establish that he is who and what he claims, even though it is all a fabrication. Most officers use a dozen or more different alias names during a career.

For the CIA, creating false identities and their supporting documentation had its origin in 1942 in the Documentation Division of the OSS Research and Development branch. Agents infiltrated by the OSS into Occupied Europe required "bullet-proof" identification papers, as the slightest whiff of duplicity could result in summary execution. OSS logbooks from October 1943 show requirements for fabricated documents such as French stamps, ID papers, and travel certificates. Agents dispatched behind German lines by OSS officer William Casey, later Director of Central Intelligence, were regular "customers" for the output of the OSS document fabrication shop in London between 1944 and 1945.[1] The London operation, manned by an assortment of craftsmen and forgers, was a field component of Stanley Lovell's OSS R&D branch and evolved in the postwar years into the Document Intelligence branch in the Operational Aids Division of the CIA's Office of Special Operations.

In 1951, the CIA's consolidation of technical and scientific work in the Technical Services Staff included the capability to manufacture documents and identity papers. The significance of documentation for Agency operations was reflected by the fact that three of TSS's six original divisions focused on some aspect of identity and documents.[2] Hundreds of CIA officers working overseas, together with every agent dispatched into Eastern Europe or China, required an alias identity along with unassailable documents to back up an airtight cover story. The alias protected the agent's true identity, while the cover legitimized his presence in the area. TSS assembled a documentation team of artists, forgers, engravers, printers, papermakers, and photographers from OSS veterans, U.S. trade schools, and selected German and Japanese artisans who had originally learned their craft while working against the United States.

Selection of appropriate cover for covert operatives was handled by a separate division within the DDP.[3] TSS, and subsequently OTS, supported cover requirements by creating and/or reproducing paper or plastic documentation that a person would normally be expected to carry, such as passports, visas, licenses, credit cards, blood donor records, stationery, membership cards, business cards, and travel documents. Paper documents were at the heart of establishing an officer's identity and legitimacy, particularly in the decades before electronic databases.[4] Historically, officially issued and printed documents carried on one's person were the standard form of identification for travelers, but in recent years, biometric identification and individual data stored on computer chips have become a required element of establishing identity.

Fabricating high-quality identity documents has always been technically difficult and unforgiving. The OSS London shop reportedly checked documents thirty times before issuance to an agent going behind German lines.[5] Minor mistakes or errors in printing, format, color, paper texture, inks, or missing security features in government-issued documents were quickly recognized in the home country. Immigration and customs officials, as well as border guards, were trained to spot false documentation; likewise, local police were keenly aware of the typical contents of a traveler's wallet or purse. Hotel receptionists, ticket agents, and bank tellers were trained to be alert for fraudulent or counterfeit documents. A suspicious document opened the door for additional investigation and questioning that, once begun, often led to an unraveling of the bearer's cover story and endangered the larger operation.

Intelligence officers working under an alias required documents that were perfect reproductions of official issuances and contained the current authentication features necessary for travel. Credit cards had to be signed with the same name and same script as the bearer's passport, driver's license, and club cards. Passports had to contain valid visas and entry and exit "chops" (rubber-stamp impressions) that corresponded to travel reflected on other documents. Dates on airline stubs or train tickets had to be consistent with dates showing entry into a country.

A production element within the OTS designed, printed, bound, laminated, and artificially aged the false documents while a separate authentication division reviewed, verified, and prepared each document before issue. The separate authentication process confirmed that every piece of an individual's document package, including any government-issued identification papers, was complete, accurate, and up to date. Before issuance, authentication officers compared the document package to data maintained in CIA's exhaustive international document inventory and archives of current and historic samples of customs and immigration forms, rubber stamps, cachets, seals, passports, and travel paperwork. To ensure that the document inventory and knowledge of international travel remained current, CIA officers or assets were sent on probes to survey travel routes, observe immigration practices, collect chops in passports at foreign border crossings, and record changes in entrance/exit procedures in countries of operational interest.

A few months before the September 11, 2001, terrorist attacks, a CIA officer had his attaché case stolen in a European country while registered as a guest at an upscale hotel near a major airport. OTS had modified the attaché case by creating an inner cavity for secreting papers that documented the officer as a resident of a neighboring country.

The attaché case came into possession of the local internal security service where the secret compartment and contents were discovered. The alias documentation was turned over to the country that had supposedly issued the documents, where an investigation determined the papers were forgeries. Suspecting that the origin of the forgery was the work of a third country, not the United States, the offended service registered a blunt objection. The third country's intelligence service knew it had no role in the fabrication, but to calm matters simply acknowledged receipt of the materials without comment.

Nearly a year later, at a conference of intelligence services cooperating on detecting terrorist forgeries, the story was related and the documents displayed. The OTS officers immediately recognized the documents as ones they fabricated. Later a friendly colleague from the third country privately commented, "We know these aren't ours. They are really good, almost perfect. Better than anything we could do. We don't know who else can do such fine work, but we thought you should know that the capability is out there." Obscured by the language of intelligence diplomacy was a professional compliment for OTS's fabrication skills.

Alias travel documents were critical to the success of a CIA sting operation in the late 1980s. The CIA needed intelligence on the capabilities of a tactical missile being designed in a country hostile to the United States. Tight security surrounded the missile program and foreign visitors were never given access to the facility where components were assembled. To attempt to gain access, a CIA agent assumed the persona of a Middle Eastern businessman and the cover of an international military equipment broker. The agent needed to find a way to get inside the building to photograph the equipment.

CIA planners contrived a scenario whereby the agent acquired information that a third country, also hostile to the United States, had interest in secretly buying several of the missile systems. The intelligence, in the form of official-appearing documents, had all been fabricated by OTS. The agent made contact with a representative from the target country for a "confidential discussion" and presented the so-called intelligence with a proposal that he broker the sale. Suspicion was high but the asset's intelligence, cover story, and identification papers were so convincing that follow-up meetings were arranged.

The asset demonstrated good faith by putting several hundred thousand dollars on the table and serious negotiations about a possible sale began. He established the position of his client that any deal would be contingent on his inspecting the goods before signing a purchase contract. Over several weeks, discussions went back and forth until the target country's senior official called for a final meeting.

The negotiations lasted three days but eventually a verbal agreement was reached and formal papers drawn up for signature. The seller agreed to the asset's demand to inspect the production facility and missile components. Then, with the contract papers on the table ready for signature, the senior negotiator abruptly called a halt. "Let me see your passport," he demanded. The asset, stunned by the request, had no choice but to comply. The negotiator slowly examined the passport page by page before handing it back with a smile. "I just wanted to make sure," he explained, "that you were in Yemen in March three years ago like you told me yesterday. I see you were. Now let us conclude our business."

The tension subsided as the asset struggled to maintain his composure. In fact, he had never been in Yemen although travel to that country had been built into his cover story. OTS had inserted the required visa into his passport, valid for the month in question. The techs had properly dated the entry and exit immigration chops, complete with scrawled initials of the immigration officers. In fact, the entire passport, visa, entry/exit stamps, chops, and signatures had been constructed by OTS. A concealed camera in the asset's briefcase silently filmed the entire drama.

Some time later, a senior DO official called the operation "one of the best of the decade." Its success hinged on the work of the OTS's document fabrication and authentication officers.

From the cover perspective, CIA employees fall broadly into two classifications: overt and undercover. The majority of CIA employees are overt, acknowledge their CIA affiliation, and each January receive W-2 forms issued by the Central Intelligence Agency. Typical of overt employees are individuals in senior management positions, those with assignments in public affairs, the Center for Study of Intelligence, congressional liaison, personnel recruitment, analytical components of the Directorate of Intelligence, and research units in the Directorate of Science and Technology. These individuals perform work that does not require a covert identity.

Within the Directorate of Operations, most employees have cover, as does any other Agency officer who participates in covert operations. Cover documents provide corroborating personal and public material to establish and support the legitimacy of a cover and a fabricated identity.

The CIA categorizes cover in two types, official and commercial. Official covers are provided by other government agencies and departments while commercial covers are acquired from private-sector companies or private individuals. The cover of any individual officer can be adapted to the operational need and ranges from "light" to "deep."

Officers may have a cover that allows them to work using their true name or they may assume an alias as part of operational cover. Two types of aliases are frequently used. The most common is a created identity. Officers are assigned randomly selected names, consistent with their ethnic appearance and supported by standard identification documents such as driver's license, credit cards, social security card, and passport. Throughout their career, officers would be issued multiple aliases and supporting documents. Since 1990, however, the created identity has become increasingly vulnerable to detection as an alias due to interconnected databases containing official and personal information about individuals. The adage "If you don't exist in cyberspace, you are probably a fraud" has become a truism that limits the long-term operational use of created identities.

Borrowed identities offer an alternative to fictitious identities but require the cooperation and temporary "disappearance" of the voluntary and cooperating donor. Borrowed identities have the advantage of possessing a verifiable personal history and require no manufactured backstopping of the individual's college attendance, work history, social connections, or forged documents. The borrowed identity also exists as a "cyber persona" since, at minimum, credit history shows up on numerous databases. Reproduction of personal ID papers is simplified by a willing donor making the material available to document specialists. OTS technicians can provide effective disguise in the form of the donor's clothing and, within reason, body appearance. Borrowed identity, normally reserved for particularly sensitive operations, was successfully employed to effect the single personal meeting that occurred between an agency officer and *TRIGON* in Moscow in 1976.

Disguise can either complement an alias or obscure the true identity of the user. The history of OTS's disguise work, like documents, began in OSS. At the formation of TSS, disguise became part of the Furnishings and Equipment Division and subsequently supported covert operations by altering the appearance of officers and agents to protect their true identity or ensure against future visual recognition. Disguise also can make an individual's appearance consistent with photo identification documents that are used to support an alias identity. Disguise conceals personal identity as wood blocks conceal microphones and transmitters.

CIA officers meeting with assets have often employed light disguise in conjunction with an alias. Such disguises may give the officer a visage that alters his true features. A light disguise might include a wig, glasses, mole, facial hair, dental appliance, or certain articles of clothing. Whether the disguise is realistic or believable is less important than the fact that it prevents the officer from being recognized later. Light disguises would typically be employed when meeting with an unknown volunteer who asks to speak to someone in intelligence. To avoid the risk of exposing an officer to someone who could be a terrorist or part of another intelligence service's dangle operation, the CIA representative would apply a light disguise before engaging the volunteer. Light disguises are also issued to members of surveillance teams to protect them from recognition by the target or by well-wishing friends whom they might inadvertently encounter while on the job.

As necessary, more elaborate disguises using full or partial facial masks could perform an ethnic or sex change to alter a person's racial or gender appearance. Among the options are padded clothing to alter body type and weight distribution, sculpted appliances that alter eye color, mouth lines, and affect speech tone, makeup and hair coloring, hand and arm "gloves" to match facial coloring, shoe lifts to add height, and torso devices to create a stooped posture. Individually and in combination, the disguise techniques can affect dramatic appearance change.[6]

For officers in need of a disguise subject to close attention and durable for hours or days, OTS specialists would spend several hours performing the transformations. These labor-intensive disguises were typically applied on individuals in high-risk situations such as illegal border crossings. Given time, the disguise specialists would alter hair color, apply facial hair, modify jaw lines, improvise dental work, create wrinkles, change complexion, or add glasses and warts to match any photographic documents and thus avoid chance recognition at a border crossing or airport checkpoint.[7]

The use of disguise to maintain secrecy is a basic means of acquiring information otherwise unavailable. It is also one of the most ancient. The Old Testament describes several disguise incidents in the history of the Hebrew people such as Jacob's deception of his father Isaac to secure the family birthright. The Chinese strategist Sun Tzu offered instructions for disguising spies in *The Art of War* with this comment: "Your surviving spy must be a man of keen intellect, though in outward appearance a fool; of shabby exterior, but with a will of iron." More recently, Shakespeare was famously fond of disguises, incorporating them into the plays, including *Twelfth Night, Measure for Measure*, and *As You Like It*.

The twenty-first-century spy's disguise must not only be flawless in outward appearance, it must also reflect his assumed identity in the array of sophisticated documentation safeguards in use throughout the world. Disguise must match a digital persona that includes holographic images and microchips containing biometric data embedded into passports and travel documents. Personal information on the Internet that is compatible with one's disguise becomes as critical to the modern spy's identity as the traditional counterfeit beard and eyeglasses.

CHAPTER 22

Concealments

The OTS concealment specialist combines the skills of a craftsman, the creativity of an artist and the illusion of a magician.

—An OTS concealment engineer

In 1586, secret correspondence to Mary Queen of Scots from the French ambassador was concealed inside barrels of beer and smuggled to her at the country estate of Chartley, England, where she was under house arrest.[1] During the American Revolution, couriers who traveled by boat, carried intelligence reports inside weighted bottles that could be dropped overboard at the threat of capture.[2] A hollowed-out lead bullet was used to conceal smaller written messages, but this was eventually replaced by a similar bullet made of silver that could be swallowed at the first sign of danger without incurring the ills of lead poisoning.[3]

OTS's laboratory for concealments grew out of the OSS Research and Development—Camouflage Division in Fort Washington, Maryland, which had produced letter drops for use by World War II agents.[4] The drops were originally made from tree limbs. The wood was split and a metal container inserted in such a fashion that the wood could be replaced and present an innocent appearance to any observer.[5] An important principle learned after receiving comments back from the field was that the drops should never be constructed of anything burnable or edible, lest they be picked up and used by some passerby needing food or fuel. Afterward, better drops were produced in various forms that included stones and old tin cans. Such drops

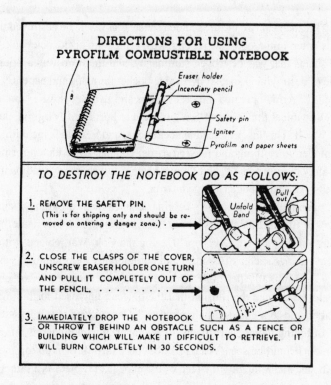

*Sensitive notes and information could be protected by use of a
Combustible Notebook. The ordinary-looking notebook contained
Pryofilm, which when ignited by an incendiary pencil, would
destroy the notebook and contents in thirty seconds, 1940s.*

were designed to be ignored by anyone not involved with the operation
and could be left at public locations, such as at a prearranged distance from
a mile-marker on a European road, for a two-way exchange of intelligence.[6]

Another research and development unit, Division 19, Miscellaneous
Weapons of the National Defense Research Committee, supported OSS's
wartime requirements and established its first lab in June 1943 at the Congressional Country Club outside Washington, D.C, operating as the Maryland Research Laboratory.[7] Under the project code-named MOTH, three
containers were created for transporting concealed secret intelligence documents with devices to destroy the contents "if opened by a person unfamiliar with its use." One device could be camouflaged inside a fountain pen or

shaving kit and held two or three folded sheets that would be destroyed thirty seconds after initiation. The second was a medium-sized notebook with bound sheets that was destroyed summarily, and the third was a briefcase capable of destroying a special insertable pocket for maps and papers.[8]

The U.S. Army operated a secret Escape and Evasion (E&E) laboratory and facility under the secret MIS-X program at Fort Hunt, Virginia, during World War II. This lab produced concealments and E&E aids including silk maps hidden in clothing and playing cards; compasses inside uniform buttons, safety razors, pencils, and fountain pens; and shortwave radios inside mess kits, baseballs, and cribbage boards.[9]

The true nature of the CIA's original concealment program, a charter function of the Technical Services Staff, was obscured by its name, "Furnishings and Equipment Division." During the Cold War, objects with unapparent cavities such as furniture and automobiles were required when it became operationally desirable to hide a person, passageway, or object. Concealment created the illusion that the object being used for hiding had no relationship to a clandestine operation.[10] Camouflage was a less-secure means of hiding than concealment; like a cover, if camouflage is removed, the contents could be seen. A large safe with a tarp thrown on top may be camouflaged and removing the tarp would expose the object as a safe. However, if a false bottom was created inside the safe and the cavity door could be opened only by manipulating a hidden latch, the safe was transformed into a concealment.

A concealment device, or CD, includes a hidden compartment to which access is obtained by mechanical decipherment of locks, hinges, and latches. The mechanical actions necessary to open a CD are normally a sequence of unnatural twists, turns, and pulls. Intelligence services used CDs to mask the entrance to tunnels or hiding places as well as for hiding spy gear. OTS categorized concealments as being active or passive.

An active CD possessed an obvious function that remained operable in addition to the internal cavity and the spy gear it housed. An example of an active CD was a fountain pen that wrote normally but contained a subminiature camera that could be operated without affecting the writing function. The writing instrument masked the presence of the concealment. Another example would be a video camera concealed inside a lamp with both devices performing their designed functions separately or simultaneously. A calculator modified with a concealment cavity for a beacon was considered active if it continued to add, subtract, multiply, and divide.

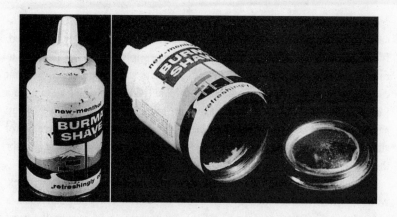

*The 1960s Burma Shave–can active concealment produced
shaving foam and masked an internal cavity. Years later such
items were copied by criminals for use in smuggling.*

A passive CD provides a cavity for concealing materials but does not
perform another function to mask its clandestine use. For example, a wooden
statue with a cavity in the base would have no function other than for display. An attaché case with a false bottom at its base or a book with a
hollowed-out cover were other examples of passive CDs that became a standard part of every case officer's and agent's operational equipment.

Concealments served five operational purposes: storing (bookcase at
home), transporting (travel purse), exchanging (a loaded dirty mitten at a
dead drop site), infiltrating (an audio transmitter inside a gift to a target),
and masking (a wine rack placed in front of the entrance to a secret passageway).

Spies with compromising equipment must secretly store and protect
the clandestine gear in their possession. One-time pads for enciphering
messages or subminiature cameras must be stored indefinitely for use at
the appropriate time. Sensitive intelligence information or documents
must be hidden until passed to the handler. CDs need to provide quick access to the equipment and information while protecting against accidental
discovery by a family member or exposure during a more threatening security search. The size of the items to be stored in the concealment and
the available method for getting the CD to the agent dictated what could
be used.

Virtually any object that offers sufficient volume can be converted into a concealment device, but the object has to fit into the user's lifestyle. The local economy in the agent's country often restricts the variety of CDs issued. In areas where consumer goods are in short supply, it may be difficult to find items that could be given to the agent for storage purposes without causing neighbors to be envious and suspicious. Constructing a CD inside a false-bottomed five-liter petrol can be an effective storage device for an agent with a car or garage, but if the country was experiencing severe petrol shortages such a luxury might be seen as out of place or be a target for theft.

In the decades before data could be stored on discs and thumb drives, wooden desks and bookcases that could conceal a four-inch briefcase, a large document, disguise items, radios, and cameras were among the most popular concealment "hosts."[11] Concealment furniture was constructed to blend in with the home decor of the user. Bookcases, in particular, were universally accepted as common furniture. They were durable and could be built with cavities throughout—at the top behind the molding, inside the shelves, in a false back, in the thickness of the sides, or with the largest cavity beneath the bottom shelf behind the skirt.

Material being exchanged by dead drop was hidden inside specially constructed CDs designed to blend in with the site's surroundings and remain unrecognized until retrieved. If the drop site was in a park, a small piece of tree limb hollowed out to hold film cassettes or a false passport would have been a typical concealment.[12] Waterproof containers, weighted with lead shot to compensate for the buoyancy, were constructed to fit inside drainpipes, toilet reservoirs, or submerged in the shallow water of a decorative pond or stream.[13] Other "natural environment" concealments resembled bricks or chunks of masonry.[14] These items were collectively known within CIA as "sticks and bricks" because, when deployed, they were indistinguishable from the original pieces in their natural environment.

Dead drop concealments normally must have no value to the society where used. Otherwise, the concealment may be collected for its assumed worth. In theory, the more repulsive a dead drop CD appeared, the more attractive its operational use. A crushed can still dripping oil, a piece of electrical cable coming out a wall with exposed wires that appeared "live," discarded bandages and medical waste, or animal excrement were unlikely to be picked up by a casual passerby.

Animal carcasses, especially decaying ones, are universally offensive and thus effective for dead drop containers.[15] OTS specialists periodically produced CDs from pigeons, rats, and an occasional roadkill. The lab animals were humanely killed, then gutted and treated to create an artificial cavity inside the stomach and chest. Some were freeze-dried and vacuum-packed in tin cans. Material intended for the agent was wrapped in aluminum foil and inserted inside the created cavity and the animal stitched back together. Before the carcass was deployed, it might be doused in Tabasco sauce as a deterrent to hungry cats roaming the streets. Pigeon carcasses were typically dropped at sites around parks and the special rats were often just left by the side of the road. To make the dead rats even more repugnant, OTS constructed rubberized "gut parts" to spill out of the carcass as it lay on the road. When deployed, the roadkill CD was intended to be retrieved quickly.

Agents needed a secure means to transport their spy gear. If recruited outside a denied area, the agent would be required to reestablish contact after they returned home and were ready to begin work. The techs found a solution by hiding one-time pads and commo schedules inside inexpensive tourist souvenirs such as statues of saints, reproductions of sculpture, and castings of famous buildings. These items could be collected in cities where the agent traveled, carried by hand, and readily explained as a tourist purchase if questioned. Low-cost items were less likely to be examined when packed in personal luggage. For greater security, these were one-time-use CDs that could not be opened without being broken to access their contents. Because there was no hidden latch or manipulation that might betray the method of opening the CD, the cavity was not likely to be detected even during a close examination.

People, as well as information often needed to be "transported."[16] During the Cold War CIA and OTS worked successfully on more than 140 "illegal movement" operations without ever losing a person.[17] OTS constructed life-supporting human concealments for defectors or escapees in the form of specially designed exfiltration crates or modified automobiles. Refrigerator boxes could house an eight-hour life support system for a person weighing up to 250 pounds and measuring up to six foot six inches tall. The less than luxurious container included items needed to support basic life and body functionality such as "piddle packs" for urination, absorbent sponges, food, water, ice packs, gel packs, a warming source, and circulating battery fans. Constraints on the internal oxygen supply usually limited the time a system could be employed.

During one exfiltration of an agent from a Soviet Bloc country, the border crossing took much longer than planned when the vehicle was held up at several checkpoints. The agent was wedged inside a concealment built in the car's trunk with virtually no room to move. Officers driving the car, although concerned about the agent's well-being, could do nothing. Finally, after several hours longer than expected, the automobile arrived at a safe location, the concealment opened and the individual pulled out. To the amazement of the officers present, the agent was smiling and seemed unperturbed by his claustrophobic adventure. When asked how he had tolerated the experience so well, the safe and grateful agent replied that he had been a tank driver in the Soviet army. As a result, he was accustomed to being a contortionist.

In a less successful operation, OTS received a requirement for a Mercedes sedan configured to conceal a man who would be driven out of Eastern Europe. The lead OTS specialist designed a concealment using space created by reducing the car's fuel tank. He worked on the project for six months to remove the original tank, replace it with a smaller one, and make other external and interior configurations to accommodate the agent. When finished, the passenger area, trunk, and underside looked factory new. The tech received unanimous acclaim for doing a first-class concealment job.

The automobile purchase had been disassociated from the Agency and the title and paperwork showed no official connection between the car and the U.S. government. The station arranged for a driver who was not aware of the intended use of the car to deliver the vehicle to Berlin. Apparently, the driver failed to heed the fuel indicator and ran out of gas while in route. He contacted the nearest Mercedes dealer because something "wasn't working" right—the tank had been full when he started the trip, yet the fuel gauge had fallen quickly and he had run out of fuel well under the normal range for the vehicle. The technician examined the car for some time and then called the driver. "Sir, you have a problem," asserted the technician as he pointed out the small tank and the cavity. The discovery immediately ended the operation but the U.S. government now owned a new limited-range Mercedes. Eventually the car became the VIP touring sedan at one of OTS's covert facilities.

Another exfiltration operation required a person be moved from a major hotel in a Middle Eastern country that was known to be under surveillance by the local service hostile to the United States. OTS techs covertly observed the comings and goings at the hotel for several days and determined that the surveillance focused exclusively on people entering and leaving but showed no interest in baggage or luggage. It was, a tech suggested, time to consult with some of OTS's Hollywood contacts who specialized in performing magic tricks. If a magician could saw his beautiful assistant in half and then have her emerge intact from a coffin half a stage away, surely he could sneak someone past a surveillance team.

A magician and his trick builders designed a dolly for rolling luggage that was loaded with varying sizes of suitcases, a steamer trunk, and an ice chest. The façade of the baggage on the dolly appeared completely realistic; each piece was designed to fit around the legs, arms, torso, and head of a person so that the agent could sit inside and be wheeled out of the hotel by the porter into a waiting van. The operation proceeded without incident, completely confounding the surveillance.

CDs are critical to infiltrating secret equipment into a facility. The "Trojan horse" is often an item desired by the target or a gift given as a gesture of goodwill that conceals a bug, a beacon, or even an explosive device.[18]

A Trojan horse operation against a communist country's ambassador in Europe exploited the diplomat's interest in a piece of sculpture he openly admired at a dinner party. The local CIA station reasoned that the sculpture, an impressive large bronze of an old farmer, might be displayed by the ambassador in his embassy conference room. The size of the piece made it an ideal host for an audio device and the batteries necessary for a long service life. The station obtained the original sculpture, but the techs could not create a hollow cavity inside the bronze and restore the original without leaving signs of alteration.

The alternative was to sculpt an identical statue and position the eavesdropping package inside the farmer's head before the final casting. By creating a forged sculpture, no visible scars would appear on the outside and the sealed bronze would limit access to the unit if examined by the embassy's technical team. OTS found an accomplished sculptor from among its artists and the final casting was declared a masterpiece; even its weight replicated that of the original. When the audio components were tested, the device accurately reproduced the room sounds using a microphone "airway" the tech had hidden inside the recesses of the farmer's mouth.

An access agent presented the statue to the ambassador during a ceremony on the annual celebration of the communist country's national day. A listening post, a block away from the embassy, recorded the event as the access agent wheeled the bronze into the embassy and made the presentation. After several minutes of listening to diplomatic social chatter, the listening post's keeper remotely switched the device off until the next morning. All waited anxiously to hear where the ambassador would display the sculpture.

When the audio was switched on again, the concealed statue continued performing splendidly. The ambassador had it sitting next to him as he conducted his daily briefing of his senior ministers. The placement could not have been better, right in the ambassador's conference room.

Congratulations for the CIA officers were, however, premature. The ambassador then announced that a treasure like this statue deserved to stand in the embassy's most prestigious location. All important visitors would pass this prominent spot at the top of the stairs leading to the ambassador's suite and thus could pause to admire the sculpture.

The announcement was terrible news for the operation. The top of the stairs would not be an area used for secret briefings and meetings. The listening post monitored the device for several weeks but obtained nothing of intelligence value. The concealment tech and sculptor had done their jobs, perhaps too well. Assessment of the ambassador had not anticipated his need to display and show off the magnificent gift.[19]

OTS created masking or camouflage for equipment whose size, location, or function precluded concealment, but which could not be left visible. Roof antennas were masked by tool or storage sheds to prevent another security service from determining the antenna's direction and configuration. A laser communication device pointing out of the window of an agent's dining room might be camouflaged as a large decorative urn. The entrance to a secret tunnel in the basement of a safe house could be masked by a bar and wine racks that slid easily aside for the tunnel access. Each masking system is designed to draw attention to the mask itself without attempting to hide the fact of its existence. Because the mask is something that would be expected in the environment, no particular curiosity is aroused and no further examination of the mask is invited.

OTS techs enjoyed James Bond movies for the beautiful women, daring men, quips, and especially for the clever ingenuity exhibited by Major Quentin Boothroyd, known as Q. They never failed to be amazed at how well Q's devices always worked in the field.[20] The gadget master for Her Majesty's Service met with Bond before each operation and issued him devices drawn from a seemingly endless array of well-designed and highly crafted gadgets. Q always anticipated needs and applied technology that pushed the boundaries of design, materials, and craftsmanship. In the character of Q, techs recognized a surprisingly realistic depiction of a scientist-craftsman-intelligence officer who shared many of their everyday problems. Q dealt with officers who were technophobic, worked with people who neither understood nor trusted technology, spent hours teaching the proper use of devices, and unsuccessfully admonished Bond to remember to return the gadget to stores when finished with it. At OTS, a real-life Q would have felt comfortable at a location known simply as "the lab."

The OTS lab, an hour's drive (or more, depending on traffic) from downtown Washington, would have been the dream of any craftsman. One tech, with a university degree in mechanical engineering, remembered thinking that the array of equipment he saw on his first day on the job was awe inspiring. In the mid-1970s the lab employed craftsmen and -women specialists for all of the hand skills needed for professional concealments—metal and automotive shop; wood, plastic, and ceramic shop; electronics; leather; fabric; glass; seamstress; bookbinding; welding; tool making; photography; drafting—and others. The lab seemed to have virtually every piece of equipment available to work on any material.

Compared to a university environment, the lab wanted for neither money nor skilled craftsmen. If the concealment tech needed a new tool or piece of equipment, he could get it or make it. Men and women with twenty or more years of experience seemed eager to pass their knowledge on to the new arrival. Not only did the lab have talented techs, it also had the necessary tools and materials, such as industrial-quality sewing machines for professional work with textiles, fabric, and leather. Thousands of dollars' worth of the finest glove and belt-weight leathers, dyes of every color, along with cutting and buffing machines, were available to make items appear "factory new" or "old as dirt." The shop could produce handbags with fine stitching and maintained a selection of needles, guards for hands, colors and weights of threads, and various types of fabrics. In the metal shop there were tools that few universities could afford, including special gauges and single-purpose cutters. The cabinetmaking shop turned out furniture, molding, or decorative boxes with the look, craftsmanship, and quality of the finest manufacturer. Exotic woods were available to match any operational need. The plastic and electronic shops were similarly equipped.

As clandestine ops in the Soviet Union accelerated, the lab devoted half of its concealment output to supporting these operations. Soviet operations were considered so sensitive that case officers demanded a different type of concealment item for every operation; and since no item was ever repeated, each required the full process from design to fabrication.[21]

Items with the largest cavities were usually wooden structures such as bookcases and desks. Techs in the 1970s designed and built all CD furniture from scratch, from unfinished raw wood to completed desk, bookcase, or bed stand.[22] Later, in the 1980s, "solid" wood products began to be replaced on the commercial market with furniture constructed from particleboard, which was less expensive, but frequently heavier and less durable. Since the techs' job was to produce CDs that blended in with other contemporary furniture, the lab shifted its construction to particleboard as well. For the OTS craftsmen this change was accompanied by a noticeable fall-off in the quality of furniture and increased operational difficulties. A case officer who issued a particleboard CD bookcase learned that after one or two moves, the tolerances and alignment of latches necessary for the CD to function properly did not hold up nearly as well as those in solid wood. The techs found adjustments almost impossible, so the only solution was to build a new piece.

Although wood was most frequently used, by no means was it the only host material employed for concealments. The lab could construct CDs in toolboxes, toasters, power supplies, large step-down transformers (2000-watt models that were available overseas), bases of small refrigerators, small air conditioners, and vehicles. The shop was equipped to work in plastics, a material that went through phases of popularity in the consumer marketplace. However, plastic generally was so light that if used as a CD for any weighty item, an explanation was required as to why this apparently lightweight plastic set of drawers seemed so heavy and solid.

The OTS concealment shop was the ultimate "form, fit, and function" business that encouraged imagination. "If you can think it, you can do it" became the unofficial motto. The techs called their lab "the greatest toy shop in the world."

Concealment techs understood that if there were an intuitive or obvious means of opening the host device, the CD would not meet the level of security required for its clandestine use. Therefore, they pursued a continuing search for a "hidden catch" or an inventive new way to keep something hidden from anyone other than the intended user. Catches and latches became the stock-in-trade of concealment specialists, since if the compartment could be opened by anyone, it was not an acceptable CD. Hinges, magnets, pins, slides, pneumatic tubes, and even old-fashioned pull bolts were used to create an array of concealed openings. Generally, to open a CD required twists, turns, or pulls in a precise combination that functioned as a form of mechanical code that had to be performed before gaining access.

The techs recognized that concealments in everyday items had to look normal, yet be easily opened by those who knew the "code." An OTS concealment tech in the 1970s remembers designing and building a concealment that required normal manual dexterity to open the device. However, word later came back to the tech that the device, though perfect in every other aspect, was not usable because the agent had nonfunctional arthritic thumbs. The case officer had not previously provided this information about the agent's limitations, and the tech never thought to ask for it. It was a lesson long remembered. Only by observing and asking probing questions during the design process could the techs devise a concealment that made the agent feel as if the device had always been part of him. The successful concealments matched the CD to the person and the CD became second nature when used and operated.

Every CD was designed to meet an anticipated level of threat. A low-threat CD for the home of a case officer might be adequate for hiding his office attaché case, which itself was a CD. At the other extreme were CDs used to transport sensitive materials across international borders where they were subjected to x-ray and magnetometer readings as well as physical examinations.

Sometimes a perfect-looking dead drop CD was not enough; it also had to pass the "smell test." In the late 1970s, the KGB determined that some of the OTS dead drop concealments made from wood to resemble tree limbs had been assembled using a type of epoxy whose odor was detectable by specially trained KGB dogs. The unfortunate success of the KGB in discovering some containers led the lab to identify the flaw in their production process and replace the epoxy with a nonscented adhesive.

The techs were ever conscious that an agent's life often depended on the skill and ingenuity they used to fabricate the CD. If spy gear inside a CD was discovered, it became prima facie evidence of espionage. Such a compromise would not only seal the fate of the agent, but could result in the detection of other agents using similar equipment and lead to the arrest of the handler.

The craftsmen of high-threat CDs worked under dual requirements that the host could always be subjected to physical search and an item's design had to fit with the agent's lifestyle and cover. When completed, the host would look and function exactly as expected in the agent's environment and the concealment cavity be inaccessible to anyone unaware of the mechanical cipher.

Detailed drawings of the concealment host and thorough documentation of concealment issuances were maintained. These records became invaluable should a CD be lost or compromised. If other CDs like it had been issued in the same country, they might need to be recalled and replaced. Once hostile counterintelligence officers learned of a CD, and how it was manipulated and opened, they could be expected to be on the lookout for similar pieces.

OTS concealment makers were masters of craft and illusion. Whimsy, smoke, and mirrors were as much the materials of CDs as wood and metal and fabric. It was expected that the techs would be masters of the fabrication skills required of their craft, but they constantly challenged each other to reach the next level by making materials do things they were not intended or expected to do. The best CDs used materials in ways not done elsewhere and possibly never done before. Fabrication skills were one part of creating an illusion; the thought process for the initial idea was of equal importance and an indispensable step in designing a device. Concealments worked because people assumed what they observed was the only reality. A person looking at a lamp could not imagine that illumination was its secondary feature. The primary function of the lamp was storage—storing the hidden camera that was taking the observer's picture. For clandestine operations, illusion and CDs worked together because people want to believe what they see. The OTS techs succeeded brilliantly in fabricating concealments from physical materials; in time, their next challenge would be to do the same with electronic software.

Clandestine Surveillance

He was now "black"—free of surveillance. Moscow was his.

—Milt Bearden in *The Main Enemy*

The word "surveillance" comes from the French *surveiller*, to watch over. The CIA broadened the definition to "watching from anywhere" and relied on TSS and its successor organizations to build and deploy special equipment for surveillance and countersurveillance operations. The CIA has used surveillance for both offensive and defensive purposes by secretly collecting information about the movement and activity of recruitment targets and using countersurveillance to protect CIA officers engaged in clandestine acts.

Surveillance operations employ stationary (fixed or static) or mobile assets as needed. "Stationary surveillance" refers to sustained observations made from fixed sites, which can be apartment buildings, cafés, airports, or intersections. The monitoring attempts to identify either the people transiting the site or the type of activity conducted at the location. The target site would be observed from an observation post manned by trained surveillance personnel using still and video camera systems. As the capability and reliability of visual surveillance equipment improved, unmanned observation posts that recorded and transmitted images to a control point significantly reduced the number of personnel required for multiple fixed-surveillance sites.[1]

"Mobile surveillance," conducted primarily by foot, automobile, or

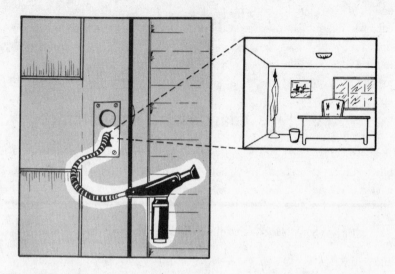

The Fiberscope is composed of a pistol-grip viewer and a flexible shaft designed to enable inspection of remote or inaccessible locations. A panoramic view was obtained by drilling a .315-inch hole in the adjacent wall or ceiling, or by sliding the tip through the target's keyhole or under the doorway, circa 1968.

airplane, tracks a person or other moving target, such as a vehicle or shipping container. OTS supplied concealed surveillance cameras, disguises, and specialized communication equipment for mobile teams. Mobile surveillance becomes particularly important when terrorists are identified and their movements need to be observed and plotted.

Surveillance photography serves dual operational purposes: to establish positive photo identification of a target and operational acts, such as meetings, exchanges of documents, and payoffs. The quality of the photography depends on selecting the right camera for the operational environment. In a stationary observation post, typically located inside a building, camouflaged cameras are prepositioned to photograph a target and can be controlled manually or remotely. Inside an apartment or hotel room with a common wall to the target, covert photographs can be taken from behind a ventilation grille, through a pinhole lens, or using a pre-installed camera "port." Images from digital cameras may be immediately transmitted to an operational base. In the early 1990s, film cameras began to be replaced by high-resolution digital cameras. At first, images were recorded to videotape and then later to digital storage media. The advantages of advanced storage capacity and digital transmission of captured images expanded the applications for photographic surveillance. Miniaturization and capacity advances in the 1990s of small, low-light video cameras allowed video concealments to employ many of the same techniques previously used by audio surveillance to hide small microphones in wood blocks, books, or office equipment.

Mobile observation posts, using surveillance cameras carried by a person on foot, or riding on a bicycle, car, train, or aircraft, add another capability for covert collection. Concealment requirements and the need to compensate for the target's movement limit the choice of cameras for mobile posts when compared to the fixed sites. At close range to the target, traditional camera systems will often be concealed beneath the user's clothing or inside a briefcase or purse. The ubiquitous presence of cell phones and their integrated imaging features fundamentally altered the nature of visual surveillance by creating the reality that any action done in public is likely to have been photographed by someone.

Historically, for intelligence photography, 35mm cameras with varying lengths of standard and telephoto lenses provided the highest possible level of detail (resolution) in still images. In 2001, with the advent of the Nikon D1X (5.9 million effective pixels), OTS accelerated its movement toward digital imaging for all photographic requirements.

At greater distances, cameras with long lenses make a target appear closer and enlarge the image on the film media. These telephoto lenses could be 500mm, 1,000mm, 2,000mm, or even longer.[2] A lens 300mm or longer is almost impossible to hand-hold without a tripod or other means of support. The commercially available Questar Seven 2,800mm lens with a 35mm camera and tripod can read the numbers on a license plate from more than two miles away with adequate lighting and favorable atmospheric conditions.[3]

Because surveillance pictures using film cameras are often taken at night during periods of little light, careful selection of available light, ultra-high-speed film, or infrared materials are required. Available-light photography requires the fastest available film and steady support. The sensitivity of commercially available ultra-high-speed film to ASA 6400 can be "push processed" to sensitivity above ASA 50,000 by manipulating the developing time, development temperature, or both.[4] At these extreme ASA levels, it is possible to photograph a subject at night illuminated by a

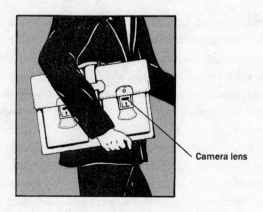

Camera lens

The spring-wound advance on the German robot camera could be remotely activated while the camera remained concealed in a leather attaché case.

single birthday candle. Infrared surveillance photography that takes place at the end of the light spectrum is not visible to the unaided eye and requires only a standard 35mm camera, infrared film, and a strobe flash unit with an infrared filter.[5]

"Tracking devices," particularly beacons, aid mobile surveillance when close surveillance is impractical or undesirable. Tracking devices are selected based on the nature of the target and operational issues. In practice, beacons are most useful in tracking inanimate objects such as vehicles and shipping containers. Success in implanting a covert beacon inside the human body, despite assertions by some that "my teeth contain government-installed devices," eluded the CIA. Contrary to certain remarkable scenes in spy movies, OTS found that human behavior, combined with technical limitations imposed by laws of physics, made "personal beaconing" practically impossible.

Beacons are considered both strategic and tactical. Tactical beacons, monitored from ground receivers, are usually located within a short range of the target; strategic beacons may be monitored from a highflying aircraft or satellites. Most clandestine beacons use a small radio frequency transmitter to broadcast a navigational signal to the surveillance team. For example, the CIA might covertly insert satellite-tracked beacons inside shipping containers of shoulder-fired missiles being transported by terrorists from Afghanistan to other locations in the Middle East and place a tactical beacon on a pickup that is hauling a few of the missiles to a terrorist safehouse.

During the Vietnam war, TSD disguised a small beacon inside imitation animal dung. Left in the foliage adjacent to a North Vietnamese Army or Vietcong campsite, the active beacon would not be noticed or disturbed because of its appearance. Attack aircraft could home in on the signal that pinpointed the site for destruction.[6] In the Middle East, beacons were deployed inside briefcases and belts to protect individuals at high risk against kidnapping. When activated by nonalerting movements of the wearer, the beacon transmitted a signal for help and provided the location of the kidnapped individual.

Software beacons were created for operations against cell phones and portable computers. With brief access to a target's laptop or cell phone, the beaconing software could be installed and, thereafter, anytime the target used his cell phone or logged onto the Internet with his laptop, his position was logged and his e-mails and conversations intercepted.

Taggants represent another means for tracking movements by identifying a targeted individual as he passes through choke points within the operational area. Similar to the plastic security tags attached to clothing that must be removed before departing a store, taggants made of chemicals, pheromones, or electronics can be remotely detected and the tagged individual identified.[7] One of the best-known taggants used by the KGB was "spydust," the chemical compound 5-(4-Nitrophenyl)-2,4-pentadien-1-al (NPPD). The Soviets' use of spydust was of such concern to CIA operatives that OTS established a special program to analyze and counter the material.

The CIA employed audio, visual, physical, forensic, and electronic technical aids to enhance the organized study and observation of targets. Throughout most of the Cold War, audio operations and satellite photography dominated technical collection successes. Audio operations were an original function of the Technical Services Staff, but initially were no more important than printing, concealments, and disguise. However, by 1960, audio technical surveillance had become TSS's top priority. Audio operations, designed to obtain positive or operational intelligence, targeted communications systems or facilities where conversations of interest might occur. Government telephone lines, official foreign missions and facilities, an office, residence, or hotel room—all were exploited by the audio techs.

Traditional landline telephones were particularly vulnerable to clandestine tapping. Almost every target individual anywhere in the world had and used a phone. The handset contained a high-quality microphone built into the mouthpiece and was connected to wires leading out of the building. TSS developed three basic systems for bugging phones in the early 1950s that remained viable for decades.

By tapping the line, both parties talking could be heard and the full conversation captured. The tap might require direct contact with the wires, or an "inductive" tap could be fitted as a collar around a line without making physical contact with the internal wires. An alternative was to modify the phone.

Normally when a telephone receiver is placed on its cradle the depressed hook switch ends the call. TSS developed a technique in the 1950s to bypass remotely the hook switch in order to use the sensitive mouthpiece microphone to listen in on all room sounds and conversation. Usually a tech required access to the telephone to make the modification but if the make and model of the targeted telephone could be obtained, a hook-switch bypass modification to an identical instrument could be made. Then, similar to a quick-plant operation, a cleaning person or service personnel could covertly exchange phones. The third basic system exploited the telephone's own current. Telephone instruments draw current from the telephone company for power to operate the unit and activate the bell or ringer. This power level was sufficient to support other bugs and listening devices in the room and eliminated the need for batteries to be replaced.

Cellular telephones are particularly vulnerable to audio attacks. Cellular conversations can be intercepted while transmitting between the nearest cell tower and the handset, or as the signal is relayed between towers to the telephone exchange. All data and conversations sent to and from a cell phone, including e-mails, videos, images, and text messages can be captured without any physical access to the phone itself. Cell phones can be located to within 100 feet by triangulating the signal strength of the cell phone with the three nearest cell towers. By integrating this geographical positioning data with a moving map display, movements of the cell phone can be monitored in near-real time.

A cell phone can also be bugged by gaining access to the instrument for the time required to swap batteries. Modified batteries containing a microphone, digital storage media, and computer chip constitute a self-contained eavesdropping system. Once the audio is captured and stored in compressed format, the microcomputer chip in the system dials a preprogrammed number and burst-transmits the stored information to a receiver. The bug automatically recharges itself when the user charges his cell phone battery.

In the 1980s, cell phones communicated using analog signals that were easily intercepted and monitored. In the 1990s, digital cellular providers began offering limited protection from amateur eavesdroppers, but fell far short of the capabilities and technical resources of intelligence services and law enforcement agencies.

A bugged olive in a martini glass served at a black-tie embassy reception might play well to movie and television audiences but such things are usually unrealistic for CIA operations. To get the "good stuff," surveillance techs installed listening systems in walls and ceilings of consulates, concealed recorders in attaché cases, and hid microphones and transmitters in apartments. They operated contact microphones to eavesdrop through the walls of hotel rooms, rigged telephones, intercepted cell calls, and bounced laser beams off windowpanes. Whether the techs left cigarette lighters with transmitters in target offices or wired microphones into a case officer's brassiere, the objective never changed—to get secret intelligence in support of national security.

For each surveillance operation, the techs selected components that work together in order to capture the audio at the target site and transmit it to the listening post. Their equipment differed markedly from the repackaged consumer-grade products masquerading as "covert electronics" and offered for sale at retail spy shops.[8] Consumer electronics normally lack the technical sophistication and reliability needed to operate in security environments where covertness is critical and climatic conditions uncontrolled and unpredictable. Compared to professional spy equipment, the consumer "spy" gadgets require excessive power, operate erratically, and emit signals that are easy to detect and intercept.

Many OTS spy electronics were the result of a collaborative development process between CIA engineers and private companies where a dedicated team of cleared contractors worked on Agency projects.[9] This model of industry–government cooperation produced components with performances that eclipsed commercial standards by decades. Among the most significant examples were rugged, sensitive audio microphones that were later made public and introduced into hearing aids and small, long-life transmitter batteries that eventually powered heart pacemakers. Charge-coupled devices

(CCDs) were used in OTS spy cameras a decade before the same technology was commercially available in digital cameras.

The latest and most sophisticated OTS audio equipment was usually reserved for targets in denied areas where hostile technical surveillance countermeasures sweep teams were the most formidable. OTS created a variety of components and eavesdropping devices, each with different characteristics and capabilities that allowed the tech to customize each system to meet the operational requirements and counter the threat.

Commercial microphones were developed in the latter quarter of the nineteenth century after Emile Berliner sold his microphone patent to the fledgling Bell Telephone Company. The world's first electronic eavesdropping system, the Turner Dictograph introduced in 1915, contained a carbon microphone, battery, and earphone. Buyers were cautioned "not to use the device for illegal or immoral purposes." Whether a microphone is located inside the mouthpiece of a telephone, or embedded in the wooden leg of a table, its purpose is the same—to convert the sounds of room noise and voices into an electrical signal.[10]

From the variety of microphones available, the techs matched the one with the most desirable characteristics to the operational requirement. Mics could be hardwired to the listening post, connected to a concealed recorder worn beneath the user's clothing, or connected to a radio-frequency transmitter. While a hardwired mic offered security advantages, the radio-frequency transmitter quickly became the most commonly used audio system because the listening post could be placed in remote locations.

Contact microphones are effective in capturing sound waves from room audio that cause every hard surface in the room, including the walls, floors, and objects, to vibrate. A sensitive contact microphone with the capability to convert vibrations into an electric signal was especially useful in operations against targets in hotel rooms when the tech had physical access to one of the adjacent rooms, or the room above or below. The tech could affix the contact microphone to the wall or floor using glue or a nail to pick up the vibrations. A special type of contact microphone, the "accelerometer," could detect vibrations of room conversation or movement through solid concrete walls up to eighteen inches thick. For opportunities that required quick reaction, OTS produced a special self-contained "motel kit," disguised inside a small toiletries case, that consisted of a contact microphone, wall adhesive, pocket amplifier with optional output to a tape recorder, and earpiece.[11] It could be packed in a briefcase or carried beneath a coat.

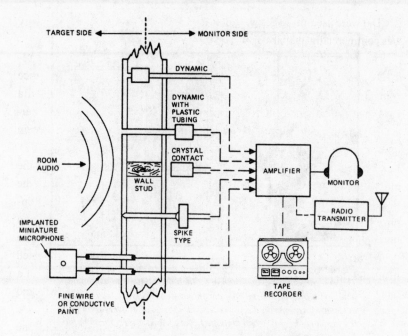

TARGET SIDE ◄─────────────► MONITOR SIDE

Illustrated are five different types of microphones intelligence services used for covert monitoring of conversations through a common wall. The degree of access to the target room and type of building construction determined which microphone was used.

The "vibro-acoustic" microphone, designed to be affixed to a reinforcing steel rod or bar inside a concrete column during building construction, could later be connected to concealed wiring that would run to a listening post. Conversations cause the concrete and rebar to vibrate and enable the vibro-acoustic sensor to capture the sounds. Multiple sensors inside the same column on different pieces of rebar could be selectively tuned by the monitor at the post to target specific conversations anywhere in a 360-degree circle around the column. However, the task of attaching the vibro-acoustic mics onto the rebar required bribing or distracting security guards at the construction site.

Pinhole mics, half the size of a pencil eraser, were a workhorse for the OTS audio techs. Whether mics were hidden behind a floorboard, inside a wall, or embedded in the base of a flowerpot, they required only a tiny (less than half a millimeter) airway to capture all of the room noises. The pinhole mic could be installed inside numerous objects or architectural

TSD developed a "Motel Kit" for surveillance of targets of opportunity.
The self-contained eavesdropping kit included contact microphones,
a battery-powered source, and earphones. The sensitive transducer
detected vibrations on the wall caused by sound or conversation in
the target room that, when amplified, were heard clearly through the
headphones, circa 1970.

features of a room. When denied access to the target room, the tech could install the mic in a common wall by drilling a pinhole too small to be noticed through the wall, floor, or ceiling.

Advances in the 1980s made it possible for OTS to design a fiber-optical microphone that operated using only light waves transmitted to it along a cable thinner than a human hair. The fiber-optical microphone defied detection by a metal detector or nonlinear junction detector and its tiny wire was easily hidden.

Directional mics were designed for operations to pick up a selected conversation from individuals standing together and talking at a social event while excluding other room noise to either side. The rifle mic, a type of directional microphone, was used in outdoor seating or smoking areas to collect conversations from a distance. The increase in smoke-free buildings turned these gathering spots into ideal target areas to collect gossip and personal information. The directional rifle mic, composed of an array of tubes of varying precalculated lengths placed in front of the sensitive microphone, filtered out extraneous sounds and reduced all noises other than those from voices in the direction of the target.

Audio played a critical role in the rescue of seventy-one people held hostage for four months by the Tupac Amaru Revolutionary Movement (MRTA) in April 1997. Fifteen armed MRTA terrorists stormed the Japanese ambassador's residence in Lima, Peru, during a diplomatic Christmas party on December 17, 1996, taking seventy-two Peruvian and foreign hostages. Several days later, when it became apparent that the hostages would be held indefinitely, the Peruvian government began infiltrating listening devices into the residence in hope of acquiring intelligence about the terrorists' intentions and the status of the hostages' well-being. Loudspeakers set up at the front of the residence to deliver messages and harass the terrorists were part of the government's attempt to pressure a surrender.

In January, one hostage, a senior Peruvian government official, suddenly assumed the persona of an isolated eccentric. He began talking incoherently and at random to various inanimate objects in ways that suggested his mental state had deteriorated. The act was a ruse; the official had knowledge of audio operations from his previous work and made a calculated guess that something in the residence could contain a bug. In fact, a religious icon did conceal a transmitter and on one afternoon the listening post monitors heard the hostage pray, "If you hear this, play 'La Cucaracha' tomorrow." Precisely at 6 AM the next morning, "La Cucaracha" blared through the loud speakers, baffling the Lima press as to why the government would use a famous Spanish Civil War song as a harassment tool.

After the musical acknowledgment, the eccentric continued talking to the icon for three months until April 22, when, minutes before the successful rescue assault, he reported that the hostages were in a relatively safe indoor area while the terrorists were in an open area playing their usual afternoon soccer match. The assault was launched, killing the fifteen MRTA revolutionaries and rescuing all but one hostage.

OTS audio techs left nothing to chance in preparation and advance planning for their operations. The complexity and the risk of the activity demanded that each phase of any technical surveillance operation be considered and documented. CIA Headquarters required that a survey and written proposal, known as "the 52-6," be prepared, submitted, and approved before an audio operation could proceed. The survey consisted of six primary elements.

The target could be a person or a facility such as a telephone line, building, room, or automobile. Methods used to operate against a target varied by the type of information sought. If the target was a briefing by a senior military attaché during his weekly staff meeting, the embassy conference room would be the place to plant the listening device. On the other hand, if the attaché was being assessed for possible recruitment, his bedroom or the telephone line he used for personal calls might be locations where his conversations would reveal an exploitable weakness. Attacking the third-floor room of a trade mission with windows overlooking a busy street would require a completely different operational plan than one to bug the general's briefing room inside a secure military base. Without a means of gaining unobserved access to the target facility, there could be no operation.

Audio operations required a thorough physical description of the site, including a viable location for a listening post. A signal "path loss" test identified any physical obstructions that would degrade the bug's transmission signal. Activity patterns of occupants were recorded. Any security and alarm systems, including the use of guard dogs, was plotted. The survey estimated the operational life of the battery in the listening device, identified the number of people, their special skills, and the type of equipment required. The techs projected the time they could be safely inside the target, the optimum date and time for the operation, a proposed escape route, individual cover requirements, and the risk of compromise. The station and Headquarters weighed in on the expected value of the information to be gained from a successful operation.

Based on what was known of the target, the techs described their plan to enter the facility and do the required deconstruction, which could involve removing baseboards, drilling, implanting devices, reconstructing damaged walls, inventorying tools, and exiting securely. The scenario also included the plan for communicating with countersurveillance teams and contingency procedures—what to do in an emergency should technical or security problems arise during the operation.

After the installation, audio devices were managed from the listening post. The survey included information about the location, equipment, and staffing of the listening post. The station had responsibility for staffing the post, manning the tape recorders, translating and producing transcripts, while OTS maintained and serviced the equipment. When an audio operation ended, the techs conducted another clandestine entry to remove the device and to restore the facility, leaving no trace of the installation. This objective was not always achievable; operational judgment would balance the risk of exposure during a reentry with the value and importance of the equipment to be retrieved.

Both audio surveillance and concealed video camera operations consisted of three primary components: the collection device, the transmission link, and the listening or observation post. Collection devices were usually a microphone or camera that would covertly acquire the information for transmission down a wire or radio-frequency broadcast to a listening post. The collector might be a subminiature microphone embedded in the woodwork, a tap placed across the telephone line, or a pinhole video camera concealed behind a dressing-room mirror. Power for the collection device came from batteries or by siphoning power from the existing electrical lines at the target location.

The transmission link sent the collected signal containing the sound or imagery from the collection device to a receiving and recording location. The configuration of the target, the cooperation of the local security service, and the distance to the listening post were all factors in determining the type of link used—hard wire, radio transmission, or a more exotic system such as laser or fiber-optics. Where the monitoring post was positioned close to the target, as in the basement of an apartment building or in the adjacent room of a hotel, a hardwired microphone to the recorder would be preferable since no over-the-air radio signal was generated. Hardwiring a microphone or video camera could also eliminate the need for a power source at the target site and made the implanted device nearly impossible to detect without x-raying the floors, furniture, and walls. However, hardwiring is usually slow to install and potentially more susceptible to accidental discovery.

For microphones hardwired to the listaning post, OTS developed special tools to aid in their invisible installation. A small, easily concealed aluminum crowbar was developed for quickly prying baseboards away from the wall to hide wires, as was a special hand-held fine-wire kit that used a razor blade to slice a small slit in a wall, insert a pair of tiny wires, and finally seal the opening using a pencil eraser. The device could lay wires across a painted surface without leaving a trace.[12] For audio installations involving damage to woodwork or walls, OTS engineers created special quick-drying putties and odorless paint to hide signs of construction. The tech could complete his installation and cover all traces of his work during a single entry into a target site.

The listening or observation post received and recorded signals from the transmission link for processing. A typical post could contain several recorders, each paired to an implanted collection device. Advances in digital recording created a virtually unlimited recording capacity.

The radio-frequency transmitter became the CIA's most frequently used device for sending a stolen signal out of a target location. Although the transmitter required batteries or another power source, its signals had an advantage in that they could be monitored anywhere within a kilometer of the installation and farther with the use of repeaters. Since the early 1970s CIA surveillance systems have included the capability for remotely turning the transmitter on and off at selected times to conserve battery power, and storing collected conversations for a remotely programmed transmission at a later time.

For the hardest targets, exotic systems were developed to collect audio via lasers, infrared light, or fiber-optic cables. More technically complex and difficult to maintain than the radio-frequency transmitters, these systems were limited in use but effective in situations where a target employed aggressive technical countermeasures to block, identify, or neutralize a radio-frequency transmission link. Transmitting signals via infrared or laser reduced vulnerability to traditional TSCM "sweep" techniques.

MI6 officer Richard Tomlinson described the difficulties experienced in an operation to bug the penthouse apartment of a suspected Russian intelligence officer in Lisbon. A loft space above the apartment provided a suitable place for hiding the small microphone, but a problem arose in linking the microphone to the recording equipment located in another apartment below. Use of a normal radio link was ruled out for technical reasons, so the alternative was to link the two areas by running a small wire "through a convoluted drainpipe that wound its way down the building."[13] Technical officers experimented with various mechanical crawlers in an effort to thread the wire through the bends of the drainpipe to no avail before hitting on the idea of using a mouse. Tomlinson describes the operation:

> Using a fishing line they could dangle the mouse, harnessed to the end of a fishing line, into the top end of the drainpipe. They would then lower it down the vertical section of the pipe to the first right-angled bend. From there the mouse could scurry along the horizontal part of the pipe to the next vertical section and so on, down to the bottom of the pipe where it could be recaptured. The wire could then be attached to the line and pulled through the pipe.
>
> Trials of the mouse-wire delivery system on the Century House

drainpipes, using three white mice borrowed from the chemical and biological weapons research establishment at Porton Down, proved reasonably successful. One mouse, nicknamed Mickey, was a natural and scampered through the pipes enthusiastically. A second, Tricky, tried to climb back up the fishing wire when dangled, but once in the pipe, was reasonably competent.[14]

Methods of clandestinely introducing a listening or photo device were as varied as the imagination of the techs. Eavesdropping devices embedded inside Trojan horse–style gifts were given to diplomats, businessmen, and other high-profile targets with the expectation that the device would be placed in an area used for important conversations.[15] The gift, described by a case officer as one that "keeps on giving," could be an engraved pen and pencil set for the target's desk, a decorative flowerpot, or a handsome globe. Two primary weaknesses of a Trojan horse operation are the inability to predict or control where the gift, with the listening device, is placed and the potential blowback on the giver should the deception be discovered.

For short-duration audio operations, OTS developed small portable eavesdropping systems embedded in functional everyday items such as lighters and disposable ballpoint pens.[16] Such a device could be attached by a member of the cleaning staff beneath a conference room table where it would not be noticed or secreted by an official visitor between the cushions of a couch. The eavesdropping device would collect and transmit room audio as long as it retained battery power or until removed or discarded. Ideally, an operation to deploy a quick plant included an advance visit to the target site to determine the best place to leave the device, identify an appropriate concealment, and determine a listening post location.

Concealments for quick-plant operations can be either tailored or generic.[17] Deployment of a generic device—say, a disposable cigarette lighter or expended ballpoint pen—requires little advance planning. OTS carried an inventory of bugged AC electrical adaptors that could be quickly installed between a lamp plug and the socket. These were available in the varied colors and styles appropriate for target countries.[18] DCI William Casey credits himself with personally deploying a generic quick plant disguised as a large needle in the office sofa of a senior Middle Eastern official during a trip abroad.[19]

The wood block represented a frequently deployed variation of a quick plant.[20] Wood blocks encased audio transmitters and were designed to be placed underneath a table or desk, or as part of a chair or a replacement for sections of chair rails and crown molding. High-quality wood blocks replicated the color and type of the wood, as well as shape of the molding and appeared as a normal part of the furniture or room design. Structural wood-blocks replaced the pieces of triangular-shaped wood that provided stability and support beneath most pieces of wooden furniture. These were unlikely to be seen and required less engineering effort to conceal beyond matching the general color of the furniture.[21] Books were a variation of wood blocks, the spine of a book providing a tailored concealment cavity for a listening device.[22] A visitor to the target location could execute a quick plant by unobtrusively replacing a specific book with a seemingly identical edition.[23]

When access inside the target location or to adjacent rooms proved impossible, more exotic systems enabled collection of audio from a distance. Laser microphones worked on the principle that a laser beam directed at an angle toward a glass window was reflected and could be captured at a listening post, compared with the original signal, and demodulated to recover audio. In the 1980s, OTS engineers developed a program that embedded a small prism inside window glass in key targets. The prism increased the sensitivity of the laser microphone and allowed OTS to control accurately the angle of reflection. With this prism system, the laser could be aimed at the window and the reflection would return along a parallel path to the LP. This eliminated the necessity for the transmitter and receiver to be in different locations and made detection more difficult.

A passive resonator concealed inside a wall or piece of furniture can be targeted with a radarlike signal transmitted from an exterior post. The reflected signal is demodulated to eavesdrop on all conversations in the room. Because its power source comes from the external signal, the device can transmit indefinitely. CIA technicians saw their first resonator in 1952 when one was discovered embedded in the carved wooden Great Seal of the United States.[24]

Inside the halls of the CIA's Original Headquarters Building, a historical display from the Directorate of Science and Technology shows the inventiveness of CIA scientists. One device, a robotic catfish named "Charlie," was designed to be indistinguishable, when viewed from the water's surface, to channel catfish commonly found in rivers around the world. It appeared so lifelike while swimming in the water, that some feared that it might be consumed by even larger predators. Charlie's mission was unspecified, but experts speculated he could be used to swim into freshwater rivers and canals to gather water samples near foreign nuclear power facilities. The mobile aquarobot could also serve as an underwater platform for eavesdropping devices.[25]

CIA officers abroad lived, worked, and operated under the constant awareness that at any time they could come under surveillance. Officer training included weeks of surveillance detection runs to develop and practice skills in recognizing and dealing with surveillance, either obvious or discreet.

Obvious surveillance was used when a foreign security service chose to send a message to an officer that his activities were being closely watched. Such surveillance could become aggressive, verging on harassment and intimidation. Tactics might include "bumper locking," in which a trailing surveillance vehicle stayed so close that its bumper actually touched the target car. On the street, surveillance watchers could walk directly in front of, behind, or adjacent to the target, staying in close proximity even in shops and buses. Slashed tires, broken windshields, and stolen car batteries conveyed the same message: "We know who you are, and whatever you are up to, we don't like it."

Aggressive actions are sometimes taken by surveillance teams to retaliate for a provocation or to thwart an operational act.[26] This happened to an active young CIA officer whose operational activities aroused suspicions of the local service. The officer received an unscheduled late-night visit at his home by the country's chief of counterintelligence. After a tense discussion, the foreign chief left behind a parting compliment coupled with an unstated warning, "Mr. Paseman, you are very good. However, I suggest the remainder of your tour should be rather boring."[27]

Discreet surveillance, while not physically intimidating, was difficult to recognize and more to be feared. Failure to detect counterintelligence watchers could lead to operational compromise and loss of an agent. Early in the 1970s OTS engineers created tiny body-worn receivers to intercept the radio transmissions of Soviet surveillance teams. These concealed receivers, unrecognized by the KGB for several years, gave CIA officers operating in Moscow a prized capability for detecting surveillance activity.

Well-trained surveillance teams, operating in familiar areas where they control the turf, will attempt to lull the officer into the false belief that he is "black" (free of surveillance). Should the officer fail to detect such surveillance and proceed to "go operational" he could unwittingly lead

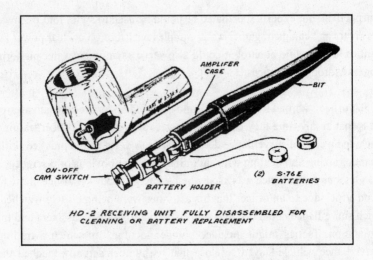

AMPLIFER CASE

BIT

ON-OFF CAM SWITCH

BATTERY HOLDER

(2) S-76E BATTERIES

HD-2 RECEIVING UNIT FULLY DISASSEMBLED FOR CLEANING OR BATTERY REPLACEMENT

The Hearing Device-2 countersurveillance device, with a neck loop antenna and body-worn receiver, allowed an officer to hear nearby hostile radio communications through bone conductivity by biting down on the pipestem.

surveillance to his agent or be caught during an operational act. Discreet Soviet surveillance played a key role in the compromise of major operations and the expulsion of CIA officers from the USSR.

Disguises offered one method of defeating the KGB's overwhelming surveillance advantages. OTS sculpted and fitted disguises for use by case officers and agents to evade surveillance and avoid recognition. Before leaving for foreign assignments, case officers were trained to apply a variety of false appearances and function normally while in disguise. Each received a "light disguise kit" tailored to the officer's gender. The kit typically included items like false mustaches and beards, hairpieces, a fake wart, planar lens eyeglasses, hair coloring, collapsible canes, reversible coats, shoe lifts, and dental appliances.[28] Some officers, whose assignment required a more elaborate disguise, received full or partial head and face disguises individually sculpted and tinted to blend fully with the wearer's skin and hair color.[29] Because surveillance teams relied heavily on visual indicators to track a target, quick changes in an officer's appearance—adding or removing a hat, letting hair down, putting on or taking off glasses, or reversing the color of a jacket—might cause surveillance to lose their target in crowds or on busy streets.

In the cat-and-mouse game between surveillance and countersurveillance, the edge traditionally went to the side controlling home territory. For the CIA this meant that they were always at a disadvantage when meeting and handling agents in high-risk or denied areas. Agents had to communicate with their handlers, and defeating surveillance was the key to their protection. Whenever OTS developed a new gadget or disguise that offered an advantage against the ever-present watchers, it would be only a matter of time before its tactical superiority was lost. There had to be a better way for agents to operate and communicate without exposing themselves to hostile surveillance and for OTS the new technology arrived in the form of digital zeros and ones.

Covert Communications

We are surrounded by a world of secret communications . . .

—Eric Cole, *Hiding in Plain Sight*

America's intelligence services (CIA, FBI, and some military elements) recruited foreign spies with the access and opportunity to procure (which is to say, steal) secret information considered vital to U.S. national security. However, without an ability to communicate securely with his handler, the spy and his purloined secrets are worthless. Spies were most vulnerable to being caught not while procuring the information, but when attempting to pass their secrets to a third party. Every agent required his own tailored covcom that fit his circumstances and the kind of information he collected. A film cassette filled with photographs of classified memos represented a different covcom problem than passing printed pages of a radar system's operating manual or the actual circuit board from a missile guidance system.[1] Information exchanges between agent and handler must be both secure and secret. Codes and ciphers provide levels of security while digital steganography hides the encrypted information in a cloak of electronic invisibility.[2]

In the last half of the twentieth century the "Holy Grail" of covcom was envisioned as a secure system of two-way, reliable, on-demand exchange of voice, text, and data 24/7 from and to any location. The message need not necessarily be encrypted, but the communication process must present a low probability of detection and interception. Once concluded, the exchange would leave no record of having occurred or any telltale electronic footprint. Such a system would be used "from anywhere to anywhere" in the world for an agent to "talk" to his handler, CIA Headquarters, or even the President of the United States.

Every CIA covcom system, from the personal meeting between handler and agent to a multimillion-dollar satellite link between an agent and the DCI, consisted of three primary segments: the field set (what the agent used either to receive or send), the transmission backbone (such as shortwave, high-frequency broadcasts that carried the message), and the receiving element. Personal meetings between agent and handler required comparatively less technology while covcom through satellites was dependent on technology.[3] Regardless of the system, each involved integrating whatever special devices were needed with sound tradecraft at every stage of system development, delivery, and, in the case of the agent, concealment of incriminating equipment. The fewer pieces of spy gear in the agent's possession and the fewer unnatural acts he had to perform operationally, the lower was his risk of being detected.[4]

Through the years, the CIA's Office of Communications, Office of Research and Development, Office of Development and Engineering, and Office of Technical Service each pursued some element of covcom's Holy Grail. Their efforts resulted in the deployment of successive generations of technically sophisticated gear that advanced one or more of the following imperatives: obtain more timely information, improve security, pack the maximum amount of information into an exchange, and deliver intelligence ever more quickly to the end user.

When selecting a covert communications system, the case officer considered factors such as the agent's lifestyle, profession, ability to travel abroad, and risk tolerance. He estimated how frequently the covcom would be used, the size and aggressiveness of the local counterintelligence service, level of surveillance directed against the handler, and the number and types of covcom systems already operating in the area. Regardless of the variables, two general categories of covcom occurred between agent and handler: personal and impersonal. Each category of covcom has advantages as well as risks.

Personal meetings between an agent and handler (often a U.S. official) represent the riskiest form of covcom. Hostile governments conducted routine surveillance of foreign diplomats under the assumption that some of them were actually intelligence officers operating under official cover. Persons suspected of having an intelligence affiliation were systematically surveilled to detect signs of clandestine activities such as clearing and filling dead drops or meeting an agent. The agent, unless already under investigation, was less likely than the American to be under surveillance, but if observed in an unauthorized meeting with a foreign official, immediately was suspect and placed under surveillance.

Despite the risk, face-to-face meetings were frequently a preferred means of communication with agents. Exchange of materials was assured, conversations could address urgent issues, conflicts could be ironed out, and the agent's morale given a boost. During personal meetings, the handler was always alert for changes in the agent's attitude, motivation, personality, and health. He was able to conduct hands-on agent training, modify requirements, change operational plans, and gauge firsthand the extent of any counterintelligence problems.[5]

Given the inherent risks, however, personal meetings in denied areas were kept to a minimum, carefully planned, and never conducted without a specific reason. The handler was always prepared for the contingencies necessary to maintain the security of the operation; meeting times, duration, and locations were selected to provide a plausible cover story for both handler and agent in case they were observed. The agenda for the meeting was scripted in advance; initial greetings were immediately followed by a standard question, "How much time do you have?" Next on the script was to agree on arrangements for the next meeting should they be interrupted.[6]

To minimize the counterintelligence exposure of an agent being spotted during a personal meeting, the CIA developed techniques known as "brief encounters." These involved a personal contact between the agent and handler, but minimized the length of time required for an exchange of material. In 1958, the CIA Chief of Station in Prague, Haviland Smith, developed the technique of a "brush contact" or "brush pass" while providing tradecraft training in New York City to a Czech agent.[7] Smith noticed that the agent was reluctant to leave his package of secrets in a dead drop for fear it would be discovered and traced back to him. As an alternative, Smith had the agent stand just inside the entrance to the Grand Central Terminal where a person entering had the option of proceeding straight ahead to the old Biltmore Hotel or turning right and descending down a flight of stairs to the subway.

Smith knew that at that point he would be momentarily out of sight of any trailing surveillance. If an agent was waiting at the top of the stairs and just inside the entrance, Smith could pass a newspaper to the agent, who would quickly turn around and head down into the subway while Smith proceeded straight ahead into the hotel. It worked so well that, in a training exercise, even when the surveillance team was looking for the move, as long as they were following Smith from behind, the pass could not be detected.[8] Only in the unlikely chance that the hostile surveillance team had somehow anticipated Smith's travel path and arrived ahead of him could the exchange be spotted. The counterintelligence surveillance team was never certain where Smith, who varied his routes and timing, was going and could not "set up" on him in advance. Case officers and techs identified similar locations in cities around the world where the same technique could be used.[9]

In a variation of the brush pass, the moving-car delivery technique allowed an agent to drop a package covertly into the handler's slowly moving vehicle through an open window.[10] The travel route selected by the handler was consistent with his normal evening routine and included a number of right-hand turns on dimly lit side streets. Following each right-hand turn, the handler's vehicle was out of sight of the trailing surveillance vehicle for a few seconds; the CIA referred to this brief window of time as being "in the gap."[11] The agent was instructed to stand in the shadows at the corner and watch for the handler's car to complete the turn. When the car was briefly out of sight or in the gap just following the turn, the driver would dim the car's lights as a signal to the agent. The agent then stepped to the curb and dropped the package through the open window. Immediately after making the exchange, the agent receded into the shadows and remained motionless until the trailing surveillance vehicle passed. A concealment cavity built into the car's dashboard or floor mat was used to hide the package until the driver and vehicle returned to a safe compound.

A higher-risk variation of the moving-car delivery occurred when the agent and handler both drove vehicles to the same traffic signal and pulled up alongside each other. With the agent's car on the right and the passenger window of the handler's car opened, the agent tossed the package into the empty seat. A moving-car exchange required thorough planning and excellent timing, but when executed properly was virtually undetectable.

Impersonal communications, those not requiring face-to-face meetings, were employed when personal meetings were excessively risky or impossible. Impersonal communications separated the agent and handler by time, space, or location.[12] During the initial phases of agent recruitment, face-to-face meetings between the case officer and target were often necessary, but would be phased out when the target accepted the clandestine nature of the relationship. The more hostile the operating environment, the greater was the need to shift to using impersonal communications to protect the agent.[13]

Impersonal communications using either dead drops or electronic devices offered advantages to the agent and handler and when properly executed were difficult for counterintelligence to detect. Dead drops avoided the necessity for the agent to possess an electronic transmission device, but required time-consuming surveillance detection runs by the handler. Conversely, electronic exchanges usually obviated the requirement for lengthy surveillance detection on runs but the technology could fail, and in its early years of use, often did. Other disadvantages to using impersonal communications included the handler being unable to directly assess the emotional and physical condition of the agent, and the communication stream being accidentally or intentionally interrupted or intercepted. Agents, as in the case of A. G. Tolkachev, could also decide the electronic gadget was unwieldy and stop using it, or that the amount of information passed during any one electronic transmission was too limited.[14]

Dead drops are the most commonly used and most secure form of impersonal communication.[15] Dead drops enable agent and handler to exchange messages, correspondence, documents, film cassettes, money, requirements, and instructions without a direct encounter. Dead drops were "timed operations" in which the dropped package remained in a location for only a short time until retrieved by either the agent or handler.

Dead drop sites are selected from locations to which both the agent and the handler have normal access. Public sites for dead drops ranged from parks and nature trails to stairwells, parking garages, and elevators. Site selection varied depending on the country in which the operation was taking place and the circumstances surrounding the agent. Examples might include a library that contained a shelf of little-used books or the door of a mosque where shoes of either the agent or handler would serve as the container for the material being exchanged. Private areas, such as social clubs and health clubs, were also used if they contained obscured areas where drops could be left without notice.

The ideal dead drop site was used only once, was in a location that could be precisely communicated to the agent, and provided speed of access for both the agent and handler. The site should also provide privacy so that it could be loaded and emptied without the agent and handler being observed.[16] Finally, the location would be selected so that both the handler and the agent had plausible reason to be at the site and in a setting where the concealment would naturally pass unnoticed. A case officer from the 1970s who handled Polish officer Ryszard Kuklinski observed, "every CIA officer serving in a denied area should have a dog." Even in areas with constant and unfriendly surveillance, the necessity of taking the dog for walks provided excellent cover for carrying out operational acts involving signal sites and dead drops.[17]

In most capital cities, such as Moscow, Vienna, Paris, Washington, and Berlin, the number of "pristine sites" that met operational requirements for dead drops was limited, since thousands of intelligence officers from different countries had worked the same areas for decades. As a result, there was continuous pressure to identify new sites for future operational use. Techs and case officers shared the responsibility to find, photograph, sketch, and maintain inventories of valid sites. The difficulty in doing so was compounded because all signal and drop sites possess the same general attributes. In response, alert counterintelligence officers could set up an observation post at likely locations and patiently wait for them to be used. Nevertheless, the value of dead drops, despite their complexity and limitations, makes them a primary tool of every professional intelligence service.

Signal sites were among several methods used to initiate a communications sequence between the agent and handler. Signals of some type usually preceded or concluded an operation and were normally linked to a specific meeting place or dead drop location. For example, a signal left at site "Alpha" may initiate a drop at site "Bravo" or a meeting at a designated park bench.

Signal sites were usually located in public places, away from the actual drop site, and positioned so that the agent or a designated observer passed them regularly. Signposts, telephone poles, bridge abutments, and mailboxes were among the sites typically used to place a signal. Visual signal marks were made using postage stamps, white adhesive tape, masking tape, colored thumbtacks, colored adhesive-backed stickers, colored chalk, lipstick, and even crushed cigarette packs. A precisely placed soft drink can is readily visible to a passing car, bus, or pedestrian and becomes an effective signal. The positioning of the tape, or the color of the thumbtack, chalk, or other signal could also send a danger signal or initiate an escape sequence.[18] Even if the chalk mark was observed, its meaning was unknown to anyone other than the agent and handler.

Usually after placing the material at a dead drop site, a signal was left to communicate that the drop had been "loaded." The person unloading the drop would then confirm the presence of the signal before proceeding to the site. Once he had retrieved the materials, or "cleared the drop," a final signal might be left to communicate that the package was safe and the operation concluded. The absence of valid signals indicated a problem and forestalled either the agent or handler from approaching the site.[19]

Priority was always given to creating a safe and secure means for the agent to both send and receive messages. Signals were forms of codes that used symbols to communicate longer meanings. Other types of signal techniques included a "car park signal" based on the direction a car was parked, its parking location, or the direction its wheels were turned, and a "window signal" that used the raised or lowered position of a window or drapes and blinds (open, partially open, or closed) to send a message. The position of a potted plant visible to passersby could also have been a signal depending on where it was positioned.

Calls placed through the public telephone system, while subject to monitoring, could be safely used to send signals. An example was a "silent call" or "dead telephone" signal that was received at the agent's home at a predetermined time. The caller, using a public phone in a nonalerting location, said nothing but remained on the line for a set number of seconds before hanging up. To the agent the call had meaning, but to anyone monitoring the agent's telephone lines the call had no significance. Even if it was traced to the public telephone, it could not be linked to a case officer. When executed carefully, and used infrequently, the silent call or other wordless signals were almost impossible for an adversary to decode.

Other impersonal exchanges may be undertaken using public systems such as the postal service, telephone, telegraph, newspapers, radio transmissions, and the Internet. Within public systems, covert communications are mixed with the billions of daily telephone calls, letters, postcards, telegrams, newspaper ads, e-mails, Web postings, and instant-messaging transmissions.

When personal meetings were required, a technique known as a "visual recognition signal" could safely send a coded message from the agent to handler prior to any personal contact. Typically, the agent would be instructed to appear at a busy intersection at a prearranged date and time wearing clothing whose color was meaningful to the handler, but not alerting to counterintelligence if he was under surveillance. Anyone aware of the operation and familiar with the agent's photo and instructions could observe from a distance to see if a properly attired individual appeared at the established time.

Secret writing has existed for at least 2,000 years and predates the establishment of the first European postal systems.[20] Letters and postcards mailed by an agent to an accommodation address outside the country of origin were commonly used throughout the twentieth century to conceal secret writing. The technique represented an early form of steganography in which the goal was to mask the existence of a communication. The CIA used three forms of secret writing: wet systems, dry systems, and microdots.[21]

Wet systems used special inks that became invisible on the paper after the writing dried; the hidden message became visible again only when a reagent matched to the ink was applied. As a simple example lemon juice was used to constitute an ink and the heat from an electric light bulb or candle as the reagent. OTS packaged dehydrated heat-sensitive inks in a variety of disguised forms. Aspirin tablets made good candidates as concealment hosts because they were commonly carried and could be stored in a medicine cabinet at the agent's home without attracting attention; when dissolved in water the pills created the ink. The agent dipped a sharpened wooden stylus or toothpick into the liquid and wrote on bond paper he had prepared by rubbing with a soft cloth in all four directions. Agents were instructed to write on paper placed on glass to prevent indentations and minimize the disturbance to the fibers in the paper. After the letter had been composed and the ink dried, the agent would again rub the paper in all four directions to eliminate any traces of the message in the paper's fibers. Later the letter would be steamed and placed inside the pages of a thick book to dry.[22] Agents would prepare a cover letter over the secret writing for mailing to an accommodation address outside of their home country.[23]

While it was most often used as an "agent-send" system, in some instances agents also received instructions by secret writing. To eliminate the complexity of developing the writing and minimize the amount of potentially incriminating reagents in an agent's possession, OTS often recommended the "scorch method." Polish military officer Ryszard Kuklinski received innocuous letters that contained hidden messages that became legible only when they were scorched with a household iron.[24] With thousands of potential combinations of inks and reagents to select from, OTS produced hundreds of such systems. In an emergency, however, diluted blood, semen, and even plain water could be used as an invisible ink.[25]

Two weaknesses of wet systems were the requirement for the agent to possess the special ink and the near impossibility of removing all traces of damage to the paper's fibers. Even if the secret ink was not detected, under close scrutiny the damage to the fibers of the paper sometimes became noticeable.

Dry systems began appearing in the late 1950s as a variation of carbon typing paper. Chemists impregnated special papers with small amounts of chemicals and bound them into common items like writing journals, endpapers of books, or the last few pages of notepaper in a checkbook. The agent would sandwich three pieces of bond paper on top of a piece of glass. The top and bottom papers were blank while the middle page was the special "carbon paper." The agent wrote the operational message on the top page and the secret chemicals were transferred onto the bottom page. These dry systems quickly became the preferred method of secret writing in the 1960s.

Former MI6 officer Richard Tomlinson described a modern dry system developed by the British service:

> Using the Pentel Rollerball pen provided by Technical and Operations Support, Secret Writing, [I] wrote up the intelligence in block capitals in the standard format of a CX [raw intelligence] report. . . . It all fitted onto one page of A4 paper from my pad of water-soluble paper. Putting the sheet faceup on the bedside locker, I laid a sheet of ordinary A4 paper over it, then on top of both of them. . . . Five minutes was enough for the imprint transfer to the ordinary A4. The sheet of water-soluble paper went into the toilet bowl, and in seconds, all that was left was a translucent scum on the surface of the water that I quickly flushed away. Back in the bedroom, I took the sheet of A4, folded it into a brown manila envelope, and taped it into the inside of a copy of the Gazzetta dello Sport.[26]

The simple but effective system eliminated the concern of the officer being detected with spy gear since the Pentel Rollerball was commercially available and uncompromising.[27] The technique allowed the agent to see what he was writing before making the offset copy. Tomlinson observed that "offset is now used routinely by MI6 officers in the field for writing up intelligence notes after debriefing agents, and is also issued to a few highly trustworthy agents, but is considered too secret to be shared with the liaison services such as the CIA."[28]

Tomlinson also described a method for developing the Pentel secret writing:

> *At the back of the pad, I ripped out the fifth-to-last page, took it to the bathroom, placed it on the plastic lid of the toilet seat, and removed a bottle of Ralph Lauren Polo Sport aftershave from my sponge bag. Moistening a small wad of cotton with the doctored cologne, I slowly and methodically wiped it over the surface of the paper. [The message] started to appear, darkening to a deep pink. Using the hotel hairdryer, I carefully dried the damp sheet, trying not to wrinkle it too much and drying away the strong smell of perfume. It now looked a normal letter, though in a slightly peculiar dark red ink.*[29]

Microdots and other reduced-image techniques represented a third form of secret writing. A microdot was an optical reduction of a page of text or photographic negative to a size that was illegible without intense magnification. Commonly defined, microdots were less than 1mm square and required optical magnification of at least $100\times$ to be read. The microdot's larger cousin, a macrodot, created with similar photographic reduction processes, was considered much less secure. Operational advantages to using microdots included their tiny size, which made it possible to conceal the dot inside a variety of hosts, and convenience of delivery to virtually anyplace in the world using public systems such as the postal service.

CIA case officers received familiarization training in microdot communications, but making and burying an operational dot required assistance from an OTS specialist possessing both the necessary equipment and practiced skills. Dots were usually a last choice for agent covcom.

There were also operational disadvantages to using microdots:

- The production and burying of a dot by the originator was exacting and time-consuming.
- Microdots were often so well concealed they were difficult for the agent to find in the host letter or document. When located, the microdot had to be carefully dug out and properly positioned for reading.

- Microdots required special optical viewers with sufficient magnification to make the message legible. Other dots had to be redeveloped before they could be read.
- Microdots were usable primarily as a one-way agent-receive system. The lack of agent photographic skills, equipment, and training for microdot preparation most often precluded their use as an agent-send system.
- Microdot preparation usually required specialized photography equipment that, if discovered in an agent's possession, would arouse suspicion of espionage.

For additional secrecy, microdots could be rendered invisible by bleaching in a small amount of diluted iodine before use; the process was reversed by redeveloping the dot after being received. The tiny bleached dot could be buried behind a postage stamp, the flap of a letter, inside the thickness of a postcard, or underneath a raised typed letter on a single sheet of paper. TSD manufactured a special "slitter" for slicing an opening on a postcard or raising a tiny flap on a piece of paper so that a microdot could be inserted.[30] Once placed inside, the dot was sealed in place with a dab of egg white, carefully rolled with the curved edge of a glass to pick up excess "glue," and placed beneath a stack of books to dry. When properly prepared a microdot could be buried so effectively that it defied detection, even when a counterintelligence service was alerted and searching for it. Often, ensuring that the microdot could be found by the agent at its destination was a greater challenge than finding ways to conceal and transport the dots.[31]

After an agent dug out his microdot, he required a reader of sufficient power to read the message. Since a microscope might appear out of place inside the living quarters of many agents, OTS issued three small, concealable microdot readers. The smallest was the "bullet" lens also known as the Stanhope lens.[32] This tiny lens, a thin glass rod (3mm × 6.8mm) slightly larger than a pencil lead, had a spherical convex curvature on one of its surfaces, and a polished plane surface on the opposite side. The microdot could be moistened with saliva and affixed on the flat side of the lens; the user held the opposite side next to the eye. The bullet lens was capable of magnifying a microdot more than thirty times.[33]

TSS bought a hundred Stanhope lenses in the early 1950s from a novelty company only to discover that they came preloaded with sexy pinup photos of American starlets. The potentially offensive images were re-

The Stanhope lens, shown against a penny, was one type of microdot reader issued by OTS.

moved before the viewer was issued to agents and the lens required no further modification. The tiny size of the bullet lens enabled it to be readily concealed and transported inside a cigarette or bottle of ink, or sewn into the seam of a jacket or dress. One case officer carried the lens in the corner of his eye. Operationally the bullet lens was so small that the care and skill required to position the dot and view the message made it unpopular with agents.

Despite its deficiencies, the bullet lens proved invaluable for some operations. In 1969, the CIA recruited a middle-aged Cantonese-speaking woman to serve as a courier into southern China. The courier, who lived in Hong Kong, had immediate family members living in Canton whom she partially supported by her modest jade and semiprecious stones shop. Among her family members was a first cousin, also a CIA asset, with connections to a prodemocracy intellectual group that produced antigovernment publications and leaflets. The CIA was supporting the effort as a covert action and used microdots embedded in personal letters and postcards as the primary means of covert communication. When the agent lost the last of his microdot readers, the covcom link collapsed.

The case officer appealed to the courier to take several new bullet lens readers, each approximately the size of grain of rice, to her cousin. Recognizing that border checks for anyone entering China were intense, with pat downs and body cavity searches common, the case officer suggested the lens be embedded into the gauze of a Band-Aid that covered an active sore on the courier's foot.

The proposal was met with scorn. From the agent's perspective, the concealment was far too dangerous because the hard lens might be felt through the gauze. If discovered, the penalty would be severe. The agent agreed to take one of the lenses but said she would think of a much better concealment by their next meeting.

A week later, as the two were meeting, the agent unexpectedly dumped on the table a catty (680 grams) of dried fish, each fish about the size of a small minnow. The tabletop was covered with dozens of little fish.

"Find it," she ordered the case officer.

"What?" he replied.

"The lens is with the fish, you find it." Her tone was not amused.

The case officer looked at the fish. All were the same, dried, and some nearly translucent. He picked up several and examined. There was no way he could even imagine where the lens might be.

"I can't."

The agent began picking up fish and rubbing each between her fingers.

After handling several, she announced, "Here it is." She peeled off the dry skin and tore open the fish. Out popped the lens.

The lens had been inserted through the mouth into the belly of fish. The expandable cavity was long and large enough to hold the lens but the lens itself was not so thick as to distort the fish's appearance.

"This will be a lot safer than your Band-Aid. They will never inspect these fish which are like what everyone takes into China," the courier asserted.

The case officer could only agree and thought of lyrics from the musical *The King and I:* "When you become a teacher, by your students you'll be taught." In this instance, however, when you become a case officer, by your asset you will be taught.

The operation was successful and the lens reached the agent as planned.

TSD produced a more user-friendly microdot viewer, the "114 Reader," that was about the size of two pencil erasers. It unscrewed so that the dot could be placed between the two halves and viewed. The superior optics and larger size of the viewer made it more popular with agents but also more difficult to conceal.

The largest of the CIA's microdot viewers was the "little telescope" (about the size of an unfiltered cigarette) with an internal telescoping section for magnification up to 150 times. The viewer was more powerful and easier for agents to use than its predecessors, but it was much larger. If detected it was clearly recognizable as a piece of spy gear, but the little telescope was still small enough to be concealed in a pack of cigarettes or a modified fountain pen.

In 1983, the CIA recruited Soviet Colonel Vladimir Mikhailovich Vasilyev in Budapest and assigned the codename *GTACCORD*. To communicate with him after his return to Moscow, OTS perfected a new technique of sending messages using a Hewlett-Packard computerized laser-engraver. The technique allowed the CIA to etch a microscopic message into the black borders of features inside the February 1983 issue of *National Geographic* magazine.[34] The hidden message in the ruled line was invisible to the unaided eye, though readable with a 30x magnifier.

The laser-engraver burned away microns of ink to leave a message that had characteristics of a microdot but did not require the additional stages of development and precise handling. By etching the message on an advertisement in a popular magazine to which *GTACCORD* had normal access, there was no link back to a specific agent should the presence of the message be detected. The secret message contained the internal commo plan for *GTACCORD* to contact CIA. It read: "Your package should always be in a waterproof wrapper placed inside a dirty, oily rag tied with string . . ."[35]

The commo plan worked. Colonel Vasilyev spied for the United States for three years until he was betrayed by CIA officers Edward Lee Howard and Aldrich Ames in 1984 and 1985. Vasilyev was arrested in 1986 and executed in 1987.[36]

One laser-engraving effort left an enduring olfactory memory with the OTS techs. In this operation, the covert message was placed on a border line of an advertisement for fancy chocolates in a gourmet magazine. The advertisement was printed with newly developed "chocolate-scented ink" and when the laser-engraver began burning the ink to embed the message, the entire OTS lab took on the smell of fresh-baked chocolate chip cookies.

Another reduced-image technique involved photography that used a

ther volumes of classic distincti·
ther inlaid with 22 karat gold.

ound for direct subscribers exac

each volu
dition rec
the center
of Arts.

ther volumes of classic distincti·
ther inlaid with 22 karat gold.
micro-engraving

ound for direct subscribers exac

each volu
dition rec
the center
of Arts.

In the 1980s, OTS developed laser-engraving for clandestine communication. This was used by CIA agent Colonel Vladimir Vasilyev in Moscow. Left: Cover of National Geographic *magazine. Right: Inside page of the magazine showing location of the line containing micro-engraving, mid-1980s.*

sensitive, fragile emulsion layer of film. This plastic wrap–like substance could be separated from the thicker cellulose base of some types of film. Called "soft film" by the KGB, it was one of the most usable methods for clandestine communication even before World War II and saw use extensively throughout the Cold War.[37]

Typically, a frame of soft film contained the image of a single page of text, which could be produced in a variety of sizes. Although a frame was much larger than a microdot and more vulnerable to detection, it was much easier for the agent to use. Larger examples could be disguised as photo protectors inside a man's wallet or the shiny coating of a postcard such as was used by George Saxe's agent in the late 1960s. Pieces of the pliable film could be rolled into tiny cylinders as small as the size of a matchstick, concealed in such varied household items as a hollow pencil or a ballpoint pen refill, or sewn into the lining of clothing, and then read using a standard magnifying glass.

Kalvar, a commercial product developed as an alternative to traditional microfilm, represented one of the OTS's most successful special films for reduced-image photography. The company that first manufactured it ceased operations in 1979 but other firms continued making Kalvar for OTS.[38] In operational use, the advantage of Kalvar was that it could be handled and processed in normal room light, did not require special chemicals, and was developed in boiling water.[39]

Ultrathin-base (UTB) films were used for the subminiature cameras provided to agents and officers for clandestine photography; the thinner backing (base) allowed a standard film cassette to contain hundreds of exposures and increased the volume of information passed in a dead drop exchange. UTB film could not withstand the rigors of passing through automated processing and developing equipment, however, and required OTS techs to hand roll, spool, and later process the exposed film at remote field photo labs. The combination of UTB film and reliable OTS subminiature cameras produced some of CIA's best Cold War intelligence.

For further enhancement of operational security for clandestine photography, TSD developed special processing film (SPR) that looked and performed exactly like a standard cassette of 35mm film. However, after the film was exposed, any attempt to develop the images by a person without knowledge of the counterintuitive steps required, would result in a completely black or transparent strip on any part of the film that was SPR treated. The advantage of SPR film was that the agent could photograph secret documents and keep the film in his camera with the knowledge that even if it should be searched and the film processed, the compromising evidence on the roll would not be discovered.

During operational meetings, both the agent and handler would make and retain written notes for reminders, specific instructions, phone numbers, and names. Because the notes were sensitive and potentially compromising, a means to destroy the notes quickly and thoroughly, if necessary, was required. OTS developed a variety of secure note-taking capabilities for protecting such information.

Water-soluble paper was produced by a small CIA-owned paper-making machine and cut and bound into forms required by operations. Visually, the special paper resembled thin copy or tracing paper although it could also be made in a variety of weights. When dropped in water or any other liquid, the paper, together with the ink or pencil markings, dissolved immediately. Splashing water on the soluble paper left an instant gooey remainder that could not be restored to recover the original writing.

The CIA agent Ryszard Kuklinski had a pad of water-soluble paper onto which he copied his exfiltration plan to study and memorize. The original plan was passed to Kuklinski on a microfilm that he kept hidden. However, by copying the plan on water-soluble paper and taping it beneath his kitchen cabinet, it was more readily accessible and Kuklinski felt confident he could destroy the information quickly if necessary by dropping it into a waiting pan of water in the kitchen sink.[40]

A case officer driving through a city to identify new dead drop and signal sites would need a way to make notes but also required a quick destruction method if stopped by local police or involved in an automobile accident. Water-soluble paper and a handy bottle of water provided the solution. If a problem arose, the water could be doused on the notes, reducing them to a mushy residue.

An alternative was flash paper, a form of nitrocellulose that burned quickly and completely with a bright flame without smoke or ash. Any printing or writing on the paper would be destroyed when ignited. Because agents and case officers often smoked, a lit cigarette could be used to ignite flash paper carrying operational notes, one-time pads, communications plans, and other sensitive material. The effectiveness of instant destruction was offset by the reality that the "flash" of the ignited flash paper would assuredly attract attention, limiting its operational use.

Another option available to case officers was the so-called more-or-less-invisible (MLI) writing instruments developed by the chemists in OTS's secret-writing program. The abundance of ballpoint pens and other plastic products in the 1960s led the scientists to coat commonly available plastic items with special chemicals. The treated items, when used as a writing instrument, left invisible traces of the chemical residue on paper that could subsequently be developed and read. While a casual observer would see nothing on the paper, professional techniques could detect the presence of secret writing.

Only imagination limited the variety of plastic that could be used. MLI chemistry could be applied to eyeglass frames, caps on ballpoint pens, plastic key fobs, credit cards, and even the plastic toothpick on commercial models of the Swiss Army knife. Case officers using an MLI device to "write" invisibly on a piece of paper could carry those safely until returning to the station where a tech developed the notes.

TSD answered a request from the Directorate of Operations in the early 1970s to provide a secure system that would allow a case officer or tech to record and store operational notes on a tape recorder. OTS modified small commercial Sony stereo tape recorders by adding an additional, or "third track" recording head. In use, the tape recorder would function normally to play tapes filled with local music in two channels of stereo. When an officer wanted to record operational notes, he activated a switch OTS had built into the recorder and turned on the secret recording head for the third track. The audio would be recorded on the tape, but on a track that was unreadable on any nonmodified tape recorder, and only an operational listener would know how to activate the switch to listen to the third track.

In a variation on the concept, the clandestine communication branch of MI6 created a similar covert system. Former MI6 officer Richard Tomlinson described it:

> *The essential feature of these gadgets is that they are noncompromising, i.e., they are identical or virtually indistinguishable from commercially available equipment. Pettle recorders were particularly ingenious. Any normal audiocassette has two tracks running parallel to each other, one for each side of the cassette. Pettle recorders exploit the unused part of the magnetic tape, which lies between the two strips. [We observed] an ordinary personal stereo, which played and recorded on both sides of the tape like an ordinary machine. But turning it upside down tripped a microswitch so that pressing the STOP and RECORD buttons together made the machine record over the central track, while pressing STOP and PLAY together made it play back the recording.*[41]

Codes and ciphers play essential roles in successful covert communications systems. A code obscures the meaning of a message of any kind by substituting words, numbers, or symbols for plaintext (the unencrypted text of the message). A single symbol could represent an idea or an entire message. The signals made with chalk, lipstick, or a thumbtack to initiate a dead drop sequence were examples of codes while the message concealed inside the dead drop had the added protection of a cipher. A cipher represented a particular type of code in which numbers and letters were systematically substituted according to a prearranged plan. Ciphers used a key to convert the plaintext message.

Probably no piece of spy gear was more often issued or more reliable than the one-time pad. OTPs, the only cipher system that was known to be theoretically unbreakable, were composed of one or more pages filled with random numbers arranged in groups of five.[42] Only two copies of an OTP were produced—one copy for the agent and one for the handler. To maintain the security of communications, the OTP page and all notes from using it were to be destroyed by the agent as soon as the working session was completed.

OTPs had great advantages and were praised by both agents and handlers. One OTS tech who ran operations in Moscow for two decades stated, "OTPs didn't let us down. They didn't leave you or the agent wondering if the communication was secure." OTPs proved to be the best covcom security available during most of the Cold War; agents had immense confidence in the security of the OTP system because they understood that even if the message was discovered, its contents would be illegible and there was no link to the agent. The difficulty with the system was that OTPs were immediately recognized as spy gear if discovered, and since only used once, they had to be constantly resupplied through dead drops.

The one-way voice link described a covert communication system that transmitted messages to an agent's unmodified shortwave radio using the high-frequency shortwave bands between 3 and 30 MHz at a predetermined time, date, and frequency contained in their communications plan. The transmissions were contained in a series of repeated random number sequences and could only be deciphered using the agent's one-time pad. If proper tradecraft was practiced and instructions were precisely followed, an OWVL transmission was considered unbreakable. The agent was able to use OWVL only to receive communications, but it had many advantages over secret writing or agent meetings. OWVL required no spy gear except a one-time pad, was generally reliable and repeatable, and precluded surveillance. As long as the agent's cover could justify possessing a shortwave radio and he was not under technical surveillance, high-frequency OWVL was a secure and preferred system for the CIA during the Cold War.[43]

The OWVL transmission consisted of a series of numbers, usually in groups of four or five. During the 1950s and 1960s, they were read by a man or woman, and in later years produced by an electronically generated voice.[44] The numbers could be spoken in any language, usually timed to begin on the hour, quarter hour, or half hour, and were repeated hours or days later on the same or a different frequency.[45]

Facilities with giant antenna farms to broadcast OWVL signals to every country of operational interest were positioned at strategic locations in the United States and abroad. The sites served the dual purpose of handling CIA staff communications traffic as well messages for agents.

In the late 1970s, OTS and the Office of Communications began upgrading the OWVL system with the development of the interim one-way link (IOWL). This used the same broadcast stations and network as OWVL, but the agent's commercial shortwave radio was replaced by a dedicated IOWL receiver. The self-contained miniature piece of spy gear was a black box about the size of a pack of cigarettes and half as deep, including the internal battery. Its size made concealment relatively easy and it could be plugged externally into a standard speaker or headphone output of a shortwave radio receiver. The primary benefit to the agent was the speed of receiving a message; the numbers were transmitted at higher speeds and then stored internally in the receiver to be recalled later. Decreasing the time an agent had to spend performing the covert activity of listening to and transcribing shortwave transmissions improved security and his efficiency; messages that previously had required the agent to listen and copy for an hour could be received in ten minutes. IOWL required the agent to possess and hide another piece of spy gear, but because it was technically equal to OWVL and offered advances in reception speed and an improvement in weak-signal reception, the system was widely deployed.

Short-range agent communications, known as SRAC systems, represented a technological revolution for covcom when OTS deployed the first units to agents inside the Soviet Union in the mid-1970s. SRAC enabled the agent and case officer to exchange information without being required to come into close proximity, or conduct a clandestine act such as loading a drop that might be observed. It also eliminated the risk of leaving sensitive material unattended in a dead drop, which might be discovered and traced back to the agent.

The original SRAC systems exchanged short-duration, encrypted radio-frequency messages of a few hundred characters in less than five seconds between two black-box transceivers. An agent carried a pocket-sized SRAC unit in his coat and "shot" his message at designated locations at any time, day or night. He did not need to know, or be concerned about, the location of the SRAC receiver, which could have been located in an embassy, a residence, or in the handbag of a lady standing in front of a department store. SRAC defeated physical surveillance by eliminating the requirement for agent and case officer ever to be in the same location. It was, however, potentially vulnerable to signal interception if an opposition service was monitoring the SRAC frequency at the time and in the area of a transmission.

Because SRAC could be initiated by the agent, the CIA then had a reliable capability to receive time-sensitive reporting and could immediately retask the agent with follow-up requirements.[46] When military tensions between Greece and Turkey were at the flash point during the 1990s, senior CIA officers credited the near-real-time agent reporting through SRAC systems with preventing war between the two countries. SRAC was the principal covcom link between the CIA and General Dimitri Polyakov when the latter actively spied for the United States during the 1970s in Moscow, and later played a critical role in Colonel Kuklinski's successful 1981 exfiltration from Poland.

The SRAC device used by Kuklinski was prepared for him under an OTS project code-named DISCUS and was known at the Warsaw station under the code name ISKRA.[47] It was described as follows:

> *The size of a pack of cigarettes, it weighed about half a pound and had a keyboard and memory. Kuklinski could type in a message at home, place the device in his pocket, and carry it somewhere else. There he could push the transmission button without removing the ISKRA from his pocket. The device had a small window through which a single line of text could be read, from an outgoing or incoming message. If he transmitted directly into the embassy, an alarm would sound in the Warsaw station. As a rule, Kuklinski was asked to leave a signal in the morning that he would transmit in the night, and an officer would take another ISKRA outside to receive the message.[48]*

The OTS SRAC systems were an early form of text messaging. In the 1980s, receive-only digital pagers were introduced to the consumer market, and later enhanced in the 1990s with the capability to both transmit and receive messages. Once text messaging over cell phones was developed, the use exploded globally, with hundreds of millions of messages being sent daily. Both pagers and cell phones offered new potential for covcom and possessed added advantage that the agent did not require dedicated spy gear to communicate. However, these systems were particularly vulnerable to counterintelligence detection if not operated with the disciplined tradecraft needed to maintain their clandestine use.

As early as the mid-1960s, the CIA recognized the potential for using satellites for agent communications. The idea was that an agent with a small handset could beam his information to an orbiting satellite, which, in turn, would relay the data to a receiving site. By combining a satellite-send system with his OWVL, the agent could transmit and receive secret intelligence inside his home country without personal contact with a CIA officer. The capability, first deployed in the late 1960s under the codename BIRDBOOK, used low-earth-orbit satellites as "bent-pipe" relays for the agents' messages.[49]

Unfortunately, field realities limited the operational use of BIRDBOOK. Agents had only a five-to-seven-minute window to "shoot" the message to the satellite as it arced across the sky. Success also depended on a clear line-of-sight transmission path as well as the precise orientation and positioning of the transmit antenna.[50] Hostile counterintelligence services learned of the system and developed means to intercept the signal and triangulate the agent's position using direction-finding techniques.

Despite its limitations, BIRDBOOK demonstrated that satellites, signal processing, and component technology could be integrated into a long-range covcom system. Over the next two decades, new generations of government and commercial satellites increased global coverage and signal processing improvements made lower power transmissions possible. Advances in electronic components, combined with an understanding of the tradecraft necessary for securely operating a satellite transmitter, addressed many of the problems at the agent's end of satellite covcom. A decade before satellite phones were available, OTS, with its industry and government partners, had created a similar covert capability for a small number of highly select CIA agents.

While appearing simple in concept, covcom systems, including the most sophisticated and technically advanced, are difficult to design and exacting in their use if they are to be employed successfully. Each technological advance, from the telegraph to the Internet, added another means of communication, but technical officers had to devise means of assuring security and covertness before the technology could be used in clandestine operations. In the end, whether the secret message is written in the disappearing ink of

Caesar's day or encoded in a radio-frequency signal transmitted by satellite, covert communication between agent and handler relies on both the technique used and confidence that the exchange cannot be detected or read by anyone except the intended parties. However, as the final decade of the twentieth century unfolded, covcom, like the other pillars of tradecraft, would be revolutionized by the electronic tidal wave of digital technology, steganography, and the Internet.

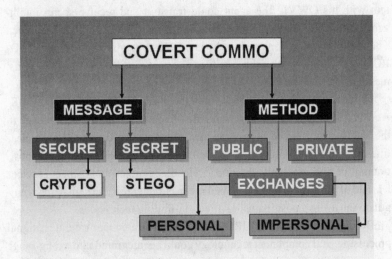

Spies and the Age of Information

The electron is the ultimate precision-guided weapon . . .

—DCI John Deutsch in Senate testimony, June 25, 1996

I n mid-December 1991, the CIA's Soviet/East European Division held its annual Christmas party. The mood was especially jubilant and attendees, including their OTS colleagues, received a campaign-style lapel button depicting the red Soviet hammer and sickle; beneath the red star were the words THE PARTY'S OVER.[1] Without media coverage, on December 31, 1991, a small detachment of Red Army soldiers marched to the Kremlin walls and replaced the red hammer-and-sickle flag of the USSR with the Russian tricolors not seen since the 1917 revolution.[2] For the CIA and OTS, their main adversary had been vanquished. A year later former DCI James Woolsey stated, "With the end of the Cold War, the great Soviet dragon was slain." Then he wryly noted that in its place the United States faced a "bewildering variety of poisonous snakes that have been let loose in a dark jungle [and] it may have been easier to watch the dragon."[3] For CIA officers, four transnational intelligence issues had emerged as competitors for intelligence resources and alongside traditional national targets such as North Korea, Cuba, Iraq, Iran, China, and Russia. These were:

- Terrorist groups and Middle Eastern Islamic extremist cells
- Proliferation of nuclear, biological, and chemical weapons

- Criminal and narcotics-trafficking cartels
- Regional instability, particularly in Africa and the Middle East

Less widely recognized at the time was the oncoming technical revolution in intelligence as the Information Age gave way to the Information Society with the accompanying creation, distribution, diffusion, use, and manipulation of information affecting global economics, politics, and cultures.[4] Digital information systems, used in the CIA for more than two decades, ceased being location specific and were now being connected and accessed throughout the world in unsecured spaces over an electronic spiderweb named the Internet. Former CIA officer James Gosler observed that because of the emergence of digital technologies in the 1990s, "the conduct of espionage had been irreversibly altered, owing in large part to the rapid expansion of global reliance on information technology."[5] These digital technologies, combined with servers, routers, and a terminal at every desk, transformed information at every level—creation, storage, processing, viewing, sharing, and transmission.[6]

Specialized subminiature cameras developed by OTS to photograph documents stored in a target's filing cabinet were of limited value when computer networks became the repositories for secrets. Gosler noted, "Clandestine photography is rapidly yielding to sophisticated technical operations that exploit these networks. Spies with authorized access to these networks—an insider—can exfiltrate more than one million pages of sensitive material inside a microelectronic memory device easily concealed within a watch, an ink pen, or even a hearing aid."[7]

Examples of the rapid obsolescence of Cold War collection devices can be found in some OTS equipment developed in the 1970s to support CIA agent Kuklinski in Poland, who had access to the Soviet war plans. Over the course of nine years, Kuklinski secretly photographed more than 25,000 pages of classified Soviet and Polish military planning and capabilities documents.[8] OTS supplied technology for the operation that included disguises, concealment devices, subminiature cameras, suicide pills, and covert communication devices.[9] Today, the technology for most of Kuklinski's communications and specialized camera equipment is obsolete and the secret documents he photographed and dead-dropped to his case officer would likely be imaged, transmitted, and disseminated in electronic form.[10]

The digital revolution did not alter the CIA's goal of clandestine collection of adversaries' secret plans and intentions. However, the role of agents fundamentally shifted from a spy who is supported by clandestine technology to the spy who supports a clandestine technical operation.[11] Spy gear had to adjust to the needs of the agent who would become an infiltrator and compromiser of computer networks rather than a reporter of information. In a sense, the technology, just like an agent, would be "recruited" to spy.

Legendary criminal Willie Sutton was once asked why he robbed banks. He responded, "Because that's where the money is." While some money remains in brick-and-mortar banks, the "mother lode" of wealth is now found in the financial cyberworld. The criminal skills and tools, and tradecraft possessed by a Sutton would be of little value in robbing a cyberbank. The same is true for intelligence collection. Over time, the location and form of secret information changed. Correspondingly, the skills, tools, partnerships, and culture of tradecraft have been forced to evolve.[12]

To uncover another country's military, political, or economic secrets, targeting an opponent's information technology can be exponentially more valuable than stealing paper documents. Thanks to new digital technologies, the covert transfer of vast amounts of information, or clandestine attacks on enemy networks no longer requires a physical presence, and can often be conducted remotely from anywhere on the globe using the Internet.[13]

Regardless of the era, sound tradecraft has always employed the best available technologies to support clandestine activities. While espionage goals and objectives remain constant, global access to digital systems and information altered time-honored methods and techniques of spying. Emerging information technologies also allowed traditional tradecraft to be applied in new ways.

The Internet and the global availability of personal database information makes the spotting of individuals with potential susceptibility to recruitment independent of geography or personal engagement. Using the Internet for assessment and as a spotting tool, intelligence services can focus on a smaller pool of potential recruits. Profession and position often indicates access to sensitive information and vulnerabilities are revealed by Internet communications and search habits.

Digital ink never fades and "private" thoughts and comments expressed in obscure publications reside permanently on the Internet as searchable public records. Whether entries are in the form of a blog, posted on a chat site, contained in a circulated e-mail, published in a book or magazine article, or transcribed from a television interview, they become available indefinitely to anyone with Internet access. Opinions and musings from one's youth might provide tantalizing clues to a person's beliefs, values, interests, and vulnerabilities, all of which are immensely valuable in the recruitment process.

Publicly accessible Internet databases enable the remote and anonymous aggregation of comprehensive personal and financial profiles. Types of information readily available include employment, profession, educational history, job-change patterns, health, marital status, address, social security number, driver's license number, income, personal debts, credit card numbers, travel patterns, favorite restaurants, lawsuits, and bankruptcies.

An examination of a computer user's database information can further reveal potential recruitment vulnerabilities. Examples include:

- Recurring purchases at a liquor store or bar might suggest problems with alcohol.
- Large expenditures at pharmacies or a hospital might reveal undisclosed health problems.
- Bankruptcy or bad credit reports could indicate financial strain.
- Travel patterns and expenditures might point to extramarital relationships.
- Frequent job changes could mask failed career expectations.
- Recreational interests in dangerous or thrill-seeking activities such as scuba diving, sky diving, or motorcycle racing, could identify a risk taker who might also be inclined to accept espionage as living even more "on the edge."

For the recruiter of spies, Internet information becomes an efficient tool for identifying targets to develop and discarding those without access or apparent vulnerabilities.

Internet accessibility to commercial databases has made the creation of effective cover and the use of disguise more problematic. The traditional identity details of address, profession, and association membership become immediately verifiable using Google or other common search tools. Because the effectiveness of cover and disguise can erode quickly under examination, light commercial cover could be compromised by a curious hotel check-in clerk with Internet access. In the hands of a counterintelligence professional, even a well-backstopped cover can be pierced by identifying anomalies and dates involved with the created identity. Because so many details of a person's identity are now publicly available, it is difficult to create sufficient supporting records to construct an individual's entire life history including records of education, credit cards, residence, family, children's schools, neighborhood associations, library cards, and driver's licenses. The amount of information needed to legitimize identity has made sustaining a cover identity over an extended period nearly impossible if a determined adversary has the ability to exploit the Internet.

Light disguises that include fake beards, mustaches, hair coloring, hats, or scars may fool the human eye, but not a camera with face-recognition software that is linked to a database. Biometric data such as iris scans, passports with memory chips, digital fingerprinting, and electronic signature matching have all emerged as new industries for commercial security and intelligence requirements.

Digital technology offers options for concealing data in forms never possible during the Cold War. The tens of thousands of pages of sensitive information collected by Kuklinski over his nine-year spy career could be compressed and stored on a memory card much smaller than a postage stamp. Embedded computer chips in toys, cameras, digital music players, calculators, watches, automobiles, and many home consumer products make it possible to alter the memory in any device to conceal secret information. An agent no longer needs to possess compromising concealment devices for hiding film, one-time pads, secret-writing chemicals, and escape instructions, since all of that information can be stored electronically in everyday devices that defy detection. The likelihood of an agent's properly concealed digital information being detected approaches zero.

The remarkable reduction in size of microphones, transmitters, and cameras since 1991 has resulted in easier concealment, less power requirements, and use of smaller batteries with a longer operational life. Tiny collection tools such as digital video cameras and microphones, small enough to be fitted on small robotic "crawlers" the size as a common cockroach, can explore, map, and exploit air-conditioning vents, drainpipes, and ventilation shafts for surveillance. It is now possible to convert any image or sound into a digital format, which can then be encrypted and transmitted instantly across the Internet or by satellite on government or commercial communication links.

For example, advanced software recognition programs can link video images to database programs that enable the surveillant to capture real-time images of license plates to build instantly a database of all vehicles and their owners passing an observation point. Such information, over time, could reveal the identities of security and intelligence personnel involved in activities near the location. Forms of "Face Trace" programs enable video images to be rapidly compared to records in distant databases for identification.

New generations of low-cost radio-frequency identification chips created for the retail industry offer an opportunity to tag an unsuspecting target by embedding a tiny chip in clothing or the sole of a shoe. These embedded passive chips can be scanned as targets pass through electronic choke points and represent a digital version of the Soviet "spy dust."

Astonishingly small, unmanned aerial vehicles with wingspans of less than one-half inch, carrying cameras and audio sensors, can be remotely piloted to surveil targets from above, or guided into a building to serve as a movable "flying bug." A Defense Advanced Research Projects Agency version is small enough to fit on a thumbnail, yet capable of carrying either audio or video sensors. Ninety percent of its internal power goes toward navigation and propulsion, while 10 percent maintains the sensors. An early CIA version of the flying device from 1976, called the *Insectothopter*, is on display inside the Agency's Original Headquarters Building alongside a prototype of an advanced Defense Advanced Research Projects Agency (DARPA) model no larger than a black horsefly.

Public awareness of Cold War tradecraft often focused around the communication techniques of brush passes, car tosses, and dead drops. Despite their sophistication and usefulness at the time, all of these techniques were vulnerable to surveillance by an alert counterintelligence service. In the United States, the arrests of Navy spy John Walker in 1985, and Aldrich Ames, a KGB mole inside the CIA, in 1994, were precipitated by their actions in communicating with their Soviet handlers.[14]

The advent of the Internet affected all of the "pillars of tradecraft," but none more so than covcom where it revolutionized clandestine communications. Criminals and terrorists, as well as intelligence services, quickly recognized that the Internet offered unprecedented capabilities to communicate with near impunity. Messages, information, and signals were transmitted in ways that appeared innocuous and defied detection by being interlaced into the burgeoning traffic transiting the Internet. As information flowed through the "Net," both the identity and location of the recipient and sender could be masked in a bewildering variety of disguises. A Cold War covcom plan that required weeks to plan, and was dangerous to execute, could be completed safely in seconds over the Internet. Encryption and steganography techniques using the latest advances in technology were developed to protect and conceal data in digital files transmitted globally.

The Internet allowed computer users, including bankers, criminals,

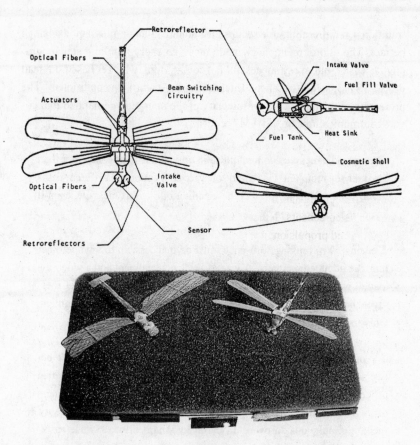

Top: Line drawing of the Insectothopter, *an early CIA attempt to develop a miniature unmanned aerial vehicle, disguised as a dragonfly, for intelligence operations, circa 1976.*

Bottom: Two prototypes of the flying Insectothopter *created by the CIA, showing variations of wing-propulsion systems, circa 1976.*

merchants, terrorists, and spies, to communicate instantly and easily, from anywhere to anywhere in the wired world. The global popularity and availability of Internet services allowed users who wished to remain undetected to blend their few messages in with billions of daily e-mails and file transfers; the hard-to-find needle in the haystack became the nearly-impossible-to-track electron in terabytes of data. Intelligence agencies recognized the potential to exploit the Internet, as they had with satellites and cell phones in earlier decades. Covert use of the Internet, however, still demanded that traditional requirements for secure and nonattributable message exchange be met. The digital technology made message encryption and steganography far easier, but every successful covcom system, whether based on dead drops, SRAC, satellites, or the Internet, had to meet four conditions.[15] The protection of the agent and the integrity of the operation demanded that covert communications be "SPAM" proof:

Secure: The message content must be unreadable to anyone other than the intended recipient. OTPs and software encryption are different paths to the same end—they protect the meaning of a covert message, even if it should be intercepted.[16]

Personal: The message presence must be inaccessible to anyone other than the intended recipient. A loaded-brick concealment and a video file loaded with digital steganography both provide a host for secret messages that would appear uninteresting and normal for their environment. Only the intended recipient would know to look inside.

Avoid traffic analysis: The existence of a communications link between the agent and handler must be hidden for the same reason that officers and agents traditionally used dead drops to preclude awareness of their covcom. There must not be any record of clandestine activity, including malicious software on the agent's hard drive, to raise suspicions about the agent during a search.[17]

Mask the existence of the fact of communication: The fact that a communication is or has occurred must remain secret. Dead drop sites would be used only once and not approached by either the agent or the case officer if suspicion of surveillance existed. Covert Internet exchanges can use remailers, cutouts, public systems, and digital dead drops for a similar objective.

The two critical components in a successful covert digital communication system are the message and its method of delivery. The message is made secure using digital encryption and secret, or invisible, using digital steganography. Both communication techniques can be used separately or together—first performing the encryption and then hiding within another file to be transmitted over the Internet.

For centuries encryption that protected information was generated by humans and early mechanical ciphers were vulnerable to being broken by other clever humans. The development of the first high-level electromechanical encryption machine took place in 1918 and produced ciphers that were, at the time, "unbreakable" by the unaided human mind alone. Though the electromechanical machines produced secure cipher text, the technology was controlled by governments with an extraordinary need for secrecy.[18] In the mid-1970s, however, strong encryption algorithms began migrating from the sole preserve of government agencies into the public domain. By the 1990s, digital encryption algorithms were widely used for protecting Internet e-commerce, mobile telephone networks, and automatic teller machines. The end of the Cold War saw the development and broad distribution via the Internet of sophisticated encryption algorithms to any user anywhere.

Phil Zimmermann is credited with developing the first version of a public encryption program, PGP (Pretty Good Privacy), in 1991. He had been a longtime antinuclear activist, and created PGP encryption to provide like-minded people with secure use of computerized bulletin board systems and messages and file storage. There was no charge for the software and the complete source code was included with all copies. PGP encryption found its way onto Usenet and from there onto the Internet. From a security standpoint, there is no publicly known method to break a PGP-generated message by cryptographic, computational means. For the first time in history government-level encryption software was available for free to anyone with access to the Internet.[19]

Intelligence services with limited financial resources soon adopted PGP and similar encryption software to create powerful and unbreakable agent covcom systems once available only to the major superpowers. The small, but aggressive, Cuban intelligence service used publicly available encryption software to communicate with its agents operating inside the United States. An advanced version of a PGP encryption program was discovered in September of 2001 during the search of the Washington, D.C. apartment of Ana Belen Montes. Montes, who the FBI code named *BLUE WREN*, was a Defense Intelligence Agency intelligence analyst for Cuban affairs, and a spy for the Cuban intelligence service.

For her covcom, Montes had been instructed to purchase a Toshiba 405CS laptop computer and was provided by her Cuban handlers, assigned to the Cuban Mission at the United Nations, with two diskettes, S-1 and R-1,

Diagram of Cuban agent Ana Belen Montes's one-way voice system for receiving encoded messages from Cuba, 2001.

for encrypting and decrypting messages. Because the possession of high-level encryption software would be alerting if Montes's laptop computer was examined forensically, digital encryption programs (PGP or similar) and one-time keys were embedded on each diskette. When receiving messages transmitted to her Sony shortwave radio by her service, she would copy and enter the ciphertext numbers into her laptop computer and insert diskette R-1 to recover the plaintext. To prepare secret information to be handed over to the Cubans she would enter the plaintext into her laptop and then use the encryption program and key embedded on diskette S-1 to convert it into ciphertext.

As long as Montes wiped her laptop hard drive after each covert use (to erase any trace of the process), and concealed her two special diskettes, the messages she was sending and receiving would have been virtually unbreakable. Despite her instructions, Montes did not wipe her hard drive after each use. As a result, during the FBI search of her apartment and computer, plaintext copies of her communications were recovered.[20] The weakness was not in the encryption software, but with the faulty tradecraft of Montes.

Once a message is encrypted, digital steganography can be used to hide it among the ones and zeros in any electronic transmission. Steganography, while not a form of encryption, protects messages by rendering them invisible. If the existence of a message cannot be discovered, its secrets are not revealed.

Publicly known digital techniques have made the use of steganography available to anyone to hide data and messages in virtually any electronic document and instantly send the secret information to anywhere on the globe over the Internet. Spies used limited digital techniques for hiding information during the Cold War. In the late 1980s, FBI Special Agent Robert Hanssen, a mole for the KGB, sent messages to his handlers on eight-inch floppy computer diskettes. Because the secrets he was selling would likely lead a trail back to him if discovered, Hanssen first encrypted the information and then concealed it on the diskettes using a technique called "40 track encryption." The little-known technical process reformatted the computer diskette and allowed data to be concealed by placing it onto specific tracks on the diskette inaccessible to the computer's internal operating system.[21] While the obscure digital techniques used by Hanssen in the 1980s were known only to so-called computer geeks, by the 1990s digital steganography programs for hiding data were readily available to anyone with Internet access.

It is possible to conceal data digitally inside music or video files in ways that make them sound and appear to be unaltered. Audio files can conceal information by altering digital bits of the file that are inaudible to the human ear. Graphic images allow the redundant bits that make up the colors to be altered in a way that appears identical to the human eye.[22] The secret messages are concealed within the bits of data. If someone does not have the original, or host file to serve as a comparison, the altered covert files with hidden messages can be very difficult to detect—especially so when combined with millions and millions of e-mails and file attachments that are sent daily over the Internet. Steganographic software uses an algorithm to embed data in a host image or sound file, and a password scheme for retrieving the information as illustrated in the graphic on page 454.[23] Professional intelligence services may use advanced steganography programs to incorporate encryption programs for additional security in case the message should be discovered.[24]

Digital technology has also reformed the classic microdot technique of using tiny pieces of film less than 1 mm square to conceal a page of text. It is now possible to create and embed large quantities of digital information

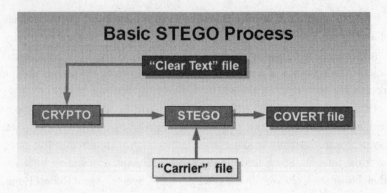

inside tiny e-mail electronic "dots." Once created, "digital dots" can be concealed in a variety of unconventional methods that defy detection. Virtually any type of digital file can be modified to conceal information, rendering the counterintelligence task not one of finding a needle in one haystack, but rather that of searching millions of haystacks without even the aid of a magnet.

Dead drops were used extensively during the Cold War as hiding places for information and money exchanges between the spy and handler, but subjected both to the risk of exposure and arrest.[25] Using the Internet, it becomes possible to create e-mail accounts to send and receive digital files and messages anonymously.

A secure digital dead drop e-mail account can be easily created from a personal laptop computer by anonymously signing up with an Internet service provider who offers a period of free access without requiring a credit card. From that newly created intermediate account, the user can log on to any similar service and create a second anonymous account to serve as the dead drop. Anyone can send digital files to the dead drop account and, with knowledge of the primary account password, content can be downloaded from anywhere in the world. To protect the security of the dead drop, users log on anonymously from the intermediate access account. America On Line and other Internet service providers also allow users to save an uncompleted e-mail or document onto the provider's hard drive to be retrieved later. This feature lets conspirators possessing the primary account password communicate by retrieving and editing the stored document without ever sending it as an e-mail or attachment. Regardless of the techniques used, for greatest security, the hard drive of the laptop would be wiped after each Internet session.

The options for covert communications using digital technology appear endless and remain a persistent problem for counterintelligence. Intelligence services anonymously establish e-mail accounts under fictional individual or business names and use them to receive coded messages and digital files from sources. The e-mail addresses, similar to a postal accommodation address, have no public association with the intelligence service and if necessary can be only once and discarded. Use of such an account would not be for agents in high-risk countries, but offers a method of anonymous communication elsewhere. A simple e-mail to a "notional account" could mask a coded communication, which would be unbreakable if used only once. For example, a Cuban agent recruited abroad and returning to Havana might send a seemingly innocuous e-mail to a friend in which he discusses his passion for stamp collecting. In reality, the "friend's" e-mail address arrives at a computer in the intelligence service and is a signal that the agent is ready to begin work. With limited use and selection of a topic consistent with the agent's lifestyle and interests, such communications defy discovery.

An unmodified computer's operating system leaves "tracks" that allow counterintelligence forensic specialists to recover plaintext copies of encrypted e-mails, regular e-mails, deleted files, cookies, temporary Internet files, Web site history, chat room conversations, instant messages, pictures viewed, recycle bin, and recent documents. Wiping the hard drive by permanently erasing its contents eliminates evidence of clandestine activity, but is often impractical for an agent using his business or family computer. As a solution, a covert operating system can be installed on a tiny concealable USB storage device smaller than the tip of one's pinkie finger. When the device is connected, the computer boots from the covert operating system inside the USB without leaving a trace of its activities of the computer's internal hard drive. The agent can then use the computer's keyboard, monitor, printer, and Internet connection without fear of leaving a forensic trail. The covert USB system is small enough to be portable and easily concealed.

The routing of voice conversations over the Internet or through any other Internet provider–based network also creates an opportunity for clandestine communication while bypassing telephone networks. Voice Over Internet Protocol (VOIP) encryption techniques scramble the conversations to render them meaningless if intercepted. Future advances in encryption techniques offer the potential to provide secure and unbreakable voice communication.

However, until encrypted VOIP communication becomes more common for businesses and individuals, the presence of such software on an agent's computer will be alerting to a counterintelligence service.

Low-cost mobile telephones, available in many countries, offer opportunities for anonymous communication. If the mobile phones are purchased for cash at randomly selected retail locations such as convenience and discount stores, there is no linkage to the user and calls made on phones with preloaded minutes cannot be traced. If the phone is discarded after one-time use, any link with the user is destroyed.

Prepaid phone cards, first introduced in the United States in the early 1980s, have grown in popularity as a cost-saving convenience for students, travelers, and spies. Telecard companies began to flourish after telephone company deregulation, when advances in satellite and communications technology created excess system capacities. Rather than allowing their global telecom systems to remain idle, the large carriers sold hundreds of millions of minutes of system usage to Telecard companies for fractions of a cent per minute. Telecard companies resell these minutes (priced at pennies a minute) on phone cards at gas stations, airports, convenience stores, and chain stores in countries around the globe. Each prepaid phone card contains a concealed PIN number and allows the user to call at no additional charge to numbers within authorized countries up to the amount of minutes designated on the card.[26]

Phone cards provide travelers with an inexpensive means of calling home, but for illicit romances, criminals, and spies they eliminate records of the calls and provide complete anonymity. If a phone card is purchased for cash at a location not under the control of the host counterintelligence service, any call made using the card is anonymous and untraceable.[27]

Cuban agent Ana Belen Montes used phone cards and digital pagers as part of a covcom system for contact with her handler at the Cuban Mission to the United Nations in New York. To contact Montes, the handler would go to a remote public pay phone in New York City and use a prepaid phone card to dial the number of a digital pager she carried and transmit his message using a three-digit number code. Montes had also been instructed to anonymously purchase a prepaid phone card and proceed to a remote public pay phone in Washington, D.C. She was to enter the phone card's 800 number, touch-tone the card's unique PIN number, and dial a telephone number linked to a digital pager carried by the Cuban intelligence officer posing as a diplomat at the United Nations. With the connection made, Montes would

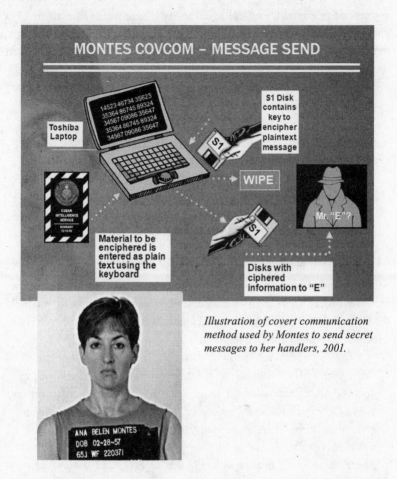

Illustration of covert communication method used by Montes to send secret messages to her handlers, 2001.

enter a three-digit code to communicate secretly with her handler. Though the CuIS system provided anonymity to the users and the calls were untraceable, her pattern of activity was alerting to FBI surveillance after she came under suspicion. Montes was known to carry a mobile phone, so there was no legitimate reason for her to seek out a remote pay phone to make a call after receiving a pager message. In this instance the covcom system was technically successful, but failed to mask the existence of the communication to well-trained counterintelligence officers.

Advancements in personal digital assistants (PDAs) in the late 1990s made it easier for information to be transmitted using SRAC techniques. FBI Special Agent Robert Hanssen, who "retired" from his role as a Russian spy when the Soviet Union collapsed in 1991, reactivated himself in 1999 and wanted to incorporate the latest digital technology for his secret communications. In a message to his handlers dated June 8, 2000, Hanssen wrote:

> One of the commercial products currently available is the Palm VII organizer. I have a Palm III, which is actually a fairly capable computer. The VII version comes with wireless Internet capability built in. It can allow the rapid transmission of encrypted messages, which if used on an infrequent basis, could be quite effective in preventing confusions if the existence of the accounts could be appropriately hidden as well as the existence of the devices themselves. Such a device might even serve for rapid transmittal of substantial material in digital form.[28]

Advances in microelectronics and personal computers added to the capabilities and effectiveness of constantly improving covcom systems. Former MI6 officer Richard Tomlinson described a system known as "Garfield the Cat" issued only to long-term and highly trusted British agents in countries such as Russia and South Africa:

> The agent writes a message on a laptop computer, then downloads it into the SRAC transmitter, a small box about the size of a cigarette pack. The receiver is usually mounted in the British Embassy and continually sends out a low-power interrogation signal. When the agent is close enough, in his car or on foot, his transmitter is triggered and sends the message in a high-speed burst of VHF. The transmitter is disguised as an innocent object. For many years "Garfield" stuffed animals were popular, as their suction feet allowed the agent to stick the transmitter on the side window of his car, giving an extra car signal as he drove past the embassy.[29]

In 2006, a spokesman for Russia's Federal Counterintelligence Service announced on Moscow television that British diplomats had been photographed servicing an "electronic dead-letter box" hidden inside a fake, hollowed-out rock in a city park. A hidden video camera planted by the FSB photographed two men identified as British diplomats as they attempted to activate their nonworking "rock." Concealed inside the rock were a receiver, transmitter, computer, and power supply designed for secret communication with a Russian agent. To activate it, the agent would secretly input his reports using only the keypad of a standard cell phone or other personal electronic device. Once prepared, the PED was placed in a transmit mode by using a combination of key strokes known only to the agent. The PED would then continuously send out a low-power signal, similar to Bluetooth technology, until it was within the range of the rock. Then, without requiring any additional actions by the agent who was carrying the PED concealed inside his coat pocket, the device would receive a confirming signal from the rock and transmit its encrypted information in a quick burst of energy. If the rock contained any stored messages for the agent, they would automatically be transmitted to his concealed PED.

The clever device updated the SRAC technology developed by OTS more than a quarter century earlier and made it safer. The agent's PED appeared unaltered apart from an undetectable software modification. Even the low-power transmission to and from the PED had a range of only thirty feet, which made it difficult to detect. The agent's handlers transmitted and received stored messages from the rock in a manner identical to that used by the agent. Once the rock had been discovered by the FSB, it was intentionally disabled to lure its caretakers, in this instance MI6 officers posted to Moscow as embassy staff members, to the rock to make repairs. The FSB announced that following the discovery of the first rock they also found a second rock concealed in a mound of snow at another location.[30]

Whether using OWVL radio in World War II, scrambled SRAC during the Cold War, or Internet steganography for the war against terrorism, every clandestine covcom system must meet five objectives:

- Break the transaction so if one end of the communication is discovered, it should not provide a link to compromise the person at the other end. The message content is secondary to the security of the agent.
- Use the best available physical or electronic hiding techniques. Covcom must always incorporate the most advanced technologies available at the time. Once a covcom system has been identified, the vulnerability of other agents being operated by the same intelligence service increases. The 1996 covcom techniques employed by Cuban agent Gerardo Hernandez and his Miami-based "Wasp" network proved helpful in recognizing similar tradecraft employed by Ana Belen Montes in 2001.[31]
- Employ strong encryption and ciphers to make the message so difficult to crack that, even if discovered, its contents are protected.

- Remain portable with capability to communicate with a variety of hardware platforms.
- Be backwards- and forwards-compatible, with sufficient flexibility to allow future security improvements to be incorporated while still allowing older communications to be read.

Jim Gosler concludes his essay "The Digital Dimension" in *Transforming U.S. Intelligence* with words of reality, caution, and optimism.[32] In Gosler's view, the reality facing the CIA is that intelligence services—allies and adversaries alike—have incorporated the "digital dimension" into their offensive and defensive operations. The caution for CIA operations is that defending against the sophisticated use of digital technology presents previously unknown and unaddressed gaps in its capabilities. The optimism is that twenty-first-century human and technical operations, when linked and mutually supportive, will solve problems otherwise intractable in either domain. A basis for such optimism can be seen in the history of OTS. The future effectiveness of American intelligence will depend on how well the lessons of that history are applied.

An Uncommon Service

by Robert Wallace

For 50 years OTS [officers] stood ready to serve whenever
and wherever America's leaders needed their talents.

—President George W. Bush, letter to CIA, August 24, 2001

John McMahon had more connections to OTS than any other senior CIA officer. In May 1973, he became its first director when TSD was moved from the Directorate of Operations to the Directorate of Science and Technology and renamed. McMahon relied on OTS for technical support as Deputy Director for Operations (1978–1981) and directed the CIA's technical programs as the Deputy Director of Central Intelligence (1981–1985). He was an informal advisor to OTS directors after leaving government service and served as the honorary chairman of the Fiftieth Anniversary Committee.

On September 7, 2001, McMahon addressed an audience of 500 retired and active CIA officers who gathered to celebrate OTS's golden anniversary in the Agency's auditorium. "Technical services [are] driven not just by advances in technology, but more importantly by the innovation of the technical officers to capture the potential of technology and bring it to application in operation," McMahon observed. Turning to how the CIA's operations officers used spy gear, McMahon said he detected a recurring fifty-year pattern: "Build it and they will try it—make it work and they will come back."

McMahon's remarks encompassed a half century of OTS contributions to clandestine operations. OTS forgers had fabricated documents for agents to infiltrate the Soviet Union in the early 1950s and created alias cyber-identities for case officers five decades later. OTS chemists made undetectable secret inks when writing was a primary means for clandestine communications, and its electrical engineers seized technology to develop radios whose millisecond transmissions were almost impossible to intercept. OTS psychologists evaluated the motivation and courage of a would-be agent. Mechanical engineers hid audio bugs in lamps to capture private conversations. These were the scientists, engineers, and craftsmen who built America's gadgets and disguises. They bugged embassies, trained saboteurs, and tracked down terrorists. When an operation went awry, they sat in foreign jails. When others were held hostage, they came to help.

From its beginning, OTS understood that supporting operations was its mission. Whether the operational requirement needed research, development, engineering, production, training, or deployment, OTS acted, driven by a philosophy of unapologetic responsiveness to national security needs. No constraints were placed on the imagination of the Agency's technical wizards. "If we can think it, we can do it. Our boundaries are the limits of what we can imagine, and, sometimes, the laws of physics," said a tech who devoted thirty-five years to OTS. This philosophy of limitless possibility produced an organization of a few thousand technical specialists who gave U.S. intelligence its decisive Cold War technical advantage and continue to equip the CIA for its battle against terrorism.

If not attempted during rush hour, the eight-mile drive from CIA Headquarters in Langley to Capitol Hill takes no more than fifteen minutes. The route, south along the George Washington Parkway across Memorial Bridge and down Constitution Avenue, provides a tourist's view of famed landmarks—the Lincoln Memorial, the Vietnam Memorial, the White House, the Washington Monument, the Jefferson Memorial, and, finally, the Capitol dome.

On such trips, I found myself moved by the familiar yet singular symbolism of these monuments to liberty. I passed those landmarks scores of times as Director of OTS to meet with the Senate Select Committee on Intelligence and the House Permanent Select Committee on Intelligence. Responsible for congressional oversight of the CIA, these committees needed to know the requirements and problems of Agency components and I was determined that they hear about them firsthand from OTS. The conversations, held in secure rooms in the Hart Senate Building or the south wing of the Capitol, were frank, detailed, and not broadcast by C-SPAN or leaked to the newspapers.

Speaking with members of Congress and their staff, I defended and advocated funding for OTS programs. In some years, I sought survival money for critical capabilities and, under other circumstances, I made the case for resources to launch programs for new capabilities. I always carried OTS toys to show the latest innovations of the office. These devices were usually small enough to put in my coat pocket and briefcase, yet impressive in illustrating the ingenuity and skill of OTS engineers, craftsman, and contractors. The toys might include dead drops or concealments, tiny batteries, prototypes of advanced beacons, or covert communications systems. Customarily, the devices were passed among the committee members and their staffs, and each person wanted to examine the component or the gadget up close.

Invariably, the committees were amazed at the small size of the devices and that otherwise ordinary objects had been modified to perform extraordinary clandestine functions. It was not difficult to imagine how Stanley Lovell and other OTS directors similarly captivated Presidents, Senators, or Congressmen with examples of ingenious devices of their era. William Donovan, it was reported, was so proud of the silenced *Hi-Standard* pistol produced by Lovell that he demonstrated it for President Roosevelt by firing into a sandbag in the Oval Office while the President talked on the phone.[1]

As remarkable as the technological progress of spy gear since World War II and the days of Penkovsky may seem, the impact of digital and materials technology on clandestine operations during my seven years at OTS was revolutionary. Paralleling the technology change was a dramatic shift in the intelligence battlefield from the dominance of a Soviet strategic threat to a demand for tactical intelligence to defeat terrorist plots and disable their weapons. In responding to post–Cold War operational requirements, OTS had adopted digital technologies that radically changed the size and capability of our equipment. By 2001, the seemingly limitless ingenuity of OTS and its contractors made many of the toys I had taken to Committee briefings in 1996 technological antiques.

On my final trip to Congress as Director of OTS in the summer of 2002, I displayed our most advanced tracking and communications equipment. My purpose was to talk about OTS's substantial role in operations against terrorism and al-Qaeda. As the committee members and staff looked over the devices, I pulled from my pocket another item that I believed spoke eloquently to the future of clandestine technical support.

"I fear that my successor OTS directors may no longer be able to show you so many neat gadgets and bugs," I said, as I held up a compact disc purchased at Radio Shack that morning. "I expect this is the spy gear you will be seeing in the future because the most significant espionage equipment will be embedded in software. This disc is spy gear, but it does not have much of a gee-whiz factor. It appears so everyday and common that its importance can be easily overlooked. We will need to learn to communicate to you and the American public that twenty-first-century digital 'spy gadgets' are as necessary as *Buster* and the T-100 camera were for their time."

Six years later, that compact disc is already obsolete.

Espionage and the Internet

The day *Spycraft* was published in 2008, we understood that the next big story of spy gear was about to break. That story would describe how the once tightly held technologies and capabilities by professional intelligence organizations were becoming broadly available to groups and individuals independent of government affiliation. For example, civilian demand for mapping, geolocating, and tracking applications of GPS technologies was as intense as that from law enforcement, intelligence, and military organizations. Having an instant personal communications device was as important to the teenager as to the combat soldier or spy. "Difficult to detect" steganography became as critical to political activists in the Middle East and China as to an agent's covert communications. Manipulating the digital environment to hack, steal data, or deny service, although complex, was simple enough that thousands of geeks in every part of the world would write applications, programs, and viruses available for anyone, anywhere to download and use for commerce, recreation, mischief, or terrorism.

While we could take momentary pride in recognizing an emerging trend, we underestimated the speed with which commercial technology would erode the technical monopoly of professional espionage organizations on clandestine capabilities. As 2008 unfolded, commercial global positioning systems; worldwide, real-time text and data communications devices; and anonymous Internet applications were being secretly integrated into a terrorist operational plan that, before the year ended, would bloody and for days hold hostage the world's second largest nation. A surprise attack would demonstrate that by the clever assembly of commercial equipment and technology, the twenty-first-century spy could operate independently of national intelligence services for equipment or training or infrastructure.

Between November 26–29, 2008, in Mumbai, India, ten youthful men who had never before been in the country used commercial devices and systems to execute one of the new century's most spectacular clandestine operations.

Like spies, terrorists employ all the fundamental elements of tradecraft, but for a different end. The classic spy seeks to acquire information secretly and communicate covertly to his handlers at a foreign intelligence service. Success in espionage is achieved by not being caught or alerting authorities that their information has been stolen. Terrorists, on the other hand, desire the notoriety of their ultimate goal—fear and terror—and often must apply similar elements of tradecraft to achieve that end.

The selection and capability of technology available over the Internet has exploded since 1999; the costs to acquire capabilities applicable for clandestine operations have decreased even more dramatically. Technologies once exclusive to national intelligence services are now available to anyone with Internet access and often at nominal cost. What made the attack on Mumbai radically different is that commercially available off-the-shelf (COTS) technologies were acquired and integrated without suspicion to enhance the severity, duration, and effectiveness of the assault.

The coordinated Mumbai attack illustrated the dramatic advance in using commercial technologies for terrorist tradecraft. On November 26, 2008, after traveling more than five hundred miles by sea from Pakistan, ten men in five two-man teams of *fedayeen* (Arabic for "redeemers") arrived undetected in an inflatable speedboat on the Mumbai waterfront. The attackers, all members of a Pakistani terrorist group, Lashkar-e-Taiba (LeT), assumed the personae of college students as they proceeded inland to launch the attack. Over the next fifty-six hours, the terrorists, armed with grenades and assault rifles, engaged eight primary targets, beginning with Mumbai's most frequented train station to luxurious five-star hotels. They killed 164 people, wounded more than three hundred, and kept the city of more than twenty million hostage while capturing the attention of the world media.

When the horror of the most sophisticated terrorist attack since 9/11 ended, a picture emerged of how commercially available technology could be used as "weapons of mass disruption" to plan, launch, and coordinate clandestine operations.

COTS equipment was the foundation of all elements of the attack and became integrated in ways that allowed the planners to remain hidden while maintaining real-time command and control over an event unfolding more than a thousand miles away. The attack required, and commercial gear enabled, that every phase—planning, reconnaissance, training, and movement until final assault—be conducted covertly to avoid detection by the counterintelligence services in India or international counterterrorism agencies.

In all, eighteen individual commercial technologies or techniques enabled the attack:

1. Internet cafés: The attack leaders remaining in Pakistan relied primarily on reconnaissance information from David Coleman Headley, who made multiple visits to Mumbai in 2007 while disguised as an American Jew. To communicate, Hadley and others used Internet cafés to anonymously access the Internet or search for unsecured Wi-Fi routers and "piggyback" on these open communication channels.

2. Long distance telephone calling cards: Low-cost plastic communications cards, when purchased anonymously, allowed the users to make untraceable phone calls.

3. Anonymous e-mail accounts: Headley and other attack planners utilized anonymous or "throwaway" e-mail accounts that were disposed of after a single use. By exploiting their ability to open an anonymous e-mail account using Gmail, Ymail, or Hotmail, communication could be conducted, and disposed of, before coming to the attention of the authorities.

4. Electronic "dead-letter boxes" or draft folders: Rather than risk traveling across international borders carrying detailed reconnaissance notes on potential targets, members of the reconnaissance teams created detailed notes in an e-mail and saved the message to the server as a draft instead of sending. Another person with Internet access, and knowing the message creator's password, could access the draft and retrieve the information without the e-mail ever being sent. This technique also eliminated the necessity of having compromising information on a user's laptop, further increasing the difficulty of the authorities to detect the operation.

5. "Hacked servers": LeT computer specialists in Pakistan hacked into servers on four continents to route communications in a manner to mask their true country of origin. The last hacked server located at an oil company in St. Petersburg, Russia, was used as the final IP address. That address was used to establish the national telephone accounts and to communicate and distribute a press release to the news services following the attacks. Russian authorities were unaware that its server had been hijacked until after the attack.

6. Google Earth: The ten attackers, training in Pakistan, had never traveled to India but were able to familiarize themselves with the topography and terrain of Mumbai using commercial software in Google Earth. The realistic 3-D satellite images acquainted them with the streets and neighborhoods they would be transiting en route to their targets.

7. GPS: Navigation of the five-hundred-mile sea voyage to Mumbai relied on skillful use of GPS. Once ashore each of the five two-man attack teams carried a Garmin Rino 120 handheld GPS system with preloaded attack coordinates.

8. Hawalas: The Middle Eastern monetary transfer system based upon honor and performance enabled operational planners to purchase secretly international cell phone numbers while evading traditional banking and financial channels.

9. Electronic online payments: Using MoneyGrams and Western Union, the planners made logistic payments in sums small enough to evade financial scrutiny.

10. Satellite phone: While at sea, a Thuraya "sat phone" allowed attackers and coordinators to communicate over the GSM network, as well as employ GPS and SMS text messaging.

11. "VoIP"—Voice over Internet Protocol: A family of communication protocols for the delivery of telephony over the Internet, served to mask the point of origin (in Pakistan) for calls between the attackers and handlers.

12. Direct inward dialing numbers: Telephone numbers as a gateway to the Internet were purchased from a U.S. company, Callphonex. This technique made it appear that calls between the handlers and attackers were from Austria.

13. Cell phones: Commercial Nokia model 1200 cell phones were purchased anonymously in Pakistan and fitted with precharged SIM cards, which had been obtained using false names in the electronic markets of Calcutta. Once the attack was under way, the terrorists used cell phones taken from victims to further confuse authorities. During the 56-hour siege, the attackers and their handlers made 284 calls and communicated for more than 995 minutes—more than 15 hours.

14. Skype: The organizers in Pakistan employed the encrypted VoIP service and software application to enhance the security of communications from Callphonex in New Jersey to Pakistan.

15. Instant messaging: Attackers and planners used SMS text messaging throughout the operation for rapid communications.

16. Crowd sourcing: During the attack, the coordinators at the command post in Pakistan monitored social network postings and tweets to garner real-time intelligence in and around the attack sites in Mumbai.

17. Exploitation of news sites: The LeT command post coordinators monitored an array of international commercial news feeds from television and the Internet as a source of intelligence that was relayed by cell phone in near real time to the attack teams.

18. Internet research: To conduct real-time counterintelligence, attackers at the Taj Mahal hotel telephoned identity information about hostages to coordinators in Pakistan while searching for the three Indian ministers reported by the news outlets to be missing at the hotel. The coordinators researched names over the Internet in an attempt to confirm the identities and unmask the ministers. Had that identification effort been successful, a hostage swap with the Indian government might have gained freedom for the attackers.

The commercial availability and covert integration of this array of technologies in advance of and during the Mumbai attack enhanced and prolonged the carnage. The success of the attack demonstrates the maturation of terrorist tradecraft in applying commercial systems as an enabler for future operations.

For American spy hunters, the Mumbai attack confirmed that commercial spy technology had widened the espionage capabilities of terrorists, and sophisticated spy tools were available to virtually anyone. The U.S. national security focus on counterterrorism and the related wars in Iraq and Afghanistan obscured an equally damaging reality that during the first decade of the twenty-first century, classic foreign espionage operations against the United States did not diminish.

On June 27, 2010, the FBI announced the arrest of ten Russian "sleeper spies" working in the United States. Some had been living in the country for more than two decades. The ten were formally charged with "carrying out long-term, 'deep-cover' assignments in the United States on behalf of the Russian Federation."

Demonstrating a naïveté about professional espionage operations, many newspapers reported the arrests as something of a spy vs. spy comedic event, disparaging the FBI operation and characterizing the ten Russian illegals as "bumbling caricatures of a bygone era. For more than a decade they used false names, dead drops, and invisible ink to gain access to the type of information more easily downloaded from the internet."

Apparently unable to comprehend the magnitude of both the long-term Russian operation and the FBI's remarkable multiyear counterintelligence investigation, press coverage turned the public attention to the young, attractive, red-headed spy, Anna Kushchenko Chapman. Her celebrity status grew when her British ex-husband produced seminude photographs of Anna that soon adorned tabloid covers around the world. Although Chapman had been in the United States only six months, her sex appeal overwhelmed any serious media inquiry into the work of the other illegals who had quietly and systematically conducted espionage operations in the United States for years.

The ten illegals were intelligence officers for the Russian Foreign Intelligence Service (Sluzhba Vneshney Razvedki, or SVR) living "normal private citizen" lives in the United States and without diplomatic protection. This type of operation is common for intelligence services that insert officers under commercial cover into select countries to perform a variety of missions. For example, during wartime, or in event of a break in diplomatic relations, illegals or non-official covered officers can remain in place to continue intelligence operations. This type of officer may handle sensitive agents, spot potential new recruits, provide logistical support, join targeted organizations or societies, and seek employment in government or other sensitive facilities.

A deciphered secret communication from the Moscow Center to illegal Richard Murphy clearly described the mission of the Russian Ten: "You were sent to USA for long-term service trip. Your education, bank accounts, car, house etc.—All these serve one goal—fulfill your main mission, i.e. to search and develop ties in policymaking circles in US and send Intel's [intelligence reports] to C [center]." The illegals' success was not judged by whom they recruited or the secrets they stole, but by whom they spotted and assessed, and the influential friends they made.

Russian illegal operations are the product of a disciplined, professional structure. Those selected to be an illegal receive the lengthy training required of all Russian intelligence officers, plus additional intensive instruction necessary for them cope with the stress of living a secret, illegal life abroad.

In almost all cases, illegals leave their birth identities behind and assume a false name and background legend before arriving in the United States. Gaps in their life legends, plus their clandestine work, make normal court-ship rituals impractical and as a result illegals often arrive as couples, have children, and do everything possible to project "normal" lives.

Spouses receive full training as illegal officers, but wives are usually operationally subordinate to their husbands and focus on providing support such as covert communications, accounting, and logistics. Among the ten illegals arrested in 2010 were four couples, one each in Massachusetts, New York, New Jersey, and Virginia. One couple, Juan Lazaro and Vicky Pelaez, had been in the United States for twenty-five years, while the others arrived in the 1990s and early 2000s.

Two illegals were "singletons": Anna Chapman in New York City, and Mikhail Semenko near Washington, D.C., were living under their true names and had received the same type of tradecraft training as the others, including agent-to-agent communications, invisible writing, and establish-ing a cover profession. Because they were single, both had active social lives. Their youth plus technical competence allowed them to work com-fortably in the evolving world of e-commerce. At the same time, their lack of intelligence maturity and condensed training time ultimately contributed to their downfall as they were lured into "false flag" meetings with FBI un-dercover officers posing as members of the Russian Consulate.

Most of the U.S. news media overlooked, or failed to recognize, the threat posed by these illegals. Portrayals of the illegals as being unproduc-tive and operationally incompetent were likely read with delight in Moscow. An image of bumbling, cartoonish Russian spies would likely diminish the American concern about future threats by the SVR.

Because the FBI had early insight into portions of the operations and could observe the illegals over several years, a remarkable level of knowl-edge about the tradecraft and equipment used by the Russian illegals was acquired. Their operational acts were planned, conducted professionally, and subject to all the vicissitudes of human and infrastructure complications.

Only two of the legal couples (the Murphys and Zottolis) met in person, and they did so on multiple occasions. Meetings between illegals are highly unusual and remain unexplained, as such lateral contact among illegals weakens security and almost never happens in the Russian operational doc-trine. The six other illegals were apparently unknown to one another prior to the mass arrest on June 27.

On four occasions between 2001 and 2005, Christopher Metsos (true name Pavel Kapustin), a senior illegal sent from Moscow Center traveling on the passport of a deceased Canadian boy, used classic tradecraft to obscure travel and activities, and directly supplied Richard Murphy with money and operational material. Metsos also cached money in 2004 at a remote location in the vicinity of Wurtsboro, New York, where Michael Zottoli uncovered it in 2006. In 2009, Murphy received money directly from a member of the Russian Consulate at the White Plains train station using a "brush meeting," or "flash pass." Afterward, he gave some of that to Zottoli at a brief meeting in Central Park. Another illegal, Juan Lazaro, who allegedly retired from active operations in 2006, met annually with a Russian intelligence officer in a South American country to receive funds.

Common to all of the illegals was their training in covert communications. They were proficient in receiving bursts of coded communications over a shortwave radio. Called "RGs" (for radiograms), these transmissions could be received either as Morse code or as a recorded voice reading a series of seemingly random numbers that are part of an encryption scheme. Juan Lazaro used a Realistic DX-440 shortwave radio to receive his communications, while others had different models of shortwave radios.

Chapman and Semenko employed more advanced covert communications (COVCOM) and initiated covert short-range communication through a private wireless network of paired laptop computers when communicating with their handlers. Creating a temporary, ad hoc wireless network between two computers allowed the transmission of data anonymously without data going over the Internet. The typical system worked by preconfiguring a laptop computer ("Laptop A") for its own private wireless local area network. This network is programmed to communicate with a second specific laptop ("Laptop B"), based on Laptop B's Media Access Control (MAC) address. To initiate communications, Laptop A transmits a signal establishing the private wireless network; when Laptop B comes within a certain physical distance of Laptop A, the two will be joined. Once the computers are joined, their private wireless network allows communications and data exchange only between the two. The data stream is encrypted and can be deciphered only with the aid of specialized decryption software.

Depending on the amount of metal in the area surrounding the transmission point, the effective range of the ad hoc network could be up to a hundred feet, but performance at that distance degraded and was not considered sufficiently reliable for operational use. Therefore, to enhance the communication range, Chapman and Semenko used USB "range extenders." Media reports would fault the illegals for using ad hoc networks because they depended on identifiable MAC addresses. Apparently unrecognized by critics, however, was that every mobile device in use in the vicinity of the transmission point was emitting a MAC address as an essential function for connecting with a network. At a busy location, such as Manhattan, there could have been hundreds of MAC addresses being transmitted at any given moment. The challenge for counterintelligence technical officers was to identify the MAC addresses of the illegals and their Russian handlers.

Using this COVCOM technique, on ten Wednesdays between January and June 2010, Anna Chapman visited coffee shops and bookstores with a tote bag that contained her laptop. On these occasions, outside, and invisible to her, was a Russian official with a laptop preconfigured to create the ad hoc network necessary to establish a communication link with her laptop. Once connected, the laptops exchanged encrypted data. In practice, however, neither Chapman nor Semenko confidently mastered this technique and both remained uncertain if their communications were being received.

Three of the illegal couples, living in Boston, Seattle, and New Jersey, employed digital steganography in their communications. Their steganography software, issued by Moscow Center, was not commercially available and less likely to be identified by standard counterintelligence forensic programs scouring the Web. FBI documents describe the technique as a software package that "permits the SVR clandestinely to insert encrypted data in images that are located on publicly available websites without the data being visible. The encrypted data can be removed from the image, and then decrypted using SVR provided software. Similarly, SVR provided software can be used to encrypt data and then clandestinely to embed the data in images on publicly available websites and blogs." The illegals used Sony FD Mavica cameras to create images of flowers that were used to mask the embedded (and encrypted) messages before being posted on the Internet.

All of the following classic and modern tradecraft, devices, and operational techniques were used by the illegals. Their identities were constructed and maintained by forged passports, false covers, and legends. Their covert communications included encrypted thumb drives, radiograms (RGs), one-time pads, encryption software, steganography software, and the Internet for communications and logistics, supplemented by secret writing, disposable cell phones, prepaid calling cards, and personal meetings. Logistical support involved the Internet, flash passes, signal sites, cache sites, precharged ATM cards from SVR funds, concealment devices, and luggage with hidden compartments for transporting money.

Operational security practices included verbal and visual "paroles" for meetings, danger signals, countersurveillance prior to personal meetings and exchanges, electronic countersurveillance from the Russian "impulse operators" who monitored known FBI communication frequencies in advance of "in flash passes," and "traps" set inside their safety-deposit boxes to detect an unexpected search.

Given the professionalism of the Russian Ten, the FBI's tightly held decade-long investigation that wrapped up the illegals, known as Ghost Stories, is impressive for its discipline, security, and vision. Ripples from the case continue as SVR illegals have been identified in other countries. The source or sources who tipped the FBI to the operation remain unidentified to the public although Russian prime minister Vladimir Putin, a former intelligence officer, stated in July 2010 that "we know who the source is." Ominously, another Kremlin spokesman was more direct when he stated, "we know who he is and where he is . . . do not doubt that a Mercader [the assassin of Trotsky in 1940] has been sent after him already."

Two weeks after their arrests, the United States deported the ten illegals to Russia in a spy swap. An eleventh operative, Christopher Metsos was arrested at Larnaca airport in Cypress on July 29, 2010, but skipped bail and is at an unknown location.

Without the benefit of exhaustive debriefings of the illegals or access to SVR records, neither the significance of the operation for Russia nor the extent of damage to U.S. security can be fully assessed. However, a senior FBI counterintelligence officer in New York City offered the perspective, "Within 6 months, Anna Chapman would have become the most dangerous Russian spy in the United States." Her attractiveness and charm opened doors for her throughout New York City and her access and contacts grew weekly.

This "Russian illegals" case will become a twenty-first-century "lessons learned" case study for counterintelligence professionals. Likely prominent among those lessons will be that foreign intelligence services are effectively integrating commercial technologies and the latest electronic advances into their covert electronic communications with the best elements of traditional tradecraft. The convergence of the new and the old poses escalating challenges to U.S. counterintelligence services.

Appendix A

U.S. Clandestine Services and OTS
Organizational Genealogy, 1941–2008

Organization	Created	Abolished
Coordinator of Information	July 1941	June 1942
Office of Strategic Services	June 1942	October 1945
Strategic Services Unit	October 1945	October 1946
Central Intelligence Group	January 1946	September 1947
Office of Special Operations	July 1946	September 1952
Central Intelligence Agency	September 1947	
Office of Policy Coordination	September 1948	September 1952
Directorate of Plans	September 1952	March 1973
–Technical Services Staff	September 1951	February 1960
–Technical Services Division	February 1960	May 1973
Directorate of Operations	March 1973	October 2005
Directorate of Research	February 1962	August 1963
Directorate of Science & Technology	August 1963	
–Office of Technical Service	May 1973	
National Clandestine Service	October 2005	

Appendix B

Selected Chronology of OTS

1942
Office of Strategic Services is established with William J. Donovan as Director. OSS establishes Research & Development Branch under Stanley P. Lovell.

1947
National Security Act establishes CIA.

1951 (September 7)
CIA Technical Services Staff (TSS) created under James H. "Trapper" Drum.

1956
First U-2 reconnaissance flight over Soviet Union.

1959
Cornerstone laid for CIA Original Headquarters Building at Langley, Virginia.

1960
First satellite photos of Soviet Union recovered. TSS renamed Technical Services Division (TSD). Three TSD audio techs arrested in Havana.

1961
Invasion of Cuba by CIA-supported Cuban exiles.

1962
Seymour Russell appointed Chief, TSD. Oleg Penkovsky arrested. Cuban Missile Crisis.

1963
TSD audio techs released from Cuban jail.

1966
Dr. Sidney Gottlieb appointed Chief, TSD. TSD relocated to former CIA Headquarters at 2430 E Street, NW, Washington, D.C.

1973 (May 4)
TSD renamed Office of Technical Service and transferred from the Directorate of Operations to the Directorate of Science and Technology.

1975
President Gerald Ford creates "Commission on CIA Activities within the United States" (Rockefeller Commission).

The Senate establishes a Select Committee to Study Governmental Operations with Respect to Intelligence Activities (Church Committee).

1978
OTS assigned responsibility for all CIA covert communications.

1985
Cornerstone laid for CIA's New Headquarters Building.

1988
OTS moves from E Street to CIA's New Headquarters Building.

1989
Berlin Wall destroyed.

1991
Collapse of Soviet Union; nation of Russia restored.

1996
Responsibility for clandestine audio operations transferred from OTS to the Clandestine Information Technology Office.

1997 (September 18)
CIA's 50th anniversary. OTS officers honored as CIA Trailblazers.

2001 (September 7)
OTS's fiftieth anniversary.

2001 (September 11)
Al-Qaeda hijacking of four commercial airliners and terrorist attack on United States.

Appendix C

Directors of OTS

TECHNICAL SERVICES STAFF, 1951–1960

Col. James H. "Trapper" Drum, September 1951–October 1952
Dr. Willis A. "Gib" Gibbons, October 1952–April 1959

TECHNICAL SERVICES DIVISION, 1960–1973

Cornelius V. S. Roosevelt, May 1959–May 1962
Seymour Russell, August 1962–March 1966
Dr. Sidney Gottlieb, March 1966–May 1973

OFFICE OF TECHNICAL SERVICE, 1973–PRESENT

John N. McMahon, May 1973–July 1974
David S. Brandwein, July 1974–June 1980
Milton C. "Corley" Wonus, June 1980–July 1984
Peter A. Marino, July 1984–September 1986
Joseph R. Detrani, December 1986–April 1989
Frank R. Anderson, April 1989–May 1991
Robert G. Ruhle, May 1991–April 1994
Robert W. Manners, February 1994–October 1996
James L. Morris, December 1996–March 1997
Patrick L. Meehan, May 1997–October 1998
Robert W. Wallace, December 1998–August 2002
Edward B. Charbonneau, August 2002–June 2003
Lawrence J. Boteler, July 2003–May 2005
Sterling K. Ainsworth, June 2005–June 2006
Anne C. Manganaro, June 2006–Present

Appendix D

CIA Trailblazers from OTS

The Central Intelligence Agency marked its 50th anniversary on September 18, 1997. As part of that anniversary, the Agency named fifty officers whose actions, example, innovation, or initiative shaped the history of the Agency. Four of these Trailblazers had significant association with Office of Technical Service. Their official citations read as follows:

DAVID E. COFFEY

Service Years: 1968–1995

Mr. Coffey's exceptional ability to solve operational problems with technology culminated in his successful creation and maintenance of an extremely sensitive—but uniquely valuable—covert communications capability. As an overseas Base Chief, his understanding of operational needs and his grasp of technology set him apart as a consummate technical operations officer and manager. His leadership significantly enhanced the integration of technical support into espionage operations. His personal commitment to excellence and teamwork did much to promote cooperative relationships between the Directorate of Science and Technology and the Directorate of Operations.

PAUL L. HOWE

Service Years: 1956–1987

Mr. Howe engineered the Agency's single greatest advance in operational photography—the ultraminiature camera. His work enabled us to photograph materials under the most difficult operational circumstances. The value of the intelligence collected solely as a consequence of the availability of this capability is beyond calculation. His intense dedication to advancing the Agency's ability to collect intelligence clandestinely has significantly contributed to the Agency's mission.

JOHN N. MCMAHON

Service Years: 1951–1986

Starting at the bottom rung of the Agency career ladder, Mr. McMahon had the distinct honor of holding leadership positions in all four Directorates, on the Intelligence Community Staff, and as Deputy Director of Central Intelligence. McMahon demonstrated extraordinary leadership, managerial savvy, decisiveness, and integrity in his many assignments, and was especially effective in dealing with the Agency's senior customers and overseers. His deep understanding of the people and substance of the intelligence profession shaped the morale of CIA's workforce and the high standards of achievement to which they aspire.

ANTONIO J. MENDEZ

Service Years: 1965–1990

Mr. Mendez is recognized for founding the development and engineering capability in the Agency's operational disguise program. His ideas led to the design and deployment of a series of increasingly sophisticated tools that enabled operations officers to change their appearance convincingly. The application of his skills to one of the Agency's highest-profile and most successful operations earned Mendez an Intelligence Star. His vision and artistic skill had a major impact on the Agency's operational capabilities in hostile environments.

Appendix E

Pseudonyms of CIA Officers Used

George Saxe Operations officer specializing in covert communications in denied areas

Ron Duncan Field technical operations officer

Ken Seacrest Technical officer specializing in denied-area operations

Tom Grant An early technical operations officer specializing in audio

Tom Linn Program manager for battery development

Stan Parker Chemist specializing in power sources chemistry

Martin Lambreth Mechanical and audio specialist

Kurt Beck Senior audio engineer and program manager

Gene Nehring Engineering manager

Greg Ford Special-devices engineering manager

Jack Knight Research and development program manager

Pat Jameson Specialist in unconventional warfare and counterterrorist covert action

Bill Parr Special missions officer

Brian Lipton Secret-writing chemist

John Orkin Electronic and explosive device evaluation engineer

Mark Fairbain Special missions manager

Frank Shumway Explosive ordnance and counterterrorism specialist

Appendix F

Instructions to Decipher the Official Message
from the CIA on page xxv

Step One: copy the numbers in the CIA message onto a large pad of graph paper, leaving approximately one inch betweens lines.

25886 14155 75126 50200

Step Two: Copy the numbers from the *TRIGON* one-time pad beneath each digit.

25886 14155 75126 50200 (message)
24765 93659 55146 09380 (OTP)

Step Three: Subtract the OTP from the "message" using "non-carrying" math.

	25886	14155	75126	50200	(message)
Subtract	24765	93659	55146	09380	(OTP)
	01121	21506	20080	51920	(deciphered)

Step Four: Separate the deciphered message into two digit numbers and convert into letters; A–Z are numbered 01 to 26 (A = 01, B = 02, C = 03, D = 04, etc.). An X (24) is used to separate sentences. The message begins:

0 1, 1 2, 1 2, 1 5, 0 6, 2 0, 0 8, 0 5, 1 9, 2 0 (grouped)
A   L   L   O   F   T   H   E   S   T   (clear text)

Author's Note: For operational use (but not in the example presented here) the first five-digit group at the top of the left-hand column was designated the "indicator group." The first five-digit group received in the OWVL transmission identified the correct page of the agent's OTP for encryption. After one-time use, the entire page would be destroyed by the agent to protect forever the security of the encrypted communication.

If you have deciphered the message successfully, it will be identical to that which appears with the Preface endnotes and you may have the skills needed for a career in the CIA. The authors invite you to learn more at: https://www.cia.gov/careers/index.html.

Glossary

Access agent—A person who facilitates contact with a target individual or entry into a facility.

Accommodation address—An address with no obvious connection to an intelligence agency, used for receiving mail containing sensitive material or information.

Active concealment—A concealment device camouflaged as an everyday item that functions in accordance to its disguise to add an additional layer of security should the device attract scrutiny. Examples of an active concealment would include a ballpoint pen capable of writing, a flashlight that lights, or a can of shaving cream that dispenses cream.

Active measures—Soviet intelligence term for highly aggressive covert action and propaganda campaigns launched against the West in an attempt to influence foreign policy or create domestic unrest. Active measures, including disinformation campaigns, were aimed at Western and Third World countries, the CIA, U.S. military, and the American public.

Agent—An individual, typically a foreign national, who works clandestinely for an intelligence service.

Agent in place—An individual employed by one government while providing secrets to another.

Alias—A false identity used to protect an intelligence officer in the field; it may be as simple as a false identification and phony business cards or as elaborate as an established career substantiated with background details and legitimate documentation.

Al-Qaeda—International radical Islamic organization established by Osama bin Laden in 1988 to unite Muslims to fight the West and expel Westerners and non-Muslims from Muslim countries. Responsible for the September 11, 2001, attack on the United States and other terrorist bombings throughout the world since the early 1990s.

Audio—The capture and recording of private conversations by electronic means. Better known as "bugs," audio devices can either be "hardwired" or transmit a signal via radio frequency, or optically using a laser or infrared light.

Bang and burn—CIA slang for personnel and operations involved with explosives, sabotage, and post-blast damage assessments.

Beacon—A device typically fastened to an object or individual that transmits a radio signal in order to track its location. The technological discipline is called beaconry.

BIGOT list—A list of names of individuals with authorized knowledge of a particularly sensitive intelligence matter.

Brush pass—A brief contact between an agent and case officer during which an exchange of physical material, such as documents, film, money, or other items, occurs.

Car toss—Similar to the brush pass, the car toss allows an agent and handler to clandestinely exchange physical material by tossing it through, or from, the open window of a car.

Case—The official record of an intelligence operation.

Case officer (also operations officer)—An intelligence officer responsible for an agent operation; responsibilities may include recruitment, instruction, as well as those of paymaster and personal advisor.

Central Intelligence Group (CIG)—The CIG, formed in 1946, was the predecessor to the Central Intelligence Agency.

Chief of station (COS)—The officer in charge at a CIA station, usually in a foreign capital.

Clear signal—A radio frequency (RF) transmission that is not masked (disguised) or encrypted.

Communicator—An intelligence officer responsible for maintaining and operating communication devices linking field stations with Headquarters.

Compartmentation—A procedure restricting knowledge of a case or operation to a small number of individuals on a "need to know" basis.

Concealment device—An object modified or fabricated to contain either a device or intelligence materials for purposes of covert storage, transport, placement within a target, or dead-dropping.

Consumer—An organization or individual who uses intelligence as part of the analysis or decision-making process. Consumers of intelligence include the military, State Department, and President.

Counterespionage—Espionage operations undertaken to detect, penetrate, and counteract foreign intelligence services.

Counterintelligence—Operations and analysis that protect information, personnel, equipment, and installations from espionage, sabotage, and terrorism.

Countersurveillance—Techniques used to detect and/or counteract hostile surveillance.

Cover—An organizational affiliation used by a person, organization, or installation to prevent identification with an intelligence service.

Covert communication (covcom)—Any technique or device used to relay data clandestinely from case officer to agent or agent to case officer.

Cryptonym—A deceptive code name assigned to an operation, individual, or case. CIA cryptonyms often included two-letter prefixes for internal identification.

Dangle—An individual or situation used as a lure to attract and identify intelligence officers of a hostile service.

DCI—Director of Central Intelligence.

DCIA—Director of the Central Intelligence Agency.

DDO—Deputy Director for Operations of CIA.

DDP—Deputy Director for Plans of CIA.

Dead drop—A method of communication between an agent and handler in which materials or devices are left unsecured in a preselected location, typically in some form of concealment. A dead drop may either be hidden from view or placed in the open in some manner that either blends in with the immediate surroundings and/or discourages close inspection by passersby.

Denied area—A term used by the CIA to describe a country or geographic region where clandestine operations are very difficult.

Diplomatic cover—An intelligence officer identified and accredited as a member of a nation's diplomatic corps.

Directorate of Operations (DO)—The CIA component responsible for conducting HUMINT espionage operations abroad. Renamed the National Clandestine Service (NCS) in 2005.

Directorate of Science and Technology (DS&T)—The CIA component responsible for applying technology and technical expertise to intelligence requirements.

Division 19—A research and development unit under the National Defense Research Committee (NDRC) that supported OSS technical requirements during World War II.

DNI—Director of National Intelligence. The position of DNI was established in 2005 for coordinating the activities of all U.S. intelligence agencies.

Doc copy (document copy)—A device or procedure related to the clandestine copying of documents with intelligence value. Doc copy devices can include small cameras, digital scanners, or media-copying procedures.

Drum's Bible—The 1951 report prepared by Colonel James H. ("Trapper") Drum, head of the Agency's Operational Aids Division, which formed the foundation for a centralized technical support organization.

Family jewels—A compilation of the most sensitive CIA operations. The phrase was first used by DCI Allen Dulles to describe his written notes on operations in the field during World War II with OSS. The term was revived in the 1970s during the U.S. Senate Church Committee investigation and applied to the compilation of CIA's activities during the 1950s and 1960s.

First Chief Directorate—KGB organization during the Cold War responsible for foreign intelligence collection. In 1994 it was renamed the *Sluzhba Vneshney Razvedki*, or Russian Foreign Intelligence Service (SVR).

Flap—An incident caused by a failed operation or error; includes both diplomatic and "public relations" repercussions resulting from an arrest of a case officer or agent and can cause internal friction and conflicts with other services.

Flaps and seals—The clandestine opening, reading, and resealing of either envelopes or packages without the recipient's knowledge.

Foreign finds—A captured clandestine device of a hostile intelligence service.

FSB (*Federalnaya Sluzhba Bezopasnosti*)—Federal Security Service of the Russian Federation responsible for internal security and counterintelligence.

GRU (*Glavnoye Razvedyvatelnoye Upravlenie* or Chief Intelligence Directorate of the General Staff)—Soviet, and later Russian, military intelligence service.

Handler—The individual, usually a case officer, who controls and directs an agent.

Hard target—A person or place assiduously guarded against espionage efforts.

High-value target—An individual or installation possessing particularly important intelligence.

HUMINT (human intelligence)—Intelligence either collected or relayed by an individual.

IED (Improvised Explosive Device)—An explosive device fashioned from disparate components and most often used in terrorism or guerilla warfare.

Illegal—A term used for Soviet and Russian intelligence officers operating abroad without benefit of "diplomatic cover." Illegals pose as legitimate residents of the target country and are protected only by a strong cover.

Intelligence requirement—Information requested by an intelligence consumer, such as the Pentagon or President.

KGB (*Komitet Gosudarstvennoy Bezopasnosti* or Committee for State Security)—The Cold War name for the Soviet Union's primary security and intelligence apparatus.

L-pill—A suicide pill or tablet issued to agents who preferred death to interrogation. Although popular in spy novels and movies, L-pills were rarely used operationally.

Legend—A carefully constructed cover for an intelligence officer.

Lemon squeezer—OTS slang for a specialist in secret writing. The term is derived from one of the oldest forms of secret writing that used lemon juice, which disappears on paper when dry and then reemerges when exposed to heat.

Listening post—A secure site at which signals from an audio operation are monitored and/or recorded. The individual charged with maintaining the site is called a keeper.

Lubyanka—KGB Second Chief Directorate Headquarters, located in the center of Moscow during the Cold War.

Main enemy—Soviet designation of the United States during the Cold War.

MI5—The British intelligence service responsible for internal security. It is comparable in some ways to the FBI, but its officers do not have the power to make an arrest.

MI6—The British foreign intelligence service also known as the Secret Intelligence Service (SIS). It is similar to the CIA.

Maryland Research Laboratory (MRL)—A wartime research laboratory run by Division 19 of the National Development and Research Committee and located on the grounds on the Congressional Country Club just outside of Washington, D.C.

Mic and wire—Tech slang for an audio operation that uses hardwiring rather than radio-frequency transmission to carry the signal from the microphone to the listening post.

Microdot—An optical reduction of a photographic negative to a size that is illegible without magnification, usually 1mm or smaller in area.

Moscow rules—The distilled wisdom for conducting clandestine operations in Moscow during the Cold War.

National Defense Research Committee (NDRC)—a civilian agency created to explore new weapons for what seemed to be America's inevitable entry into World War II.

NIACT (night action)—An indicator used in CIA messages or cables requiring an immediate response regardless of the time of receipt.

NKVD (*Narodnyy Komisariat Vnutrennikh Del* or The People's Commissariat for Internal Affairs)—Forerunner of the KGB from 1934 until 1946.

Non-official cover—A CIA officer operating under a cover that has no connection to the United States government.

One-time pad (OTP)—Groups of random numbers or letters arranged in columns, used for encoding and decoding messages. Since the codes are used only once, a properly employed OTP is theoretically unbreakable.

One-way voice-link (OWVL)—A broadcast over a shortwave radio frequency containing a ciphered message to an agent.

OPC (Office of Policy Coordination)—CIA element responsible for covert action until 1952. See OSO.

Open source—Any publicly available information.

OSO (Office of Special Operations)—CIA element responsible for clandestine collection of intelligence. OSO combined with the Office of Policy Coordination in August 1952 to form the Directorate of Plans.

OSS (Office of Strategic Services)—America's World War II intelligence organization.

OTS (Office of Technical Service)—The office in the DS&T that provides technical support to clandestine operations.

Overhead platform—Satellite or airplane capable of gathering intelligence through either interception of electronic signals or photography.

Paper mill—A counterfeiting operation specializing in creating realistic-appearing "intelligence documents" in post–World War II Europe. Paper mills were responsible for the dissemination of countless false or misleading pieces of intelligence.

Persona non grata—Latin for "unwelcome person." Intelligence officers working under diplomatic cover who were caught in the act of espionage were declared persona non grata and ordered to leave the country.

Pocket litter—Commonly carried items, such as credit cards, driver's licenses, receipts, and matchbooks that contribute to establishing a cover or legend.

Recruitment—The process of enlisting a potential agent to spy.

Rezidentura—The Russian intelligence station inside a foreign mission.

Roll-up—The capture of an agent or intelligence officer that shuts down an operation.

Safe house—A location used for clandestine meetings and assumed temporarily safe.

Second Chief Directorate—KGB organization responsible for internal security and counterintelligence operations.

Secret writing (SW)—Use of special inks or chemically treated "carbon papers" to produce a hidden message.

Seventh Chief Directorate—KGB organization during the Cold War responsible for surveillance.

Short-range agent communication (SRAC)—A device that allows agent and officer to communicate clandestinely over a limited distance.

Signal site—A covert means of communication using a nonalerting signal, such as a chalk mark on a lamppost, to either initiate or terminate a clandestine act.

Signals intelligence (SIGINT)—Intelligence gathered from the interception of either electronic emissions or transmissions.

Silent call—An operational signal in which the agent or intelligence officer places a call from an anonymous phone and then hangs up after a predetermined amount of time without speaking.

SIS (Secret Intelligence Service)—See MI6.

Special Operations Executive (SOE)—A British Special Forces organization charged with sabotage operations and support of underground forces during World War II.

Station—A forward-deployed operational office of the CIA.

Station chief—See Chief of station.

Surveillance detection run/route (SDR)—A planned route taken by an agent or handler prior to conducting a clandestine act in hostile territory, designed to identify or elude surveillance.

Target—A location, thing, facility, organization, or person against which an intelligence or counterintelligence operation is directed.

Technical Operations Officer (TOO)—A DS&T officer providing direct technical support to clandestine operations.

Temporary duty (TDY)—A field assignment of short duration.

Tradecraft—The techniques, technology, and methodologies used in covert intelligence operations. Tradecraft applies to both the procedures, such as surveillance detection routes, as well as the use of devices in covert audio surveillance and agent communications.

Tray rocker—OTS slang for one working in clandestine photography. The name derives from the way in which prints were developed in trays of chemicals.

TSCM—Technical Surveillance Countermeasures.

TSD—Technical Services Division (1960–1973), predecessor to OTS.
TSS—Technical Services Staff (1951–1960), predecessor to OTS.
Walk-in—A volunteer who approaches an intelligence agency for the purpose of espionage.
Wood block—An audio device concealed within a block of wood that can then be attached to a piece of furniture or an architectural detail within the targeted room.

SELECTED INTELLIGENCE ACRONYMS

AA	Accommodation address
COS	Chief of station
DCI	Director of Central Intelligence
DCIA	Director of Central Intelligence Agency
DDO	Deputy Director for Operations
DDP	Deputy Director for Plans
DDS&T	Deputy Director for Science and Technology
DNI	Director of National Intelligence
FSB	*Federalnaya Sluzhba Bezopasnosti* (Federal Security Service of the Russian Federation)
GRU	*Glavnoye Razvedyvatelnoye Upravleine* (Chief Intelligence Directorate of the General Staff)
HUMINT	Human intelligence
IED	Improvised explosive device
KGB	*Komitet Gosudarstvennoy Bezopasnosti* (Committee for State Security)
LP	Listening Post
MRL	Maryland Research Laboratory
NDRC	National Defense Research Committee
NIACT	Night action
NKVD	*Narodnyy Komisariat Vnutrennikh Del* (The People's Commissariat for Internal Affairs)
NOC	Non-official cover
OSO	Office of Special Operations
OSS	Office of Strategic Services
OTP	One-time pad
OTS	Office of Technical Service
OWVL	One-way voice link
SDR	Surveillance detection run/route
SIGINT	Signals intelligence
SIS	Secret Intelligence Service
SOE	Special Operations Executive
SRAC	Short-range agent communication
SW	Secret writing

TDY	Temporary duty
TOO	Technical Operations Officer
TSCM	Technical Surveillance Countermeasures
TSD	Technical Services Division
TSS	Technical Services Staff

Notes

PREFACE

1 CIA's senior ranks were reshuffled in 1995. When John Deutsch moved from Deputy Secretary of Defense to DCI, he brought with him Nora Slatkin as the Executive Director. Dave Cohen, previously Associate Deputy Director for Intelligence, became DDO and Dr. Ruth David from Sandia National Laboratories became DDS&T. George Tenet was appointed CIA Deputy Director in July 1995.

2 These components and functions are presented in detail in the unclassified OTS fiftieth anniversary booklet, "The Central Intelligence Agency's Office of Technical Service, 1951–2001" by Benjamin B. Fischer, 2001.

3 "Budget weenie" is not an authorized CIA occupational title. The more formal designations are budget analyst, financial officer, or resource manager.

4 All of these capabilities were critical to agent operations and covert action programs but all were strapped for resources until 1999. OTS power-sources scientists solicited support from other government agencies to save the battery program. Programs such as the effort to understand and counteract the use of "spy dust" by the Soviets to track CIA officers were closed. Consideration was given in 1994 to closing, due to cost, the OTS counterterrorism training and explosive test range. Other OTS programs, such as disguise, were reduced to survival status and development of new tracking systems was limited to a handful of projects. In 1997, the Executive Director intervened to increase OTS's covcom budget but the first significant new resources in a decade did not come until a 1999 supplemental congressional appropriation funded CIA's intensified counterterrorism efforts.

5 William Hood, *Mole* (New York: W. W. Norton & Company, 1952), 11–16.

6 Charles E. Lathrop, *The Literary Spy* (New Haven, Connecticut: Yale University Press, 2004), 279.

7 Richard Helms, *A Look over My Shoulder—A Life in the Central Intelligence Agency* (New York: Random House, 2003), vii.

Deciphered message from page xxv: All of the statements of fact, opinion, or analysis expressed are those of the authors and do not reflect the official positions or views of the CIA or any other U.S. Goverment [sic] agency. Nothing in the contents should be construed as asserting or implying U.S. government authentication of information or Agency endorsement of the authors' views. This material has been reviewed by the CIA to prevent disclosure of classified information.

INTRODUCTION

1 Allen Dulles, *The Craft of Intelligence* (Guilford, CT: Lyons Press), 51.
2 "Robo/Nano Spies and More," http://www.defensemedianetwork.com/stories/robonano-spies-and-more/.
3 Ibid.
4 John Keegan, *Intelligence in War: Knowledge of the Enemy from Napoleon to al Qaeda* (New York: Alfred A. Knopf, 2003), 24.
5 See: http://www.whatsabyte.com/.
6 See: http://www.geeks.com/techtips/2006/techtips-26nov06.htm.
7 See: http://www.minduploading.org/.
8 See: http://www.elon.edu/e-web/predictions/150/1930.xhtml.
9 See: http://www.makeuseof.com/tag/8-spectacularly-wrong-predictions-computers-internet/.

CHAPTER ONE – MY HAIR STOOD ON END

1 The Medal of Honor, often called the Congressional Medal of Honor, is the highest award for valor in action against an enemy force that can be bestowed on an individual serving in the U.S. Armed Forces.
2 Donovan was initially appointed as director of the OSS's predecessor, the COI (Coordinator of Information) on July 11, 1941. COI's name was changed to OSS on June 13, 1942. See: http://www.cia.gov/cia/publications/oss/art03.htm.
3 Stanley P. Lovell, *Of Spies & Stratagems* (Englewood Cliffs, New Jersey: Prentice Hall, Inc., 1963), 21.
4 Ibid., 17
5 Ibid.
6 Ibid, 21.
7 A little over a year later, FDR created the OSRD, which took over weapons research and created another group, Division 19, within the NDRC. Benjamin B. Fischer, *The Journal of Intelligence History* (Nuremberg, Germany, Vol. 2, Number 1, Summer 2002), 16.
8 Ibid., 21.
9 Ibid.
10 After accepting the position, Lovell wore two hats; he retained his position in the NDRC while heading R&D for OSS.
11 Fredric Boyce and Douglas Everett, *SOE: The Scientific Secrets* (Phoenix Mill, England: Sutton Publishing Limited, 2003), 5–6.
12 Joseph Persico, *Roosevelt's Secret War* (New York: Random House, 2001), 114.
13 Ibid., 187.
14 Anthony Cave Brown, *Wild Bill Donovan: The Last Hero* (New York: Times Books, 1982), 301.

15 Ibid.

16 Norman Polmar and Thomas B. Allen, *Spy Book: The Encyclopedia of Espionage* (New York: Random House, 1998), 408.

17 Corey Ford, *Donovan of the OSS* (Boston: Little, Brown & Company, 1970), 135–136.

18 Ibid.

19 Bradley F. Smith, *The Shadow Warriors: O.S.S. and the Origins of the C.I.A.* (New York: Basic Books, 1983), 171.

20 Brown, *Wild Bill Donovan,* 236.

21 Ibid., 185.

22 Michael Warner, *The Office of Strategic Services: America's First Intelligence Agency* (Washington, D.C.: Central Intelligence Agency, 2000), 8.

23 Boyce and Everett, *SOE,* Appendix A.

24 Lovell, *Of Spies & Stratagems,* 22.

25 For an illustration and details see: H. Keith Melton, *OSS Special Weapons & Equipment: Spy Devices of WWII* (New York: Sterling, 1991), 95.

26 Lovell, *Of Spies & Stratagems,* 42. For images and a description of the *Firefly* see: Melton, *OSS Special Weapons & Equipment,* 85, and Donald B. McLean, *The Plumber's Kitchen: The Secret Story of American Spy Weapons* (Wickenburg, Arizona: Normount Technical Publications, 1975), 167–171.

27 For images and a description of the *Limpet* see: Melton, *OSS Special Weapons & Equipment,* 58–59, and McLean, *The Plumber's Kitchen,* 229–232.

28 Melton, *OSS Special Weapons & Equipment,* 97.

29 Ford, *Donovan of the OSS,* 170; William Stevenson, *A Man Called Intrepid* (New York: Harcourt Brace Jovanovich, 1976), 114; Stuart Macrae, *Winston Churchill's Toyshop* (New York: Walker and Company, 1972), 7–11.

30 For illustrations and details see: Melton, *OSS Special Weapons & Equipment,* 67–68, and McLean, *The Plumber's Kitchen,* 183–186.

31 Ibid, Melton, 65–66, and McLean, 81–105.

32 For images and a description of the kit for disguising *Explosive Coal,* see: Melton, *OSS Special Weapons & Equipment,* 70–71.

33 McLean, *The Plumber's Kitchen,* 137–142.

34 H. Keith Melton, *The Ultimate Spy Book* (New York: DK Publishing, 1996), 32; Ivan V. Hogg, *The New Illustrated Encyclopedia of Firearms* (Secaucus, New Jersey: Wellfleet Press, 1992), 220.

35 For images and a description of the *Liberator* see: Melton, *OSS Special Weapons & Equipment,* 34–35.

36 Woolworth was a popular "five and dime" store during World War II.

37 Lovell, *Of Spies & Stratagems,* 40. In reference to the special pistol, Lovell recounted a colorful incident in which Donovan fired the weapon in the White House in the presence of President Roosevelt to demonstrate its flashless and silent characteristics. When Gary Powers was shot down in the U-2 aircraft over the USSR on May 1, 1960, he was armed with an OSS .22 caliber silenced pistol.

38 For images and a description of the *Stinger* see: Melton, *OSS Special Weapons & Equipment,* 29.

39 For images and a description of the *Matchbox Camera* see: Melton, *OSS Special Weapons & Equipment,* 103–104.

40 Joseph E. Persico, *Piercing the Reich* (New York: Viking, 1979), 26–31.

41 Ibid, 28.

42 Warner, *The Office of Strategic Services,* 33.

43 Lovell, *Of Spies & Stratagems,* 56–57. For images and a description of *Who Me?* see: Melton, *OSS Special Weapons & Equipment,* 83. The *Who Me?* formula was a mineral oil solution that used as its active ingredients skatol (from baby diarrhea), *n*-butyric acid, *n*-valeric acid, and *n*-caproic acid. Also see: McLean, *The Plumber's Kitchen,* 177–178.

44 Jack Couffer, *Bat Bomb: World War II's Other Secret Weapon* (Austin, Texas: University of Texas Press, 1992), 4–7.

45 McLean, *The Plumber's Kitchen,* 62.

46 Couffer, *Bat Bomb,* 113–120.

47 McLean, *The Plumber's Kitchen,* 62.

48 Ibid.

49 Ibid, 61–63.

50 Lovell, *Of Spies & Stratagems,* 84–85.

51 Ibid.

52 Brown, *Wild Bill Donovan,* 745.

53 David Bruce, Memo to General William Donovan, May 8, 1943. Declassified records of the OSS, MORI ID # 24190.

54 Persico, *Roosevelt's Secret War,* 337.

55 Lovell, *Of Spies & Stratagems,* 86. Lovell refers to this, or a similar project, as "Campbell."

56 Center for the Study of Intelligence, *Office of Strategic Services 60th Anniversary Special Edition* (Washington, D.C.: Central Intelligence Agency, June 2002), XI.

57 Ibid., 11.

CHAPTER TWO – WE MUST BE RUTHLESS

1 Ford, *Donovan of the OSS,* 302.

2 Ibid., 303.

3 Thomas F. Troy, *Donovan and the CIA: A History of the Establishment of the Central Intelligence Agency* (Frederick, Maryland: University Publications of America, 1981), 282.

4 Ford, *Donovan of the OSS,* 314.

5 Donovan had hoped to retain OSS or a civilian intelligence service based on the OSS structure after the war. Indeed, he had submitted a proposal to President Roosevelt for such a service, presumably with the intention of heading it. Initially, Truman rejected the idea of a follow-on service, claiming that it would turn into an "American Gestapo." One measure of the change of perception about the threat posed by the USSR in the immediate aftermath of WWII was the sudden shift in Truman's thinking during 1946.

6 Ford, *Donovan of the OSS,* 312.

7 Ibid., 314.

8 Persico, *Roosevelt's Secret War,* 448

9 Peter Grose, *Gentleman Spy: The Life of Allen Dulles* (New York: Houghton Mifflin, 1994), 273.

10 Ibid., 11.

11 Ibid.

12 OSO had responsibility for foreign intelligence collection, counterintelligence, covert action, and technical support. OPC had responsibility to conduct paramilitary and psychological operations.

13 Grose, *Gentleman Spy,* 13.

14 Ibid.

15 Benjamin B. Fischer, "The Central Intelligence Agency's Office of Technical Service, 1951–2001" (Washington, D.C.: Central Intelligence Agency: Center for the Study of Intelligence, 2001), 13.

16 The name TSS was born from bureaucratic infighting. Components in the DDP that ran operations carried the designation "division." Drum's initial proposal for a technical "division" met with stern objection from the other division chiefs because this would imply an operational rather than a support role for technical services. "Staff" became the acceptable alternative. Fortunately for TSS, Drum's first suggestion for a name, the Material Assistance and Development Office, was also rejected. Inevitably, the staff would have been called the "MAD" techs.

CHAPTER THREE – THE PENKOVSKY ERA

1 The message originated in Moscow. A CIA officer had first written the text in longhand, then, using a one-time pad, he converted the text into what appeared to be a series of random letters. These were given to the local communicator and the coded message was fed through an electronic encryption machine before being transmitted to Langley. Because the message was first enciphered by hand and then by machine, the term "superencipherment" encompassed the full process.

2 Ronald Kessler, *Inside the CIA* (New York: Pocket Books, 1992), 178.

3 See Robert Louis Benson and Michael Warner (editors), *VENONA: Soviet Espionage and the American Response, 1939–1957* (Washington, D.C.: National Security Agency and Central Intelligence Agency, 1996).

4 Jerold L. Schecter and Peter S. Deriabin, *The Spy Who Saved the World* (New York: Charles Scribner's Sons, 1992), 348.

5 The operation ran from April 1961 until August 1962. For a concise description, see: Polmar and Allen, *Spy Book,* 490–493.

6 Schecter and Deriabin, *The Spy Who Saved the World,* 92–93.

7 Ibid., 411.

8 Ibid., 340.

9 Ibid., 337.

10 Ibid., 262.

11 Ibid., 413.

12 Jacob had run a surveillance detection route (SDR) that made a circuitous route through Moscow culminating at a bookstore that he entered through one door and exited from another. Called "dry cleaning" at the time of the operation, the term has been replaced with the less colorful term SDR. Fifteen years later CIA officers were equipped with hidden earpieces to monitor the transmissions of KGB surveillance teams, but Jacob had no such advantage.

13 Schecter and Deriabin, *The Spy Who Saved the World,* 307.

14 Ibid., 394.

15 Ibid., 301.

16 Ibid., 365–366.

17 Viktor Suvorov, *Aquarium: The Career and Defection of a Soviet Military Spy* (London: Hamish Hamilton, 1985), 1–4.

18 Schecter and Deriabin, *The Spy Who Saved the World,* 377.

19 Ibid., 95, 351.

20 David C. Martin, *Wilderness of Mirrors* (New York: Harper & Row, 1980), 90.

21 Christopher Andrew and Vasili Mitrokhin, *The Sword and the Shield: The Mitrokhin Archive and the Secret History of the KGB* (New York: Basic Books. 1999), 182.

22 Schecter and Deriabin, *The Spy Who Saved the World,* 159.

23 Ibid., 248.

24 Ibid., 280.

25 Ibid., 330.

26 Ibid., 29.

27 Ibid., 334.

CHAPTER FOUR – BEYOND PENKOVSKY

1 I Samuel 19:18–42 relates the story of how Jonathan, son of Israel's King Saul, communicated covertly to David through shooting of arrows to specific locations.

2 Schecter and Deriabin, *The Spy Who Saved the World,* 320–321.

3 Ibid., 394.

4 Ibid., 184

5 The Model IIIs was an improvement over the original Minox, which had been made of stainless steel throughout World War II. This new, postwar version of the classic spy camera was light since it was made of aluminum and featured a better lens. The suffix "s" indicated that the camera could be used for synchronized flash, though in the world of espionage this feature was seldom employed.

6 The operational file on each staff member contained the available data that could be gleaned from previous assignments abroad and with information from the KGB's network of Soviet nationals who regularly reported their suspicions to their contacts about their American colleagues. Information about the newly arriving diplomat's age, marital status, hobbies, education, and official position created the KGB's "profile of the individual." It was only those activities that were "out of profile" that would result in special attention from the "watchers." For example, a newly arrived midlevel employee seen having lunch regularly with more senior staff might draw interest. The Soviets working at the U.S. Embassy had been screened and approved by the KGB. Many spoke excellent English (sometimes without even a trace of an accent), and did the bulk of the actual work for the Embassy when dealing with other Soviets who wanted to apply for a travel visa or to emigrate to the United States. They also were "fixers" that enabled the Embassy to cut through the Byzantine Soviet bureaucracy, as well as cooks, drivers, cleaning staff, gardeners, and even building maintenance personnel. They were seen as so essential to the smooth running of the Embassy that they became ubiquitous. Too frequently their nationality was forgotten and they were treated as "friends and colleagues" by many of the American personnel stationed at the Embassy despite awareness that they reported regularly to the KGB. Even seemingly innocuous patterns of who sat next to each other in the cafeteria, or whose wives were chatting closely at official functions, were considered as they attempted to unmask officers.

7 The Second Directorate is responsible for internal security.

8 Andrew and Mitrokhin, *The Sword and the Shield*, 185.

9 Ibid.

10 At Soviet embassies throughout the world, the spouses and dependents of diplomatic and intelligence personnel filled the required administrative and support jobs.

11 Ronald Kessler, *Moscow Station* (New York: Charles Scribner's Sons, 1989), 68, 106.

12 As a result of his exposure, the officer received thorough annual medical checkups in the following years. No physical harm from the radiation was ever identified.

13 Allen Dulles, *The Craft of Intelligence* (New York: Harper & Row, 1963), 192.

14 Fifteen years later, in 1970, after most of CIA had relocated to the Langley Headquarters, East and South Buildings were occupied by TSD. The tech found himself assigned to the office formerly occupied by the DCI. He removed one of the acoustic tiles above where he imagined the DCI's desk might have sat. Hanging by a now unconnected wire was a single DD-4 microphone, apparently missed when the recording system was dismantled.

15 Grose, *Gentleman Spy,* 308.

16 This was not the only unusual assignment the young officer would receive. Dulles also sent him to Princeton University's Institute for Advanced Studies in an attempt to secure the services of Albert Einstein for the agency.

17 Martin, *Wilderness of Mirrors,* 61–62.

18 Throughout the late 1950s, several TSS engineers worked on or were assigned to the large technology programs such as the U-2, but the programs were neither managed nor owned by TSS.

19 Jeffrey T. Richelson, *A Century of Spies: Intelligence in the Twentieth Century* (New York: Oxford University Press, 1995), 25.

20 Grose, *Gentleman Spy,* 391.

21 Schecter and Deriabin, *The Spy Who Saved the World,* 101.

22 Ibid.

23 Helms, *A Look Over My Shoulder,* 105. Compared to satellites and spy planes, agents were far less costly. During his time as an agent for the Americans in the 1950s, Colonel Pyotr Popov was paid an estimated $4,000 a year (approximately $25,000 adjusted for inflation) and provided intelligence on GRU and KGB operations in both Europe and the United States. A single satellite would pay the salaries of ten thousand agents like Popov.

24 In 1962, the formation of the Directorate of Research consolidated high-altitude reconnaissance and satellite programs along with CIA-sponsored research in the new directorate. However, TSD and its technical support to operations responsibility remained under the Directorate of Plans.

CHAPTER FIVE – BRING IN THE ENGINEERS

1 Charles E. Lathrop, *The Literary Spy* (New Haven, Connecticut: Yale University Press, 2004) 339.

2 The TOO's referred to this as "the tech culture." Mention is also made by Grose, *Gentleman Spy,* 389.

3 Grose, *Gentleman Spy,* 155–156.

4 When used in reference to TSD and OTS, "research and development" or "R&D" means "applied research and development." The term "development and engineering" is usually more descriptive of the type of work done by TSD and OTS. The TSD/OTS-sponsored R&D focused on technologies and development processes that would lead to production of a product, device, or capability that could be used in clandestine operations. TSD/OTS R&D programs aimed toward a two- to five-year payoff—the shorter the better.

5 Undated CIA brochure, "Directorate of Science & Technology: People and Intelligence in the Service of Freedom," page 3. The CIA's Directorate of Research, established in 1962, preceded the DS&T by one year.

6 TSD would be part of the operations directorate until 1973. In a CIA reorganization that year, TSD was moved to the Directorate of Science and Technology and renamed the Office of Technical Service (OTS).

7 *Cambridge Dictionary of Science and Technology* (New York: Cambridge University Press, 1990), 632.

8 Sergo A. Mikoyan, "Eroding the Soviet 'Culture of Secrecy,' *Studies in Intelligence*, No. 11, Central Intelligence Agency, 2001.

CHAPTER SIX – BUILDING BETTER GADGETS

1 Benjamin Weiser, *A Secret Life: The Polish Officer, His Covert Mission, and the Price He Paid to Save His Country* (New York: Public Affairs, 2004). 221–223.

2 Within ten years, TSD had developed several electronic short-range agent communication (SRAC) devices from this idea.

3 An accommodation address is most commonly a street or post office box address not associated with an intelligence service or a government agency.

4 The process, while complex for the novice, was well-known within the photographic industry. "Stripping film," was a commercial product. Intelligence services made direct contact prints on the emulsion as part of a covert microphotography process. "Bleaching" the completed emulsion was a step in standard microdot concealments that the KGB refined in the early years of the Cold War. If an agent had the necessary technical aptitude to perform the procedures, affixing "bleached" emulsion of varying sizes onto postcards became a common technique for covert communications.

5 The instructions provide an exemplar of the detail and complexity necessary to provide technical training to an agent through impersonal communications. The exemplar does not convey the precise methodology or actual text of the operational message.

CHAPTER SEVEN – MOVING THROUGH THE GAP

1 In 1963, during McCone's tenure as DCI, the Directorate of Research was renamed the Directorate of Science and Technology.

2 The DDP had been redesignated as the Directorate of Operations (DO) in 1973.

3 Despite the abrupt change, McMahon's engaging personality, and leadership skill earned the lasting respect of OTS during his fourteen-month tenure as

Director. McMahon served as Honorary Chairman of the OTS fiftieth anniversary committee in 2001. However, for many senior DO and OTS officers, Schlesinger's separating TSD from the operations directorate was viewed as a historic mistake.

4 Colby was sworn in as DCI in September. He served the CIA in both the Korean and Vietnam conflicts. President Reagan later tapped a fourth OSS veteran, William Casey, as DCI in 1981.

5 Seymour Hersh, *The New York Times*, December 22, 1974. The mail-opening program's crypt was HGLINGUAL.

6 OTS's role in the CIA mail-opening program is described in chapter 15.

7 U.S. House of Representatives, Special Subcommittee on Intelligence of the Committee on Armed Services, 1974; U.S. Senate, Select Committee to Study Governmental Operations with Respect to Intelligence Activities, 1975, 1976. U.S. Senate Select Committee on Intelligence and the Subcommittee on Health and Scientific Research of the Committee on Human Resources, 1977; Rockefeller Commission Report to the President on CIA Activities within the U.S., 1975.

8 "Family jewels" was an ironic nod to Allen Dulles who used the term to refer to a personal notebook during WWII that contained the names of his most important agents.

9 For pictures and technical details of the "Dart Gun," see: Melton, *CIA Special Weapons & Equipment*, 22.

10 Helms, *A Look over My Shoulder*, 431.

11 CCDs had also been among the technologies that allowed satellites to image targets and transmit the "pictures" back to earth in real time. The first of these satellites was the KH-11, launched in 1976. Before the KH-11, film capsules of photographs taken by satellite cameras were jettisoned from the satellite and parachuted to earth.

CHAPTER EIGHT – THE PEN IS MIGHTIER THAN THE SWORD (AND SHIELD)

1 Milton Bearden and James Risen, *The Main Enemy: The Inside Story of the CIA's Final Showdown with the KGB* (New York: Random House, 2003), 37.

2 Various aids were improvised by the agent to assist in ensuring the proper distance between lens and document as well as centering the camera. Knitting needles and threads of the proper length could be used as reference points. While these could be used if the agent was assured of privacy while photographing, they could be alerting and the objective would be to provide the agent sufficient training and confidence to operate the camera without any other aids. Over time, with improved lens design, the focusing tolerance expanded.

3 In OTS jargon, the pen was known as an "active concealment" because the concealment functioned as the product it represented, in this case a writing instrument.

4 Comments by policy officials on intelligence are usually offered in informal exchanges with officers who present the information and passed to senior Agency managers. The fact that the feedback reached a working-level case officer, like Saxe, was unusual and indicative of the significance of the information.

5 The L-pill, concealed in a pen identical to the one that housed the agent's subminiature camera, was passed to *TRIGON* during a single hourlong clandestine meeting with a CIA officer in Moscow in 1976. After *TRIGON's* arrest and suicide, the KGB produced the concealment pen said to have contained the L-pill.

6 Richelson, Jeffrey T., *A Century of Spies: Intelligence in the Twentieth Century* (New York: Oxford University Press, 1995), 343.

7 Kalvar, a commercial product developed as an alternative to traditional microfilm, represented a commercial product that OTS could apply to reduced-image photography. The original company ceased operations in 1979 but other firms continued making the product. For operational use, Kalvar had the advantage that it could be handled and processed in normal room light and developed in boiling water without requiring special chemicals.

8 Wording is based on a translation of the purported note on display in the FSB museum in Moscow.

9 Igor Peretrukhin, *Agent Code Name—TRIANON (Agenturnaya Klichka— TRIANON)* (Moscow: Tsetrpoligraf, 2000), 217–218.

10 Polmar and Allen, *Spy Book,* 362.

CHAPTER NINE – FIRE IN THE ARCTIC

1 Bearden and Risen, *The Main Enemy,* 10–11.

2 The original stump is on display inside the FSB museum in Moscow. A replica is displayed inside the International Spy Museum in Washington, D.C.

3 The TOO approach to providing technical support to operations officers contrasted with the OSS model. In 1947, OSS printed a catalog of available spy gear that could be ordered as needed. Following World War II and prior to the TOO program, TSS and TSD operated primarily as a supply and on-call service. If sustained technical support was needed for an operation, a TSS officer would be sent on TDY for that specific purpose. TOO represented a different philosophy that said a properly cross-trained officer with technical aptitude could make an ongoing contribution to the full range of an office's operations. The forward-deployed TOO could provide expertise in his principal technical area and working-level technical support from his cross-training in other areas. Further, he would have immediate and direct access to the experts in every OTS discipline when those were required. The TOO became integral to operational planning and execution in locations of their assignments.

4 The first SRR-100 models were approximately $3/4 \times 2\frac{1}{2} \times 3\frac{1}{2}$ inches. The key design dimension was thickness because the unit would likely be worn in a shirt pocket or the inside pocket of a man's suit coat.

5 The basic technology to intercept surveillance transmissions was widely understood. Soviets stationed in the United States had once used Bearcat scanners purchased at the local Radio Shack to pick up FBI transmissions. Decades later, some KGB technicians would still exhibit a genuine fondness for the stores and merchandise, praising the reliability and quality of the products such as batteries, wires, and other parts they had been able to obtain.

6 Victor Sheymov, *Tower of Secrets: A Real Life Spy Thriller* (Annapolis, Maryland: Naval Institute Press, 1993).

7 Barry G. Royden, "Tolkachev, A Worthy Successor to Penkovsky," *Studies in Intelligence,* Vol 47:3, Central Intelligence Agency, 2003, 12.

CHAPTER TEN – A DISSIDENT AT HEART

1 Driven by the character of Q, Major Boothroyd, who was not in the Bond novels, but added to the movies, spy gadgets became an instant hit. Q's equipment was impressive on the movie screen, often defying the laws of physics. A more realistic portrayal of spy gadgetry was seen in *Mission: Impossible*, where every device had to be based upon technology available at the time. For more information see: Danny Biederman, *The Incredible World of SPY-FI* (San Francisco: Chronicle Books, 2004).

2 Case officers were not encouraged by the fact that a great many of the elements required in technical collection violated some of the basic tenets of denied area tradecraft. For instance, installation of a technical device could involve an officer remaining at the target site for extended periods of time. In virtually every other operational procedure in denied areas, dead drops, signals, brush passes, car tosses, and the expanding arsenal of SRAC technology was aimed at minimizing the time of the operational act. Technical operations, such as installing a tap or emplacing a sensor, could require longer times at the target site to attach, adjust, and test the system.

3 Bearden and Risen, *The Main Enemy,* 194.

4 Polmar and Allen, *Spy Book,* 508–510; Richelson, *A Century of Spies,* 422. Polyakov is also known under the FBI codename *TOPHAT* and CIA codename *BOURBON*.

5 Polmar and Allen, *Spy Book,* 509. A "brush pass" allows for an imperceptible exchange of a small "package" such as a message or film cassette, to take place between two people as they "brush" past each other in a public area. There is no outward sign of recognition between the parties involved.

6 Communications between agents and case officers divides into two general systems known as "agent send" and "agent receive." Due to the difficulty of concealing spy gear, agents were limited in the type of covert communications (covcom) equipment they could possess. As a result, agents had fewer options in sending messages—secret writing, in limited cases microdots, and dead drops. Options for agent receive systems included those plus OWVL, microprinting, and "blind" newspaper placements. Satellites introduced the option of long-range electronic covcom and BUSTER introduced short-range electronic covcom options.

7 A "mole" is a serving intelligence officer who is secretly working for another intelligence service.

8 David Wise, *Spy: The Inside Story of How the FBI's Robert Hanssen Betrayed America* (New York: Random House, 2002), 20–24.

9 Ibid., 193.

10 Ibid., 193–194. In a 2008 interview, Polyakov's son stated the execution occurred in 1988.

11 Polmar and Allen, *Spy Book,* 509.

12 Royden, "Tolkachev, A Worthy Successor to Penkovsky," *Studies in Intelligence,* 47:3, Central Intelligence Agency, 2003, 5.

13 Ibid., 12.

14 Schecter and Debriabin, *The Spy Who Saved the World,* 5, 25, 28.

15 "Dangles" appear to be legitimate volunteers, but are actually being controlled by another intelligence service.

16 The prepared text was often in the form of a personal letter containing information designed to avoid attracting unwanted attention and scrutiny from Soviet postal censors. The content was written in another person's handwriting so even if censorship detected secret writing, the handwriting would not incriminate the agent.

17 Soviet postal censors used the word "perlustration" for the examination of mail to detect secret writing and microdots. Soviet censors detected secret writing by swabbing across the item with a "cocktail" of chemical reagents designed to expose the hidden content. Examination of the item after it was received by OTS would detect traces of the swabbing. For CIA use, secret writing systems developed by OTS were tested against such "cocktails" before being approved for operational use.

18 Royden, "Tolkachev, A Worthy Successor to Penkovsky," 10.

19 The location in which the car was parked, direction it was facing, and other simple variables could be used as the signal.

20 Personal meetings in a denied area are inherently dangerous for the agent and avoided if at all possible. Though the agent may be unknown to counterintelligence, the case officer is always subject to being surveilled and may unknowingly lead surveillance to the meeting. Dead drops are a form of "impersonal communication" in which the agent and handler are separated by time, but not space. Personal meetings do, however, provide the handler with the important advantage of assessing the agent's mental state and verifying that operational instructions are understood.

21 Information has more value when the adversary does not realize that it has been "lost." As such, secretly copying documents is almost always preferred to taking the original document.

22 The KGB and other intelligence services recommended Minox cameras for their agents. U.S. Navy Warrant Officer John Walker, a mole for the KGB, was trained in the use of a Minox Model-C camera for "doc copy" during a trip to Vienna in the late 1960s. His technical instructions are still valid: use B/W Plus-X Pan film (ASA 125), shutter speed at 1/100th, distance to document eighteen inches, and even illumination with a 75–100 watt bulb placed at a 45-degree angle to the document.

23 In 1938, the original "Riga" Minox camera could be concealed in a man's closed fist. Postwar Minox models (II and III) were just as small, but often necessitated the use of a separate light meter which also had to be concealed. In 1958 Minox incorporated an internal light meter for the first time into the slightly larger Model-B. Though still a "pocketable" camera, the Minox "B" and later models would continue to add features and size. In 1981 Minox introduced its smallest and lightest camera series, the "EC," but its fixed-focus lens (three feet to infinity) was unusable for document photography. Regardless of the Minox camera being used, they were not designed for covert use and the act of "doc copy" would be obvious to anyone observing the user.

24 Tolkachev was instructed to be home from 6:00 PM till 8:00 PM on the evening of the date that corresponded to the number of the month; 1 January, 2 February, 3

March, 4 April, etc., and "cover" (stand by to answer) his phone. The call would be disguised as a "wrong number" wherein the caller would ask for one of three names. Each name was linked to a prearranged dead drop site: OLGA, ANNA, or NINA. If the caller asker for VALERIY it would trigger a personal meeting at a prearranged location exactly one hour following the call. Each month, on a date that equaled the number of the month plus fifteen—21 June, 22 July, 23 August—Tolkachev was further instructed to be at a prearranged site at a specific time and wait for five minutes. If his regular handler did not meet him, he was provided with a "parole" (a recognition signal and password) to authenticate the identity of the person sent to meet him.

25 An additional advantage of using the unmodified commercially available Pentax ME camera was that it was not a piece of "tradecraft" equipment and there was a plausible explanation for him to have it in his apartment. Conversely, possession of a noncommercial subminiature "doc copy" camera was proof of espionage.

26 Tolkachev was also provided with parked-car signals (PCS) which would confirm receipt of a transmission by the direction in which his car was parked. The CIA also parked cars on routes frequented by Tolkachev in a similar prearranged PCS to signal to the agent.

27 Tolkachev used the updated OTS-provided demodulator to receive the ciphered message. At the predetermined time and date a ten-minute-long transmission would take place that could include both real and dummy messages. To keep the KGB guessing about the messages the airwaves were often filled with dummy messages; only the real agent would know the date, time, and frequency for the message intended for him. The newly developed demodulator was connected to the radio and captured the message as it was received. The agent could then later recall it and scroll it across the screen of the demodulator unit. The first three digits of the message contained an indicator that told Tolkachev if the following message was intended for him. If so, he could scroll out the reaming portion of the message, which could be as long as 400 five-letter groups. Tolkachev would then use his OTP to decipher the message. Tolkachev attempted to monitor the IOWL transmission, but was unable to do so because of the lack of privacy in his apartment. Shortwave transmissions usually took place at night when atmospheric conditions provided greater transmission ranges and clearer signals, but this also conflicted with the times his family was in the apartment. As a result, subsequent transmissions were moved to the daytime hours when Tolkachev could arrange to be home. Unfortunately his institute's change in security regulations eliminated trips away from the office during working hours and in December of 1982 Tolkachev returned all of his IOWL equipment to his handler.

28 Royden, "Tolkachev, A Worthy Successor to Penkovsky," 27.

29 Unbeknownst to Tolkachev, the plan was modeled after the CIA's first successful exfiltration of an agent from the USSR three years earlier. Victor Sheymov, a Soviet communications security specialist, and his wife and young daughter were hidden in the back of a van and secretly transported from a site near Leningrad to freedom in May 1980. The daring escape story is told in *Tower of Secrets* by Victor Sheymov. Then, an almost identical exfiltration plan was used in the summer of 1985 by MI6 (British Intelligence) to rescue their agent, KGB Col. Oleg Gordievsky, from inside the USSR. (see: Gordievsky, Oleg, *Next Stop Execution: The Autobiography of Oleg Gordievsky* ([London: Macmillan, 1985]).)

30 Royden, "Tolkachev, A Worthy Successor to Penkovsky," 33.
31 Royden, "Tolkachev, A Worthy Successor to Penkovsky," 5.

CHAPTER ELEVEN – AN OPERATION CALLED CKTAW

1 Pronounced "See Kay Taw," this operational code name would have no meaning to anyone who had not been briefed on the activity. CK initials referred to the Soviet/East European Division that ran the operation.

2 Jeffrey T. Richelson, *The Wizards of Langley: Inside the CIA's Directorate of Science and Technology* (Boulder, Colorado: Westview Press, 2001), 239.

3 Ibid., 29.

4 Bearden and Risen, *The Main Enemy*, 28.

5 The specialized antennas and broad-spectrum radio monitors used from a listening post continuously searched the airwaves for three reasons: (1) to gain positive intelligence, (2) to monitor police and counterintelligence frequencies to identify levels of surveillance activity, and (3) to spot transmissions that might indicate the presence of hidden listening devices transmitting from within the Embassy. Once a signal of interest was spotted on the cathode ray display, every effort was made to locate and identify the source and purpose of the transmission. If the signal had intelligence value it would be tagged and recorded, otherwise the monitoring equipment ignored it and continued its search for new and unrecognized signals.

6 Bearden and Risen, *The Main Enemy*, 28.

7 Early satellites captured images on film that was jettisoned and recovered as it parachuted to earth. The film was then flown to a facility to be processed and analyzed. Depending on where the film was recovered, the process from satellite to analysis could take a week or longer. Real-time satellites, however, capture images digitally and then transmit them to ground stations where they are relayed back to intelligence headquarters for immediate analysis.

8 OTS designed a new type of "secure room" that improved the confidence of the CIA that their operational discussions were protected from KGB eavesdropping. The special room, including chairs and tables, was constructed entirely of clear plastic to expose any electronic listening devices, or "bugs." In theory, it was comparable to the fictional "cone of silence" from the 1960s television show *Get Smart*.

9 Edward Lee Howard was one of the officers who trained on the mock-up site at "The Farm."

10 Time in the manhole became an important consideration for every entry. Sufficient time had to be allocated to do the necessary work, but longer times meant greater risk.

11 See: Polmar and Allen, *Spy Book,* 529. The threat of "tagging" was a genuine concern. The KGB's infamously aggressive program code-named METKA, used a variety of covert tracking substances and techniques, the best-known was dubbed "spy dust." Discovered in the early 1970s, the use of spy dust was made public in the mid-1980s by the U.S. Ambassador to the USSR. The chemical substance, when placed on door handles in cars or on other common objects, allowed the KGB to track those who touched the compound. When analyzed, the mysterious substance was found to be nitrophenyl pentadien (NPPD) and luminol.

12 The capabilities of the counterintelligence services in countries covered by SE

Division (USSR, East Germany, and the other Soviet Bloc countries in Eastern Europe) posed increased risks for Agency operations. Case officers required additional training for that environment.

13 An SDR is a route of travel to the place where an operation will occur, including stops and varied modes of transportation and is selected to reveal surveillance to the case officer without him having to appear to be looking. The security of Moscow operations demanded that the case officer make an absolute determination that he was free of surveillance before conducting the clandestine act.

14 Choke points referred to locations where vehicles or pedestrians are required to merge as they move from one to another area such as the only bridge connecting two sections of a city across a river. Anyone going from one to the other section must cross the bridge. Surveillance teams establish positions at choke points knowing that their target will eventually be compelled to pass through.

15 A "near field" receiver used a specially detuned antenna to ignore any transmission other than those very close to the receiver.

16 KGB surveillance teams often communicated nonverbally using a series of clicks that were created by "keying their microphones" with the radio control unit carried in their pants or jacket pocket. With this technique they avoided possible detection that might happen if seen speaking into the microphone sewn under the lapel of their surveillance clothing.

17 According to a former member of the Seventh Directorate, surveillance teams were known as *Naruzhnoye Nablyudeniye* or the "NNs". Two different team configurations were employed depending on the target. For routine surveillance the team consisted of six officers, a team leader, and three cars. For special targets and suspected CIA officers the team was increased to eight officers, a team leader, and three cars. The additional officers were added in case the target was seen in contact with an unknown individual whereby they would detach from the main team and continue to follow the unidentified suspect.

18 The Russian clothing constituted a "light disguise" that affected external changes in appearance such as style and color of clothing and shoes, hats, wigs, beards and moustaches, eyeglasses, walking canes and heel lifts that could be adopted quickly. Light disguises were primarily most effective at a distance.

19 Counterintelligence services are usually more interested in identifying the spy than the case officer. Arresting a foreign intelligence officer is less important to the KGB than the opportunity to identify a possible traitor.

20 The history of the Penkovsky case is detailed by Schecter and Deriabin in *The Spy Who Saved the World*.

21 Because one never knew what products might appear in the market from day to day, Soviet women carried an empty bag thinking "perhaps" scarce items would be available.

22 One Moscow chief would not allow officers to use surveillance receivers during their first months in-country. He wanted the officer's observation and detection skills tested and proven lest the technology become a substitute for awareness and intuitive judgment.

23 During Operation GOLD (Berlin, 1955–56) the KGB had protected their underground communications lines by placing them inside airtight cables that had been pressurized with nitrogen gas. Any penetration of the cable would lower the pressure and alert the KGB communication technicians. To overcome this KGB

safeguard, the CIA constructed a "tapping chamber" around an underground section of the cable that was pressurized before the cable was opened and the taps placed on the lines. Because the pressure inside the "tapping chamber" was the same as that inside the cable, the alarm did not sound.

24 H. Keith Melton, *CIA Special Weapons and Equipment: Spy Devices of the Cold War* (New York: Sterling Publishing, 1993), 37. A standard 35mm camera loaded with Kodak high-speed-infrared 2481 film and utilizing a flash unit fitted with an infrared filter over the lens (Kodak Wratten gelatin filters nos. 87, 87C, 88A, or 89B) allowed photographs to be taken in complete darkness without betraying the use of the flash.

25 The remote interrogation allowed the CIA to transmit a signal to a transceiver built into CKTAW. It would then automatically reply with a signal indicating the operational status of the unit. The "tamper indicated" signal was sent if the CKTAW device had been tampered with, or compromised. If the CIA officer received this signal (or no signal) after the device was "interrogated," the operation would be aborted.

26 Bearden and Risen, *The Main Enemy*, 29.

27 Ibid., 30.

28 During an interview with coauthor Keith Melton in 1997 in Moscow, Vitaly Yurchenko stated that his formal rank was that of a naval Commander, not Colonel.

29 Ronald Kessler, *Escape from the CIA: How the CIA Won and Lost the Most Important KGB Spy Ever to Defect to the U.S.* (New York: Pocket Books, 1991), 45.

30 Ibid., 47.

31 David Wise, *The Spy Who Got Away: The Inside Story of Edward Lee Howard, the CIA Agent Who Betrayed His Country's Secrets and Escaped to Moscow* (New York: Random House, 1988), 19.

32 Bearden and Risen, *The Main Enemy*, 83–84.

33 Wise, *The Spy Who Got Away*, 40.

34 Bearden and Risen, *The Main Enemy*, 83–85.

35 Wise, *The Spy Who Got Away*, 59.

36 Ibid., 59–60.

37 Bearden and Risen, *The Main Enemy*, 86.

38 Ibid.

39 Ibid.

40 Wise, *The Spy Who Got Away*, 137.

41 Ibid., 138–139.

42 Kessler, *Escape from the CIA*, 184.

43 Wise, *The Spy Who Got Away*, 186–187.

44 Ibid.,188.

45 Ibid., 113.

46 Ibid., 192.

47 Though polygraph examinations are not admissible in court, the FBI uses them routinely as an investigation tool and a way for a suspect to "prove his innocence." Howard, however, had a history of failing polygraph examinations and never considered submitting to the testing.

48 Wise, *The Spy Who Got Away*, 62.

49 Ibid., 199.

50 Ibid., 204.

51 Ibid., 204–205.

52 Bearden and Risen, *The Main Enemy,* 115.

53 Wise, *The Spy Who Got Away,* 213.

54 Bearden and Risen, *The Main Enemy,* 83–85.

55 Ibid., 84.

56 The CIA later learned that Howard had met Soviet intelligence officers during the fall of 1984 and again during the Spring of 1985. See: http://www.nacic.gov/pubs/misc/screen_backgrounds/spy_bios/edward_howard_bio.html.

57 In May 1985, Aldrich H. Ames, a CIA counterintelligence officer, began spying for the USSR and also revealed the CKTAW operation.

58 Nikolai Brusnitsyn, *Openness and Espionage* (Moscow: Military Publishing House, 1990). Soviet officials gave copies of the article to members of the U.S. delegation the Strategic Arms Limitations Talks in Geneva.

59 Ibid., 32.

60 Krassilnikov, *The Phantoms of Tchaikovsky Street Prizraki c Ulitsy Chaykovskogo* (*The Phantoms of Tchaikovsky Street*) (Moscow: GEYA Iterum, 1999).

61 Ibid., 179–187.

62 Ibid.

CHAPTER TWELVE – COLD BEER, CHEAP HOTELS, AND A VOLTMETER

1 "What happens in the field stays in the field" was often repeated and applied, except when the stories were too good not to be retold at Headquarters. These stories became part of OTS's culture and lore.

2 TDY is a government acronym for "temporary duty" and refers to assignments, usually less than 180 days, away from an employee's home area.

3 "Tech hotels" were not exclusively the culture of OTS. Officers from the Office of Communications, who managed the Agency's communications networks around the world, earned a similar reputation for knowing where to find cheap rooms.

4 See: U.S. Department of State web page: http://moscow.usembassy.gov/embassy/embassy.php?record_id = spaso.

5 Ibid.

6 The common listening devices of the time were phone taps or microphones hidden in ceilings and walls and hardwired to a manned listening post.

7 Sanche de Gramont, *The Secret War: The Story of Espionage since World War II* (New York: Putnam, 1962), 411. See also U.S. Department of State Web site: http://moscow.usembassy.gov/embassy.

8 George F. Keenan, *Memoirs: 1925–1950* (New York: Pantheon, 1967), 189.

9 At the time, audio operations that were hardwired were state-of-the-art. In most cases they required only that a microphone be planted, with wires leading away to a nearby listening post. In this way, the listeners were assured of secure lines and a steady power source from the post or the target building's own power. The unit could be turned off and on at will. It was only later, when technology had developed sufficiently to provide small, reliable transmitters, that "wireless" audio operations came into being.

10 See: http://www.cnn.com/SPECIALS/cold.war/episodes/01/interviews/beria/ for

excerpts of an interview with Sergo Beria, the son of Lavrenty Beria (chief of the NKVD—the Soviet secret police) who participated in the eavesdropping operations at Tehran and Yalta.

11 Ibid.

12 Gary Kern, "How 'Uncle Joe' Bugged FDR," *Studies in Intelligence,* Vol. 47:1, 2003, 19–31.

13 Peter Wright, *Spycatcher* (New York: Viking, 1987), 20.

14 Ibid.

15 Peter Wright, *The Spycatcher's Encyclopedia of Espionage* (Port Melbourne, Victoria: William Heinemann Australia, 1991), 238. and *Spycatcher* (New York: Dell, 1987), 26, 28–29.

16 See: Melton, *Ultimate Spy* (New York: DK, 2002), 104, for a diagram of "the Thing," photos, and a description of its operation.

17 Wright, *Spycatcher,* 78–79.

18 Ibid.

19 Wright, *The Spycatcher's Encyclopedia of Espionage* (Port Melbourne, Victoria, Australia: William Heinemann Australia, 1991), 212–213.

20 Albert Glinsky, *Theremin: Ether Music and Espionage* (Urbana and Chicago: University of Illinois Press, 2000), 273.

21 The developer, Soviet scientist Léon Sergeyevich Termen, first caused the filament inside an incandescent light bulb to resonate as a microphone in 1943 before perfecting the eavesdropping device inside the carved wooden seal in 1945. In 1947 Termen subsequently developed a system to eavesdrop on foreign embassies in Moscow using infrared light beams aimed at "points of architectural resonance" such as windowpanes. For this accomplishment he was awarded the Stalin Prize, 1st Class, the equivalent at the time to the Nobel and Pulitzer Prizes combined. For more on Termen see: Glinsky, *Theremin.*

22 Glinsky, *Theremin,* 263–264.

23 "Finds" are systems, components, and devices made and used by a foreign (non-U.S.) intelligence service for clandestine operations and usually returned to the United States for examination and analysis. These include any spy gear such as communications, surveillance, and forgery equipment as well as special weapons and improvised explosives.

24 For examples of OSS equipment see: Melton, *OSS Special Weapons & Equipment.*

25 Michael Riordan and Lillian Hoddeson, *Crystal Fire* (New York: W. W. Norton, 1997), 211–212.

26 "Contractors" is a generic term that describes private sector persons or companies that provide goods and services to the CIA, including OTS.

27 Audio would be a priority of OTS until 1996 when the audio program was moved to the Clandestine Information Technology Office (CITO). CITO existed until 2000 when most of its functions were absorbed by the Information Operations Center that drew staff from both the DS&T and the DO.

28 OTS development or procurement programs usually had a nondescriptive name that served two purposes. First, all of the program's contracts and financial activities would carry the designation to assure controls, tracking, and audit functions could be performed. Secondly, the names offered a layer of security and compartmentation when programs were being discussed. EARWORT would

mean nothing to one not briefed about the name and the activity. To ask, "What does EARWORT mean?" would convey that the individual had not been granted access to the program.

29 "Listening post" refers to the location, usually a safe house near the location under surveillance, where the covert audio feed is received, recorded and initially evaluated. Listening posts were commonly staffed by speakers of the targets' native language and equipped with headphones, amplifiers, and recorders that would capture the audio of operational relevance. The best of the "transcribers" or "monitors" could also provide cultural and emotional interpretation of the conversations they heard.

30 All OTS audio operations required submission of a formal "survey" before the operation could commence. The survey included detailed information about the target, purpose, planned operational activities, equipment to be used, and anticipated risk level.

31 In the 1980s, many advertisements would no longer be "blind" and the CIA would be identified as the employer. For an example see: Melton, *CIA Special Weapons and Equipment*; 45.

32 Despite the many engineering and scientific achievements of TSD and OTS, the "tinkerer" reputation followed the techs. In 1996, more than three decades after Scoville's remark and twenty-three years after OTS had become an office in the Directorate of Science and Technology, the Deputy Director of Operations referred to OTS as "my blue-collar guys" at a DS&T senior staff meeting. The Deputy Director for Science and Technology, who was also present, offered no objection to the characterization.

33 Something that was "jerry-rigged" meant that it had been cobbled together quickly, usually from available parts; such solutions were often intended for use only in the short term.

34 Amtorg offices had long provided cover for Soviet intelligence officers. See: William R. Corson and Robert T. Crowley, *The New KGB: The Engine of Soviet Power* (New York: Quill, 1986), 296, and Andrew and Mitrokhin, *The Sword and the Shield,* 186–187. When Robert Hanssen sought to sell secrets to the GRU in 1979–80 he did so through the NYC office of Amtorg. See David Wise, *Spy: The Inside Story of How the FBI's Robert Hanssen Betrayed America* (New York: Random House, 2002), 21.

35 A "front" company is or appears to be a legitimate firm whose visible image has no association with an intelligence service. However, the company supports or serves some clandestine activities. Front companies are commonly used by every intelligence organization.

CHAPTER THIRTEEN – PROGRESS IN A NEW ERA

1 See: http://www.militaryradio.com/spyradio/tsd.html for a picture of the early transmitters, named SRT for "surveillance radio transmitter."

2 "Sweeping" an environment with detectors for indicators of electronic, RF, and magnetic signatures of microphones, recorders, and transmitters is done by specialists known as TSCM (technical systems countermeasures teams). By later adding a remote on/off switch operated at the listening post, the transmitter could be turned off when a TSCM team was about to search the room. Once the

transmitter quit sending the signal, detection was much more difficult. Later, remote on/off would also be used as a means of putting the transmitter in "sleep mode" to save battery power. The longer the batteries operated the less frequently entries into the target would be necessary to replace them.

3 See: Melton, *CIA Special Weapons and Equipment*, 67.

4 Helms, *A Look Over My Shoulder,* 128.

5 For details and photos of "bugs" in the SRT series see Pete McCollum's clandestine communications website: http://www.militaryradio.com/spyradio/tsd.html

6 Critical characteristics for radio frequency audio transmitters intended for covert use include (1) reliability, primarily a function of the device's design, components, and power supply; (2) concealability, primarily a function of device size and configuration; and (3) detectability, primarily a function of the intended or unintended signals generated and the materials from which the device is constructed. The same three characteristics are critical for every other component of an audio surveillance system such as the microphones, wires, connectors, batteries, and recording devices. Finally, reliability, concealability, and detectability are standards by which the operational utility of the fully integrated and operating system is judged.

7 Mallory since evolved into the well-known Duracell Company.

8 In addition to powering surveillance devices, batteries were vital to other OTS espionage equipment for covert communications, tracking beacons and signaling devices. Any gadget with electronics required a power source and, in most instances, that meant some type of battery.

CHAPTER FOURTEEN – THE AGE OF BOND ARRIVES

1 Philip Agee, *CIA Diary: Inside the Company* (New York: Stonehill, 1975), cover and end flap. The case held a tracking beacon, not an audio device.

2 The continuing miniaturization of circuits followed "Moore's Law," an observation made in 1965 by Gordon Moore, cofounder of Intel. The "law" observed that the number of transistors per square inch on integrated circuits had doubled every year since the integrated circuit was invented, and Moore predicted that this trend would continue for the foreseeable future.

3 Richelson, *The Wizards of Langley*, presents an organizational history of ORD.

4 Richelson, *The Wizards of Langley,* 147.

5 See: http://www.thesmokinggun.com/archive/ciacats2.html for a redacted TSD memorandum about the Acoustic kitty project.

6 Both OTS and ORD experimented with other unconventional ideas using animals for intelligence collection. Ravens were tested as winged couriers to deposit audio devices on windowsills, though ambient noise made this idea impractical.

7 Melton, *CIA Special Weapons and Equipment*, 62.

8 For another description of the *Backscatter Gauge* in use see: F. W. Rustmann, Jr., *CIA, INC.: Espionage and the Craft of Business Intelligence* (Washington, D.C.: Brassey's, 2002) 62.

9 Antonio J. Mendez in *The Master of Disguise: My Secret Life in the CIA* (New York: Morrow, 1999), provides an account of his career as an OTS operational disguise specialist.

10 George Gardner, *Picks, Clicks, Flaps and Seals: A Monograph on Surreptitious Entry* (unpublished monograph), 1944, 5. This rare manual was the primary "surreptitious entry" manual used by the OSS and later by TSD. "George Gardner" is most likely the nom de plume for Willis George, the senior OSS "entry expert" and postwar author of *Surreptitious Entry: The Sensational Story of a Government Agent Who Picked Locks and Cracked Safes in the Service of His Country* (New York: Appleton-Century, 1946).

11 Ibid., 5.

12 The lock-picking course included a "final" exam requiring the student to pick open sixty different locks in sixty minutes. "It was a tough course," one tech noted, "I passed only because the generous instructor included several simple suitcase and luggage locks."

13 Because a tech might not have an idea of the types of locks he would encounter inside a target, he would be forced to bring with him as many types of tools as possible, sometimes carried in a small black bag. The FBI historically referred to these as "black bag operations."

14 Images and an operational description of the kit appear in: Melton, *CIA Special Weapons and Equipment*, 75.

15 After a "woods metal" (commercially sold as Cerebun) copy of the key had been made, it was placed in a key-cutting machine to copy the "key cuts" onto a stronger key blank. This key was then available to be used operationally. For images and a description of the kit see: Melton, *CIA Special Weapons and Equipment*, 76.

16 For images and descriptions of TSD lock picking kits, see: Melton, *CIA Special Weapons & Equipment*, 73, and Melton, *Ultimate Spy*, 114–115.

17 Details of HTLINGUAL are presented in the U.S. Senate, *Supplementary Detailed Staff Reports on Intelligence Activities and the Rights of Americans, Final Report, Book III*. 1976.

18 For a detailed description of the TSD "Flaps and Seals" course, see: Mendez, *The Master of Disguise*, 72–76.

19 Gardner, *Picks, Clicks, Flaps and Seals*, 93. The "dry" process involved separating the two sides of the glued flap with an ivory tool, and required more training and practice. The "steaming," which softened the glue with steam to allow it to be opened, was easier, but more hazardous. There was always the possibility that the steam would affect the script, and if the envelope was tinted, the dye could run or change color.

20 Ibid.

21 See: U.S. Senate, *Supplementary Detailed Staff Reports on Intelligence Activities and the Rights of Americans, Final Report, Book III*.

22 Ibid.

23 This decision, the amount of time techs would be in the target facility, went to the heart of operational planning, affecting the activities of both the tech and the case officers. Sometimes this was moot due to the target. A vacant building may offer unrestricted time. An occupied office building usually dictated a "quick plant" job. In other examples, a three-hour installation in an apartment could be accomplished by a ruse that temporarily drew the occupants out for the evening. However, that would also likely mean the audio would be a single bug with no redundancy and left no margin for installation complications or errors. A multiday installation actually

could become more complicated, requiring separate operations to assure vacancy of premises, sustained countersurveillance, and logistical supplies for the audio team. Advantages included a less pressured schedule, opportunity to emplace multiple bugs, time to test system performance, and margin to correct errors.

24 Using more than one microphone and recording each "channel" separately allowed the signals to be individually filtered and amplified to achieve a result comparable to turning one's head to hear a sound or conversation coming from a different part of a noisy room. "Audio steering" was accomplished by increasing or decreasing the amplification of the differing sound channels to focus on specific conversations.

25 The efficiency of a device, component, or system in electronics and electrical engineering is defined as useful power output divided by the total electrical power consumed (a fractional expression).

26 The 1980 Moscow Olympics that were boycotted by the United States ended the operation. Late in the spring of 1980, the Soviet government ordered a general clean up of Moscow prior to the Olympics. The shacks were declared public eyesores and razed. The wood block and transmitter along with the table were buried in an unknown dumpsite.

27 Oleg Kalugin, *The First Directorate: My 32 Years in Intelligence and Espionage* (New York: St. Martin's Press, 1994), 261.

28 Victor Cherkashin, *Spy Handler: Memoir of a KGB Officer* (New York: Basic Books, 2005), 194.

29 Based on interview with former KGB communications security officer living in the West.

30 Ibid.

31 By the 1990s, classical audio operations had become engulfed in a tidal wave of digital information technology.

CHAPTER FIFTEEN – GENIUS IS WHERE YOU FIND IT

1 Dulles, *The Craft of Intelligence,* 68.

2 Fischer, *The Journal of Intelligence History,* 2, Summer 2000, 16.

3 Amy Knight, *Beria: Stalin's First Lieutenant* (Princeton, New Jersey: Princeton University Press, 1993), 138.

4 Ibid.,136, and verified by author interview with former KGB officer.

5 Knight, *Beria,* 106.

6 Simon Sebag Montefiore, *Stalin: The Court of the Red Tsar* (New York. Alfred A. Knopf, 2003), 503.

7 Knight, *Beria,* 136.

8 Interview with former Soviet security officer.

9 John Markoff, James Early obituary, *The New York Times,* January 19, 2004.

10 ARPA: The Advanced Research Projects Agency was founded in February of 1958 as a research branch of the Department of Defense. The name was changed to Defense Advanced Research Projects Agency (DARPA) in 1972. In 1993 the name was changed back to ARPA and then back to DARPA in 1996. The agency is credited with development of the Internet.

CHAPTER SIXTEEN – CONSPICUOUS FORTITUDE, EXEMPLARY COURAGE IN A
CUBAN JAIL

1 Nathan Miller, *Spying for America: The Hidden History of U.S. Intelligence* (New York: Paragon House, 1989), 179.

2 Nathan Nielsen, "Our Men in Havana," *Studies in Intelligence,* Vol 23:1, Central Intelligence Agency, 1988, 1.

3 "Pocket litter" includes all secondary and incidental items that individuals normally carry in wallets and purses. Some pieces of pocket litter, such as library cards, credit cards, and blood donation cards, while not constituting official identification papers, are expected to be consistent with passports, driver's licenses or other government-issued identification documents. Pocket litter of a tourist/businessman might include business cards, club membership cards, laundry receipts and movie ticket stubs. This type of pocket litter created by TSD carried the alias name of the user consistent with the alias official identification documents.

4 Nielsen, "Our Men in Havana," 3.

5 Hugh Thomas, *Cuba or the Pursuit of Freedom* (New York: Da Capo Press, 1988), 1,219.

6 Ibid.

7 Ibid., 1,257.

8 Nielsen, "Our Men in Havana," 3.

9 Grose, *Gentleman Spy,* 495–496.

10 Walter E. Szuminski, *Our Man in Havana: TDY Hell* (unpublished monograph), 5. See also National Oceanic & Atmospheric Administration (NOAA) Web site: http://www.noaa.gov/.

11 Ibid., 3.

12 Walter E. Szuminski with Edward Mickolus, *Temporary Duty Hell: Our Man in Cuba's Jails* (an unpublished monograph, 2001), 17.

13 Sound tradecraft required the team to exit the elevator at a different floor than the apartment they would enter to mislead anyone attempting to surveil their movements by watching elevator stops. In addition, all members of the team would have been alert for surveillance during their travel to and from the apartment building.

14 The SRT-3 was the CIA's first all-transistor transmitter receiver. For information on the ST-2A, the predecessor of the SRT-3, see Peter McCollum's web page: http://www.militaryradio.com/spyradio/tsd.html

15 "Clear" signals were not protected by masking or encryption. If intercepted, the signal could be monitored, understood, and traced.

16 "Sweep teams" in the 1960s located "bugs" using special radio receivers to identify the clandestine transmissions. By remotely switching the transmitter off at the first indication that the room might be "swept," the post keeper eliminated the signal that would have betrayed the hidden eavesdropping device.

17 Szuminski, *Our Man in Havana,* 9.

18 Thomas, *Cuba or the Pursuit of Freedom,* 1,295.

19 Nielsen, "Our Men in Havana," 2.

20 Prior to creation of the CIA in 1947, the FBI had responsibility for U.S. intelligence operations in Caribbean and Central and South America countries.

21 Thomas, *Cuba or the Pursuit of Freedom,* 1,460.

22 Ibid., 1,297.

23 Ibid.

24 Ibid., 847.

25 *Cambridge World Gazetteer: A Geographical Dictionary* (New York: Cambridge University Press), 157.

26 *The New York Times*, September 2, 1925.

27 Mary Bosworth (editor), *Encyclopedia of Prisons and Correctional Facilities*, Volume 2 (Thousand Oaks, California: Sage Publications, 2005), 663–665.

28 *Illustrated London News*, February 13, 1932.

29 Szuminski, *Our Man in Havana: TDY in Hell*, 36–38.

30 Grose, *Gentleman Spy,* 519.

31 Ibid.

32 Polmar and Allen, *Spy Book,* 163–164.

33 A mixture of gasoline or alcohol and soap that is poured into a bottle, tightly corked, with a cloth fuse wrapped around the outside. The cloth was ignited and the bottle thrown. Upon impact with the target, the bottle breaks and the gasoline ignites.

34 Polmar and Allen, *Spy Book,* 166.

35 Ibid., 167.

36 Craig R. Whitley, *Spy Trade: The Darkest Secrets of the Cold War* (New York: Times Books, 1994), 54–55; Brown, *Wild Bill Donovan,* 579.

37 Whitley, *Spy Trade,* 432. Also see: James B. Donovan, *Strangers on a Bridge: The Case of Colonel Abel* (New York: Atheneum, 1964).

38 Szuminski with Mickolus, *Temporary Duty in Hell: Our Man in Cuba's Jails,* 86–87.

39 Christ's recommendation enumerated hardships endured including months of no sunlight, a year without any correspondence from family, constant communist propaganda as well as living in filthy, disease-ridden conditions. Christ described the two officers' emotional stability, focus on positive opportunities such as studying Spanish, teaching English, lecturing on capitalism and democracy, maintaining personal standards of cleanliness and decency and assisting fellow prisoners who were sick or mentally vulnerable. As a result both left Cuba "35–40 pounds lighter but without mental or emotional aberrations."

40 The DIC recognizes Agency employees who, like the three techs, performed "a voluntary act or acts of exceptional heroism involving the acceptance of existing dangers with conspicuous fortitude and exemplary courage." The CIA has recognized twenty-six employees with the Distinguished Intelligence Cross award as of August 2005.

41 Peter Weyden, *Bay of Pigs: The Untold Story* (New York: Simon & Schuster, 1979), 35.

42 Benson, a paratrooper in World War II, jumped with Marshal Tito partisans behind German lines into Yugoslavia and is credited with assisting in the evacuation of more than 200 downed allied airmen, political escapees, and partisans. He was awarded the Bronze Star for World War II service and then served with the CIA in China and Greece. On April 5, 1962, he received, posthumously, the CIA Intelligence Star for "a voluntary act of courage performed under hazardous conditions."

43 Saxitoxin is a thousand times more deadly than a typical synthetic nerve gas such as sarin; a dose of 0.2 milligrams would be lethal for the average male. For more information on saxitoxin, see the article by Neil Edwards, School of Chemistry, Physics, and Environmental Science at the University of Sussex at Brighton: http://www.chm.bris.ac.uk/motm/stx/saxi.htm. The CIA's remaining inventory

of saxitoxin was provided to the National Institute of Health (NIH) in 1975 on the premise that it could be "extremely valuable for medical research on diseases of the nervous system and for our understanding of how the nervous system normally works." See: Ritchie J. Murdoch, Ph.D., D.Sc., *Yale Medicine*, Fall 1975, and also at: http://www.med.yale.edu/external/pubs/ym_fw0001/archives.htm. A television documentary, *The History Detectives*, aired on June 27, 2005, stated that the needles were produced by Fort Detrick machinist Milton Frank.

44 Martin, *Wilderness of Mirrors,* 120.

45 Ibid., 121.

46 Ibid.

47 Ibid.

48 Ibid., 121–122.

49 U.S. Senate, Select Committee to Study Governmental Operations with Respect to Intelligence Activities, *Alleged Assassination Plots Involving Foreign Leaders, An Interim Report*, November 20, 1975, 80.

50 U.S. Senate, *Alleged Assassination Plots Involving Foreign Leaders,* 71.

51 Ibid., 72.

52 Ibid., 72.

53 Ibid., 72.

54 David Atlee Phillips, *The Night Watch* (New York: Atheneum, 1977), 91.

55 Warren Hinkle and William Turner, *The Fish Is Red: The Story of the Secret War Against Castro* (New York: Harper & Row, 1981), 30–31, and U.S. Senate, *Alleged Assassination Plots Involving Foreign Leaders*, 73.

56 David Wise and Thomas B. Ross, *The Espionage Establishment* (New York: Random House, 1970), 130.

57 U.S. Senate, *Alleged Assassination Plots Involving Foreign Leaders*, Senate Report Number 94-465, November 20, 1975, 85.

58 Ibid., 85–86.

59 Ibid., 88–89.

60 "Silenced" pistols and rifles are never completely "silent." The purpose of the "suppressor" is to reduce the sound when the weapon is fired and make it harder to pinpoint the direction of the shot. Pistols are relatively much easier to suppress than rifles.

61 U.S. Senate, *Alleged Assassination Plots Involving Foreign Leaders*, 90.

62 John Ranelagh, *The Agency: The Rise and Decline of the CIA* (New York. Touchstone. 1987), 210–211, 336–345.

63 Lumumba was eventually ousted from his post in the Congolese government and taken into protective custody by United Nations guards; he escaped and was then captured and executed by his Congolese enemies. See: Martin, *Wilderness of Mirrors,* 124.

64 Ranelagh, *The Agency,* 358.

CHAPTER SEVENTEEN – WAR BY ANY OTHER NAME

1 Stanley Karnow, *Vietnam: A History* (New York: Penguin, 1997), 214.

2 Ibid., 211

3 Civil Air Transport (CAT) was the successor to Claire Chennault's Flying Tigers, and the predecessor to "Air America." In June 2001 the CIA issued a Unit

Citation Award in recognition of all who served with Civil Air Transport and its secret successor, Air America, which ended operations in 1976. See: http://www .air-america.org/newspaper_articles/france_honors_cat.shtml

4 Ranelagh, *The Agency,* 419.

5 Karnow, *Vietnam,* 212.

6 Helms, *A Look Over My Shoulder,* 310.

7 Ibid., 230.

8 Ranelagh, *The Agency,* 419. Ranelagh notes that covert CIA involvement in Vietnam began in 1954 when DCI Allen Dulles sent Colonel Edward Lansdale to Saigon with these objectives.

9 "Covert action" is an operation designed to influence governments, events, organizations, or persons in support of foreign policy in a manner that is not necessarily attributable to the sponsoring power; it may include political, economic, propaganda, or paramilitary activities.

10 John L. Plaster, *SOG: A Photo History of the Secret Wars* (Boulder, Colorado: Paladin Press, 2000), 17–18.

11 John L. Singlaub, *Hazardous Duty* (New York: Summit, 1991), 293.

12 The junk's unexpected speed amazed the crew. Years later, the same OTS engineer raised eyebrows around Mobile, Alabama, when testing a new a high-speed, high-lift vessel in the harbor. People stopped their cars to watch the test craft outpace traffic on the adjacent highway.

13 For details and images of the Metascope see: Melton, *CIA Special Weapons and Equipment,* 36–37.

14 For pictures and a description of an "Anti-disturbance device" see: Melton, *CIA Special Weapons and Equipment,* 88.

15 Plaster, *A Photo History of the Secret Wars,* 217.

16 John L. Plaster, *SOG: The Secret Wars of America's Commandos in Vietnam* (New York: Simon & Schuster, 1977), 22–23.

17 For images and details of the RS-6 see: Melton, *CIA Special Weapons and Equipment,* 47.

18 For images and details on *Dust Powder* (B-3) see: Melton, *CIA Special Weapons and Equipment,* 99.

19 For images and details on *Puppy Chow* see: Melton, *CIA Special Weapons and Equipment,* 115.

20 For images and details on the *Document Copying Attaché Case* see: Melton, *CIA Special Weapons and Equipment,* 43.

21 These were an early version of lightweight, condensed, dehydrated, vacuum-packed foods used by backpackers and mountain climbers.

22 Plaster, *SOG: A Photo History of the Secret Wars,* 17–18, and Melton, *OSS Special Weapons & Equipment,* 36.

23 Francis Gary Powers was piloting a CIA U-2 aircraft when downed over the USSR on May 1, 1960. His survival weapon was an OSS .22 caliber silenced Hi-Standard pistol.

24 See: Melton, *OSS Special Weapons & Equipment,* 34–35.

25 For images and a reproduction of its issue instruction sheet, see: Melton, *CIA Special Weapons and Equipment,* 12–13.

26 For images and details on the *Stinger* see: Melton, *CIA Special Weapons and Equipment,* 14–15.

27 For photos, history, and firing details of the MBA *Gyrojet* pistol see: http://www
 .smallarmsreview.com/pdf/Gyrojettest.PDF. An MBA *Gyrojet* Rocket Carbine
 was shown in the 1967 James Bond film *You Only Live Twice.*
28 Plaster, *SOG: A Photo History of the Secret Wars,* 161.
29 Plaster, *SOG: The Secret Wars of America's Commandos in Vietnam,* 77. See
 also http://www.sfalx.com/moh/sisler_george_SF.html.
30 *Concise Dictionary of World History* (New York: Macmillan Publishing Company. 1983), 593.
31 During the Vietnam War, the CIA proprietary airline Air America flew a variety
 of missions in the Far East. These missions ranged from covert CIA operations
 to overt air transportation contracted by the Republic of Vietnam and various
 U.S. government agencies. At one time it was the largest airline in the world
 based on the number of aircraft it operated.
32 See: http://www.signonsandiego.com/uniontrib/20050224/news_lz1n24france
 .html).
33 For an image and details of the *Rubber Airplane* see: Melton, *CIA Special Weapons and Equipment,* 102.
34 William M. Leary, "Robert Fulton's Skyhook and Operation Coldfeet," *Studies
 in Intelligence,* 38:5 Central Intelligence Agency, 1994, 99–110. Also see: http://
 www.cia.gov/csi/studies/95unclass/Leary.html.
35 Ibid.
36 Interview, Jim Morris, Spring 2005.
37 James B. and Sybil B. Stockdale. *In Love and War* (New York: Bantam, 1985). 135.
38 Ibid., 136.
39 Ibid., 128. In 1966 POWs in North Vietnam were allowed to send and receive a
 single letter each month.
40 Ibid., 140.
41 Ibid., 144.
42 Ibid., 192.
43 Ibid., 194. The clear "decal-like thing" was a thin piece of photographic emulsion
 with reduced writing; it resembled the Kalvar process used a decade later.
44 Ibid., 193–194.
45 Ibid., 196–197.
46 Ibid., 199–200.
47 Ibid., 207–209.
48 Ibid.
49 Ibid., 209–211.
50 *Concise Dictionary of World History* 553.
51 Bob Woodward, *Veil: The Secret Wars of the CIA 1981–1987* (New York: Simon
 & Schuster, 1987), 192.
52 Ibid., 189.
53 Duane R. Clarridge, *A Spy for All Seasons: My Life in the CIA* (New York: Scribner, 1997), 269–270.
54 Ibid., 263.
55 Ibid. The chain gun, whose fire could penetrate the armor of Soviet T-55 tanks,
 was developed by the Army for the Bradley fighting vehicle.
56 The original *Cigarette* was designed by Don Aronow and named after a classic
 boat he once owned that was reputed to have been used by a "bootlegger" during

Prohibition. Bootleggers used high-speed boats to smuggle shipments of whiskey from Canada into the northeastern United States. Even if the boats were spotted by the U.S. Coast Guard, the "rum runners" depended on the speed of their craft to outrun their pursuers. See: "How a Kid From Brooklyn Put Go-Fast Boats on the Map," *Power & Motor Yacht,* July 2000.

57 Clarridge, *A Spy for All Seasons,* 271–272.

58 Ibid., 274.

59 A copy of the text of the Goldwater letter can be found at http://homepage.ntl-world.com/jksonc/docs/US-mining-nicaragua-harbors.html

60 Clarridge, *A Spy for All Seasons,* 277.

CHAPTER EIGHTEEN – CON MEN, FABRICATORS, AND FORGERS

1 The country's name changed from the Central African Republic to the Central African Empire in 1976 and back again in 1979.

2 *Cambridge World Gazetteer: A Geographical Dictionary* (New York: Cambridge University Press, 1988), 122.

3 *The New York Times,* November 5, 1996.

4 *The Concise Columbia Encyclopedia* (New York: Columbia University Press, 1983), 151.

5 *The New York Times,* June 13, 1987.

6 Interview with Dr. David A. Crown, 2005.

7 *The Concise Columbia Encyclopedia,* 151.

8 Crown interview.

9 Dr. David Crown holds a doctorate degree in criminology from University of California—Berkeley and had compiled a distinguished career before joining the CIA. During the 1950s, he served as a U.S. Counterintelligence Corps Special Agent with field experience in Europe and then as Assistant Director of the U.S. Postal Inspection Service Identification Laboratory in San Francisco.

10 *The Liberia Official Gazette,* Vol. L, 2.

11 Crown interview.

12 Ibid.

13 Ibid.

14 *The New York Times,* December 5, 1977.

15 *The New York Times,* June 13, 1987.

16 *Cambridge World Gazetteer,* 122.

17 *The New York Times,* January 13, 1987.

18 *The New York Times,* November 5, 1996.

19 Dr. Robert Managhan, "Trends in African Forgeries," *Studies in Intelligence,* 19:1, 1975, 14.

20 Crown interview.

21 Ibid.

22 Ibid.

23 Managhan, "Trends in African Forgeries," 14.

24 Crown interview.

25 *The Liberia Official Gazette.*

26 Helms, *A Look over My Shoulder,* 93.

27 Ibid., 95.

28 Ibid., 99. Rosters containing the names of the fabricators, known as "burn lists," were circulated among the Allied intelligence services as a means of limiting the damage caused by the perpetrators and maintaining integrity of intelligence information. Fifty years later, similar mechanisms, called "watch lists," were created to identify terrorists, while unverifiable, but seeming plausible "hoax" data manufactured by unknown sources permeated the Internet.

29 Ibid.,110.

30 Managhan, *Trends in African Forgeries,* 13.

31 Peter Deriabin, *Watchdogs of Terror* (New Rochelle, NY: Arlington House, 1972), 94.

32 James Ridgeway, *Blood in the Face* (New York: Thunder's Mouth Press, 1990), 30, 32.

33 Ibid., 32.

34 U.S. House of Representatives, Hearing Before the Subcommittee on Oversight of the Permanent Select Committee on Intelligence, House of Representatives, Ninety-Sixth Congress (February 19, 1980), 6.

35 U.S. Senate, Hearing Before the Subcommittee to Investigate the Administration of the Internal Security Act and Other Internal Security Laws of the Committee on the Judiciary United States Senate (June 2, 1961), 6.

36 U.S. House of Representatives, Hearing Before the Subcommittee on Oversight of the Permanent Select Committee on Intelligence, House of Representatives, Ninety-Sixth Congress (February 19, 1980), 65.

37 U.S. Senate, Hearing Before the Subcommittee to Investigate the Administration of the Internal Security Act and Other Internal Security Laws of the Committee on the Judiciary United States Senate (June 2, 1961), 6.

38 Ibid., 22.

39 Ibid., 18.

40 Kalugin, *The First Directorate,* 137.

41 David A. Crown, "Political Forgeries in the Middle East," *Studies in Intelligence,* 22:2, Central Intelligence Agency, 1978, 10.

42 Ibid.

43 Ibid., 9.

44 Managhan, *Trends in African Forgeries,* 14.

45 Ibid., 12.

46 Ibid., 11.

47 Ibid.

48 Ibid., 211.

49 U.S. House of Representatives, Hearing Before the Subcommittee on Oversight of the Permanent Select Committee on Intelligence, House of Representatives, Ninety-Sixth Congress (February 19, 1980), 69.

50 *The Journal of Intelligence History* (1:1, 2001), 62.

51 Andrew and Mitrokhin, *The Sword and the Shield,* 238.

52 Ibid., 224.

53 Ibid.

54 David A. Spetrino, "Aids Disinformation," *Studies in Intelligence,* 32:4, Central Intelligence Agency, 1988, 10.

55 Ibid., 11.

56 Ibid.

57 Ibid., 9.

58 Ibid., 12.

59 *The Washington Post*, January 25, 2005.

60 Ibid.

61 Markus Wolf, *Man Without a Face* (New York: Public Affairs, 1997), 289.

62 Ibid.

63 Ibid., 290. Wolf writes, "We left [Sudan] in 1971 and never returned."

64 Crown interview.

65 Soon after the assassinations, a variety of credible reports surfaced that linked the late PLO Chairman, Yassir Arafat, directly to the killings of Noel and Moore. However, the U.S. Department of Justice concluded in 1986 that it lacked the evidence to bring an indictment against Arafat and argued further that if such evidence existed the potential for compromise of national security information would likely preclude it from being disclosed. See: David A. Korn, *Assassination in Khartoum* (Bloomington, Indiana: Indiana University Press, 1993), 245–247.

66 National Commission on Terrorist Attacks upon the United States, *The 9/11 Commission Report* (New York: W.W. Norton, 2004). "Monograph on 9/11 and Terrorist Travel," Chapter 3, 1.

67 Ibid., 22.

68 Crown interview.

69 National Commission on Terrorist Attacks upon the United States, "Monograph on 9/11 and Terrorist Travel," Chapter 3, 1.

70 "Redbook 1986" booklet published by OTS.

71 National Commission on Terrorist Attacks upon the United States, "Monograph on 9/11 and Terrorist Travel," Chapter 3, 1.

72 Ibid.

73 Ibid., 12.

74 Ibid., 22.

CHAPTER NINETEEN – TRACKING TERRORIST SNAKES

1 Walter Laqueur, *The New Terrorism: Fanaticism and the Arms of Mass Destruction* (New York: Oxford University Press, 1999), 3–5.

2 Ibid., 11.

3 *Concise Dictionary of World History,* 336.

4 Jessica Stern, *The Ultimate Terrorists* (Cambridge, Massachusetts: Harvard University Press, 1999), 6.

5 Ibid., 13.

6 Ammonium nitrate and diesel fuel combine to produce an explosive of the type used in the 1993 attack against the World Trade Center and again in April 1995 against the Alfred P. Murrah Federal Building in Oklahoma City.

CHAPTER TWENTY – ASSESSMENT

1 The importance and difficulty of selecting the "right" people for intelligence missions was one of the important lessons the CIA learned from OSS. The OSS recruitment experience was compiled and published by the OSS Assessment

Staff in *Assessment of Men: Selection of Personnel for the Office of Strategic Services* (New York: Rinehart & Company, 1948).

2 Jerrold M. Post, "The Anatomy of Treason," *Studies in Intelligence*, 19:2 Central Intelligence Agency, 1975, 37.

3 Ibid., 36.

4 OSS found this to be an immediate problem. Since recruiters for OSS were not allowed to name the organization for which the individual would be working and did not know in what capacity the recruit would be working, the "pitch" talked about "mysterious, exciting overseas assignments." This attracted "the bored, the pathologically adventuresome, the neurotically inclined to danger, and psychopaths." The latter have a particular ability to make good short-term impressions. Structured assessment attempted to identify and weed out those who would be a danger to themselves, others, and the mission. See: Donald W. MacKinnon, "The OSS Assessment Program," *Studies in Intelligence*, 23:3, Central Intelligence Agency, 1979, 22–23.

5 David Wise, *Nightmover: How Aldrich Ames Sold the CIA to the KGB for 4.6 million* (New York: HarperCollins, 1995), 114.

6 Golytsin, a KGB major in the First Chief Directorate, defected in December 1961. Nosenko, a Soviet security officer, defected in 1964. Both had access to sensitive counterintelligence information about worldwide Soviet operations. They offered explosive and contradictory information particularly surrounding the KGB's relationship with Lee Harvey Oswald and the assassination of President Kennedy. See: Ranelagh, *The Agency*, 404–409, 563–568, and Martin, *Wilderness of Mirrors*, 151–158, 173–176, for detailed accounts of the two cases.

7 See: J. F. Winne and J. W. Gittinger, "An Introduction to the Personality Assessment System," *Journal of Clinical Psychology*. Monograph Supplement No. 38, April 1973. The PAS as used by OTS was an adaptation of the Wechsler Adult Intelligence Scale that had been developed by psychologist David Wechsler. Measurements along the PAS scale were designed to predict an individual's behavior in various situations.

8 To OTS officers and many "old hands" in operations, Jeffrey Richelson's work, *The Wizards of Langley,* on the history of the CIA's Directorate of Science and Technology, incorrectly bestows the wizard title on engineers and scientists. Ask a case officer or a tech, "Who are the wizards?" and he or she will likely reply, "Those are the shrinks in OTS."

9 In the mid-1990s, after the collapse of the USSR, requirements for graphological operational assessments decreased to the point that the service no longer required a full-time staff in OTS. Graphology has, nevertheless, been growing as a personnel service for U.S. companies for applicant screening and job interviews. See: "Deciphering the Handwriting on the Wall," *Washington Post,* October 17, 2004.

10 James Van Stappen, "Graphological Assessment in Action," *Studies in Intelligence,* 3:4, Central Intelligence Agency, 1959, 49–58.

11 Keith Laycock, "Handwriting Analysis as an Assessment Aid," *Studies in Intelligence* (Washington, DC: Central Intelligence Agency, 1959) vol 3:3 (1959), 27.

12 E. A. Rundquist, "The Assessment of Graphology," *Studies in Intelligence*, 3:3, Central Intelligence Agency, 1959, 45–51.

13 Former DDP and DCI Richard Helms in *A Look over My Shoulder,* 426, commented that the studies "proved a useful extension of the routine diplomatic and

military reporting." Former DDO and DCI William Colby in *Honorable Men—My Life in the CIA* (New York: Simon & Schuster, 1978), 335, cited "psychological advice on how to handle alien agents" as an important TSD capability. OTS deployed psychologists to field bases to support requirements at CIA offices throughout the world. At Headquarters OTS devoted one or more full-time operational psychologists to handle the caseloads in high demand operational components such as the Soviet and Far East Divisions and the Counterterrorism Center.

14 John Waller, "The Myth of the Rogue Elephant Interred," *Studies in Intelligence*, 22:2 Central Intelligence Agency, 1978, 6.

15 Allen Dulles, "Brain Warfare," speech to the National Alumni Conference of the Graduate Council of Princeton University, Hot Springs, VA, April 10, 1953.

16 See: DCI Stansfield Turner's 1977 testimony. The DCI grouped MKULTRA's 149 subprojects into three categories: (1) Research into behavior modification, drug acquisition, and testing and clandestine administration of drugs; (2) financial and cover mechanisms for each of the subprojects; (3) subprojects, of which there were thirty-three, funded under the MKULTRA umbrella but unrelated to behavioral modification, drugs, or toxins. Polygraph research and control of animal activity were examples offered. The process to completely phase out all of the MKULTRA projects required several years.

17 Michael Edwards, "The Sphinx and the Spy: The Clandestine World of John Mulholland," *Genii: The Conjurors Magazine*. April 2001, see: http://www.frankolsonproject.org/Articles/Mulholland.html

18 Ibid.

19 Letter to Dr. Sidney Gottlieb, Central Intelligence Agency, MKULTRA document 4-29, April 20, 1953.

20 Ibid.

21 Edwards, "The Sphinx and the Spy."

22 Mulholland letter to Sidney Gottlieb, Central Intelligence Agency, MKULTRA document 19-2, November 11, 1953.

23 Memorandum for the Record, *Project MKULTRA, Subproject 34,* Central Intelligence Agency, MKULTRA document 34–46, October 1, 1954.

24 Edwards, "The Sphinx and the Spy."

25 Memorandum for the Record, *Project MKULTRA, Subproject 34,* Central Intelligence Agency, MKULTRA document 34–46, October 1, 1954.

26 Memorandum for the Record, *Definition of a Task under MKULTRA, Subproject 34,* Central Intelligence Agency, MKULTRA document 34–39, August 25, 1955.

27 Memorandum for the Record, *MKULTRA, Subproject 34,* Central Intelligence Agency, MKULTRA document 34–29, June 20, 1956.

28 Edwards, "The Sphinx and the Spy."

29 Memorandum for the Record, Central Intelligence Agency, MKULTRA document 8312, March 26, 1959.

30 In 1962, Dr. Gottlieb, who had been Chief of R&D for TSD, was promoted to Deputy Chief/TSD under Seymour Russell. Richard Krueger replaced Dr. Gottlieb as Chief of R&D for TSD but was not initially briefed on any of the MKULTRA projects, which continued to report to Gottlieb. Following the IG

report, however, Krueger was "read into the program" and developed a process for phasing out over three years all remaining projects. Three years were required to close down the projects through orderly steps that would not expose the covert relationships or compromise the security of the participating institutions and individuals as well as fulfill government contractual obligations to the parties.

31 Marks, *The Search for the Manchurian Candidate*, 219.

32 Waller, "The Myth of the Rogue Elephant Interred," 6–7.

33 U.S. Congress, Senate, Select Committee to Study Government Operations with Respect to Intelligence Activities. *Foreign and Military Intelligence Final Report, Book 1.* 94th Congress, 2nd Sess., April 26, 1976.

34 The new MKULTRA records were potentially explosive for two separate reasons. First, the fact that they had not been discovered when sought by the Church Committee investigation could have pointed to a CIA "cover-up." Second, the additional records had the potential for containing significantly new information about MKULTRA experiments and operational plans. Despite the hundreds of thousands of words written about MKULTRA, most of the CIA's documentation about the program was destroyed in 1972–1973 at the direction of DCI Richard Helms. As recounted to the author by a TSD officer who was involved with the destruction, Dr. Gottlieb returned to TSD headquarters in late 1972 or early 1973 and advised his senior staff that the Director has ordered all MKULTRA records destroyed. It was a verbal order; there would be no memo. There followed some discussion about the advisability of destroying all records, particularly the documentation of research procedures and scientific results. Gottlieb responded that the directive had been unambiguous—all the records were to be destroyed. In the following days, MKULTRA project and operational records held by TDS were systematically shredded. Subsequently, remaining documentary information about MKULTRA has been subject to numerous FOIA requests and released to the public with some redactions of information judged to require continued classification.

35 Center for the Study of Intelligence, "An Interview with Richard Helms," *Studies in Intelligence*, 25:3, Central Intelligence Agency, 1981, 21.

36 U.S. Senate, Select Committee on Intelligence and the Sub Committee on Human Resources, *Project MKULTRA, The CIA's Program of Research in Behavioral Modification.* 95th Congress, 1st Sess., August 3, 1977.

37 See: Marks, *The Search for the 'Manchurian Candidate*, for a detailed account of the CIA research into human behavior based on official documents declassified under the Freedom of Information Act and released after publication of the 1975 Rockefeller Commission and 1976 Church Committee reports.

38 Bart Barnes, "Obituary, Sidney Gottlieb," *Washington Post,* March 11, 1999, B.05. The opening sentence of the obituary identifies Gottlieb as "the former chief of the Central Intelligence Agency's technical services division who in the '60s directed CIA mind-control experiments, including the administration of drugs and LSD to unwitting humans . . ." Not mentioned is that the experimentation cited was authorized, limited, and ended by the mid-'50s.

39 Ted Gup, "The Coldest Warrior," *The Washington Post Magazine,* December 16, 2001.

40 Center for the Studies of Intelligence, "An Interview with Former General Counsel John S. Warner," *Studies in Intelligence,* 22:2, Central Intelligence Agency, 1978, 49.

Chapter Twenty-One – Cover and Disguise

1 Interview with Carl A. Strahle. Strahle was one of the printers in the London OSS document shop. See also Christof Mauch, *The Shadow War Against Hitler* (New York: Columbia University Press, 1999), 179ff, and Persico, *Piercing the Reich,* 23–26. Casey would become Director of Central Intelligence in 1981.

2 The "intelligence division," "graphic arts reproduction division," and the "furnishings and equipment division" each produced identity-related materials. The other three original TSS divisions were organized to support agent communications, audio surveillance, and research and development.

3 "Cover" refers to the assumed or created identity, occupation and background of an intelligence operative. An operative can be assigned an occupational cover in his true name or can be placed "under cover" in an alias identity. The "backstopping" of cover refers to all the personal and official documentation as well as confidential arrangements made with government or private organizations to verify the operative's legitimacy.

4 Documenting the alias identities of officers and agents has remained a core mission of OTS throughout its history.

5 Strahle interview.

6 During disguise presentations, particularly for middle-aged audiences, OTS briefers had one assured laugh line. "Here in the OTS disguise shop we specialize in making you look older and fatter. If you want to look younger and thinner, we recommend you talk with the Office of Medical Service." There was truth in the humor. No matter how tight the corset, body weight did not change. No matter the direction of push, flesh would not move more than an inch or two. Graying the hair of a thirty-five-year-old and adding a salt-and-pepper beard could make a twenty-year difference. Coloring jet-black the graying hair of a paunchy fifty-year-old produced little more than the appearance of a middle-aged guy unsuccessfully dealing with his midlife crisis.

7 For a more detailed treatment of "exfiltrations" and disguise, see: Mendez, *The Master of Disguise.*

Chapter Twenty-Two – Concealments

1 David Kahn, *The Codebreakers* (New York: Macmillan, 1967), 122.

2 Center for the Study of Intelligence, "Intelligence in the War of Independence" monograph (Washington, D.C.: Central Intelligence Agency, undated), 33.

3 Ibid.

4 Anthony Cave Brown (editor), *The Secret War Report of the OSS* (New York: Berkley, 1976), 76–77.

5 Ibid., 77.

6 Ibid.

7 McLean, *The Plumber's Kitchen,* 11.

8 Ibid., 239–242.

9 For more on MIS-X see: Lloyd R. Shoemaker, *The Escape Factory: The Story of MIS-X* (New York: St. Martin's Press, 1990).

10 David Crawford, *Volunteers: The Betrayal of National Defense Secrets by Air Force Traitor*s (Washington, D.C.: Air Force Office of Special Investigations, 1988), 24.

11 See: Melton, *CIA Special Weapons and Equipment*, 113, for photographs of a concealment desk.

12 See: Melton, *Ultimate Spy*, 162, for photographs.

13 See: Melton, *Ultimate Spy*, 161, for photographs.

14 Mendez, *The Master of Disguise*, 224.

15 The concealment techs could prepare virtually any type of wild or domesticated animal as "host carcasses" for dead drops. The DO case officers, however, were squeamish and agreed to proceed only with pigeons and rats. Mendez, *The Master of Disguise*, 224–225, presents more examples.

16 An "exfiltration" is a clandestine operation to move an officer, defector, or agent across international borders without the knowledge of any hostile security service.

17 Mendez, *The Master of Disguise*, 140.

18 The al-Qaeda terrorist organization used a Trojan-horse concealment to assassinate Ahmed Shah Massoud, the leader of the Afghanistan's anti-Taliban Northern Alliance in Afghanistan, on September 9, 2001. Two al-Qaeda operatives posing as journalists secured an audience with Massoud. One carried a video camera packed with explosives. During the "interview" the explosive was detonated and Massoud killed.

19 Rustmann, *CIA, INC.*, 53–56.

20 Actor Desmond Llewelyn became Q beginning with *From Russia with Love* in 1963 and continued in the Bond movies until his accidental death in a car accident in 1999.

21 On occasion, senior CIA executives would visit the lab and be treated to a "show and tell" to demonstrate the value of concealments. During one such visit, a lab engineer proudly displayed a laptop computer that had been configured as an active concealment device for use by a Soviet agent. The electronics for the intelligence function had been masterfully integrated into the overall unit and even when the computer was disassembled, the fact that any modification had been made was not apparent or visible. The executive visitor asked, "How long did this take?" "Two hundred work days," replied the tech. "How many did you make?" "Only this one." "Well, if you made a hundred, you can be a lot more efficient," came the response. The operational imperative for one-of-a-kind concealments for Soviet operations seemed lost to the visitor.

22 For images and a description of a concealment desk see: Melton, *CIA Special Weapons and Equipment*, 113.

Chapter Twenty-Three – Clandestine Surveillance

1 A target staying at a little-used hotel might be surveilled from OP's setup temporarily to monitor the hotel's entrances. Covert video cameras could be concealed in parked cars outside each hotel entrance and feed "live video" to the watcher team set up in an adjacent hotel. Conversely, a "long-term" stationary OP might

be established for the purpose of photographing all individuals entering and leaving a radical mosque that is a known transit point for new recruits departing Europe for terrorist training in the Middle East.

2 There are two types of telephoto lenses, refractive and reflex mirror (catadioptric). Refractive lenses (as in a telescope) are usually much larger than "mirror" lenses which have a system of mirrors and lenses to fold up the optical path causing the light passing through the instrument to do so in a zigzag fashion and greatly reduce the physical size and length of the unit. The compactness of a mirror lens is often desirable for surveillance photography. "Fast" lenses have larger areas of glass to gather more light, but are more difficult to conceal. "Slow" lenses are easier to hide, but require longer exposure times to take acceptable images and are vulnerable to vibration. See: Raymond P. Siljander, *Fundamentals of Physical Surveillance: A Guide for Uniformed and Plainclothes Personnel* (Springfield, Illinois: Charles C. Thomas, 1977), 182.

3 Telephoto lenses amplify any vibrations present and require firm support of the lens and camera assembly. The longer the lens on the camera, however, the more difficult it is to use and the more likely is a loss of image quality. Inclement weather, dust, and haze can significantly degrade the quality of the image. See: Siljander, *Fundamentals of Physical Surveillance,* 195.

4 Each doubling of film speed represents a doubling in the film's sensitivity to light. "Push processing" allows film to be exposed at higher than rated ASA levels and developed using special processes to artificially "push" the ASA sensitivity to match exposure levels. It is possible to "push process" a commercially available ASA 6400 film to ASA 12800, ASA 25600, or even higher and still take an acceptable photograph of a target to produce a positive identification.

5 A conventional strobe flash unit covered with a Kodak Wratten 87C filter emits light in the infrared spectrum at wavelengths from 750 to 900 millimicrons. For an example of clandestine infrared photography using this technique see: Melton, *CIA Special Weapons and Equipment,* 37.

6 Melton, *CIA Special Weapons and Equipment,* 71.

7 Pheromones, chemicals secreted by an animal, especially an insect, are used as aids for surveillance tracking.

8 A well-known "spy shop" in New York City in the 1980s advertised repackaged, commercial-grade products, using unsubstantiated claims of technical capability that bordered on the unbelievable, such as: "Tell if any phone call anywhere in the world is bugged or recorded using this all-in-one briefcase counterspy kit!" Even more amazing was its "graduated" concept of pricing that produced catalogs and pricelists in English, Spanish, and Arabic. The same equipment was in all catalogs, only the prices changed; the prices in Spanish were double those in the English, and the Arabic version was four times higher!

9 Contractors and employees who have successfully undergone background investigations are provided with varying levels of security clearances in order to be able to work with the CIA.

10 In 1827, Sir Charles Wheatstone coined the phrase "microphone." See: http://inventors.about.com/od/mstartinventions/a/microphone.htm. When sound waves contact a microphone, they cause the thin flexible internal diaphragm to vibrate. These vibrations are converted into an electrical signal, which varies in voltage,

amplitude and frequency in an analog of the original sound. See: http://www. edinformatics.com/inventions_inventors/microphone.htm.

11 For photographs and technical descriptions see: Melton, *CIA Special Weapons and Equipment*, 65.

12 For photographs and information on the fine-wire kit see: Melton, *CIA Special Weapons and Equipment*, 67, and Melton, *Ultimate Spy*, 102.

13 Richard Tomlinson, *The Big Breach: From Top Secret to Maximum Security* (Moscow: Narodny Variant Publishers, 2000), 104.

14 Ibid., 104–105.

15 Rustmann, *CIA, INC.*, 54.

16 For a photograph of a disposable "quick plant" writing pen, see: Melton, *Ultimate Spy*, 103.

17 Rustmann, *CIA, INC.*, 57.

18 Melton, *Ultimate Spy*, 96, 105.

19 Bob Woodward, *Veil*, 147.

20 A "wood block" is an audio eavesdropping device usually consisting of a microphone, transmitter, and batteries built inside a hollow section of wooden molding or part of a table support or chair leg. Such devices are intended to be quickly exchanged with their identical counterpart inside the target location.

21 For photographs of "wood blocks" inside modified furniture components see: Melton, *Ultimate Spy*, 105.

22 See: Melton, *Ultimate Spy*, 103.

23 Experience taught that the ideal book to be swapped was on the top shelf where it was harder for the target to reach and less likely to be read and examined.

24 Glinsky, *Theremin*, 273; see also: Melton, *Ultimate Spy*, 104, for photos of the Great Seal and diagrams of the resonator.

25 See: Associated Press article by Ted Bridis "CIA Gadget Museum Showcase Robot Fish, Pigeon Camera, Tiger Dropping Microphone" on http://www.mindfully.org/Technology/2003/CIA-Museum26dec03.htm.

26 A "provocation" would be an act by the officer to elude surveillance such as getting on, and then immediately off a subway car, or speeding through an urban area to "lose" the trailing vehicle.

27 Floyd L. Paseman, *A Spy's Journey: A CIA Memoir* (St. Paul, Minnesota: Zenith Press, 2004), 61.

28 Planar eyeglass lenses are noticeably thick, but do not provide any optical magnification. The contents of a portable disguise kit issued to a CIA officer going abroad can be seen in Melton, *CIA Special Weapons and Equipment*, 105, and Melton, *Ultimate Spy*, 131.

29 The sculpted facial disguises were remarkably lifelike. OTS received assistance from Oscar-winning mask designer John Chambers (*Planet of the Apes*) to help create disguises for intelligence officers. See: by Michael E. Ruane, "Seeing is Deceiving," *Washington Post*, February 15, 2000.

CHAPTER TWENTY-FOUR – COVERT COMMUNICATIONS

1 An agent requires the ability to both receive and send covert communications. A shortwave radio with a one-time pad is an example of an "agent-receive"

system. Secret writing carbons and accommodation addresses represent an "agent-send" system.

2 Steganography is defined as "covered writing" or the art of communicating in a way that masks the very existence of the communication. See: Eric Cole, *Hiding in Plain Sight: Steganography and the Art of Covert Communication* (Indianapolis, Indiana: Wiley Publishing, 2003) for a detailed description of the uses of steganography for clandestine communication. Cole is a former CIA officer who specialized in the development of secure communication systems.

3 Through the 1970s, personal meetings also became increasingly technology dependent. Electronic signaling and nonattributable telephone calls replaced chalk marks and lipstick smears to trigger clandestine meetings. Case officers wore earpieces to listen for surveillance transmissions to determine if they were being followed. Identity-altering clothing and accessories were among other technical tools applied to assure the security of "low tech" personal meetings.

4 "Unnatural acts" appear out-of-the-ordinary or suspicious when observed and serve as a "flag" to alert counterintelligence. For example, making a large chalk X on a telephone pole or repeatedly looking over one's shoulder while walking down a street are uncharacteristic actions and could prompt further attention from security or law enforcement officials.

5 Crawford, *Volunteers,* 26.

6 Ibid., 27.

7 Weiser, *A Secret Life,* 74.

8 Ibid.

9 For visual examples of sophisticated criminal "brush passes" see the 1973 movie *Harry in Your Pocket.*

10 Weiser, *A Secret Life,* 81.

11 Ibid., 75.

12 Crawford, *Volunteers,* 27.

13 Victor Suvorov, *Inside Soviet Military Intelligence* (New York: Macmillan, 1984), 119.

14 Crawford, *Volunteers,* 28.

15 British intelligence (MI6) refers to them as "dead letter boxes" (DLBs).

16 CIA officer Aldrich Ames, a mole for the KGB and Russian Intelligence Service (SVR), sharply complained to his handler about the size of the dead drop site (code name PIPE) they had selected for his use along a horse path in Maryland's Wheaton Regional Park. Ames communicated that he needed more money and estimated that the size of the drainpipe used for the dead drop would accommodate up to $100,000. See: Wise, *Nightmover,* 220.

17 Weiser, *A Secret Life,* 58.

18 Before the FBI arrested Ames on February 21, 1994, they attempted to lure a SVR officer into a trap by leaving a horizontal chalk mark on the side of a U.S. Postal Service mailbox located at the corner of R Street and 37th Street in Washington, D.C. Unbeknownst to the FBI, however, the SVR had changed the location of the signal site. The horizontal mark left by the FBI meant nothing and the SVR didn't respond. See: Wise, *Nightmover,* 272–273.

19 Crawford, *Volunteers,* 30. The most famous instance of a botched signal occurred on May 19, 1985, when KGB spy John Walker was participating in a compli-

cated signal and drop sequence in rural Montgomery County, Maryland. The KGB officer arrived in the general area and left a soda can at the base of a stop sign to signal his presence. Walker left a second can at the base of another stop sign as a signal to the KGB. Walker saw the can left by the KGB officer and proceeded to leave his secret documents at the base of a telephone pole some distance away. Unbeknownst to the KGB, the FBI had Walker under surveillance and made the mistake of removing the signal can he left at the stop sign. When the KGB officer could not see Walker's can to confirm that he was in the area, he followed his instructions and aborted the operation. After leaving the secret documents, Walker proceeded to a second drop location where the KGB was to have left money for him. He found nothing and when he returned to retrieve the documents, they were gone as well. Walker was arrested early the next morning at the Ramada Hotel in nearby Rockville, Maryland, and the KGB officer departed the country for the Soviet Union the following day. See: Jack Kneece, *Family Treason, The Walker Spy Case* (New York: Stein and Day, 1986), 109–123.

20 Polmar and Allen, *Spy Book*, 573, describes secret writing. *Spy Book* attributes to Ovid in "Art of Love" counsel that one's missive "can escape curious eyes when written in new milk [then] touch it with [charcoal] dust and you will read."

21 Joseph B. Smith, *Portrait of a Cold Warrior: Second Thoughts of a Top CIA Agent* (New York: G.P. Putnam's Sons, 1976), 130–131.

22 Ibid., 131.

23 Accommodation addresses differed from dead drops in that once a letter was dropped in a mailbox both the agent and the handler lost control of the message. An effective AA would have a constant stream of business or personal correspondence, letters, and postcards going and coming. In some countries, a post office box or a "letter-drop" provided an adequately safe and convenient type of accommodation address. A message can be communicated through an AA without SW if the type, style, or color of a postcard is, in itself, the signal. For example, a recruited agent who has just returned from a posting abroad could signal to the CIA his willingness to begin his clandestine work by mailing a specific type of postcard to an innocuous AA. For the most sensitive agents, AAs were used only once.

24 Weiser, *A Secret Life,* 23.

25 During World War I German agents traveling to the United States impregnated articles of clothing with their secret inks. To recover the ink, a scarf or a shoelace would be soaked in distilled water.

26 Tomlinson, *The Big Breach,* 82–83.

27 Ibid., 65.

28 Ibid., 65–66.

29 Ibid., 127. The chemical formulation for the reagent was not specified by Tomlinson.

30 See: Melton, *Ultimate Spy,* 150.

31 A microdot communication to an agent would normally carry several identical dots buried in different locations. One longtime microdot user explained, "Three dots to an agent was the minimum. The first he wouldn't find. He would find the

next but would drop it or a gust of wind from an open window would blow it away. Hopefully he would find and read the third one."

32 The original "bullet" lens was invented by Charles Stanhope in London in the late 1700s. This small magnifying lens (the "Stanhope") was used originally in the textile industry to count the number of cotton fibers in a field, and was further refined by Henry Coddington in 1830. The Stanhope lens was used well into the 1800s and later saw popular use in the making of "peeps" for viewing tiny "girlie pictures" that were sold at carnivals and sideshows, and even for a miniature version of "The Lord's Prayer."

33 Ibid.

34 The original magazine is displayed inside Moscow's FSB Counterintelligence Museum.

35 Wise, *Nightmover*, 259–260. The message continued to describe each signal site and dead drop to be used in Moscow. Signal site ZVONOK was accessed by boarding a number 10 trolley bus traveling toward Krymsky Most. The agent was to get off at the fifth stop and locate a specific phone booth where he would make his mark, a 10 cm Cyrillic "R" on the building wall to the left of the phone booth and drainpipe. The mark was to be made waist high using black crayon or red lipstick so that it could be easily read from a passing vehicle. The CIA would acknowledge receipt of Vasilyev's signal by parking a car with the license plate number D-004 opposite the Lenin Central Museum.

36 Ibid.

37 Cellulose base film, both cellulose nitrate and cellulose acetate, was the primary choice for creating "soft films."

38 Xidex Corporation acquired Kalvar Corporation in March of 1979 and three months later closed the New Orleans plant and fired the production personnel. See: http://www.keypointconsulting.com/downloads/pub_Event_Studies.pdf .

39 To produce "soft film" a frame of Kalvar was placed between two pieces of glass together with a developed negative containing the message for the agent (emulsions sides together). The glass plate was exposed to a 500 watt lamp for 40 to 50 seconds and then held with tweezers and dipped into boiling water for two seconds. As it cooled, the emulsion was carefully peeled away from the backing and allowed to dry. The resulting image on "soft film" was ready to be camouflaged and passed to the agent.

40 Weiser, *A Secret Life*, 66.

41 Tomlinson, *The Big Breach*, 66.

42 Messages were created by adding the random series of numbers (the "keys") on the designated page of the OTP to the plaintext message. The person receiving the ciphered message subtracts the random numbers (found on his matching copy of the OTP) to recover the original message.

43 It needed to be a good-quality shortwave receiver capable of single sideband reception.

44 These are often referred to outside the CIA as "numbers stations" or "counting stations." For more information on "spy numbers stations" and an opportunity to listen-in on sample transmissions, go to: http://www.spynumbers.com/enigma .html

45 A 150 five-number message would contain 750 numbers. It was possible, but less common, for transmissions to also be sent using phonetic language where letters

were "spoken" (alpha, bravo, charlie, delta, foxtrot, etc.). Most messages were usually fixed at a length of 150 five-number groups, but could be longer. If the message was shorter than 150 groups, additional numbers would be added as "pads or "filler" at the end.

46 "Time-sensitive" information reported events or circumstances of immediate significance. If not received by the intelligence service quickly, the reporting rapidly lost its value.

47 Weiser, *A Secret Life,* 215, 229.

48 Ibid., 229–230.

49 "Bent-pipe" refers to the satellite's limited role in receiving and relaying the signal without any processing. In essence, the signal from the agent bounced off the satellite to the ground receiving station.

50 The Russian Federal Service (FSB) museum in Moscow displays an attaché case, labeled as the BIRDBOOK system, which is filled with electronics and has a transmitting antenna built into the lid.

CHAPTER TWENTY-FIVE – SPIES IN THE AGE OF INFORMATION

1 Bearden and Risen, *The Main Enemy,* 522–523.

2 Ibid.

3 R. James Woolsey, in testimony before the Senate Select Committee on Intelligence, February 2, 1993, just before his installation as DCI. The colorful metaphor provided a "sound-bite" justification for his view that substantial intelligence resources were still needed in the post–Cold War era.

4 A term applied to the period where movement of information became faster than physical movement, more narrowly applying to the 1980s onward. The Information Age also heralded the era when information was a scarce resource and its capture and distribution generated competitive advantage. Microsoft became one of the largest companies in the world based on its influence in creating the underlying mechanics to facilitate information distribution.

5 James Gosler, "The Digital Dimension," from *Transforming U.S. Intelligence,* Jennifer Sims and Burton Gerber (editors) (Washington, D.C.: Georgetown University Press, 2005), 96.

6 Ibid.

7 Ibid.

8 Weiser, *A Secret Life,* 158.

9 Gosler, "The Digital Dimension," 100.

10 Ibid.

11 Ibid., 101.

12 Ibid.

13 Construction of the Internet began in 1969 with the Advanced Research Projects Agency Network (ARPANET) by academic researchers under the sponsorship of the Advanced Research Projects Agency (ARPA). Two decades later, the Internet became publicly accessible with a worldwide system of interconnected computer networks. The "Net" connected thousands of smaller commercial, academic, domestic, and government networks, creating an interlinked "World Wide Web" that provided varied information and services including online chat, electronic mail, and instant messaging.

14 Walker was arrested on May 20, 1985, shortly after leaving secret documents for the KGB at a dead drop location in rural Montgomery County, Maryland. The warrant authorizing the search of the Ames residence was made possible by the discovery in their household trash of a yellow Post-it note referencing a covert meeting with the Russian Intelligence Service to be held in Bogotá, Columbia.

15 Decision Support Systems, Inc., *Secure Communications Operational Tradecraft; "How Not To Be Seen,"* January 11, 2002, website: http://www.metatempo .com/SecureCommo.PDF.

16 For even greater protection, the agent may choose to superencipher the message using an OTP first, and then enciphering it again using a "strong and proven" encryption program such as PGP (Pretty Good Privacy). See: http://web.mit. edu/network/pgp.html. Such protection, properly employed, would be slow and cumbersome to use, but would result in an "unbreakable message."

17 "Malware" (malicious software) includes programs for data encryption, digital steganography, password "cracking," and "hacking." Possession of such software, while not illegal, may become a basis for suspicion if detected during an examination of the agent's computer and hard drive.

18 The first electromechanical encryption machine was developed and patented by Edward Hebern in 1918.

19 A free version of PGP can be downloaded from http://www.pgpi.org/. Advanced commercial versions of PGP are available from http://www.pgp.com/.

20 FBI Affidavit for the Arrest of Ana Belen Montes; September 2001, pg. 8. The complete affidavit can be downloaded at http://www.fbi.gov/pressrel/pressrel01/ 092101.pdf.

21 Robert Hanssen 2001 Arrest Affidavit No. 86-87. Hanssen's technique was so effective that the KGB was unable to locate the message on the diskette. A month later, on March 28, 1988, he sent a letter to them with the simple instruction "Use 40 TRACK MODE." See: http://www.cicentre.com/Documents/DOC_Hanssen _Affidavit.htm.

22 Cole, *Hiding in Plain Sight,* 5.

23 Ibid.

24 Hundreds of stego programs are commercially available over the Internet and allow data to be hidden in a variety of file formats using familiar graphic interfaces found on the Windows operating systems.

25 Dead drops are still in use and difficult to detect when used only once. FBI Special Agent Robert Hanssen was arrested on February 18, 2001, after loading a dead drop with information for the SVR at Foxstone Park in Vienna, Virginia.

26 See: http://www.callingcards.com/ for pricing and coverage of international phone cards.

27 The assumption is made that retail outlets that sell phone cards near foreign embassies and missions are likely to be under surveillance and stocked with special phone cards that allow all calls to be monitored and traced.

28 Robert Hanssen 2001 Arrest Affidavit, No. 131.

29 Tomlinson, *The Big Breach,* 66–67.

30 See news articles: http://news.scotsman.com/international.cfm?id = 115032006;

http://www.dlmag.com/news/12/01-24-2006/418-britain-russia-and-a-spyrock.
html.

31 FBI Affidavit for the Arrest of Ana Belen Montes; September 2001, pg. 11.

32 Gosler, "The Digital Dimension," 110.

EPILOGUE – AN UNCOMMON SERVICE

1 Lovell, *Of Spies and Stratagems,* 40–41.

Selected Bibliography

Agee, Philip, *CIA Diary: Inside the Company* (New York: Stonehill, 1975).

Andrew, Christopher and Vasili Mitrokhin, *The Sword and the Shield: The Mitrokhin Archive and the Secret History of the KGB* (New York: Basic Books, 1999).

Assessment of Men: Selection of Personnel for the Office of Strategic Services (New York: Rinehart & Company, 1948).

Barnes, Bart, "Obituary, Sidney Gottlieb," *The Washington Post,* March 11, 1999.

Bearden, Milton and James Risen, *The Main Enemy: The Inside Story of the CIA's Final Showdown with the KGB* (New York: Random House, 2003).

Benson, Robert Louis, and Michael Warner, *VENONA: Soviet Espionage and the American Response, 1939–1957* (Washington, D.C.: National Security Agency and Central Intelligence Agency, 1996).

Biederman, Danny, *The Incredible World of SPY-Fi* (San Francisco: Chronicle Books, 2004).

Bosworth, Mary (editor), *Encyclopedia of Prisons and Correctional Facilities,* Volume 2 (Thousand Oaks, California: Sage Publications, 2005).

Boyce, Fredric and Douglas Everett, *SOE: The Scientific Secrets* (Phoenix Mill, England: Sutton Publishing Limited, 2003).

Brown, Anthony Cave (editor), *The Secret War Report of the OSS* (New York: Berkley, 1976).

Brown, Anthony Cave, *Wild Bill Donovan: The Last Hero* (New York: Times Books, 1982).

Brunsnitsyn, Nikolai, *Openness and Espionage* (Moscow: Military Publishing House, 1990).

Cambridge Dictionary of Science and Technology (New York: Cambridge University Press, 1990).

Cambridge World Gazetteer: A Geographical Dictionary (New York: Cambridge University Press, 1988).

Carl, Leo D., *International Dictionary of Intelligence* (McLean, Virginia: International Defense Consultant Services, 1990).

Center for the Study of Intelligence, "Intelligence in the War of Independence" (monograph) (Washington, D.C.: Central Intelligence Agency, undated).

Center for the Study of Intelligence, "An Interview with Former General Counsel John S. Warner," *Studies in Intelligence,* 22:2, Central Intelligence Agency, 1978.

Center for the Study of Intelligence, "An Interview with Richard Helms," *Studies in Intelligence,* 25:3, Central Intelligence Agency, 1981.

Center for the Study of Intelligence, *Office of Strategic Services 60th Anniversary Special Edition XI* (Washington, D.C. Central Intelligence Agency, June 2002).

Central Intelligence Agency, "Directorate of Science & Technology: People and Intelligence in the Service of Freedom" (Washington, D.C.: Central Intelligence Agency, printed circa 2003, undated).

Cherkashin, Victor, *Spy Handler: Memoir of a KGB Officer* (New York: Basic Books, 2005).

Clarridge, Duane R., *A Spy for All Seasons: My Life in the CIA* (New York: Scribner, 1997).

Colby, William, *Honorable Men: My Life in the CIA* (New York: Simon & Schuster, 1978).

Cole, Eric, *Hiding in Plain Sight: Steganography and the Art of Covert Communication* (Indianapolis, Indiana: Wiley Publishing, 2003).

The Concise Columbia Encyclopedia (New York: Columbia University Press, 1983).

Concise Dictionary of World History (New York: Macmillan Publishing Company, 1983).

Corson, William R. and Robert T. Crowley, *The New KGB: The Engine of Soviet Power* (New York: Quill, 1986).

Couffer, Jack, *Bat Bomb: World War II's Other Secret Weapon* (Austin, Texas: University of Texas Press, 1992).

Crawford, David, *Volunteers: The Betrayal of National Defense Secrets by Air Force Traitors* (Washington, D.C.: Air Force Office of Special Investigations, 1988).

Crown, David A., "Political Forgeries in the Middle East," *Studies in Intelligence,* 22:2, Central Intelligence Agency, 1978.

Deriabin, Peter, *Watchdogs of Terror* (New Rochelle, NY: Arlington House, 1972).

Donovan, James B., *Strangers on a Bridge: The Case of Colonel Abel* (New York: Atheneum, 1964).

Dulles, Allen, "Brain Warfare," speech to the National Alumni Conference of the Graduate Council of Princeton University, Hot Springs, VA, April 10, 1953.

Dulles, Allen, *The Craft of Intelligence* (New York: Harper & Row, 1963).

Earley, Pete, *Confessions of a Spy: The Real Story of Aldrich Ames* (New York: Berkley, 1998).

Edwards, Michael, "The Sphinx and the Spy: The Clandestine World of John Mulholland," *Genii: The Conjurorsí Magazine,* April 2001.

Fischer, Benjamin B., "The Central Intelligence Agency's Office of Technical Service, 1951–2001," (OTS 50th anniversary booklet) (Washington, D.C.: Central Intelligence Agency, 2001).

Fischer, Benjamin B., *The Journal of Intelligence History* (Nuremburg, Germany) 2:1, Summer 2002.

Ford, Corey, *Donovan of the OSS* (Boston: Little, Brown & Company, 1970).

Gardner, George, *Picks, Clicks, Flaps and Seals: A Monograph on Surreptitious Entry* (unpublished monograph), 1944.

George, Willis, *Surreptitious Entry: The Sensational Story of a Government Agent Who Picked Locks and Cracked Safes in the Service of His Country* (New York: Appleton-Century, 1946).

Glinsky, Albert, *Theremin: Ether Music and Espionage* (Urbana and Chicago: University of Illinois Press, 2000).

Gordievsky, Oleg, *Next Stop Execution: The Autobiography of Oleg Gordievsky* (London: Macmillan, 1995).

Gosler, James, "The Digital Dimension," in *Transforming U.S. Intelligence*, Jennifer Sims and Burton Gerber, editors (Washington, D.C.: Georgetown University Press, 2005).

de Gramont, Sanche, *The Secret War: The Story of Espionage since World War II* (New York: Putnam, 1962).

Grose, Peter, *Gentleman Spy: The Life of Allen Dulles* (New York: Houghton Mifflin, 1994).

Gup, Ted, "The Coldest Warrior," *The Washington Post Magazine,* December 16, 2001.

Helms, Richard, *A Look over My Shoulder—A Life in the Central Intelligence Agency* (New York: Random House, 2003).

Hersh, Seymour, "Huge CIA Operation Reported in U.S. Against Antiwar Forces, Other Dissidents in Nixon Years," *The New York Times,* December 22, 1974.

Hinkle, Warren and William W. Turner, *The Fish Is Red: The Story of the Secret War Against Castro* (New York: Harper & Row, 1981).

Hogg, Ivan V., *The New Illustrated Encyclopedia of Firearms* (Secaucus, New Jersey: Wellfleet Press, 1992).

Hood, William, *Mole* (New York: W. W. Norton & Company, 1952).

Kahn, David, *The Codebreakers* (New York: Macmillan, 1967).

Kalugin, Oleg, *The First Directorate: My 32 Years in Intelligence and Espionage* (New York: St. Martin's Press, 1994).

Karnow, Stanley, *Vietnam: A History* (New York: Penguin, 1997).

Keenan, George F., *Memoirs: 1925–1950* (New York: Pantheon, 1967).

Kern, Gary, "How 'Uncle Joe' Bugged FDR," *Studies in Intelligence,* 47:1, Central Intelligence Agency, 2003.

Kessler, Ronald, *Escape from the CIA: How the CIA Won and Lost the Most Important KGB Spy Ever to Defect to the U.S.* (New York: Pocket Books, 1991).

Kessler, Ronald, *Inside the CIA* (New York: Pocket Books, 1992).

Kessler, Ronald, *Moscow Station* (New York: Charles Scribner's Sons, 1989).

Kneece, Jack, *Family Treason, The Walker Spy Case* (New York: Stein and Day, 1986).

Knight, Amy, *Beria: Stalin's First Lieutenant* (Princeton, New Jersey: Princeton University Press, 1993).

Korn, David A., *Assassination in Khartoum* (Bloomington, Indiana: Indiana University Press, 1993).

Krassilnikov, Rem, *Prizraki c Ulitsy Chaykovskogo* [*The Phantoms of Tchaikovsky Street*] (Moscow: GEYA Iterum, 1999).

Laqueur, Walter, *The New Terrorism: Fanaticism and the Arms of Mass Destruction* (New York: Oxford University Press, 1999).

Lathrop, Charles E., *The Literary Spy* (New Haven, Connecticut: Yale University Press, 2004).

Laycock, Keith, "Handwriting Analysis as an Assessment Aid," *Studies in Intelligence,* 3:3, Central Intelligence Agency, 1959.

Leary, William M., "Robert Fulton's Skyhook and Operation Coldfeet," *Studies in Intelligence,* 38:5, Central Intelligence Agency, 1994.

Lovell, Stanley P., *Of Spies & Stratagems* (Englewood Cliffs, New Jersey: Prentice Hall, 1963).

Macrae, Stuart, *Winston Churchill's Toyshop* (New York: Walker and Company, 1972).

MacKinnon, Donald W., "The OSS Assessment Program," *Studies in Intelligence*, 23:3, Central Intelligence Agency, 1979.

Managhan, Robert, "Trends in African Forgeries," *Studies in Intelligence,* 19:1, Central Intelligence Agency, 1975.

Marks, John, *The Search for the Manchurian Candidate* (New York: W. W. Norton & Company, 1979).

Martin, David C., *Wilderness of Mirrors* (New York: Harper & Row, 1980).

Mauch, Christof, *The Shadow War Against Hitler* (New York: Columbia University Press, 1999).

McLean, Donald B., *The Plumber's Kitchen: The Secret Story of American Spy Weapons* (Wickenburg, Arizona: Normount Technical Publications, 1975).

Melton, H. Keith, *CIA Special Weapons and Equipment: Spy Devices of the Cold War* (New York: Sterling Publishing, 1993).

Melton, H. Keith, *OSS Special Weapons & Equipment: Spy Devices of WW II* (New York: Sterling Publishing, 1991).

Melton, H. Keith, *The Ultimate Spy Book* (New York: DK Publishing, 1996).

Mendez, Antonio J., *The Master of Disguise: My Secret Life in the CIA* (New York: Morrow, 1999).

Mikoyan, Sergo A., "Eroding the Soviet 'Culture of Secrecy,'" *Studies in Intelligence,* No. 11, Central Intelligence Agency, 2001.

Miller, Nathan, *Spying for America: The Hidden History of U.S. Intelligence* (New York: Paragon House, 1989).

Montefiore, Simon Sebag, *Stalin: The Court of the Red Tsar* (New York: Alfred A. Knopf, 2003).

Munro, David (editor), *Cambridge World Gazetteer: A Geographical Dictionary* (New York: Cambridge University Press, 1988).

National Commission on Terrorist Attacks upon the United States, *The 9/11 Commission Report* (New York: W. W. Norton, 2004).

Nielsen, Nathan, "Our Men in Havana," *Studies in Intelligence,* 32:1, Central Intelligence Agency, 1988.

Olson, James M., *Fair Play: The Moral Dilemmas of Spying* (Washington, D.C.: Potomac Books, 2006).

Paseman, Floyd L., *A Spy's Journey: A CIA Memoir* (St. Paul, Minnesota: Zenith Press, 2004).

Peretrukhin, Igor, *Agent Code Name—TRIANON* (*Agenturnaya Klichka-TRIANON*) (Moscow: Tsentrpoligraf, 2000).

Persico, Joseph E., *Piercing the Reich* (New York: Viking, 1979).

Persico, Joseph E., *Roosevelt's Secret War* (New York: Random House, 2001).

Phillips, David Atlee, *The Night Watch* (New York: Atheneum, 1977).

Plaster, John L., *SOG: A Photo History of the Secret Wars* (Boulder, Colorado: Paladin Press, 2000).

Plaster, John L., *SOG: The Secret Wars of America's Commandos in Vietnam* (New York: Simon & Schuster, 1977).

Polmar, Norman and Thomas B. Allen, *Spy Book: The Encyclopedia of Espionage* (New York: Random House, 1998).

Post, Jerrold M., "The Anatomy of Treason," *Studies in Intelligence*, 19:1, Central Intelligence Agency, 1975.

Ranelagh, John, *The Agency: The Rise and Decline of the CIA* (New York: Touchstone, 1987).

Richelson, Jeffrey T., *A Century of Spies: Intelligence in the Twentieth Century* (New York: Oxford University Press, 1995).

Richelson, Jeffrey T., *The Wizards of Langley: Inside the CIA's Directorate of Science and Technology* (Boulder, Colorado: Westview Press, 2001).

Ridgeway, James, *Blood in the Face* (New York: Thunder's Mouth Press, 1990).

Riordan, Michael and Lillian Hoddeson, *Crystal Fire* (New York: W. W. Norton, 1997).

Rockefeller, Nelson A. (chairman), *Report to the President,* Commission on CIA Activities within the United States, United States Government, June 1975.

Royden, Barry G., "Tolkachev, A Worthy Successor to Penkovsky," *Studies in Intelligence,* 47:3, Central Intelligence Agency, 2003.

Ruane, Michael E., "Seeing Is Deceiving," *Washington Post*, February 15, 2000.

Rundquist, E. A., "The Assessment of Graphology," *Studies in Intelligence,* 3:3, Central Intelligence Agency, 1959.

Rustmann, Jr., F.W., *CIA, INC.: Espionage and the Craft of Business Intelligence* (Washington, D.C.: Brassey's, 2002).

Schecter, Jerold L. and Peter S. Deriabin, *The Spy Who Saved the World* (New York: Charles Scribner's Sons, 1992).

Sheymov, Victor, *Tower of Secrets: A Real Life Spy Thriller* (Annapolis, Maryland: Naval Institute Press, 1993).

Shoemaker, Lloyd R., *The Escape Factory: The Story of MIS-X* (New York: St. Martin's Press, 1990).

Siljander, Raymond P., *Fundamentals of Physical Surveillance: A Guide for Uniformed and Plainclothes Personnel* (Springfield, Illinois: Charles C. Thomas, 1977).

Singlaub, John L., *Hazardous Duty* (New York: Summit, 1991).

Smith, Bradley F., *The Shadow Warriors: O.S.S. and the Origins of the C.I.A.* (New York: Basic Books, 1983).

Smith, Joseph B., *Portrait of a Cold Warrior: Second Thoughts of a Top CIA Agent* (New York: G. P. Putnam's Sons, 1976).

Spetrino, David A. "Aids Disinformation," *Studies in Intelligence,* 32:4, Central Intelligence Agency, 1988.

Stevenson, William, *A Man Called Intrepid* (New York: Harcourt Brace Jovanovich, 1976).

Stern, Jessica, *The Ultimate Terrorists* (Cambridge, Massachusetts: Harvard University Press, 1999).

Stockdale, James B. and Sybil B. Stockdale. *In Love and War* (New York: Bantam, 1985).

Suvorov, Viktor, *Aquarium: The Career and Defection of a Soviet Military Spy* (London: Hamish Hamilton, 1985).

Suvorov, Victor, *Inside Soviet Military Intelligence* (New York: Macmillan, 1984).

Szuminski, Walter E., *Our Man in Havana: TDY in Hell* (unpublished monograph), undated.

Szuminski, Walter E. with Edward Mickolus, *Temporary Duty in Hell: Our Man in Cuba's Jails* (unpublished monograph), 2001.

Thomas, Hugh, *Cuba or the Pursuit of Freedom* (New York: Da Capo Press, 1988).

Tomlinson, Richard, *The Big Breach: From Top Secret to Maximum Security* (Moscow: Narodny Variant Publishers, 2000).

Troy, Thomas F., *Donovan and the CIA: A History of the Establishment of the Central Intelligence Agency* (Frederick, Maryland: University Publications of America, 1981).

U.S. House of Representatives, Permanent Select Committee on Intelligence. *Soviet Active Measures.* 97th Congress, 2nd Sess., July 1982.

U.S. House of Representatives, Permanent Select Committee on Intelligence, Subcommittee on Oversight. *Soviet Covert Action (The Forgery Offensive).* 96th Congress, 2nd Sess., February 1980.

U.S. House of Representatives, Select Committee on Intelligence. *U.S. Intelligence Agencies and Activities: Committee Proceedings.* 94th Congress, 2nd Sess., January, February 1976.

U.S. House of Representatives, Special Subcommittee on Intelligence of the Committee on Armed Services. *Inquiry into the Alleged Involvement of the Central Intelligence Agency in the Watergate and Ellsberg Matters.* 91st Congress, 1st Sess., May, June, July 1973, and February, March, June, July 1974.

U.S. Senate, Select Committee on Intelligence and the Subcommittee on Health and Scientific Research of the Committee on Human Resources. *Project*

MKULTRA, The CIA's Program of Research in Behavioral Modification. 95th Congress, 1st Sess., August 3, 1977.

U.S. Senate, Select Committee to Study Governmental Operations with Respect to Intelligence Activities. *Alleged Assassination Plots Involving Foreign Leaders, An Interim Report.* 94th Congress, 1st Sess., November 20, 1975.

U.S. Senate, Select Committee to Study Governmental Operations with Respect to Intelligence Activities. *Foreign and Military Intelligence, Final Report Book I.* 94th Congress, 2nd Sess., April 26, 1976.

U.S. Senate, Select Committee to Study Governmental Operations with Respect to Intelligence Activities. *Intelligence Activities and the Rights of Americans, Final Report, Book II.* 94th Congress, 2nd Sess., April 26, 1976.

U.S. Senate, Select Committee to Study Governmental Operations with Respect to Intelligence Activities. *Supplementary Detailed Staff Reports on Intelligence Activities and the Rights of Americans, Final Report, Book III.* 94th Congress, 2nd Sess., April 26, 1976.

U.S. Senate, Select Committee to Study Governmental Operations with Respect to Intelligence Activities. *Supplementary Detailed Staff Reports on Foreign and Military Intelligence, Final Report, Book IV.* 94th Congress, 2nd Sess., April 26, 1976.

U.S. Senate, Select Committee to Study Governmental Operations with Respect to Intelligence Activities. *Unauthorized Storage of Toxic Agents.* 94th Congress, 1st Sess., September 1975.

Van Stappen, James, "Graphological Assessment in Action," *Studies in Intelligence,* 3:4, Central Intelligence Agency, 1959.

Waller, John, "The Myth of the Rogue Elephant Interred," *Studies in Intelligence,* 22:2, Central Intelligence Agency, 1978.

Warner, Michael, *The Office of Strategic Services: America's First Intelligence Agency* (Washington, D.C.: Central Intelligence Agency, 2000).

Weiser, Benjamin, *A Secret Life: The Polish Officer, His Covert Mission, and the Price He Paid to Save His Country* (New York: Public Affairs, 2004).

Westerfield, H. Bradford (editor), *Inside CIA's Private World: Declassified Articles from the Agency's Internal Journal, 1955–1992* (New Haven, Connecticut: Yale University Press, 1995).

Weyden, Peter, *Bay of Pigs: The Untold Story* (New York: Simon & Schuster, 1979).

Whitley, Craig R., *Spy Trade: The Darkest Secrets of the Cold War* (New York: Times Books, 1994).

Willis, George, *Surreptitious Entry: The Sensational Story of a Government Agent Who Picked Locks and Cracked Safes in the Service of His Country* (New York: Appleton-Century, 1946).

Winne, J. F. and J. W. Gittinger, "An Introduction to the Personality Assessment System," *Journal of Clinical Psychology.* Monograph Supplement No. 38, April 1973.

Wise, David, *Nightmover: How Aldrich Ames Sold the CIA to the KGB for 4.6 Million* (New York: HarperCollins, 1995).

Wise, David, *Spy: The Inside Story of How the FBI's Robert Hanssen Betrayed America* (New York: Random House, 2002).

Wise, David, *The Spy Who Got Away: The Inside Story of Edward Lee Howard, the CIA Agent Who Betrayed His Country's Secrets and Escaped to Moscow* (New York: Random House, 1988).

Wise, David and Thomas B. Ross, *The Espionage Establishment* (New York: Random House, 1970).

Wolf, Markus, *Man Without a Face* (New York: Public Affairs, 1997).

Woodward, Bob, *Veil: The Secret Wars of the CIA 1981–1987* (New York: Simon & Schuster, 1987).

Wright, Peter, *Spycatcher* (New York: Viking, 1987).

Wright, Peter, *The Spycatcher's Encyclopedia of Espionage* (Port Melbourne, Victoria, Australia: William Heinemann Australia, 1991).

Memorandum for the Record, *Project MKULTRA, Subproject 34,* Central Intelligence Agency, MKULTRA document 34-46, October 1, 1954.

Memorandum for the Record, *Definition of a Task under MKULTRA, Subproject 34,* Central Intelligence Agency, MKULTRA document 34-39, August 25, 1955.

Memorandum for the Record, *MKULTRA, Subproject 34,* Central Intelligence Agency, MKULTRA document 34-29, June 20, 1956.

Acknowledgments

The authors acknowledge with deepest appreciation the overwhelming and gratifying support we received from more than a hundred active and retired CIA officers in preparing this history. We conducted dozens of interviews and received correspondence from many others who devoted their careers to the Office of Technical Service or related engineering and development offices. Other case officers and operations managers who used the equipment, disguises, and alias documents OTS produced offered significant insights into processes that tightly wove operations and technology together. Virtually every request for assistance, whether for an interview, illustrative stories, verification of information, or photographs received the traditional OTS response: "So what can I do to help?"

The leadership of the Technical Service Retirees Association facilitated our contact with its members. Bruce Bixby, Dave Gokey, Tom Herring, Jim Joyce, Jerry Lee, Karl Muenzenmayer, Ray Parrack, and John Tredwell have devoted significant personal time to preserving OTS history and traditions through the TSRA and were especially helpful in identifying techs and case officers for us to contact.

Significant contributors to chapters on OTS's "early years" include John Aalto, Andy Anderson, Tom Beale, Howard Gamertsfelder, Cleo Gephart, Lyle Greeno, Norm Jackson, Irv Kemp, Dick Krueger, Hugh Montgomery, Al Schumann, Pauline Sypolt, Elsie Szuminski, Wally Szuminski, and Glenn Whidden.

Episodes and adventures from the middle decades of OTS history were recounted by Lynn Ashe, Bob Barron, Rosemary Capuzzi, David Coffey, Dick Corbin, Sam David, Phil Dean, Walt DeGroot, Jack Finarelli, Stuart H., Chris Hsu, Charles Janak, Dick Kessler, Andre Kesteloot, Ed Levitt, Ron Looney, the late Bob Ruhle, Sue Ruhle, Marti Shogi, Scotty Skotzko, Bob Stevens, Bob Swadell, Tom Twetten, Pat Wartell, Charlie Schuilla, Elisabeth Wilton, Judy Wonus, and Jon.

For information on more recent decades, Don Bailey, Dave Banks, Harlene Barton, Dan Bradley, Jack C., Roy Combs, Jim Cotsana, Ivan Danzig, Janet Donahue, Forrest Fleming, Bill Geary, Connie Geary, Thomas E. Gebbie, Bob Hart, Diane James, Leo Labaj, Lois Lees, Ellen Martin, Randy Mays, and Iris Stansfield all offered fresh insights and personal experiences.

Many others deserve mention but cannot be named due to current duties, cover, or other considerations. They know of their contribution and with justifiable pride can say to their families, "You know, I had a hand in this as well."

We owe special recognition to three career OTS officers whose deaths preceded *Spycraft*'s publication. Each of these officers, in spite of serious health conditions and pain, made themselves available for extended interviews, relating with honesty and pride their years of service to America through the CIA.

Arthur "Mick" Donahue supported CIA's covert action programs for forty years, from the Vietnam War to the war against terrorism. Although Mick had been retired for several years when 9/11 occurred, OTS management recognized immediately that his skills and experience were again required. So did Mick. Before night fell on September 11, 2001, Mick had contacted OTS offering to help rebuild our covert action capabilities, as he had done in the 1960s and the 1980s. "When a war is over, we always shut down covert action," Mick correctly observed, "and a few years later it's needed again. Good thing there are still a few of us around who know how to do it." Mick was engaged the next day and continued until his health prevented him from working.

Paul Howe, one of fifty CIA officers recognized in 1997 as Agency Trailblazers, thrived on working "under the radar." Paul's modesty was exceeded only by the remarkable engineering achievements for which he is rightly credited. Among the most significant were the T-100 subminiature camera and its subsequent models that became, arguably, the CIA's most effective Cold War intelligence-collection devices. Over his career, Paul combined his technical expertise with a unique skill of harnessing the capabilities of private contractors to produce generations of covert communications equipment that consistently outpaced our adversaries' technical counterintelligence capabilities. When we concluded our final interview, Paul summed up his career with characteristic modesty: "Well, I did what I could. I think it helped."

Sol Kurtzman lived in Washington nearly fifty years and never lost his

New York accent or demeanor. Sol was among the early professional engineers in the Technical Services Staff and remained with the office until the end of the Cold War. Sol told of the TSS's determined struggle to establish a reputation as "the place" in CIA where technical solutions to operational requirements would be found no matter how "impossible" the problem. Sol's reputation for cajoling and prodding engineers to create smaller, more reliable, less power-hungry clandestine devices matched his personal uncompromising standard for technical excellence. I was unaware of the seriousness of Sol's declining health when he asked in late 2006 if he could read a draft of *Spycraft*. After several days, Sol returned the manuscript pages with critical, positive, and invariably insightful annotations. His admonishment that OTS "engineers deserve as much ink as the ops guys" resulted in Chapter 15, which we dedicate to Sol's memory.

Without the contributions of Mick, Paul, and Sol, three OTS giants and American patriots, *Spycraft* could not have been written.

Several friends assisted us in obtaining artifacts, providing photographs, or validating information regarding operations of other intelligence services. These included Michael Hasco; Dan Mulvenna, retired RCMP Security Service officer; Gerald "Jerry" Richards, retired FBI Special Agent and Soviet tradecraft specialist; and Peter Earnest, Executive Director, International Spy Museum. Additional valuable contributions were made by Pete Burns, Chase Brandon, Brian Kelley, Jim LeCroy, Bill Mosebey, Jonna Mendez, Tony Mendez, Pat Merriweather, Harry Price, and the history preservationists Nick Benigsen, Lyle Hunger, and Mr. "X" and friends. The CIA curator Toni Hiley and her assistant Carolyn Reams facilitated our access to images from the CIA's museum collection. Through the generosity of Richard Lovell, we acquired papers of his father, the late Stanley Lovell, who directed research and development for the Office of Strategic Services.

Hayden Peake, author, historian, and curator of the CIA's Historical Intelligence Collection, is the dean of intelligence bibliophiles. His wise counsel, literary criticism, and encouragement proved invaluable. Former CIA Chief Historian Ben Fischer has been a friend and contributor to this project from its inception. Danny Biederman, Dr. David Crown, Jack Downing, and Bill Mulligan urged us to persevere when it seemed the project might not succeed. Critiques by Jim Gosler, Richard Lawrence, and Lou Mehrer on early drafts provided helpful commentary on the text from perspectives outside the OTS family. As demands to devote more time to *Spycraft* increased, Paul Johnson, former Director of the CIA's Center for the Study of Intelligence, and Nick Dujmovic, head of the CIA's oral history program, graciously offered schedule flexibility that allowed me to complete this manuscript while concurrently fulfilling my CSI assignments.

The CIA's publication review staff, particularly Paul B. and Kate M., information review officer Suzanne Fleischauer, and Publication Review Board member Larry Boteler worked professionally with the authors to resolve potential issues of classification. Herb Briick assisted us in obtaining, under the Freedom of Information Act, several historical documents and images seen for the first time in *Spycraft*. Michael Morell, the CIA's Associate Deputy Director, encouraged our efforts through prompt and considered adjudication of policy questions that arose during the prepublication process.

We are indebted to Brian Tart, president of Dutton Books, and Mitch Hoffman, our initial editor, for tolerance during the two-year CIA review process and their confidence that this work would eventually receive publication approval. The *Spycraft* story is better told due to the editorial counsel of Dutton editor Stephen Morrow and his assistant, Erika Imranyi. The cover, photos, and images reflect the creative talents of the Dutton art department. Dan Mandel, our literary agent from Sanford Greenberger, directed us through the necessary business processes required for a work of this magnitude. Mark Zaid's legal perspective provided constructive options for dealing with the CIA's official review bureaucracy. Randy Bookout and Al Cumming, from the staff of Senate Select Committee on Intelligence, took a welcome and special interest in the progress of the project.

The enduring patience of our families over the past five years is matched only by their loyal, unflagging support. Their uncompensated assistance cannot be calculated in dollars. We simply offer "thank you" a million times over to Mary Margaret Wallace, Kristen Melton, and Melissa Suzanne for the hours spent reading and rereading dozens of drafts, fact checking, telling us when we were boring, making us explain government acronyms, running errands, typing transcripts, and the hundred other tasks we took for granted. They must have done it for love.

Finally, to every reader, we are grateful for your interest in *Spycraft* and hope the story of OTS will renew your confidence in America's intelligence institutions and in the men and women who devote their careers to this service.

Index

Note: Page numbers in *italics* refer to illustrations.